THE MAMMILLARIA HANDBOOK

PREFACE

I am delighted and honoured to be able to write a preface on the occasion of a further reprint of this classic work on what has long remained one of the most popular genera of the family Cactaceae.

Following the similarly classic base for modern study of the family as a whole in the 1920s, Britton & Rose's *The Cactaceae*, Craig's work in the mid-1940s distilled the knowledge of the genus Mammillaria for the first time with the sort of detail essential for their proper appreciation and identification. It remains the bridge between older works including this genus and more modern publications, and still rides alongside the latter as essential reading for any serious student or collector of this wonderfully diverse genus, as a constant source of reference.

Indeed it was my constant source of study for many years as an invaluable base on which to build my own work, culminating in the publication in 1981 of my book, *Mammillaria - a Collector's Guide*. It is an invaluable guide to earlier literature on the subject. Dr Craig did his research thoroughly, and the book abounds with such references, as well as his own experience of growing, and clearly loving, the many species covered. The contemporary photographs of many of the species discovered and described at that time are immensely valuable as an indication of what exactly those names applied to in terms of actual plants.

In the years that Craig's work has not been available, much study has taken place within the two Societies formed for the study of this genus, the Mammillaria Society based in the UK but with nearly 50 per cent foreign membership, and the similar organisation based in West Germany. This reprint is long overdue, and will once again make available to the many enthusiasts for this genus a cornerstone for their study.

John Pilbeam 1989

Fig. 1. *Mammillaria guerreronis* in Zapilote Canyon, Guerrero, Mexico.

THE
MAMMILLARIA
HANDBOOK

WITH DESCRIPTIONS,
ILLUSTRATIONS, AND KEY TO THE SPECIES
OF THE GENUS MAMMILLARIA
OF THE CACTACEAE

By

ROBERT T. CRAIG, D.D.S.

Reprinted by
EP PUBLISHING LTD., and JOHNSON REPRINT CORPORATION
1963, 1965 & 1973

LOFTHOUSE PUBLICATIONS
1989

ABBEY GARDEN PRESS : PASADENA
1945

Reprinted by
LOFTHOUSE PUBLICATIONS
29 Ropergate, Pontefract, West Yorkshire. WF3 1LG

Under the auspices of the
MAMMILLARIA SOCIETY OF GREAT BRITAIN

ISBN 0900370 98 X

Originally
PRINTED IN PASADENA, CALIFORNIA, U.S.A.
BY
THE ABBEY GARDEN PRESS
P. O. BOX 101

Printed in Great Britain at The Bath Press, Avon

DEDICATED TO

MY WIFE

INA Y. CRAIG

WHO PARTICIPATED IN

THE PLEASURES AND HARDSHIPS

OF OUR EXPLORATIONS AND FIELD WORK

IN THE DESERTS OF THE SOUTHWEST AND MEXICO

COLLABORATOR

MR. ERNEST SHURLY

LONDON, ENGLAND

PREFACE

THE FASCINATION EXERTED by the cactus family upon so great a number of people is not difficult to understand. So different are these odd plants from other members of the plant kingdom, except certain Euphorbias and the equally fantastic Mesembryanthemums—whose mimicry is purely accidental—that they are found to attract attention even from people least affected by the aesthetic aspect of the merely beautiful flowers. When as a child of tender years on the unbroken prairie of extreme western Kansas, the writer was shown some "pincushion" cacti but was warned not to touch them, although he might "look at the pretty flowers." Years later he became acquainted with other Mammillarias on the rocky hills of southern New Mexico, and was impressed by the beauty of the diminutive *Mammillaria lasiacantha* and related forms. Some years later his interest was renewed when hardy plants of the same genus were seen repeatedly on the sun-baked mountain sides of Guatemala.

In Guatemala, just as in the United States, people with a taste for unusual house plants gather the several native Mammillarias and install them in their patios. Of all the infinitely varied yet always immediately recognizable members of the cactus family, scarcely any others are as well adapted to intimate culture as these appropriately named "pincushion" cacti. They may—at least if one does not value them too highly—be neglected for weeks and months, left unwatered, or even thrown untended upon a window sill, to resume normal growth when soil or a minimum quantity of water is again supplied. On the other hand, they respond gratefully to kind treatment. Their slender attachment to the soil facilitates easy collection in their native habitats. Their mass is so slight that most of them may be enjoyed by those with the most limited quarters; and best of all, they may be taken in hand without fear of the painful results that follow too close a contact with many of the other cacti.

Taking all their traits into consideration, their beauty, their infinite diversity and their puzzling and intricate variation, they form one of the most fascinating and curious groups of cacti, and many people would place them at the very top of the list. They are entertaining to horticulturists and taxonomists alike, but to the latter they are often most irritating because of those very inconsistencies that endear them to their growers.

That the genus is a perplexing one to divide into species, no one can deny. In this respect it is equaled perhaps only by the Opuntias, which possess so few of the praiseworthy characters of the Mammillarias and do provide the most diabolical representatives of the cactus family. No two authorities seldom agree upon the number of species of Mammillarias or the manner in which they should be separated as there are so many closely related forms.

Dr. Craig, although actively engaged in his practice of dentistry as a vocation, has taken time to indulge in an interesting avocation as well. Like many other persons, he became interested in plant collecting as a hobby but he carried it a little farther in one particular field until it became quite an involved project. Some years ago at the university before taking up his dental studies, he laid a background in the fundamental concepts of scientific evaluation of details and accuracy in recording observations, which he was later able to apply to a very good advantage in his study of these plants.

The account of the Mammillarias presented on the following pages might be called monumental if one does not wish to avoid that obnoxious word, which is applied so indiscriminately to other objects besides mortuary momentos. However, this is really monumental in the great amount of care and investigation involved in its preparation, a thankless labor to be appreciated only by those who have attempted, though on a smaller scale, something of the same nature.

Here in a single volume and in a single series is presented the most comprehensive account that ever has been attempted on a single group of cacti. Of course, this is partly the result of substantial additions to the knowledge of the cacti in recent years because of the increased attention given to them, particularly on the Pacific coast where climatic conditions permit their continued normal growth under cultivation. Extensive demands for new plants for culture have fostered wide exploration by persons interested in cacti alone, so that the dry mountain sides of Mexico, where Mammillarias most abound, are being combed in all accessible areas for further novelties. Yet one who appreciates the infinite variety of arid vegetation and the narrow limits of distribution of some of its elements, will know the possibilities of finding new Mammillarias are not exhausted.

This volume contains far more than a mere supplement to previous treatment of this genus. This is evident from the ample, detailed and methodical descriptions that are present, in contrast to the annoyingly brief ones usually provided. The nomenclature and the identification of the literally myriad of published names in this vast genus have been verified, apparently with a care not previously attempted. And last and perhaps even most important, the species are mostly all illustrated, and many with excellent new photographs. It goes without saying that in cacti—as well as in most other groups of plants—a glance of a few moments at a good illustration tells more than hundreds of words.

Dr. Craig's splendid account of the Mammillarias is an admirable example of what can and should be done for all groups of cacti, and for a great many other plants of horticulture or purely scientific interest. There are a few such treatments, but the emphasis still has to be placed on the word *few*. This one should give renewed impetus (if any were needed) to critical study and horticultural interest in this most popular group of cacti, whose cultivation and study can give pleasure whether one commands the most elaborate facilities for growth or only very simple equipment.

It is superfluous to express words of praise for this volume that bears the clear marks of careful preparation. It is necessary only to open the book at random, as Edward L. Green used to remark, "Use your eyes man! Use your eyes!"

PAUL C. STANDLEY
Chicago Natural History Museum

FOREWORD

THE MAMMALLARIA SECTION of the Cactaceae has interested more collectors than any other genus of the entire cactus family. This popularity with the many and diversified collectors is due in part to the adaptability of its many species to various localities and to different methods of culture. In the warmer climates, they thrive well in out-of-doors gardens where they can remain throughout the year. Under less favorable climatic conditions, they do very well in pot culture and they seldom attain a size so great as to present a problem to those with limited in-doors space. Because of this popularity, it is highly desirable that the available information concerning them be assembled and summarized for the most advantageous use.

The object of this monograph is to bring up to date, as near as possible, all the available information on this very interesting group of plants. Previously the most comprehensive treatment of this genus was that published in 1923 by Britton and Rose in "The Cactaceae." This extensive work brought together for the first time in the English language a wealth of information which previously had not been available. They accomplished this in spite of the fact that unsettled political conditions and various revolutions in Mexico prevented much extensive field study except in restricted areas and at times it was nearly impossible for foreigners to travel even between some of the centers of population. The unfriendly attitude of some of the Indian tribes made it inadvisable to do much exploring or any plant hunting in the more remote sections of the country. Since the establishment of a stable government some years ago, the conditions are decidedly more favorable for travel and exploration. With the improvement of the automobile highway systems throughout that nation, the presence of foreigners has greatly increased and any danger is largely non-existent. With these opportunities at hand we have endeavored to correlate the work of Britton and Rose with the recent additions to the literature as well well as with data from our own explorations.

Since 1923, many new species have been discovered and described, as well as much additional data concerning the older species; this material has become available through the efforts of various explorers and investigators who have worked in the native habitat of the plants. About twelve years ago the author first began a specialized study of this particular genus by gathering the descriptions of the species that have been published since 1923. In order to associate them with the previously known species, an attempt was made to fit them into the "Key to the Species" of Britton and Rose. Many difficulties were encountered because of inadequate descriptions of the older species as reported in recent publications and to some extent to the discrepencies between the "Key" and the descriptions.

In order to lay a better foundation for study, a new approach was made to the problem in which our investigations consisted of three main phases of endeavor. First, the original descriptions of all published species were obtained and these were used as the basis for our own descriptions. This was necessary because much of the essential data had not been included in the works of recent authors and some of the later interpretations of the original descriptions had altered the intent of the original authors in that they had been made to fit some plant for which they were not intended.

Second, although some of the original descriptions were often extremely meager in essential details, there were sometimes subsequent publications by the same or other authors that often added much valuable information. On the other hand, later authors sometimes made errors which were perpetuated by subsequent authors who failed to refer to the original description thus causing much confusion. Hence a survey of the literature that has been published between 1676 and 1943 was made and all recorded information, which did not conflict with the original description, was added to our growing descriptions. These were being built along the same general outline so as to have them all as nearly uniform as possible and to make it easier to compare the various species. This was a considerable task because most of the literature had been written in Latin or German with some of it in French, Spanish, Danish and Dutch; all of which had to be translated into English before it could be used.

The third phase of the investigation was the supplementing of the previous works with data from our study and comparison of plants collected in the field, whenever possible from the type localities, and from information reported to us by other explorers, collectors and dealers who have actually observed the plants in their native habitats. This field study necessitated considerable travel on our part and there still remains much to cover as the transportation facilities in large sections of Mexico are still not available except by the very primitive means of either on foot or by horseback which is very time consuming. Much new territory has been explored during the last few years and doubtlessly there will be much more in the near future with the construction of the many new roads which will permit automobile travel into places that were not possible to reach previously.

This field work, which was begun in 1932 and most of which was done in company with my wife, has taken us over a large part of the desert areas of the southwestern states of California, Arizona, and parts of New Mexico and Texas, and across the international border into Mexico. We dipped into the northern part of the peninsula of Baja California, but for the most part we have relied upon the collections and plant information as reported to us by such collectors as Mr. Howard Gates, Mr. George Lindsay, Mr. Edgar Baxter and others who have covered much of the territory on the entire length of the peninsula to Cape San Lucas.

We have made six trips across the border into the state of Sonora. In 1933 we explored the area in the northwestern part of the state in the neighborhood of Sonoyta. In 1935, crossing the border at Nogales, collections were made southward to the vicinity of Hermosillo and down to the coast at Guaymas and especially at San Carlos Bay which is the type locality of several species. In 1936, in company with Mr. and Mrs. John Hilton, a trip was made to the southeastern part of the state to the McCarty ranch at Guirocoba in the Alamos district where we made collections and obtained the extensive one that had been made by Mr. and Mrs. Howard Gentry during the preceeding spring and summer. On our return trip we collected in the Baca Tete Mountains; at that time it bordered on the dangerous side due to the hostile nature of the Yaqui Indians. It was certainly not a healthy trip to make as these natives have a very strong dislike for anyone who trespasses their domain and particularly if he does any digging among the rocks, whether it is looking for mineral or just mere cactus plants. In 1937 we entered the northeastern part of the state through the border town of Naco near Douglas and followed the Sonora River down to Hermosillo. This was a trip which is not advisable to make by automobile (we had been strongly advised against making it) as it is little more than a trail. Between Arispe and Ures it took us fourteen hours to make fifteen miles and none of that time was spent in plant collecting but in digging the car out of the seemingly bottomless sand, trying to avoid the sink holes and negotiating the car over the rocks in the river bottom as we found it easier to drive in the river than to try to follow a practically nonexistent road. Connecting with the main highway at Hermosillo, we proceeded on to Guirocoba ranch. In 1938 a short trip was made into the northern part of the state in the neighborhood of Nogales. In 1939 an expedition was formed in company with Mr. George Lindsay to explore the Sierra Tarahumara in Chihuahua. From Guirocoba, Sonora, which is the end of the automobile "trail", we went by horseback into Sinaloa, thence northward into the higher mountains of Chihuahua. This was a very enlightening trip in many ways inasmuch as part of it was in practically virgin territory. We not only had the opportunity to study the most interesting Tarahumara Indians but were able to make some valuable botanical collections. Some new species, as well as additional information on previously known species, were obtained in not only the field of cacti but also on several succulents as well as on other plant families. It has always been our great regret that we had to turn back where we did because of lack of time as we felt that we were just on the edge of a territory that doubtlessly contained much interesting material.*

In both 1941 and 1942 we crossed the southwestern states of California, Arizona, New Mexico, and entered Mexico by the way of Laredo, Texas, and traversed the Pan-American Highway to Mexico City. Using the capital as a hub, we made trips into various states: northeast to Queretaro, west to Michoacan, southwest to Guerrero, south to Oaxaca, and southeast to Puebla. These two trips likewise revealed several new species as well as considerable much needed information about some of the older species from that section of the country. In all, our field work has added much information not recorded in the literature as well as a better conception of the various plants in general.

ACKNOWLEDGEMENTS

The extent of this work would not have been possible if it had not been for the extensive cooperation that various persons have so generously rendered and to them is herewith extended the most sincere appreciation and gratitude. In historical research, Mr. Ernest Shurly of London, England, has contributed a large part of the data on the historical development of this genus. He has assembled for the first time the most complete records of the published data on all the species. He has copied in chronological order for each species nearly everything that has been published in the various books and journals since a species was first described. These data have added much to our knowledge of each of the species as it has supplied much information that was not in the original description. It also has shown by whom errors were first made and how they have been perpetuated by subsequent authors. Previously much of these data have not been available to the average investigator as there are only a very few copies of some of these very old books and many of them are not in this country. We also obtained other valuable assistance by means of the microfilm photographs through the Library of Congress at Washington, D.C., of some of the very old and priceless books which otherwise are not available to the general public.

In the contribution of plant material, various collectors and explorers have been very cooperative in supplying us with plants and information. Sr. and Sra. Ferdinand Schmoll of Cadereyta de Montes, Queretaro, Mexico, have been most generous in supplying us not only with plants but also from their store of information about the plants from the central plateau region and more particularly from the states of Queretaro and Guanajuato.

Mr. Howard Scott Gentry who has explored and collected extensively in the states of Sonora, Sinaloa, Chihuahua, as well as in Baja California, has been very helpful in the collection of material for us. In 1936, while he was making a botanical survey of the Rio Mayo watershed in Sonora and Chihuahua, he collected for us a very representative group of plants from that region, which included several new species and numerous intergrading varieties. We had the misfortune to loose much of this material because of a very severe and prolonged freezing weather shortly after our return with the plants. He is certainly to be commended for his explorations in this practically virgin territory.

Mr. George Lindsay of Lakeside, California, has repeatedly brought us material from his many trips along both the Mexican west and east-coastal regions and from as far south as Oaxaca and particularly from Baja California, where he has explored extensively.

---

*Cact. Succ. Journ., 15:47, 1943.

It was with him that the expedition was made from Sonora through Sinaloa into the Sierra Tarahumara in Chihuahua in 1939. His writings on his expeditions into Mexico and his studies of its flora have added much to our knowledge of the plants from these regions.

Mr. Howard Gates of Corona, California, has been most generous with his plant material and his intimate knowledge of the flora of Baja California, which he gathered on his several trips down the full length of that long and almost roadless peninsula. Mr. Edgar Baxter of Santa Barbara, California, has contributed material from Mazatlan and southward along the coast of Sinaloa.

Mr. William Hertrich, curator of the Huntington Library at San Marino, California, has aided materially in extending to us the use of the facilities of the library and gardens as well as for his contributions of photographs and other material.

Much valuable data on the flowering habits and other plant details have been contributed by Mr. Robert Peebles of Sacaton, Arizona; Mr. Robert Kelly of Temple City, California; Mr. Robert S. Woods of Azusa, California; Mr. A. Wilhite of Corona, California; Mr. Jack Whitehead of the University of California Botanical Garden at Berkeley.

For botanical and technical suggestions and criticisms, we are indebted to Mr. Harry Johnson of Hynes, California; Dr. Ira Wiggins of Stanford University; Dr. Carl Epling of the University of California at Los Angeles; Dr. Philip Munz of Pomona College.

R. T. CRAIG
Baldwin Park, California.

September 1, 1944

CONTENTS

FRONTISPIECE	iv
PREFACE	ix
FOREWORD	xi
INTRODUCTION	1
KEY TO THE SPECIES	7
SPECIES OF MAMMILLARIA	17
UNCLASSIFIED SPECIES	299
LITTLE KNOWN SPECIES	323
SPECIES KNOWN BY NAME ONLY	343
NAMES TO BE EXCLUDED	346
ASSOCIATED GENERA	351
APPENDIX	359
BIBLIOGRAPHY	364
INDEX	371

THE MAMMILLARIA HANDBOOK

INTRODUCTION

DISTRIBUTION

THE GENUS MAMMILLARIA is predominantly of Mexican distribution but it also has a very wide range in the neighboring regions north of the equator. The southern-most limits appear to be in Colombia and Venezuela in South America which are represented by two known species and a doubtful one. In the Central American countries two species are to be found only in Guatemala and Honduras. To the east in the West Indies two species are somewhat widely distributed on the various islands. Westward from the islands of the Caribbean it is next encountered along the coast in the south-central part of Texas which is its easternmost limits in the United States. From there it extends north-westward through Texas to southern New Mexico and then westward through Arizona to California.

Mammillarias are quite generally distributed over most of Mexico with the exception of the more tropical regions although a few are found even in the damper climates. The greatest distribution appears to be in three main areas: first, in the more arid sections of the central plateau region, especially in the states of Hidalgo, Queretaro, Guanajuato and San Luis Potosi; second, in the north-western part of the mainland in the states of Sonora, Chihuahua and Sinaloa; and thirdly, on the peninsula of Baja California and the adjacent islands. These apparent areas of greater distribution may be accounted for by the fact that in these general areas there have been more intensive botanical explorations and commercial exploitations than in most of the other sections where transportation facilities are not available at present. As these unexplored areas are being opened to travel, many new species are being discovered and doubtlessly many more new ones as well as distribution data on previously known species will be uncovered in the future. There are still large areas in north-central, central-western, south-western and southern Mexico that as yet have hardly been touched by the exploring botanist and doubtlessly these potential areas may contain many interesting and different species.

The types that are found in South America as far as we know now, are endemic to that region and likewise in the West Indies one of the species is found only on the islands while one of the other species has a very closely related form that is found in Texas and north-eastern Mexico. The species that are found in the southwestern United States are for the most part also found in northern Mexico. The Central American species are very closely related to and probably intergrade with the species found in the neighboring Mexican states.

Some species appear to be very limited in their distribution as they have as yet not been reported from any locality other than in a relatively small area of their original discovery. On the other hand other species are found over a large area of several hundred miles wide and still they maintain their peculiar characteristics intact. Some species with apparently very limited distribution may in fact have a much wider range but the information available at present, limits it to a small area because of insufficient explorations.

In the following descriptions of the various species, the information listed under "Distribution" refers to the state and the country in which they are normally found. The "Type locality" refers to the particular place in that state from which was obtained the material that was used in preparing the original description. Often, many of the earlier descriptions did not cite any particular locality other than only "Mexico" which has led to much confusion and duplication of descriptions of the same species by different authors.

When there is no type locality given in the original description but when it is known that plants of this genus have been found in a definite locality that information is listed herein —under the heading of "Type Locality" as "reported" from that particular place.

Much of our present information on the specific localities has been supplied by recent explorers and commercial dealers and to them much credit is due for this most valuable datum.

NOMENCLATURE

The genus *Mammillaria* in the *Cactaceae* was proposed by Haworth (Pl. Succ., 177, 1812) in the description of *M. simplex* (cf. *M. mammillaris*). Stackhouse previously in 1809 had used this generic division for a genus of *Algae*. Britton & Rose (Cact., 4:65, 1923) on the strength of this priority use by Stackhouse proposed the name *Neomammillaria* to take the place of *Mammillaria* which they considered as a homonym, but this change was not generally accepted internationally. Inasmuch as Stackhouse's genus was never recognized or came into general use, the International Botanical Congress of 1930 recognized the genus *Mammillaria* in the *Cactaceae* as official and with full generic rank as *nomen conservanda*.

Various spellings of this name have been used by different authors, namely: *Mammilaria* by Tory and Gray (Flora, 1:553) and *Mamillaria* by many of the early authors as well as Engelmann, Schumann, Berger et al but the accepted spelling now is *Mammillaria*.

The author is herein following the practice that has been used for a long time by most zoologists and many botanists in the elimination of the use of all capital letters from the specific names regardless of whether they refer to proper nouns, geographical places, or descriptive adjectives. This procedure was adopted in interest of simplicity and uniformity as there is no specific rule regarding the capitalzation of some names and not of others.

For the most part we are listing the species synonyms as given by Britton & Rose (Cact., 4, 1923). Most of them have been checked and were found in most cases to be reasonably accurate.

SPECIES VARIATION

The variation-limits of the characteristics of any species are often not well defined. Some species are so decidedly distinct in their characteristics that they stand alone and there are no other species that are very similar to them, while on the other hand there are others that appear to be a part of a series with progressively changing characteristics. In the latter type it is difficult to draw any distinctive line of demarkation between the various forms presented in the series but the extremes are so decidedly different in many respects that there is no question about their distinction. Also within what is generally considered a species, often there are variations that are somewhat distinct from the typical form but still not enough to justify a separate species. This is examplified by the wide variation that is sometimes found in a single species in a very limited area of one hillside. It is hard to believe that the several variations that might be found within a few hundred feet of one another could be called different species just because of a little heavier armament or an extra spine or so but still they do appear to be different. Without the knowledge of that habitat variation and careful investigation and comparison, they might readily be considered as different species.

In some species with a wide distribution, it has been our observation in the field that there is often a marked evidence of geographical variation. The slight changes, that take place in the typical characteristics, often become more pronounced as the distance from the type locality increases. These changes may continue to the extent that the original plant has so changed that it has actually become a different species. On the other hand we have observed in some species a greater change in a very limited area than other species will exhibit over several hundred miles. Some species may appear to look different in different localities because of the difference in growth habit, spine coloration, or spine length but when the basic characteristics are compared, they are found to coincide with the type. These are often given varietal rank to designate the slight differences. Likewise the nature of the terrain in which they are found will influence the growth and coloration factors, especially in areas of greater moisture, more fertile soil, and greater protection by the other flora.

In the past, some investigators have been inclined to erect as new species any slight variation from the type. This has led to much confusion and long synonymy-lists for some of the species. In contrast to this, other writers have grouped a number of closely related forms under a single name even though they often exhibit distinct differences in their basic characteristics. The treatment used in this monograph has been to try to avoid either extreme and to judge the species wherever possible from our own or other investigators field observations.

The theory has been advanced by some workers that the *Cactaceae* is of a comparatively recent development and it is still in the process of its evolution, that is, the characteristics of the species are still in the formative stage, and as a result they have not become stable. The geographical variations and the environmental factors have their influence on the development of the various species but on some of them to an apparently greater extent than on others. In the final analysis it is the major specific characteristics that determine a species and not the minor variable differences and when all phases of the matter are taken into consideration, our conception of a species is more or less an artificial means of specifying a plant of certain characteristics in order to differentiate it from another. If the plants over a large area could be systematically checked, it would probably reveal that many of our species are only progressive variations of a comparatively few species and that many so-called species are only the steps in between the more outstanding types. Inasmuch as no person could cover the entire areas of the distribution of this genus and very few people interested in these plants ever could have the opportunity to observe all the variations in nature, we will continue to designate the various species until some better classification can be devised.

CLASSIFICATION

The arrangement of the various species into groups in order that specimens may be better and more easily identified and classified, must still be based on more or less purely artificial dividing criteria. Two main systems have been used by recent authors. The Berger modification of Schumann's system, which has been used by some recent authors, divided all the species into ten named groups in which the dividing factors were, to some extent, quite indefinite and flexible. The Britton and Rose system arranged the classification characteristics in a more comparative system of keys with quite tangible separation factors. By using the basic characteristics of a plant, this system is more workable and makes easier the classification and identification of plants. The system used herein is a modification of the Britton and Rose system. When more detailed data are available on the seeds of all of the species, this may be a valuable basis for a classification scheme.

CHARACTERISTICS

The **shape** or growth habits of a plant cannot be used very conclusively as a separation factor between the species except in the marked extremes. This is especially true of cultivated plants but there are some species in their natural habitat that have very definite growth habits such as some of the small elongated spreading species like *M. viperina* as contrasted with the large globular types that form large solid compact mounds like *M. compressa*. Whether a plant is always simple or cespitose, is also a most variable factor for even in its native habitat there may be conditions that may favor one type rather than another; and certainly under cultivation with often unnatural conditions, it is a most unreliable character.

The crestate forms are not considered herein as they are not typical and represent an abnormal growth.

The location of the **branching** is also not a constant characteristic with each species and particularly whether it is from the base or from the side of the body of the plant. On the other hand that type of dichotomous branching or division of heads at the apex is more or less characteristic of certain groups of species and it is seldom, if ever, found in other groups unless there has been some injury to the growing center.

The tubercles are arranged in spirals downward from the apex in both clockwise and counter-clockwise directions. There appears to be a general grouping of these spiral arrangements that are fairly consistent throughout the entire genus. They are arranged in a regular series of progressive mathematical steps with a definite relationship within each series as well as between the various series. In the first series there are 3 spirals in one direction and 5 in the other with a difference of 2 between the two rows; in the next series there are 5 and 8 spirals with a difference of 3; in the next there are 8 and 13 spirals with a difference of 5. It will be noted that if the amount of the difference between the spirals of the first two series (3 and $5=2$) and (5 and $8=3$) are added together, the sum will be equal to the amount of the difference between the spirals of the third group (8 and $13=5$). The next series is 13 and 21 with a difference of 8 or the sum of the differences of the two preceeding series. The same is true of the next series of 21 and 34 with a difference of 13 which is likewise equal to the sum of the two preceeding series, etc.

SERIES	DIFFERENCES
3 and 5	2
5 and 8	3
8 and 13	5 (2+3)
13 and 21	8 (3+5)
21 and 34	13 (5+8)
34 and 55	21 (8+13)

There have been a few odd series reported but these are not the usual and may represent some abnormal condition or intermediate step in the evolution of the genus. Those reported are: 7 and 12, 10 and 16, 11 and 18, 16 and 26. Each species usually has a more or less definite and constant arrangement but it may step to the next higher series in larger specimens. As yet there has been no explanation for this unique mathematical relationship but further study may reveal a group relationship.

The **sap** that is found in the tubercles has been used as one of the major dividing factors in this genus. One fairly well defined group of species has a nearly clear to amber-colored or so-called **watery sap** while the other group has a whitish or so-called **milky sap**. There is also a small intermediate group of species that has a **semi-milky sap** in that they display both types of sap under different conditions.

The tubercle is used herein as the basis for the determination of the type of sap of the plant instead of the central body core as was proposed by Schumann, Berger, and others. From the standpoint of the average plant fancier, it is doubtful if he will cut up his prized plant just to determine whether it contains a watery or a milky sap. A small cut into a tubercle in an inconspicuous place will usually supply the desired information and it will have no ill effects on the plant because usually the sap readily coagulates and heals over the cut. Care must be exercised nevertheless, because once we had a very milky plant in active growing condition that actually bled to death from a small cut in a tubercle. The sap continued to flow from the cut until it finally died regardless of our efforts by various means to stop it. Even though this is a very unusual experience, still it is one of those things that can happen.

The plants in the semi-milky group contain a sap that is nearly clear when the plant is in a dormant or shrunken condition but when it fills out and resumes growth, the sap becomes more milky although sometimes not to the extent that is found in the true milky group. This group will usually display a scant milky sap in the inner body core even while in the dormant state even though it cannot be demonstrated in the tubercle.

The watery sap is usually associated with all the plants with the hooked central spines (except *M. strobiliana* and *M. uncinata* in the milky group and *M. guerreronis* var. *subhamatam*, *M. rekoi*, and *M. zapilotensis* in the semi-milky group) as well as with some of those with straight central spines. It is generally associated with seeds that are mostly black but some of them have been reported as brownish. The milky sap is associated only with those with straight central spines (except *M. strobiliana* and *M. uncinata*) and always with brown seeds. The semi-milky sap is associated with both the straight and the hooked central spines and only with brown seeds.

In many species the **axils** between the bases of the tubercles contain more or less whitish wool. When present, it is usually more abundant at the apex and in the flowering and fruiting area. In some species it may persist for several seasons but it usually disappears after fruiting. It may vary in amount from only a dot of a hardly discernable woolly mat to an abundant flocculent mass that sometimes nearly covers the tubercles. The texture of it varies from very fine silky thread-like to coarse hair-like to stiff bristle-like or even to almost spine-like. The distinction between hairs and bristles is very often difficult to determine as it is usually only a question of degree. The presence or absence of wool has not been used herein as a distinguishing factor in the separation of species because its presence is not always consistent as it often depends upon whether or not the plant is in an active growing condition or whether it is in the flowering or fruiting season. The presence of bristles is usually more consistent but not always, however, when present, they will usually persist even after the wool has disappeared. A plant, that was grown under the more adverse conditions of its native habitat, will sometimes exhibit very pronounced types of bristles but when they are transplanted to more favorable conditions or are raised from seeds, the robustness of these bristles diminishes or they may disappear entirely. Although the presence or absence of bristles has been used as a dividing factor in the key, it should be remembered that this characteristic may be a little difficult to determine for certain in those plants that have been grown in cultivation for some time.

The **spines** are arranged in various positions within the areole at the apex of the tubercle. For convenience of description, they are classified as **central** and **radial** spines. Some species bear only central spines and others bear only radial spines, while most species have both. In the species with only one type, it is sometimes debatable whether the spines are truly central or truly radial but in most cases their nature is fairly well established. In some specimens in which only the radial spines are presumed to be present, there will often be found what appears to be a central spine but which in fact is often only a more or less centrally placed radial spine. However, it is not uncommon in those species which ordinarily have only radial spines to find what are unquestionably central spines on the very robust specimens but they are not typical of the species. This condition is found more often in the group with the milky sap and has caused much misunderstanding of some of the species when only a single specimen is at the disposal of the investigator. The variation in spine size and structure is often influenced very strongly by the environmental conditions as well as by the hereditary factors, hence the spine count as a species separation factor is in many cases a purely artificial division and individual members of a species will often overlap these arbitrary limits. The central spines may be either straight, bent or hooked at the tip but the radial spines are nearly always straight or at the most only slightly bent or tortuous in their over-all length.

The color of the **flower** is usually rather constant but sometimes it varies somewhat although the other characteristics of the plant may be typical. This variation may represent a distinct variety or sometimes it may be a geographical variation. The size of the flowers in so many of the species is so small that it is necessary to examine them with a magnifying glass to determine the finer characteristics of the various parts of the flower.

Even though most of them are small, there are a few species that are very conspicuous by their size and the color of the bloom as *M. wrightii* and *M. wilcoxii* from Arizona and New Mexico and *M. beneckei*, *M. balsasoides* and *M. zephyranthoides* from southern Mexico. The structure of the flower in most species is fairly constant even though there may be considerable variation in the spine arrangement so it has been used to some extent in the separation of the species. The dried perianth of the flower remains firmly attached to the fruit in some species while in others it is easily detached. The full significance of this characteristic has not been fully ascertained because data on many of the species are at present not available but it may be a valuable characteristic in the grouping of the species.

The **seeds** have often been inadequately described as only either brown or black but there are many other important details other than color that are characteristic of the species. In some, the color has been further qualified as reddish, reddish brown, brownish black, olive-gray and dark brownish gray. It has been our observation that if the seeds are properly washed to remove the adhering pulp their true color will more nearly be revealed. For example, Englemann reported that the seeds of *M. barbata* were brown. When we obtained the seeds of this species they did appear to be dull brown but when they were washed of the dried adhering pulp it was revealed that they were in fact a glossy black. References to the color of the seeds from immature fruit is also often misleading.

The appendage or corky base of the seeds is present in varying degrees in several species. It is present to only a limited degree in such species as *M. guelzowiana* and *M. longiflora* while it probably reaches its greatest development in *M. phellosperma*. This characteristic has been used as a basis for a different generic classification but inasmuch as it merges into this genus through the intermediate forms, it is being retained in this genus. Here again is one of those debatable and "hard to draw the line" questions.

The shape and other characteristics of the seeds may be a better basis for the grouping of the species than that based on the morphology of the other parts of the plant but inasmuch as all of these data are seldom available to the average collector, it has not been used herein.

KEY TO THE SPECIES

A. Tubercles with milky sap, seeds brown.
  B. Central spines absent.
    C. Radial spines less than 10.
      D. Bristles in axils.
        E. Outer perianth-segments entire . . . . 1. *M. pyrrhocephala*
        EE. Outer perianth-segments ciliate.
          F. Inner perianth-segments reddish.
            G. Tubercles not angled . . . . 2. *M. compressa*
            GG. Tubercles angled.
              H. Tubercles arranged in 8 and 13 spirals . . . . 3. *M. seitziana*
              HH. Tubercles arranged in 13 and 21 spirals . . . 4. *M. polyedra*
          FF. Inner perianth-segments greenish yellow.
            G. Tubercles not sharply angled.
              H. Tubercles arranged in 8 and 13 spirals . . . . 5. *M. confusa*
              HH. Tubercles arranged in 13 and 21 spirals . . . 6. *M. karwinskiana*
            GG. Tubercles angled to tip.
              H. Tubercles narrow at base, 4-5 mm. . . . . . 7. *M. knippeliana*
              HH. Tubercles wide at base, 8 mm. or more . . . . 8. *M. praelii*
      DD. No bristles in axils.
        E. Tubercles not angled to apex.
          F. Inner perianth-segments pink to reddish . . . . . 9. *M. ortegae*
          FF. Inner perianth-segments cream to yellow.
            G. Outer perianth-segments entire . . . . 10. *M. winteriae*
            GG. Outer perianth-segments ciliate.
              H. Stigma-lobes green . . . . 11. *M. zahniana*
              HH. Stigma-lobes pink . . . . 12. *M. magnimamma*
        EE. Tubercles angled to apex.
          F. Inner perianth-segments yellowish green . . . . . 13. *M. peninsularis*
          FF. Inner perianth-segments reddish . . . . 14. *M. crocidata*
          FFF. Inner perianth-segments whitish.
            G. Tubercles broad pyramidal . . . . 15. *M. roseo-alba*
            GG. Tubercles slender . . . . 16. *M. lloydii*
    CC. Radial spines more than 10 . . . . 17. *M. albiarmata*
  BB. Central spines present.
    C. Central spines hooked.
      D. Bristles in axils . . . . 18. *M. strobiliana*
      DD. No bristles in axils . . . . 19. *M. uncinata*
    CC. Central spines straight.
      D. Radial spines usually absent (or occasionally deciduous bristles)
        E. Bristles in axils . . . . 20. *M. mendeliana*
        EE. No bristles in axils.
          F. Tubercles angled.
            G. Inner perianth-segments pink . . . . 21. *M. carnea*
            GG. Inner perianth-segments yellow . . . . 22. *M. melispina*

Key to the Species

FF. Tubercles terete.
  G. Tubercles short conic . . . . 23. *M. orcuttii*
  GG. Tubercles elongate . . . . 24. *M. polythele*
DD. Radial spines present.
  E. Radial spines less than 10.
    F. Bristles in axils.
      G. Tubercles angled.
        H. Inner perianth-segments yellowish.   25. *M. woburnensis*
        HH. Inner perianth-segments reddish.
          I. Central spines 2 . . . . 26. *M. polygona*
          II. Central spines 4-6 . . . . 27. *M. tenampensis*
      GG. Tubercles not sharply angled.
        H. Inner perianth-segments yellowish.
          I. Central spines 0-1 . . . . (5) *M. confusa centrispin.*
          II. Central spines 1-2.
            J. Perianth-segments acute . . . . . . 28. *M. petrophila*
            JJ. Perianth-segments acuminate . . . . 29. *M. eichlamii*
        HH. Inner perianth-segments reddish.
          I. Tubercles terete . . . . 30. *M. collinsii*
          II. Tubercles pyramidal . . . . 31. *M. mystax*
    FF. No bristles in axils (or occasionally scattered ones).
      G. Tubercles sharply angled to tip.
        H. Inner perianth-segments yellowish.
          I. Central spines 1.
            J. Radial spines 4-5 . . . . 32. *M. flavovirens*
            JJ. Radial spines 6-8 . . . . 33. *M. bocensis*
          II. Central spines 2-4.
            J. Radial spines 3-4 . . . . 34. *M. sempervivi*
            JJ. Radial spines 6-8 . . . . 35. *M. obscura*
        HH. Inner perianth-segments reddish.
          I. Radial spines 3-5.
            J. Stigma-lobes carmine . . . . 36. *M. bucareliensis*
            JJ. Stigma-lobes yellow . . . . 37. *M. vagaspina*
          II. Radial spines 6-9.
            J. Outer perianth-segments serrate . . . . 38. *M. craigii*
            JJ. Outer perianth-segments entire . . . . 39. *M. melanocentra*
      GG. Tubercles obscurely angled at tip.
        H. Inner perianth-segments greenish to yellowish.
          I. Central spines 1.
            J. Areoles with additional subradial spines . . 40. *M. nivosa*
            JJ. Areoles without additional subradial spines
              K. Outer perianth-segments ciliate.
                L. Tubercles arranged in 5 and 8 spirals. 41. *M. dawsonii*
                LL. Tubercles arranged in 8 and 13 spirals 42. *M. gatesii*
              KK. Outer perianth-segments entire . . . 43. *M. baxteriana*
          II. Central spines 2-4 . . . . 44. *M. brandegeei*
        HH. Inner perianth-segments reddish.
          I. Stigma-lobes reddish.

  J. Central spines 1-2.
   K. Outer perianth-segments ciliate . . .   45. *M. sartorii*
   KK. Outer perianth-segments entire . . .   46. *M. zuccariniana*
  JJ. Central spines 4 . . . .   47. *M. bachmannii*
 II. Stigma-lobes yellowish to green.
  J. Radial spines 3-7.
   K. Tubercles arranged in 8 and 13 spirals .   48. *M. phymatothele*
   KK. Tubercles arranged in 13 and 21 spirals   49. *M. hamiltonhoytae*
  JJ. Radial spines 8-10 . . . .   50. *M. scrippsiana*
EE. Radial spines 10-20 (see below for EEE).
 F. Bristles in axils.
  G. Inner perianth-segments reddish.
   H. Stigma-lobes rose.
    I. Radial spines 10 . . . .   51. *M. esseriana*
    II. Radial spines 16-20 . . . .   52. *M. geminispina*
   HH. Stigma-lobes yellow to greenish.
    I. Radial spines 10-15 . . . .   53. *M. evermanniana*
    II. Radial spines 15-20.
     J. Perianth-segments whitish with reddish
      mid-line . . . .   54. *M. saetigera*
     JJ. Perianth-segments reddish purple with
      darker mid-line . . . . . .   55. *M. standleyi*
  GG. Inner perianth-segments yellow to white.
   H. Radial spines 10-14 . . . .   56. *M. lindsayi*
   HH. Radial spines 18-20 . . . .   57. *M. ritteriana*
 FF. No bristles in axils (or occasionally scattered ones).
  G. Tubercles more or less angled.
   H. Inner perianth-segments predominantly reddish.
    I. Tubercles not sharply angled . . . . .   58. *M. sonorensis*
    II. Tubercles angled to tip . . . .   59. *M. hertrichiana*
   HH. Inner perianth-segments white to yellowish,
    some with reddish mid-line.
    I. Central spines 1-2.
     J. Radial spines 10-15.
      K. Outer perianth-segments entire . . .   60. *M. hemisphaerica*
      KK. Outer perianth-segments ciliate.
       L. Inner perianth-segments greenish . .   61. *M. macdougalii*
       LL. Inner perianth-segments white with
        reddish mid-line . . . .   62. *M. gummifera*
     JJ. Radial spines 15-20.
      K. Inner perianth-segments white . . .   63. *M. pachyrhiza*
      KK. Inner perianth-segments cream to pink
       to reddish.
       L. Outer perianth-segments entire . .   64. *M. applanata*
       LL. Outer perianth-segments serrate . .   65. *M. johnstonii*
    II. Central spines 4-7.
     J. Outer perianth-segments entire.
      K. Central spines porrect . . . . . .   66. *M. arida*
      KK. Central spines horizontal . . . . .   67. *M. pseudocrucigera*

JJ. Outer perianth-segments serrate.
    K. Central spines strongly subulate . . . 68. *M. gigantea*
    KK. Central spines more slender and shorter
        (65) *M. johnstonii guaymen*
GG. Tubercles nearly terete.
  H. Central spines 1-2.
    I. Tubercles arranged in 8 and 13 spirals . . . 69. *M. gaumeri*
    II. Tubercles arranged in 13 and 21 spirals.
      J. Tubercles short, 5 mm. or less . . . . . 70. *M. xanthina*
      JJ. Tubercles elongate, 10-12 mm. long . . . 71. *M. tesopacensis*
  HH. Central spines 3-5.
    I. Inner perianth-segments white to yellow . . 72. *M. mammillaris*
    II. Inner perianth-segments orange to reddish.
      J. Inner perianth-segments orange . . . . 73. *M. zeyeriana*
      JJ. Inner perianth-segments purplish pink . . 74. *M. petterssonii*
EEE. Radial spines more than 20.
  F. Bristles in axils.
    G. Radial spines 20-30.
      H. Stigma-lobes reddish . . . . 75. *M. brauneana*
      HH. Stigma-lobes greenish.
        I. Central spines 1-4.
          J. Inner perianth-segments pink.
            K. Tubercles nearly terete.
              L. Central spines acicular . . . . . 76. *M. hahniana*
              LL. Central spines subulate . . . . . 77. *M. bravoae*
            KK. Tubercles angled . . . . 78. *M. woodsii*
          JJ. Inner perianth-segments whitish . . . . 79. *M. chionocephala*
        II. Central spines 6 . . . . 80. *M. mayensis*
    GG. Radial spines more than 30.
      H. Tubercles angled to tip . . . . 81. *M. neopotosina*
      HH. Tubercles terete, at least so at the tip.
        I. Inner perianth-segments yellowish . . . . 82. *M. parkinsonii*
        II. Inner perianth-segments reddish . . . . . 83. *M. klissingiana*
  FF. No bristles in axils.
    G. Radial spines 20-30.
      H. Central spines 1-2.
        I. Outer perianth-segments entire . . . . . 84. *M. heyderi*
        II. Outer perianth-segments ciliate . . . . . 85. *M. infernillensis*
      HH. Central spines 4-6 . . . . 86. *M. formosa*
    GG. Radial spines more than 30 . . . . 87. *M. morganiana*
AA. Tubercles with semi-milky sap, seeds brown (see below for AAA).
  B. Central spines straight.
    C. Stigma-lobes purple . . . . 88. *M. crucigera*
    CC. Stigma-lobes yellow to green.
      D. Flowers tubular . . . . 89. *M. vaupelii*
      DD. Flowers campanulate . . . . 90. *M. rekoiana*
  BB. Central spines mostly straight but in some varieties occasionally
    hooked . . . . 91. *M. guerreronis*

Key to the Species

BBB. Central spines strongly hooked.
    C. Radial spines to 20 . . . . 92. *M. rekoi*
    CC. Radial spines 30 . . . . 93. *M. zapilotensis*
A. Tubercles with watery sap, seeds mostly black but a few brown.
  B. Central spines absent (occasionally present in very robust specimens)
    C. Radial spines less than 10.
      D. Bristles in axils, tubercles elongate.
        E. Outer perianth-segments serrate . . . . 94. *M. camptotricha*
        EE. Outer perianth-segments entire . . . . 95. *M. albescens*
      DD. No bristles in axils, tubercles not elongate.
        E. Radial spines ascending . . . . 96. *M. kewensis*
        EE. Radial spines horizontal . . . . 97. *M. durispina*
    CC. Radial spines 10-20 (see below for CCC).
      D. Inner perianth-segments reddish.
        E. Outer perianth-segments entire . . . . 98. *M. napina*
        EE. Outer perianth-segments serrate . . . . 99. *M. lanata*
      DD. Inner perianth-segments yellowish.
        E. Branches easily detached, seeds black . . . . . . 100. *M. fragilis*
        EE. Branches not easily detached, seeds brown . . . . 101. *M. elongata*
    CCC. Radial spines more than 20.
      D. Bristles in axils.
        E. Bristles stout, 7-8 . . . . 102. *M. humboldtii*
        EE. Bristles only occasional.
          F. Inner perianth-segments red . . . . 103. *M. viperina*
          FF. Inner perianth-segments whitish . . . . . . 104. *M. lenta*
      DD. No bristles in axils.
        E. Spines plumose . . . . 105. *M. plumosa*
        EE. Spines not plumose.
          F. Radial spines less than 30 . . . . 106. *M. aureilanata*
          FF. Radial spines more than 40.
            G. Outer perianth-segments entire . . . . . . 107. *M. herrerae*
            GG. Outer perianth-segments serrate.
              H. Axils with wool and hair . . . . 108. *M. schiedeana*
              HH. Axils naked . . . . 109. *M. lasiacantha*
  BB. Central spines present.
    C. Central spines hooked.
      D. Radial spines less than 10.
        E. Central spines 1 . . . . 110. *M. criniformis*
        EE. Central spines 3-4 . . . . 111. *M. wildii*
      DD. Radial spines 10-20 (see below for DDD).
        E. Bristles in axils.
          F. Seeds brownish.
            G. Inner perianth-segments bright yellow . . . . 112. *M. aurihamata*
            GG. Inner perianth-segments white . . . . . . 113. *M. leucantha*
          FF. Seeds black.
            G. Central spines only 1 . . . . 114. *M. capensis*
            GG. Central spines usually more than 1.
              H. Stigma-lobes reddish to orange.
                I. Tubercles soft in texture.

                    J. Inner perianth-segments red, small . . . 115. *M. erythrosperma*
                    JJ. Inner perianth-segments yellow, very large 116. *M. balsasoides*
                II. Tubercles firm in texture.
                    J. Axillary bristles 1-3 . . . . 117. *M. fraileana*
                    JJ. Axillary bristles 15-20 . . . . 118. *M. verhaertiana*
        HH. Stigma-lobes greenish (see below for HHH).
            I. Outer perianth-segments entire.
                J. Tubercles firm in texture.
                    K. Tubercles terete . . . . 119. *M. dioica*
                    KK. Tubercles angled . . . . 120. *M. armillata*
                JJ. Tubercles soft in texture.
                    K. Tubercles arranged in 5 and 8 spirals . 121. *M. glochidiata*
                    KK. Tubercles arranged in 13 and 21 spirals . 122. *M. pygmaea*
            II. Outer perianth-segments ciliate . . . . . 123. *M. angelensis*
        HHH. Stigma-lobes white.
            I. Outer perianth-segments entire . . . . . 124. *M. icamolensis*
            II. Outer perianth-segments serrate . . . . 125. *M. pubispina*
    EE. Bristles only occasionally present in axils
        (see below for EEE).
        F. Tubercles firm in texture.
            G. Tubercles arranged in 5 and 8 spirals, body slender 126. *M. occidentalis*
            GG. Tubercles arranged in 8 and 13 spirals, body robust 127. *M. swinglei*
        FF. Tubercles soft in texture . . . . 128. *M. trichacantha*
    EEE. No bristles in axils.
        F. Stigma-lobes orange to red.
            G. Stigma-lobes dark red.
                H. Tubercles elongate . . . . 129. *M. mainae*
                HH. Tubercles short . . . . 130. *M. fasciculata*
            GG. Stigma-lobes orange . . . . 131. *M. beneckei*
        FF. Stigma-lobes yellow to green.
            G. Central spines usually 1.
                H. Outer perianth-segments entire.
                    I. Inner perianth-segments white with red
                        mid-stripe . . . . 132. *M. carretii*
                    II. Inner perianth-segments orange-yellow . . . 133. *M. surculosa*
                HH. Outer perianth-segments ciliate.
                    I. Tubercles elongate and flattened . . . . 134. *M. zephyranthoides*
                    II. Tubercles short and conic . . . . 135. *M. hutchisoniana*
            GG. Central spines 3-6.
                H. Tubercles more or less firm in texture.
                    I. Outer perianth-segments ciliate.
                        J. Central spines 1-3 . . . . 136. *M. sheldonii*
                        JJ. Central spines 4-6 . . . . 137. *M. neocoronaria*
                    II. Outer perianth-segments entire.
                        J. Radial spines 11-15, tubercles arranged in
                            8 and 13 spirals . . . . . . 138. *M. goodridgei*
                        JJ. Radial spines 15-20, tubercles arranged in
                            13 and 21 spirals . . . . 139. *M. blossfeldiana*

Key to the Species

        HH. Tubercles soft in texture.
            I. Inner perianth-segments yellow . . . . . 140. *M. crinita*
            II. Inner perianth-segments pink to purple.
                J. Outer perianth-segments ciliate.
                    K. Stigma-lobes 11, yellow; radial spines
                        12-14 . . . . 141. *M. wrightii*
                    KK. Stigma-lobes 7, green; radial spines
                        14-22 . . . . 142. *M. wilcoxii*
                JJ. Outer perianth-segments entire.
                    K. Inner perianth-segments purple . . . 143. *M. zeilmanniana*
                    KK. Inner perianth-segments white to pink
                      with reddish mid-stripe
                        L. Seeds black . . . . 144. *M. schelhasii*
                        LL. Seeds brownish black . . . . 145. *M. rettigiana*
DDD. Radial spines over 20.
    E. Bristles in axils.
        F. Inner perianth-segments reddish.
            G. Radial spines 20-25.
                H. Central spines mostly 2 . . . . 146. *M. umbrina*
                HH. Central spines 4-6 . . . . 147. *M. solisii*
            GG. Radial spines 30 . . . . 148. *M. multiformis*
        FF. Inner perianth-segments white to yellow.
            G. Seeds brownish.
                H. Central spines 2 . . . . 149. *M. erectohamata*
                HH. Central spines 4 (see below for HHH).
                    I. Radial spines 20-25 . . . . 150. *M. knebeliana*
                    II. Radial spines 40-60 . . . . 151. *M. phellosperma*
                HHH. Central spines 7-9 . . . . 152. *M. multihamata*
            GG. Seeds black.
                H. Central spines smooth . . . . 153. *M. gilensis*
                HH. Central spines pubescent.
                    I. Radial spines smooth . . . . 154. *M. hirsuta*
                    II. Radial spines pubescent.
                        J. Tubercles 4-5 mm. long . . . . . . 155. *M. longicoma*
                        JJ. Tubercles 8-15 mm. long . . . . . . 156. *M. kunzeana*
    EE. Bristles only occasional in axils (see below for EEE).
        F. Inner perianth-segments red . . . . 157. *M. bombycina*
        FF. Inner perianth-segments yellow . . . . 158. *M. seideliana*
    EEE. No bristles in axils.
        F. Radial spines 20-30.
            G. Inner perianth-segments reddish, or at least with
                reddish mid-stripe.
                H. Central spines pubescent.
                    I. Tubercles soft in texture . . . . 159. *M. scheidweileriana*
                    II. Tubercles firm in texture . . . . . . 160. *M. mercadensis*
                HH. Central spines smooth.
                    I. Central spines 1 . . . . 161. *M. insularis*
                    II. Central spines more than 1.

Key to the Species

     J. Flower tube elongate . . . . 162. *M. longiflora*
     JJ. Flower tube not elongate.
       K. Outer perianth-segments entire . . . 163. *M. posseltiana*
       KK. Outer perianth-segments ciliate.
         L. Stigma-lobes green . . . . . . 164. *M. microcarpa*
         LL. Stigma-lobes white to rose . . . . 165. *M. weingartiana*
  GG. Inner perianth-segments white to yellow.
    H. Central spines pubescent.
      I. Central spines usually 1 . . . . 166. *M. bocasana*
      II. Central spines 4-5 . . . . 167. *M. painteri*
    HH. Central spines smooth.
      I. Tubercles arranged in 8 and 13 spirals . . . 168. *M. boedekeriana*
      II. Tubercles arranged in 13 and 21 spirals . . 169. *M. sinistrohamata*
 FF. Radial spines 30 or more.
  G. Outer perianth-segments entire . . . . 170. *M. moelleriana*
  GG. Outer perianth-segments serrate.
    H. Central spines 1-2.
      I. Inner perianth-segments white to cream . . 171. *M. gasseriana*
      II. Inner perianth-segments reddish.
        J. Radial spines acicular, 50-60 . . . . . 172. *M. barbata*
        JJ. Radial spines setaceous, 60-80 . . . . 173. *M. guelzowiana*
    HH. Central spines 4-8 . . . . 174. *M. jaliscana*
CC. Central spines occasionally hooked (see below for CCC).
 D. Radial spines less than 20.
  E. Bristles in axils . . . . 175. *M. colonensis*
  EE. No bristles in axils . . . . 176. *M. guirocobensis*
 DD. Radial spines 20-40 (see below for DDD).
  E. Bristles in axils.
    F. Seeds black . . . . 177. *M. phitauiana*
    FF. Seeds brown . . . . 178. *M. haehneliana*
  EE. No bristles in axils . . . . 179. *M. oliviae*
 DDD. Radial spines to 75 . . . . 180. *M. magallanii*
CCC. Central spines straight.
 D. Radial spines absent (or at the most, only fine deciduous hairs).
  E. Tubercles terete . . . . 181. *M. hidalgensis*
  EE. Tubercles more or less angled . . . . 182. *M. tetracantha*
 DD. Radial spines present.
  E. Radial spines less than 10.
    F. Bristles in axils.
      G. Central spines 1-2 . . . . 183. *M. decipiens*
      GG. Central spines 7-11 . . . . 184. *M. viereckii*
    FF. No bristles in axils.
      G. Radial spines 6, subulate . . . . 185. *M. kelleriana*
      GG. Radial spines 8-10, acicular . . . . . . 186. *M. fertilis*
  EE. Radial spines 10-20 (see below for EEE).
    F. Bristles in axils.
      G. Central spines 1-2 . . . . 187. *M. picta*
      GG. Central spines 4-6.

Key to the Species

        H. Outer perianth-segments serrate.
           I. Stigma-lobes green . . . . 188. *M. ruestii*
           II. Stigma-lobes pink to tan . . . . 189. *M. rhodantha*
        HH. Outer perianth-segments entire . . . . . . 190. *M. phaeacantha*
    FF. Only occasional bristle in axils (see below for FFF).
        G. Inner perianth-segments whitish . . . . . . 191. *M. inaiae*
        GG. Inner perianth-segments reddish.
           H. Stigma-lobes greenish.
              I. Dried perianth missing on fruit . . . . . 192. *M. sphacelata*
              II. Dried perianth persisting on fruit . . . . 193. *M. mazatlanensis*
           HH. Stigma-lobes whitish . . . . 194. *M. wiesingeri*
FFF. No bristles in axils.
    G. Central spines 1-2.
        H. Inner perianth-segments reddish.
           I. Stigma-lobes reddish.
              J. Outer perianth-segments ciliate . . . . 195. *M. perbella*
              JJ. Outer perianth-segments entire . . . . 196. *M. lesaunieri*
           II. Stigma-lobes greenish (see below for III, IIII
              and IIIII)
              J. Tubercles 4 angled . . . . 197. *M. haageana*
              JJ. Tubercles ovoid . . . . 198. *M. amoena*
           III. Stigma-lobes white.
              J. Radial spines 10-12.
                  K. Central spines 1 . . . . . . (95) *M. napina centrispina*
                  KK. Central spines 2 . . . . 199. *M. mundtii*
              JJ. Radial spines 16-18 . . . . 200. *M. collina*
           IIII. Stigma-lobes golden . . . . 201. *M. donatii*
           IIIII. Stigma-lobes uncertain . . . . 202. *M. albilanata*
        HH. Inner perianth-segments white to yellow.
           I. Radial spines 12-14 . . . . (97) *M. fragilis centrispina*
           II. Radial spines 16-18 . . . . 203. *M. echinaria*
    GG. Central spines 4-7.
        H. Inner perianth-segments reddish.
           I. Tubercles more or less angled.
              J. Radial spines stiff acicular . . . . . . 204. *M. graessneriana*
              JJ. Radial spines setaceous . . . . 205. *M. hoffmanniana*
           II. Tubercles conic.
              J. Stigma-lobes deep pink . . . . 206. *M. fuliginosa*
              JJ. Stigma-lobes yellow.
                  K. Outer perianth-segments entire . . . 207. *M. ochoterenae*
                  KK. Outer perianth-segments serrate . . . 208. *M. pringlei*
        HH. Inner perianth-segments white (or with reddish
           mid-line).
           I. Stigma-lobes green . . . . 209. *M. discolor*
           II. Stigma-lobes white . . . . 210. *M. albicans*
           III. Stigma-lobes yellow . . . . 211. *M. esperanzaensis*
EEE. Radial spines more than 20.
    F. Bristles in axils.
        G. Radial spines 20-30.

Key to the Species

        H. Central spines 1-2.
            I. Inner perianth-segments yellow . . . . . 212. *M. halbingeri*
            II. Inner perianth-segments red . . . . . . 213. *M. conspicua*
        HH. Central spines 4-6 (see below for HHH).
            I. Inner perianth-segments deep pink . . . 214. *M. nunezii*
            II. Inner perianth-segments whitish.
                J. Outer perianth-segments ciliate . . . 215. *M. ortiz rubiona*
                JJ. Outer perianth-segments entire . . . 216. *M. neopalmeri*
        HHH. Central spines 12-15 . . . . 217. *M. spinosissima*
    GG. Radial spines more than 30.
        H. Central spines 1-4 . . . . 218. *M. albicoma*
        HH. Central spines more than 4.
            I. Stigma-lobes red . . . . 219. *M. candida*
            II. Stigma-lobes yellow to green.
                J. Central spines reddish to brownish . . . 220. *M. multiceps*
                JJ. Central spines white to pale yellow . . . 221. *M. prolifera*
FF. No bristles in axils.
  G. Radial spines 20-30.
    H. Central spines 1-6.
      I. Inner perianth-segments reddish.
        J. Stigma-lobes reddish.
          K. Tubercles more or less angled . . . 222. *M. calacantha*
          KK. Tubercles conic.
            L. Central spines 2 . . . . 223. *M. pseudoperbella*
            LL. Central spines 4-6.
              M. Inner perianth-segments carmine 224. *M. celsiana*
              MM. Inner perianth-segments purplish 225. *M. fuscata*
        JJ. Stigma-lobes yellow to greenish.
          K. Outer perianth-segments entire . . . 226. *M. elegans*
          KK. Outer perianth-segments serrate.
            L. Flower tubular . . . . 227. *M. columbiana*
            LL. Flower campanulate . . . . . 228. *M. yucatanensis*
      II. Inner perianth-segments yellow.
        J. Tubercles more or less angled.
          K. Spines pubescent . . . . 229. *M. eriacantha*
          KK. Spines smooth . . . . 230. *M. densispina*
        JJ. Tubercles conic . . . . 231. *M. vetula*
    HH. Central spines 10-15 . . . . 232. *M. schmollii*
  GG. Radial spines more than 30.
    H. Inner perianth-segments white to yellow.
      I. Central spines 1-8.
        J. Spines thin acicular to setaceous.
          K. Stigma-lobes linear . . . . 233. *M. baumii*
          KK. Stigma-lobes revolute . . . . . 234. *M. radiaissima*
        JJ. Spines stout acicular . . . . 235. *M. microhelia*
      II. Central spines 8-11 . . . . 236. *M. droegeana*
    HH. Inner perianth-segments rose to purple.
      I. Outer perianth-segments entire . . . . . 237. *M. microheliopsis*
      II. Outer perianth-segments serrate . . . . . 238. *M. pottsii*

Fig. 2. *Mammillaria pyrrhocephala* x 0.8

1. Mammillaria pyrrhocephala Scheidweiler

Mammillaria pyrrhocephala Scheidweiler, Allg. Gartenz., 9:42, 1841.
Mammillaria maletiana(*) Cels, Portef. Hort., 2:222.
Mamillaria pyrrhocephala donkelaeri Salm-Dyck, Cact. Hort. Dyck. 1849, 17 & 121, 1850.
Cactus pyrrhocephalus Kuntze, Rev. Gen. Pl., 1:261, 1891.
Neomammillaria pyrrhocephala Britton & Rose, The Cactaceae, 4:99, 1923.
Mammillaria maletiana pyrrhocephala Schelle, Kakteen, 337, 1926.
Mammillaria maletiana pyrrhocephala fulvolanata Schelle, Kakteen, 337, 1926.
Mammillaria pyrrhocephala maletiana (Cels) Borg, Cacti, 354, 1937.
Mammillaria pyrrhocephala fulvolanata (Hildm.) Borg, Cacti, 354,1937.
Mammillaria pyrrhocephala confusa (B. & R.) Borg, Cacti, 354, 1937.

BODY simple and cespitose from base and sides, cylindric, with apex sunken, to 8 cm. high. TUBERCLES closely set in 13 and 21 spirals, greenish to nearly bluish, 4-5 angled, keeled ventrally, with milky sap, 10 mm. long, 6-7 mm. wide at base. AREOLES oval, 2-3 mm. wide, with white to tan wool, later becoming naked. AXILS with white or brownish wool and long bristles. CENTRAL SPINES none (rarely 1, to 3 mm. long, straight, black, later becoming gray with black tip, erect, often it is only a more centrally placed radial spine). RADIAL SPINES 4-6, 2-4 mm. long, upper longer, all subulate, straight to slightly recurved, very dark reddish brown with lighter red base, strongly ascending. FLOWERS funnelform, 20 mm. long. *Outer perianth-segments* pale olive-green below, bright red above, with brownish ventral-stripe, short-lanceolate, tip acuminate, margins entire. *Inner perianth-segments* deep pink mid-line, pale pink margins, lanceolate, tip acuminate, margins entire, recurved. *Filaments* deep rose. Anthers pink with tan pollen.

(*) Frequently mispelled with *ll* and *tt* but *maletiana* is the form used in the original as the plant was named after M. Malet.

Style yellowish to whitish below. *Stigma-lobes* 4-6, greenish to tannish pink. FRUIT pink below to light green above, clavate, 20x6 mm., with dried perianth persisting. SEEDS light tan, curved pyriform with lateral hilum near base, nearly smooth, 1 mm.

Distribution: Hidalgo, Oaxaca (?), Mexico.

Type locality: Real del Monte, Hgo.

Illustration: Fig. 2 is from a photograph of a plant sent to us by Sr. F. Schmoll of Cadereyta, Qro. The illustration in Britton & Rose, The Cactaceae, 4:103, Fig. 100 as *Neomammillaria pyrrhocephala* is not referrable here but it is probably nearer one of the varieties of *Mammillaria confusa*.

FIG. 3. *Mammillaria compressa*

2. Mammillaria compressa DeCandolle

Mammillaria compressa DeCandolle, Mem. Mus. Hist. Nat. Paris, 17:112, 1828.
Mammillaria subangularis DeCandolle, Mem. Mus. Hist. Nat. Paris, 17:112, 1828.
Mammillaria triacantha DeCandolle, Mem. Mus. Hist. Nat. Paris, 17:113, 1828.
Mammillaria cirrhifera Martius, Nov. Act. Nat. Cur., 16:334, 1832.
Mammillaria angularis Link & Otto in Pfeiffer, Enum. Cact., 12, 1837.
Mammillaria cirrhifera angulosior Lemaire, Cact. Gen. Nov. Sp., 95, 1839.
Mammillaria longiseta Mühlenpfordt, Allg. Gartenz., 13:346, 1845.
Mamillaria cirrhifera longiseta Salm-Dyck, Cact. Hort. Dyck. 1849, 18, 1850.
Mamillaria plinthimorpha Jacobi, Allg. Gartenz., 24:92, 1856.
Mamillaria squarrosa Meinshausen, Wochenschr. Garten. Pflanz., 2:116, 1859.
Cactus cirrhifer Kuntze, Rev. Gen. Pl., 1:260, 1891.
Cactus compressus Kuntze, Rev. Gen. Pl., 1:260, 1891. Not Salisbury 1796.
Cactus longisetus Kuntze, Rev. Gen. Pl., 1:260, 1891.
Cactus squarrosus Kuntze, Rev. Gen. Pl., 1:261, 1891.
Cactus subangularis Kuntze, Rev. Gen. Pl., 1:261, 1891.
Cactus triacanthus Kuntze, Rev. Gen. Pl., 1:261, 1891.
?*Mammillaria angularis fulvispina* Schumann, Gesamtb. Kakteen, 576, 1898.
Mammillaria angularis longiseta Salm-Dyck in Schumann, Gesamtb. Kakteen, 576, 1898.

Mamillaria angularis triacantha Salm-Dyck in Schumann, Gesamtb. Kakteen, 576, 1898.
Mamillaria angularis compressa Schumann, Gesamtb. Kakteen, 577, 1898.
Mamillaria oettingenii Zeissold, Monatsschr. Kakt., 8:10, 1898.
Mamillaria kleinschmidtiana Zeissold, Monatsschr. Kakt., 8:21, 1898.
Neomammillaria compressa Britton & Rose, The Cactaceae, 4:90, 1923.
Mammillaria compressa compressa (Schum.) Borg, Cacti, 352, 1937.
Mammillaria compressa fulvispina (Schum.) Borg, Cacti, 352, 1937.
Mammillaria compressa longiseta (Salm-Dyck) Borg, Cacti, 353, 1937.
Mammillaria compressa triacantha (Salm-Dyck) Borg, Cacti, 353, 1937.
Mammillaria compressa rubispina Borg, Cacti, 353, 1937.

FIG. 4. *Mammillaria compressa*

BODY simple and cespitose to form large mounds of 1 meter wide, individual heads globose to clavate-cylindric, to 20 cm. high, 5-8 cm. wide. TUBERCLES closely set in 8 and 13 spirals, firm in texture, light bluish gray-green, compressed 4-sided at base, ovate above, more or less bluntly angled, keeled ventrally, with milky sap, 4-6 mm. long, 8-15 mm. wide at base. AREOLES oval, small, with white wool in youth. AXILS with dense white wool, and strong white bristles. CENTRAL SPINES none. RADIAL SPINES 4-6 (sometimes with 1-3 very short accessory spines from the lower part of the areole), very unequal and variable in length, 20-70 mm. long, upper shorter, lower longer and heavier, all stout acicular to subulate, often angled, straight to very tortuous, short ones stiff, longer ones somewhat flexuous, all smooth, white to reddish, later becoming gray with brown tip, strongly ascending. FLOWERS campanulate, 10-15 mm. long, March. *Outer perianth-segments* wide purplish brown mid-stripe, narrow tannish pink margins, lanceolate, tip acuminate, margins ciliate to finely serrate. *Inner perianth-segments* deep purplish pink narrow mid-stripe, rose-red margins, linear-oblong, tip acute to acuminate, margins entire. *Filaments* very pale pink. *Anthers* pale greenish yellow. *Style* tan. *Stigma-lobes* 4-6, pale pinkish tan, linear. FRUIT bright red, clavate, 23x7 mm., with dried perianth persisting. SEEDS light brown, curved globular-pyriform with lateral hilum at base, nearly smooth to slightly wrinkled.

Distribution: Hidalgo, Queretaro, San Luis Potosi; Mexico.
Type locality: None given but abundant near Ixmiquilpan and Tasquillo, Hgo.
Illustrations: Fig. 3 is from a photograph taken of a plant near Tasquillo. Fig. 4 is

from a photograph of a plant sent to us by Sr. F. Schmoll of Cadereyta, Qro. Fig. 5 is from a photograph of a plant collected for us by Mr. George Lindsay.

This species was found by us in great abundance in the hills in the more desert sections of the central plateau region, especially in the neighborhood of Ixmiquilpan, Zimapan, and Actopan. It forms beautiful clusters of often over a meter in diameter with its

FIG. 5. *Mammillaria compressa*

light green body and long cream colored spines. When this is studded all over with the bright red fruits, it is certainly a sight to long remember. The spine arrangement is extremely variable in the length of the spines and different specimens appear to be different species but when the various forms are all placed together, the grading of one into another can easily be seen.

M. rufispina Allnut (Cactus, 131, 1877) probably is referrable here but the very scant description makes it uncertain. *M. subcirrhifera* Foerster (Handb. Cact., 234, 1846) is only a name which was referred by Schumann to *M. angularis*.

3. Mammillaria seitziana Martius

Mammillaria seitziana Martius in Pfeiffer, Enum. Cact., 18, 1837.
Mamillaria foveolata Mühlenpfordt, Allg. Gartenz., 14:372, 1846.
Cactus foveolatus Kuntze, Rev. Gen. Pl., 1:260, 1891.
Cactus seitzianus Kuntze, Rev. Gen. Pl., 1:261, 1891.
Neomammillaria seitziana Britton & Rose, The Cactaceae, 4:83, 1923.

BODY simple and later branching from base, globose to cylindric, with apex deeply sunken, to 9 cm. wide. TUBERCLES arranged in 8 and 13 spirals, firm in texture, green, pyramidal, angled to tip, with milky sap, 10-14 mm. long, 7 mm. wide at base. AREOLES round, 3 mm., with white wool in youth, later becoming nearly naked and sunken. AXILS with wool and bristles (bristles do not appear in new growth in cultivation but are found in the older axils in imported plants). CENTRAL SPINES none. RADIAL SPINES 4-6, length variable, lower longest to 30 mm., lateral shortest 3-4 mm., all stout acicular, straight to recurved, stiff, smooth, flesh-colored to yellowish brown, with black tip, strongly ascending, lateral ones more dorsal. FLOWERS funnelform, 25 mm. long, 15 mm. wide. *Outer perianth-segments* olive to brownish light pink with browner red mid-line only near the tip, very narrow clear margins, linear-lanceolate, tip acute, margins ciliate. *Inner perianth-segments* pale rose-white, red mid-line, linear, tip acute to acuminate, margins ciliate near the base, entire above. *Filaments* white to very pale pink.

FIG. 6. *Mammillaria seitziana* x 1

Anthers pale yellow. *Style* nearly white at base, medium dark pink above. *Stigma-lobes* 6-7, light cream to nearly white, pink line ventrally, 1.5-2 mm. long, overtop anthers 2 mm. FRUIT unknown. SEEDS brown.

Distribution: Hidalgo, Mexico.

Type locality: Ixmiquilpan and also reported from Zimapan, Hgo.

Illustration: Fig. 6 is from a photograph of a plant sent to us by Sr. F. Schmoll of Cadereyta, Qro.

"*M. senckeana* and *M. senckei* are two names listed as syononyms of this species but we do not find that they have ever been published." Britton & Rose, Cact., 4:83.

Ehrenberg (Linnaea, 349, 1847) gives the type locality as Oaxaca and not Hidalgo but we question this change.

4. **Mammillaria polyedra** Martius

Mammillaria polyedra Martius, Nov. Act. Nat. Cur., 16:326, 1832.
Mammillaria villifera Otto in Pfeiffer, Enum. Cact., 18, 1837.
Mammillaria polytricha Salm-Dyck, Allg. Gartenz., 10:289, 1842.
Mammillaria polytricha tetracantha Salm-Dyck, Allg. Gartenz., 10:290, 1842.
Mammillaria polytricha hexacantha Salm-Dyck, Allg. Gartenz., 10:289, 1842.
Mammillaria polytricha laevior Salm-Dyck in Labouret, Monogr. Cact., 105, 1853.
Mammillaria polytricha scleracantha Labouret, Monogr. Cact., 105, 1853.
Cactus polyedrus Kuntze, Rev. Gen. Pl., 1:261, 1891.
Cactus villifer Kuntze, Rev. Gen. Pl., 1:261, 1891.
Cactus polytrichus Kuntze, Rev. Gen. Pl., 1:261, 1891.
Neomammillaria villifera Britton & Rose, The Cactaceae, 4:102, 1923.
Neomammillaria polyedra Britton & Rose, The Cactaceae, 4:102, 1923.

Fig. 7. *Mammillaria polyedra* x 0.8

BODY simple and cespitose, globular to cylindric with sunken apex, heads to 10 cm. wide and high, clumps to 30 cm. wide. TUBERCLES widely separated in 13 and 21 spirals, very firm in texture, deep green, flattened pyramidal, sharply angled, more so at the base, flattened dorsally, strongly keeled ventrally, with milky sap, 12 mm. long, 6-11 mm. wide at base, wider laterally. AREOLES sunken eliptical to inverted pyriform, 2-3 mm. wide, with dense wool in apex, soon becoming naked. AXILS with dense white wool in youth and in flowering and fruiting areas, also with white bristles to length of tubercles, but appearing only in fruiting and older axils. CENTRAL SPINES none. RADIAL SPINES 2-6, mostly 4, unequal length, 1-3 uppers 3-4 mm., lower one to 25 mm. long, all straight or slightly curved, upper slender acicular, lower acicular to slender subulate, all stiff, smooth, black to purplish brown in youth, later becoming grayish ivory with purplish tip, strongly ascending. FLOWERS funnelform, 20 mm. wide, 15-25 mm. long. May to June. *Outer perianth-segments* reddish tan mid-stripe to somewhat greenish above, very pale yellowish tan to nearly white margins, linear-lanceolate, tip acuminate, margins lightly ciliate. *Inner perianth-segments* pale pink to rose, darker mid-line, with apex darker and somewhat tannish, darker ventrally, linear-elliptical, tip acute, margins entire or with ends split. *Filaments* white to very pale yellow, numerous. *Anthers* yellow. *Style* white to very pale greenish. *Stigma-lobes* 6-8, greenish white to yellowish green, 2 mm. long, overtop anthers 2-3 mm. FRUIT scarlet, wide clavate, very juicy, 10x20 mm., with dried perianth not persisting. SEEDS reddish brown, curved pyriform with lateral hilum, finely pitted, 1.7 mm. long. ROOTS fibrous.

Distribution: Oaxaca, Mexico.

Type locality: None given but Bravo and Lindsay both report it from near Oaxaca City, O.

Illustration: Fig. 7 is from a photograph of a plant collected for us by Mr. George Lindsay.

"*Mammillaria anisacantha* Hortus first appeared as a synonym of *M. polyedra anisacantha* Salm-Dyck (Hort. Dyck. 1844, 11, 1845) and then as a synonym *of M. polyedra laevior* Salm-Dyck (Cact. Hort. Dyck. 1849, 17, 1850) ; neither of the varieties was here described, but the latter was briefly characterized by Labouret. *M. scleracantha* is cited from Monville's Catalogue of 1846, but we have not seen this publication; it does occur as a synonym of *M. polyedra scleracantha* in Labouret's Monograph, 105, 1853." Britton & Rose Cact., 4:103.

M. villifera Otto is being referred here because the separation of the two species by Britton & Rose by one more or less spine is hardly a strong differenciating factor and because Ehrenberg in 1847 reports finding it in Oaxaca, the distribution area of *M. polyedra*. It was often compared by early writers with *M. mystax* Hort. (Not *M. mystax* Martius.) We have not been able to obtain any definite data on the Hort. type.

M. polygona Zuccarini in Pfeiffer (Enum. Cact., 17, 1837) is often referred here but it was never described. The name was later used for a different plant.

Schumann, followed by Boedeker, reported the presence of central spines but the original description did not contain any such reference to them and our imported plants did not present any evidence of any. There is a probability that a plant now and then will present a central spine but it is not typical. We found a similar condition in the closely related species of *M. karwinskiana* and *M. confusa*.

Mammillaria subpolyedra Salm-Dyck

Mammillaria subpolyedra Salm-Dyck, Dyck. Verz. Gart., 343, 1834.
Mammillaria jalapensis Pfeiffer, Enum, Cact., 17, 1837. Nomen.
Cactus polyedrus Kuntze, Rev. Gen. Pl., 1:261, 1891.
Neomammillaria subpolyedra Britton & Rose, The Cactaceae, 4:105, 1923.

This species is doubtlessly referrable to the above species because the differences as listed in the original description as "less prominent angles of the tubercles, darker color of the spines, inner perianth-segments fewer in number and more obtuse" could well come within the variation of any species. Britton & Rose for some unknown reason referred it to the group with the non-milky sap. Until further exploration can establish more pronounced differences, it will be considered here as synonymous with or at the most a possible variety of *M. polyedra*.

5. Mammillaria confusa (B. & R.) Orcutt

Neomammillaria confusa Britton & Rose, The Cactaceae, 4:102, 1923.
Mamillaria confusa Orcutt, Cactography, 7, 1926.

BODY simple and cespitose by dichotomous and basal branching, flattened globose to cylindric with sunken apex, to 15 cm. high, to 10 cm. wide. TUBERCLES widely separated in 8 and 13 spirals, firm in texture, dark green in youth, becoming light grayish olive-green, somewhat 4-sided, not angled, keeled ventrally, with milky sap, 10 mm. long, 8-9 mm. wide at base. AREOLES inverted pyriform, sunken, with dirty cream colored wool only in youngest. AXILS with white wool and 10-12 tortuous white bristles which are longer than the tubercles. CENTRAL SPINES usually none (but sometimes 1-2, very strong subulate. See varieties.) RADIAL SPINES 4-6, 5-30 mm. long, lower longer, all acicular to strong subulate, straight to curved, smooth, stiff, in youth black, becoming tan-cream with brown tips, later chalky, strongly ascending but later becoming nearly horizontal. FLOWERS funnelform, 15-20 mm. long, 10 mm. wide, March to fall months. *Outer perianth-segments* pale green below, bright red tapering mid-stripe above, black tip, tannish olive-green to pale yellow margins, lanceolate, tip acuminate, margins ciliate. *Inner perianth-segments* very pale greenish white to very pale greenish yellow, pale fine reddish mid-line, linear-lanceolate, tip acuminate, margins mostly entire (some lightly ciliate). *Filaments* white. *Anthers* yellow. *Style* white. *Stimga-lobes* 4-6, pale green, 1.5 mm. long, overtops anthers 2 mm. FRUIT red, clavate, 20x5 mm., with dried perianth persisting. SEEDS light tan, curved pyriform, with lateral hilum near the base, 1.1x0.5 mm., faintly roughened. ROOTS fibrous.

Distribution: Oaxaca, Mexico.

Type locality: None given but collected near Oaxaca, O.

FIG. 8. *Mammillaria confusa* var. *conzattii* x 1

a. var. **conzattii** (B. & R.) comb. nov.

Neomammillaria conzattii Britton & Rose, The Cactaceae, 4:103, 1923.
Mamillaria conzattii Orcutt, Cactography, 7, 1926.

Radial spines 4-5, acicular, 5-9 mm.

Distribution: Oaxaca, Mexico.

Type locality: Cerro San Felipe and also reported from Monte Alban, O.

Illustration: Fig. 8 is from a photograph of a plant we collected at Monte Alban near Oaxaca, O., in 1942.

The original description calls for "3 white stigma-lobes" but in all the material that we collected, we found no pure white ones but some were quite pale green.

FIG. 9. *Mammillaria confusa* var. *centrispina*

b. var. **centrispina** var. nov.

Spinis centralibus 1-2, subulatis validis, ad 25 mm. longis, rectis aut curvis, porrectis; spinis radialibus 4-6, ad 15 mm. longis, rectis, aut curvis.

Central spines 1-2, strong subulate, to 25 mm. long, straight to curved, smooth, porrect; radial spines 4-6, to 15 mm., straight to curved.

Distribution: Oaxaca, Mexico.

Type locality: Between Oaxaca and Milta, O.

Illustration: Fig. 9 is from a photograph of a plant collected by us at the type locality in 1942. The illustration in Britton & Rose (Cact., 4:103, Fig. 100) as *Neomammillaria pyrrhocephala* is referrable here instead of to the latter species.

This variety is very close to *M. collinsii* and may be synonymous with it, but in order to be certain, additional data is necessary on the plants from San Geronimo, O., which is a couple of hundred miles to the southeast.

c. var. **robustispina** var. nov.

Spinis radialibus 4-5, acicularibus validis ad subulatis validis, 5-9 mm. longis.

Radial spines 4-5, strong acicular to strong subulate, 5-9 mm. long.

Distribution: Oaxaca, Mexico.

Type locality: Between Oaxaca and Mitla, O.

Illustration: Fig. 10 is from a photograph of a plant collected by us at the type locality in 1942.

This species is quite variable in the spine structure and arrangement but the flowers are quite constant. The three varieties intergrade one with another and the different characteristics are often found on the same plant. We collected a number of specimens in the hills around Oaxaca City, and on the way to Mitla which is 25 miles to the south-east and they were very inconsistent in the spine characters. Specimens from the same site ranged from acicular to very subulate spines and some exhibited the heavy central spines in young plants and others did so only in the very large plants while other large plants

Fig. 10. *Mammillaria confusa* var. *robustispina*

did not show any central spines at all. We visited Prof. C. Conzatti in Oaxaca City and he directed us to the hills where *M. conzattii* could be found and we noted the same variations.

6. Mammillaria karwinskiana Martius

Mammillaria karwinskiana Martius, Nov. Act. Nat. Cur., 16:335, 1832.
Mammillaria fischeri Pfeiffer, Allg. Gartenz., 4:257, 1836.
Mammillaria karwinskiana flavescens Zuccarini in Pfeiffer, Enum. Cact., 19, 1837.
Mammillaria virens Scheidweiler, Allg. Gartenz., 9:43, 1841.
Mammillaria karwinskiana virens Salm-Dyck, Hort. Dyck. 1844, 10, 1845.
Cactus fischeri Kuntze, Rev. Gen. Pl., 1:260, 1891.
Cactus karwinskianus Kuntze, Rev. Gen. Pl., 1:260, 1891.
Cactus virens Kuntze, Rev. Gen. Pl., 1:261, 1891.
Neomammillaria karwinskiana Britton & Rose, The Cactaceae, 4:95, 1923.

BODY simple and cespitose, branching from base and by dichotomous division, globular to oval cylindric with rounded apex. TUBERCLES arranged in 13 and 21 spirals, firm in texture, glossy, bright to bluish to dark green, pyramidal, angled below, nearly terete at apex, with milky sap, 6-8 mm. long, 7-8 mm. wide at base. AREOLES oval, sunken, to 2.5 mm. wide, with white wool in youth, becoming naked very soon. AXILS with white wool and ivory tortuous bristles with brown tip that are nearly as long as the tubercles. CENTRAL SPINES usually none (but occasionally 1, to 25 mm. long, strong subulate, straight to recurved, porrect or dorsally recurved. See variety). RADIAL SPINES 4-6, upper and lower 10-30 mm. long, lateral 4-16 mm. long, all heavy acicular to subulate, nearly straight to slightly recurved, stiff, at first blood-red, later becoming chalky white below, with dark reddish brown to black tip, ascending with sometimes one nearer the center. FLOWERS funnelform, 20 mm. long, 15 mm. wide. *Outer perianth-segments* 10-16, base pale greenish yellow, above purplish with reddish greenish purple back stripe, pale greenish yellow margins, linear-lanceolate, tip acuminate, margins ciliate. *Inner perianth-segments* white base, red to carmine mid-line above, white to cream-colored wide margins, oblong, tip acute to short aristate, margins entire. *Filaments* yellowish white. *Anthers* sulphur-yellow, small, oval. *Style* whitish to very pale yellowish cream. *Stigma-lobes* 4-6, pale yellowish green to yellow, linear, overtop anthers 1-2

Fig. 11. *Mammillaria karwinskiana* x 0.7

mm. FRUIT red, clavate, to 20 mm. long, juicy, with dried perianth persisting. SEEDS light brown, curved pyriform with lateral slanting hilum, slightly rugose, 1x0.6 mm.

Distribution: Puebla, Oaxaca (?); Mexico.

Type locality: None given but reported from Puebla-Tehuacan, P. and from Mitla, O.

Illustration: Fig. 11 is from a photograph of a plant collected by us near Tehuacan, Puebla.

This species is quite variable in its spine arrangement. In some specimens, the spines are all small and quite slender acicular, while in others they are very robust and sometimes a strong central spine is present. Schumann, followed by Boedeker, included the presence of central spines in their description of the species, but inasmuch as the original description did not contain any such reference, we are referring this variation to the variety *centrispina* Salm-Dyck. Mr. George Lindsay collected specimens for us in 1937, between Tehucan and Oaxaca City, which had no central spines but the specimens that we collected between Tehucan and Puebla in 1941 and 1942 were more variable and presented scattered central spines on the more robust specimens. The types reported from Mitla, Oaxaca, by Miss Helia Bravo may be referrable to *M. confusa* a very closely related species.

a. var. **centrispina** (Pfeiffer) Salm-Dyck

Mammillaria centrispina Pfeiffer, Allg. Gartenz. 4:258, 1838.
Mamillaria karwinskiana centrispina Salm-Dyck, Hort. Dyck. 1844, 10, 1845.
?*Mamillaria closiana* Roumey, Bull. Soc. Bot. France, 2:372, 1855.
Cactus centrispinus Kuntze, Rev. Gen. Pl., 1:260, 1891.

The central spines are usually present in only those specimens in which the spines are of the more robust type. They are not always present in all of the areoles of the same plant and vary in the same plant from a slightly heavier and often darker colored radial spines, in practically the same plane as the radials, to a very robust one which is definitely centrally placed. Here again it is difficult to absolutely define the limits of a species where the members of the same plant cluster will vary so greatly.

Fig. 12. *Mammillaria knippeliana* x 1

7. **Mammillaria knippeliana** Quehl.

Mamillaria knippeliana Quehl, Monatsschr. Kakteenk., 17:59, 1907.

BODY simple, later branching, with slightly sunken apex, 5-6 cm. high, 4-5 cm. wide. TUBERCLES arranged in 13 and 21 spirals, pyramidal, 4-sided to apex, rounded dorsally, with milky sap, 8 mm. long, 5 mm. wide at base. AREOLES round, 2 mm., with white wool but becoming naked very soon. AXILS with white wool which persists for some time, and white tortuous bristles. CENTRAL SPINES none. RADIAL SPINES 6, sometimes only 4, to 30 mm. long, lower longest, all acicular, straight, stiff, smooth, white with blood-red or brown tip, ascending. FLOWERS funnelform, isolated, not far from top, 15 mm. long, 10 mm. wide. *Outer perianth-segments* red with yellowish margins in outer scales, becoming yellow with coppery reddish mid-stripe, lanceolate, tip acute, margins finely ciliate. *Inner perianth-segments* straw-yellow, with red tip, linear-lanceolate, tip acute, margins finely ciliate. *Filaments* white. *Anthers* straw-yellow. *Style* white. *Stigma-lobes* 6, bright green. FRUIT red, elongate globular, flattened, 9x5 mm., with dried perianth persisting. SEEDS dull tan, curved pyriform with lateral hilum, faintly reticulate, 1.2x0.6 mm. ROOTS fibrous.

Distribution: Morelos, Mexico. (Schmoll).

Type locality: None given but reported from near Cuernavaca (Schmoll).

Illustration: Fig. 12 is from a photograph of a plant obtained from Sr. F. Schmoll of Cadereyta, Qro.

The description is based on the original description and a subsequent flower description by the same author and from plants sent to us by Sr. Schmoll. The original description gives the tubercles as having a milky sap only on deep wounding, which may have been due to a more or less dormant condition of the original plants. Our imported plants that answer the description otherwise do have a milky sap in the tubercles when in a growing condition.

Britton & Rose refer this species to *M. karwinskiana* but our observations are that there is enough difference between the two to justify a distinct species. It is probably an intermediate or connecting link between *M. karwinskiana* and *M. praelii*.

FIG. 13. *Mammillaria praelii* x 1

8. **Mammillaria praelii** Mühlenpfordt

Mamillaria praelii Mühlenpfordt, Allg. Gartenz., 14:372, 1846.
Mamillaria inclinis Lemaire, Illustr. Hort., 5: Misc. 9, 1858.
Cactus praelii Kuntze, Rev. Gen. Pl., 1:261, 1891.
Neomammillaria praelii Britton & Rose, The Cactaceae, 4:96, 1923.

BODY simple and cespitose from base, globose to short cylindric with sunken apex, heads 11 cm. high, 8 cm. wide. TUBERCLES closely set in 13 and 21 spirals, firm in texture, pale to dark bluish green, pyramidal, 4-6 angled to tip, keeled ventrally, with milky sap, 10 mm. long, 6-9 mm. wide at base. AREOLES oval, 2-3 mm., with short white wool, later becoming naked. AXILS with dense white to brownish wool and 15-18 bristles as long as the tubercles. CENTRAL SPINES none. RADIAL SPINES 4-5 (mostly 4), 2-8 mm. long, lateral longer, all subulate, straight, stiff, smooth, yellowish cream to brown in youth, later becoming white with brown tip, horizontal to somewhat ascending in cross formation. FLOWERS funnelform, 15-20 mm. long, 10 mm. wide. *Outer perianth-segments* brown-red mid-stripe, greenish yellow margins, lanceolate to ovate, tip obtuse to apiculate, margins ciliate to serrate. *Inner perianth-segments* greenish yellowish white, with reddish mid-stripe, lanceolate, tip obtuse, margins entire to lightly ciliate. *Filaments* pale greenish yellow to yellowish white. *Anthers* sulphur-yellow. *Style* pale greenish yellow to yellowish white. *Stigma-lobes* 3-6, green. FRUIT red, clavate, 20x5 mm., with dried perianth persisting. SEEDS brown, curved pyriform with lateral hilum near base, faintly rugose, 1.2x0.5 mm. ROOTS fibrous.

Distribution: Guatemala to Oaxaca (?) Mexico.

Type locality: None stated.

Illustration: Fig. 13 is from a photograph of a plant in our collection.

This species, like *M. karwinskiana* and *M. confusa,* usually do not have any central spines but occasionally a plant is found in which they are present.

a. var. **viridis** Salm-Dyck

Mamillaria viridis praelii Salm-Dyck, Cact. Hort. Dyck. 1849, 16, 1850.
Mamillaria viridis Salm-Dyck, Cact. Hort. Dyck. 1849, 116, 1850.
Cactus viridis Kuntze, Rev. Gen. Pl., 1:261, 1891.

Central spines 1 (rarely 2), 6 mm. Radial spines 5-6. Flowers pale yellow, with red mid-line.

Fig. 14. *Mammillaria ortegae*

9. **Mammillaria ortegae** (B. & R.) Orcutt

Neomammillaria ortegae Britton & Rose, The Cactaceae 4:83, 1923.
Mamillaria ortegae Orcutt, Cactography, 8, 1926.

BODY simple, short clavate with sunken apex, 5-8 cm. wide. TUBERCLES arranged in 8 and 13 spirals, pyramidal, obscurely angled, keeled ventrally, with milky sap, 8-10 mm. long, broader at base. AREOLES oval, with scant white wool only in youngest. AXILS with white wool but no bristles. CENTRAL SPINES none. RADIAL SPINES 3-4, mostly 4, (sometimes 1-2 small additional spines or bristles, perhaps deciduous), 6-10 mm. long, stout acicular, straight, stiff, smooth, yellowish tan, ascending. FLOWERS unknown but reported to be reddish. FRUIT reddish, clavate, 10 mm. long. SEEDS light brown, small, angled.

Distribution: Sinaloa, Mexico.

Type locality: None stated.

Illustration: Fig. 14 is a reproduction of an illustration, Fig. 76, in Britton & Rose (Cact., 4:84) as *Neomammillaria ortegae*.

We have not been able to recollect this species to complete the description.

Mammillaria bergii Miquel

Cactus (Mammillaria) bergii Miquel, Comm. Phyt., 103, 1838.

BODY simple, somewhat globose. TUBERCLES yellowish green, conic pyramidal, somewhat rhomboidal below, with milky sap, 7-8 mm. long, 6-7 mm. wide at base. AREOLES round, with wool in youth, later becoming naked. AXILS with wool. CENTRAL SPINES none. RADIAL SPINES 4, 12-17 mm. long, upper longer, all pale grayish yellow to chalky reddish, in cross formation, divergent porrect. FLOWERS campanulate, 20 mm. long and wide. *Outer perianth-segments* 8, rose-green below, greenish yellow to reddish above, linear-lanceolate, tip acute, margins entire. *Inner perianth-segments* 12, bright rose, reddish mid-line, margins white transparent, tip obtuse, margins entire. *Filaments* rose. *Anthers* yellow. *Style* pale rose. *Stigma-lobes* 3, olive-green. FRUIT and SEEDS unknown.

Distribution: Mexico.

Type locality: Unknown.

This species is known to us only from the description as no type locality is recorded. It may be referrable to the above species which is very incomplete regarding the flower characteristics. If the missing information of the type is found to correspond to this description, the older name will have priority.

FIG. 15. *Mammillaria winteriae*

10. Mammillaria winteriae Boedeker

Mammillaria winteriae Boedeker, Monatsschr. Deutsch. Kakt. Ges., 1:119, 1929.

BODY simple, flattened globular with sunken apex, 20-30 cm. wide. TUBERCLES arranged in 8 and 13 spirals, light to bluish green, 4-angled to tip, rounded dorsally, keeled ventrally, lateral sides somewhat sunken, with milky sap, 15 mm. long, 15-25 mm. wide at base. AREOLES round to nearly square, 2 mm. wide, with some wool in youth but soon becoming naked. AXILS with scant wool at first but becoming densely woolly but no bristles. CENTRAL SPINES none. RADIAL SPINES 4, upper and lower to 30 mm., lateral to 15 mm., all straight or slightly bent, stout acicular, stiff, enlarged base, yellowish gray to faint reddish above, with brownish tip, strongly ascending in cross formation. FLOWERS funnelform, lateral, 30 mm. long, 25 mm. wide. *Outer perianth-segments* brownish red mid-stripe, wide yellowish white margins, oblong to moderate lanceolate, tip acuminate, margins entire, 10-15x4-5 mm. *Inner perianth-segments* yellowish white with whitish margins and pale sulphur-yellow mid-stripe with pale rose mid-line, oblong to moderate lanceolate, tip acuminate, margins entire, 10-15x4-5 mm. *Filaments* white below, pale violet-rose above. *Anthers* yellow. *Style* yellowish white. *Stigma-lobes* 5-9, greenish yellow, later bright yellow, overtop anthers. FRUIT pink, clavate. SEEDS bright reddish brown, slightly oblong globular with lateral deep seated small hilum, laterally compressed, 1 mm. ROOTS tuberous.

Distribution: Nuevo Leon, Mexico.

Type locality: Near Monterrey.

Illustration: Fig. 15 is a reproduction of an illustration in Monatsschr. Deutsch. Kakt. Ges., *1*:119 as *Mamillaria winteriae*.

FIG. 16. *Mammillaria zahniana*

11. **Mammillaria zahniana** Boedeker & Ritter

Mamillaria zahniana Boedeker & Ritter, Monatsschr. Deutsch. Kakt. Ges., **1**:120, 1929.

BODY simple, depressed globose with somewhat sunken apex, 10 cm. wide, 6 cm. high, almost entirely sunken into ground when dormant. TUBERCLES arranged in 8 and 13 spirals, slightly glossy, dark leaf-green, pyramidal, 4-angled to faintly rounded, keeled ventrally, rounded dorsally, with milky sap, 20 mm. long, 20 mm. or more wide at base. AREOLES round, 3 mm. wide in youth but becoming smaller, with white wool only in youngest. AXILS with very faint amount of wool, no bristles. CENTRAL SPINES none. RADIAL SPINES 4, upper three to 8 mm. long, lower one to 15 mm., all subulate, straight, stiff, smooth, base enlarged, faint whitish transparent horn-color, with short blackish tip, ascending, lateral two directed more dorsally. FLOWERS funnelform, 30 mm. long, 25 mm. wide. *Outer perianth-segments* pale greenish yellow, whitish margins, red ventral-stripe, spatulate, tip acute, margins ciliate, 13x4 mm. *Inner perianth-segments* sulphur-yellow, darker yellow near tip (never reddish), margins paler yellow, lanceolate, tip acuminate, margins ciliate, 15x4 mm. *Filaments* yellowish white. *Anthers* citron yellow. *Style* white. *Stigma-lobes* 8-10, bright green, 4 mm. long, overtop anthers. FRUIT red, clavate. SEEDS reddish brown, curved pyriform with lateral hilum at base, roughened, hardly 1 mm.

Distribution: Coahuila, Mexico.

Type locality: Saltillo.

Illustration: Fig. 16 is a reproduction of an illustration in Monatsschr. Deutsch. Kakt. Ges., 1:121 as *Mamillaria zahniana*.

Habitat: In gravel soil at an altitude of 1600-2000 meters.

FIG. 17. *Mammillaria magnimamma*

12. Mammillaria magnimamma Haworth

Mammillaria magnimamma Haworth in Till, Phil. Mag., 63:41, 1824.
Mammillaria divergens DeCandolle, Mem. Mus. Hist. Nat. Paris, 17:113, 1828.
Mammillaria gladiata Martius, Nov. Act. Nat. Cur., 16:336, 1832.
Mammillaria ceratophora Lehmann, Allg. Gartenz., 3:228, 1835.
Mammillaria recurva Lehmann in Pfeiffer, Enum. Cact., 15, 1837.
Mammillaria hystrix Martius in Pfeiffer, Enum. Cact., 21, 1837.
Mammillaria ehrenbergii Pfeiffer, Allg. Gartenz., 6:274, 1838.
Mammillaria microceras Lemaire, Cact. Aliq. Nov., 6, 1838.
Mammillaria deflexispina Lemaire, Cact. Aliq. Nov., 6, 1838.
Mammillaria arietina Lemaire, Cact. Aliq. Nov., 10, 1838.
Mammillaria versicolor Scheidweiler, Bull. Acad. Sci. Brux., 5:494, 1838.
Mammillaria conopsea Scheidweiler, Bull. Acad. Sci. Brux., 5:496, 1838.
Mammillaria gladiata aculeis rectis Scheidweiler, Bull. Acad. Sci. Brux., 6:93, 1839.
Mammillaria gladiata aculeis minimis Scheidweiler, Bull. Acad. Sci. Brux., 6:93, 1839.
Mammillaria gladiata spinis longissimis Scheidweiler, Bull. Acad. Sci. Brux., 6:93, 1839.
Mammillaria centricirrha Lemaire, Cact. Gen. Nov. Sp., 42, 1839.
Mammillaria centricirrha macrothele Lemaire, Cact. Gen. Nov. Sp., 42, 1839.
Mammillaria neumanniana Lemaire, Cact. Gen. Nov. Sp., 53, 1839.
Mammillaria conopsea longispina Scheidweiler, Bull. Acad. Sci. Brux., 6:92, 1839.
Cactus magnimamma Salm-Dyck in Steudel, Nom. ed 2, 1:246, 1840.
Mammillaria subcurvata Dietrich, Allg. Gartenz., 12:232, 1844.
Mammillaria magnimamma arietina Salm-Dyck, Hort. Dyck. 1844, 12, 1845.
Mammillaria diadema Mühlenpfordt, Allg. Gartenz., 13:346, 1845.
Mammillaria krameri Mühlenpfordt, Allg. Gartenz., 13:347, 1845.
Mamillaria foersteri Mühlenpfordt, Allg. Gartenz., 14:371, 1846.
Mamillaria grandicornis Mühlenpfordt, Allg. Gartenz., 14:372, 1846.

Mamillaria bockii Foerster, Allg. Gartenz., 15:50, 1847.
Mamillaria pazzani Steiber, Bot. Zeit., 5:491, 1847.
Mamillaria divaricata Dietrich, Allg. Gartenz., 16:210, 1848.
Mamillaria hopferiana Link, Allg. Gartenz., 16:329, 1848.
Mamillaria glauca Dietrich in Link, Allg. Gartenz., 16:330, 1848.
Mamillaria centricirrha hopferiana Salm-Dyck, Cact. Hort. Dyck. 1849, 17, 123, 1850.
Mamillaria megacantha Salm-Dyck, Cact. Hort. Dyck. 1849, 123, 1850.
Mamillaria megacantha rigidior Salm-Dyck, Cact. Hort. Dyck. 1849, 18, 124, 1850.
Mamillaria lactescens* Meinshausen, Wochenschr. Gartn. Pflanz., 2:117, 1859.
Mammillaria falcata Hort. ex Foerster, Handb. Cact., ed. 2, 345, 1885.
Mammillaria gebweileriana Haage in Foerster, Handb. Cact. ed. 2, 358, 1885.
Mammillaria schmidtii Sencke in Foerster, Handb. Cact. ed. 2, 376, 1885.
Mammillaria krameri viridis Haage in Foerster, Handb. Cact. ed. 2, 372, 1885.
Mammillaria krausei Rebut, Catal., 7, 1886.
Cactus bockii Kuntze, Rev. Gen. Pl., 1:260, 1891.
Cactus centricirrhus Kuntze, Rev. Gen. Pl., 1:260, 1891.
Cactus conopseus Kuntze, Rev. Gen. Pl., 1:260, 1891.
Cactus diadema Kuntze, Rev. Gen. Pl., 1:260, 1891.
Cactus divergens Kuntze, Rev. Gen. Pl., 1:260, 1891.
Cactus ehrenbergii Kuntze, Rev. Gen. Pl., 1:260, 1891.
Cactus foersteri Kuntze, Rev. Gen. Pl., 1:260, 1891.
Cactus gladiatus Kuntze, Rev. Gen. Pl., 1:260, 1891.
Cactus glaucus Kuntze, Rev. Gen. Pl., 1:260, 1891.
Cactus krameri Kuntze, Rev. Gen. Pl., 1:260, 1891.
Cactus lactescens Kuntze, Rev. Gen. Pl., 1:260, 1891.
Cactus megacanthus Kuntze, Rev. Gen. Pl., 1:260, 1891.
Cactus microceras Kuntze, Rev. Gen. Pl., 1:260, 1891.
Cactus hystrix Kuntze, Rev. Gen. Pl., 1:260, 1891.
Cactus divaricatus Kuntze, Rev. Gen. Pl., 1:261, 1891. Not Lamarck 1783.
Cactus neumannianus Kuntze, Rev. Gen. Pl., 1:261, 1891.
Cactus pazzanii Kuntze, Rev. Gen. Pl., 1:261, 1891.
Cactus recurvus Kuntze, Rev. Gen. Pl., 1:261, 1891.
Cactus versicolor Kuntze, Rev. Gen. Pl., 1:261, 1891.
Cactus subcurvatus Kuntze, Rev. Gen. Pl., 1:261, 1891.
Mamillaria centricirrha magnimamma Schumann, Gesamtb. Kakteen, 582, 1898.
Mamillaria centricirrha divergens Schumann, Gesamtb. Kakteen, 582, 1898.
Mamillaria centricirrha bockii Schumann, Gesamtb. Kakteen, 582, 1898.
Mamillaria centricirrha recurva Schumann, Gesamtb. Kakteen, 582, 1898.
Mamillaria centricirrha krameri Schumann, Gesamtb. Kakteen, 582, 1898.
Mammillaria krameri longispina Haage, Cact. Cult., 134, 1900. Nomen.
Mammillaria centricirrha amoena Schelle, Handb. Kakteenk., 266, 1907.
Mammillaria centricirrha arietina Schelle, Handb. Kakteenk., 266, 1907.
Mammillaria centricirrha boucheana Schelle, Handb. Kakteenk., 266, 1907.
Mammillaria centricirrha ceratophora Schelle, Handb. Kakteenk., 266, 1907.
Mammillaria centricirrha conopsea Schelle, Handb. Kakteenk., 266, 1907.
Mammillaria centricirrha cristata Schelle, Handb. Kakteenk., 266, 1907.
Mammillaria centricirrha deflexispina Schelle, Handb. Kakteenk., 266, 1907.
Mammillaria centricirrha destorum Schelle, Handb. Kakteenk., 266, 1907.
Mammillaria centricirrha de tampico Schelle, Handb. Kakteenk., 266, 1907.
Mammillaria centricirrha diacantha Schelle, Handb. Kakteenk., 266, 1907.
Mammillaria centricirrha diadema Schelle, Handb. Kakteenk., 266, 1907.
Mammillaria centricirrha divaricata Schelle, Handb. Kakteenk., 266, 1907.
Mammillaria centricirrha ehrenbergii Schelle, Handb. Kakteenk., 267, 1907.
Mammillaria centricirrha falcata Schelle, Handb. Kakteenk., 267, 1907.
Mammillaria centricirrha foersteri Schelle, Handb. Kakteenk., 267, 1907.
Mammillaria centricirrha gebweileriana Schelle, Handb. Kakteenk., 267, 1907.
Mammillaria centricirrha gladiata Schelle, Handb. Kakteenk., 267, 1907.
Mammillaria centricirrha glauca Schelle, Handb. Kakteenk., 267, 1907.
Mammillaria centricirrha globosa Schelle, Handb. Kakteenk., 267, 1907.
Mammillaria centricirrha grandidens Schelle, Handb. Kakteenk., 267, 1907.
Mammillaria centricirrha guilleminiana Schelle, Handb. Kakteenk., 267, 1907.
Mammillaria centricirrha hystrix Schelle, Handb. Kakteenk., 267, 1907.
Mammillaria centricirrha hystrix grandicornis Schelle, Handb. Kakteenk., 267, 1907.
Mammillaria centricirrha hystrix longispina Schelle, Handb. Kakteenk., 267, 1907.
Mammillaria centricirrha jorderi Schelle, Handb. Kakteenk., 267, 1907.
Mammillaria centricirrha krameri longispina Schelle, Handb. Kakteenk., 267, 1907.
Mammillaria centricirrha krausei Schelle, Handb. Kakteenk., 267, 1907.
Mammillaria centricirrha lactescens Schelle, Handb. Kakteenk., 267, 1907.
Mammillaria centricirrha longispina Schelle, Handb. Kakteenk., 267, 1907.
Mammillaria centricirrha megacantha Schelle, Handb. Kakteenk., 267, 1907.
Mammallaria centricirrha microceras Schelle, Handb. Kakteenk., 267, 1907.

*Here was referred *M. neumannii glabrescens* Regel in Foerster, Handb. Cact., ed. 2, 370, 1885.

Mammillaria centricirrha monstii Schelle, Handb. Kakteenk., 267, 1907.
Mammillaria centricirrha moritziana Schelle, Handb. Kakteenk., 267, 1907.
Mammillaria centricirrha neumanniana Schelle, Handb. Kakteenk., 267, 1907.
Mammillaria centricirrha nordmannii Schelle, Handb. Kakteenk., 267, 1907.
Mammillaria centricirrha obconella Schelle, Handb. Kakteenk., 267, 1907.
Mammillaria centricirrha pazzanii Schelle, Handb. Kakteenk., 267, 1907.
Mammillaria centricirrha scheideana Schelle, Handb. Kakteenk., 267, 1907.
Mammillaria centricirrha schmidtii Schelle, Handb. Kakteenk., 267, 1907.
Mammillaria centricirrha spinosior Schelle, Handb. Kakteenk., 267, 1907.
Mammillaria centricirrha subcurvata Schelle, Handb. Kakteenk., 267, 1907.
Mammillaria centricirrha valida Schelle, Handb. Kakteenk., 267, 1907.
Mammillaria centricirrha versicolor Schelle, Handb. Kakteenk., 267, 1907.
Mammillaria centricirrha viridis Schelle, Handb. Kakteenk., 267, 1907.
Mammillaria centricirrha zooderi Schelle, Handb. Kakteenk., 267, 1907.
Mammillaria centricirrha zuccariniana Schelle, Handb. Kakteenk., 267, 1907.
Neomammillaria magnimamma Britton & Rose, The Cactaceae, 4:77, 1923.
Mammillaria magnimamma bockii Borg, Cacti, 345, 1937.
Mammillaria magnimamma divergens Borg, Cacti, 345, 1937.
Mammillaria magnimamma krameri Borg, Cacti, 345, 1937.
Mammillaria magnimamma recurva Borg, Cacti, 345, 1937.

BODY simple and cespitose from the base and forming large clusters, individuals are deep-seated so as to be almost level with the ground, flat topped with more or less sunken apex. TUBERCLES arranged in 8 and 13 spirals, firm in texture, grayish green to dark bluish green, 4-sided, but not sharply angled, with milky sap, 10 mm. long, 10-12 mm. wide at base. AREOLES sunken diamond shaped, with white wool in youth. AXILS with white wool especially in the flowering area. CENTRAL SPINES usually absent, but occasionally present as a more centrally placed radial spine. RADIAL SPINES 3-5, quite variable and unequal in length, upper ones shorter and straighter, lower one or two 15-25 mm. long and recurved, all subulate, smooth, horn-colored with black tip, somewhat ascending. FLOWERS 20-25 mm. long and wide. *Outer perianth-segments* dirty cream-colored margins, ventrally a wide tannish red mid-stripe, lanceolate, tip acute, margins serrate. *Inner perianth-segments* dirty cream-color, fine reddish mid-line, oblong, tip acute, margins entire. *Filaments* white to rose above. *Anthers* yellowish. *Style* nearly white. *Stigma-lobes* 5-7, tannish pink, 2 mm. long, overtop anthers. FRUIT carmine red, clavate, 20 mm. long, with dried perianth persisting. SEEDS dull brown, curved pyriform with lateral hilum, roughened but not pitted, 1.6x0.7 mm.

Distribution: Hidalgo, San Luis Potosi, Federal District; central Mexico.

Type locality: None given but common on the central plateau.

Illustration: Fig. 17 is a reproduction of an illustration, Fig. 77, in Britton & Rose (Cact., 4:78) as *Neomammillaria magnimamma.*

This species is quite common throughout the central plateau region. We collected specimens near Pachuca, Actopan, and the Pedregal near Mexico City and observed it in many places elsewhere. It forms fairly large clusters but they are usually nearly level with the ground and as such they are often overlooked because the body and the more or less tuberous root is so sunken in the ground. It is quite variable in its spine arrangement, which to some extent, has been responsible for the long list of synonyms. Slight variations have been described by authors who have never had the opportunity to observe the habitat variations but have erected species on often a single specimen with minor differences.

M. cirrosa Poselger (Allg. Gartenz., 21:94, 1853) and *M. cirrhosa* Schumann (Gesamtb. Kakteen, 582, 1898) are not referrable here because it was described as having 2-3 elongated central spines.

M. grandidens is a name listed by Schumann (Gesamtb. Kakteen, 582, 1898) as a synonym of *M. centricirrha* but we have not been able to find that it was ever described.

M. tetracentra Otto in Foerster (Handb. Cact., 214, 1848) and *Cactus tetracanthus* Kuntze (Rev. Gen. Pl., 1:260, 1891) which was referred here by Britton & Rose is very probably referrable to *M. tetracantha* because of the color of the flower.

M. nordmanniana was only mentioned by Rebut (Catal., 7, 1896) but this name was later used as variety in *M. centricirrha nordmanniana* by Schelle (Handb. Kakteenk., 266, 1907).

We are listing the synonyms of this species as given by Britton & Rose with only minor revisions as we have not attempted to edit them because we feel that much more field work is needed to straighten out some of the probable errors.

FIG. 18. *Mammillaria peninsularis* x 1

13. Mammillaria peninsularis (B. & R.) Orcutt

Neomammillaria peninsularis Britton & Rose, The Cactaceae, 4:85, 1923.
Mamillaria peninsularis Orcutt, Cactography, 8, 1926.

BODY simple and cespitose with flattened top, deep seated, 4 cm. wide. TUBERCLES arranged in 5 and 8 spirals, firm in texture, pale bluish green, sharp 4-angled, keeled ventrally, rounded dorsally, with milky sap, 8 mm. long, 4-5 mm. wide. AREOLES oval, with very slight trace of wool only in youngest. AXILS with long white wool in youth, later becoming naked. CENTRAL SPINES none to very occasionally one more centrally placed radial spine. RADIAL SPINES 4-8, upper one sometimes nearly central, 6 mm. long, all stout acicular to slender subulate, straight, stiff, smooth, pale yellow with brown tips, ascending. FLOWERS funnelform, tube to top of tubercles, near center but in old axils, lasts for several days, 15 mm. long. *Outer perianth-segments* green base, reddish brownish green mid-stripe, pale green margins, linear-lanceolate, tip acuminate, margins serrate. *Inner perianth-segments* greenish yellow, reddish brown mid-line near the tip, but fading below, lanceolate, tip acuminate, margins serrate. *Filaments* greenish white. *Anthers* deep yellow. *Style* bright green. *Stigma-lobes* 6, bright green, linear, 1 mm. long, overtop anthers. FRUIT unknown. SEEDS probably brown.

Distribution: Baja California, Mexico.

Type locality: Cape San Lucas.

Illustration: Fig. 18 is from a photograph of a plant collected for us by Mr. Howard Gates at the type locality.

FIG. 19. *Mammillaria crocidata* x 0.5

14. **Mammillaria crocidata** Lemaire

Mammillaria crocidata Lemaire, Cact. Aliq. Nov., 9, 1838.
Mammillaria webbiana Lemaire, Cact. Gen. Nov. Sp., 45, 1839.
Mammillaria uberimamma Monville in Labouret, Des Cactees, 120, 1853.
Mammillaria webbiana longispina Jacobi, Allg. Gartenz., 24:83, 1856.
Cactus crocidatus Kuntze, Rev. Gen. Pl., 1:260, 1891.
Cactus webbianus Kuntze, Rev. Gen. Pl., 1:261, 1891.
Mammillaria centricirrha uberimamma Schelle, Handb. Kakteenk., 267, 1907.
Neomammillaria crocidata Britton & Rose, The Cactaceae, 4:87, 1923.

BODY simple to sometimes branching, globose to cylindric with sunken apex, 7-8 cm. wide, to 10 cm. high. TUBERCLES arranged in 13 and 21 spirals, firm in texture, dark bluish green, pyramidal, 4-angled in youth, becoming somewhat terete later, with milky sap, 6-10 mm. long, 12 mm. wide at base. AREOLES very small, oval, with thick white flocculent wool, later becoming gray and persisting. CENTRAL SPINES none. RADIAL SPINES 2-4, (often several fine accessory hairs), upper 8 mm. long, lower 10-17 mm. long, all stout acicular to slender subulate, straight, rigid, rose-white with apex dark purple to nearly black, divergent in cross formation, ascending. FLOWERS campanulate, in crown, 15 mm. long. *Outer perianth-segments* greenish white, linear-lanceolate, tip acuminate, margins finely ciliate. *Inner perianth-segments* carmine-red, purplish mid-stripe, paler margins, linear-lanceolate, tip acuminate, margins entire. *Filaments* white below, pink to purplish above. *Anthers* pale yellow. *Style* green below, rose above. *Stigma-lobes* 4-5, reddish, spreading. FRUIT greenish turning to light pink, clavate, 18x5 mm., with dried perianth not persisting. SEEDS light dull brown, curved pyriform with lateral hilum, lightly rugose, 1.6x1 mm.

Distribution: Central Mexico; Hidalgo and Queretaro.

Type locality: None stated but reported from Real del Monte, Sierra Taro, and Mesa de la Magdalena; also from near Tecozanbla (Schmoll).

Illustration: Fig. 19 is from a photograph of a plant obtained from Sr. F. Schmoll of Cadereyta, Qro.

Britton & Rose gives the radial spines as 6-7 but the original description (and most other authors) report as 2 to mostly 4. Our imported plant follows the original.

Fig. 20. *Mammillaria roseo-alba* x 0.8

15. Mammillaria roseo-alba Boedeker

Mammillaria roseo-alba Boedeker, Monatsschr. Deutsch. Kakt. Ges., 1:87, 1929.

BODY simple, flattened globular with slightly sunken apex, 6 cm. high, to 18 cm. wide. TUBERCLES arranged in 8 and 13 spirals, dull grayish bluish green, pyramidal, angled to tip, flattened dorsally, keeled ventrally, with milky sap, 10-15 mm. long, 8 mm. wide at base. AREOLES diamond shaped, somewhat depressed, to 3 mm. wide, but later smaller, with white wool only in youngest, very soon becoming naked. AXILS with very scant white wool in youth. CENTRAL SPINES none. RADIAL SPINES 4-6, upper 1-2 smaller and often deciduous, 1 more or less subcentral, dorsal and ventral ones to 15 mm. long, dark brown to black and heavier, lateral ones 5-6 mm. long, cream with black tip, all straight to somewhat recurved, stout acicular to thin subulate, somewhat ascending, dorsal one nearly porrect. FLOWERS campanulate, 25 mm. wide, 15 mm. long. *Outer perianth-segments* light green base, tannish light green to rose-brown mid-stripe, orange tinge near tip, pale greenish ivory margins, linear-lanceolate, tip acute, margins finely ciliate, 15x3 mm. *Inner perianth-segments* white, pink to rose mid-line, linear, tip acute and often split, margins entire, 20x2 mm. *Filaments* white to pale pink. *Anthers* citron-yellow. *Style* rose to pale pink. *Stigma-lobes* 5-7, greenish yellow. FRUIT red, curved clavate, 15 mm. long. SEEDS dull brownish yellow, short curved pyriform with lateral hilum, faintly roughened but not pitted, 1.4x0.8 mm.

Distribution: Tamaulipas, Mexico.

Type locality: Progresso near Victoria.

Illustration: Fig. 20 is from a photograph of a plant obtained from Mexico.

This species has been offered in the trade as *Mammillaria rosealeuca*.

FIG. 21. *Mammillaria lloydii*

16. Mammillaria lloydii (B. & R.) Orcutt

Neomammillaria lloydii Britton & Rose, The Cactaceae, 4:89, 1923.
Mamillaria lloydii Orcutt, Cactography, 7, 1926.

BODY simple, flattened globular to cylindric with sunken apex, 6-7 cm. wide. TUBERCLES closely set in 8 and 13 spirals, firm in texture, dark green, 4-angled to tip, with milky sap, 6 mm. long, 5 mm. wide at base. AREOLES small, oval, with wool only in youth. AXILS with wool in flowering area but only slightly otherwise. CENTRAL SPINES none. RADIAL SPINES 3-4, upper and lower 6 mm. long, lateral shorter, upper one often subcentral, all subulate, straight, stiff, smooth, upper red to dark brown, others whitish, all ascending. FLOWERS wide funnelform, 12-15 mm. wide. *Outer perianth-segments* reddish brown tapering mid-stripe, very pale green margins, lanceolate, tip acute, margins serrate. *Inner perianth-segments* white with tinge of red, darker mid-stripe, lanceolate, tip apiculate, margins entire, spreading. *Filaments* pale pink above, paler below. *Anthers* yellow. *Style* pale pinkish above, paler below. *Stigma-lobes* 6, pale tan. FRUIT pink, short clavate, 8x5 mm., with dried perianth missing. SEEDS light tan, curved pyriform with lateral hilum near base, faintly wrinkled, 0.7x0.4 mm.

Distribution: Zacatecas, Mexico.

Type locality: None given.

Illustration: Fig. 21 is a reproduction of an illustration, Fig. 82, in Britton & Rose (Cact., 4:89) as *Neomammillaria lloydii*.

Fig. 22. *Mammillaria albiarmata* x 1

17. Mammillaria albiarmata Boedeker

Mammillaria albiarmata Boedeker, Jahrb. Deutsch. Kakt. Ges., 1:67, 1935-6.

BODY simple, flattened globular with slightly sunken apex but not woolly, 40 mm. wide, 15 mm. high. TUBERCLES arranged in 13 and 21 spirals, semi-flabby in texture, grass-green, 4-sided at base, pyramidal cylindric, terete above, with milky sap, 5-6 mm. long, 3 mm. wide at base. AREOLES round to oval, to 1.5 mm. wide, with white to cream wool only in youngest, soon becoming naked. AXILS naked or with very slight woolly tuft. CENTRAL SPINES none. RADIAL SPINES 20-25, 2-5 mm. long, upper ones shortest, lower ones longer, all very fine acicular, straight, smooth, stiff, with base not enlarged, white to creamy white, orange-pink at tip, soon horn colored, horizontal, in older areoles the spines are somewhat pectinate. FLOWERS wide funnelform, 20 mm. long and wide, April. *Outer perianth-segments* brownish olive-green wide tapering midstripe, narrow pale greenish white margins, linear-lanceolate, tip acute to acuminate, margins finely serrate, 5-8x2 mm. *Inner perianth-segments* creamy white, pale brownish pink to pale rose fine mid-line, linear, tip acuminate, margins finely serrate, 3 mm. wide. *Filaments* very pale pink to almost white. *Anthers* pale yellow. *Style* very pale pink to almost white. *Stigma-lobes* 4-6, bright greenish yellow, becoming white ventrally, 1 mm. long, overtop anthers 1-2 mm. FRUIT reddish pink, elongate globular, 10x5 mm. SEEDS brown, pyriform with ventral hilum, surface pitted, 1.2x0.9 mm. ROOTS thick tuberous, deep seated.

Distribution: Coahuila, Mexico.

Type locality: Near Saltillo.

Illustration: Fig. 22 is from a photograph of a plant sent to us by Sr. F. Schmoll of Cadereyta, Qro.

Boedeker reports that the perianth-segments have serrated margins but those on our plants were nearly entire.

Fig. 23. *Mammillaria strobiliana*

18. Mammillaria strobiliana Tiegel

Mammillaria strobiliana Tiegel in Moeller, Deutsch. Gartner. Zeit., 48:329, 1933.

BODY simple, conical with slightly sunken apex, 6 cm. wide. TUBERCLES arranged in 13 and 21 spirals, dark green, sharply angled pyramidal, curved, with milky sap, 10-12 mm. wide. AREOLES round, 2 mm. wide, with white wool in youth. AXILS with white hairy wool and an occasional isolated white bristle. CENTRAL SPINES 1, 12 mm. long, subulate, often hooked, ashen-gray, porrect. RADIAL SPINES 4-5, 6 mm. long, upper shorter, all heavy acicular, ashen-gray with darker tip, later becoming whitish, somewhat ascending. FLOWERS in crown near top, 18 mm. wide. *Outer perianth-segments* smudgy yellowish white, faint greenish mid-stripe ventrally, lanceolate, tip acuminate, margins ciliate. *Inner perianth-segments* very pale greenish yellow to smudgy white, often with maroon mid-line, lanceolate, tip acute, margins ciliate to entire. *Filaments* white to reddish above. *Anthers* dirty yellow. *Style* whitish. *Stigma-lobes* 4-5, pale greenish yellow, 1.5 mm. long. FRUIT carmine-red, clavate, 20x5 mm., with dried perianth persisting. SEEDS light tan, curved pyriform with lateral hilum at base, slightly roughened, 1 mm.

Distribution: Oaxaca, Mexico.

Type locality: None given.

Illustration: Fig. 23 is a reproduction of an illustration in Moeller, Deutsch. Gartner Zeit. 48:367 as *Mammillaria strobiliana*.

This species probably may be related to one of the *M. confusa* variations or *M. collinsii* group inasmuch as the similarity of the body, flower, spine count, and general distribution. This group of plants from Oaxaca varies so widely in spine arrangement but still they are so similar in other characteristics that it is very possible that they are all very closely related.

The illustration is through the courtesy of Mr. William Hertrich of the Huntington Garden of San Marino, California.

Fig. 24. *Mammillaria uncinata*

19. **Mammillaria uncinata** Zuccarini

Mammillaria uncinata Zuccarini in Pfeiffer, Enum. Cact., 34, 1837.
Mammillaria depressa Scheidweiler, Bull. Acad. Sci. Brux., 5:494, 1838.
Mammillaria uncinata spinosior Lemaire, Cact. Gen. Nov. Sp., 96, 1839.
Mammillaria uncinata rhodacantha Dietrich, Allg. Gartenz., 18:185, 1850.
Cactus depressus Kuntze, Rev. Gen. Pl., 1:260, 1891.
Cactus uncinatus Kuntze, Rev. Gen. Pl., 1:261, 1891.
Neomammillaria uncinata Britton & Rose, The Cactaceae, 4:140, 1923.

BODY simple, globose, rounded above, with somewhat sunken apex, 3-8 cm. high, 5-10 cm. wide. TUBERCLES closely set in 8 and 13 spirals, firm in texture, dark bluish green, pyramidal with angled base and terete above, with milky sap, 7-10 mm. long, 8-11 mm. wide at base. AREOLES with white wool in youth but becoming naked soon. AXILS with white wool in youth, later becoming naked. CENTRAL SPINES 1, (occasionally 2-3), 10-12 mm. long, heavy acicular to subulate, hooked, smooth, stiff, flesh-colored with darker tip, porrect. RADIAL SPINES 4-7, 5-6 mm. long, upper curved, lower straight, 4 in cross formation, 2 upper accessory and smaller, upper flesh-colored, lower white with black tip, all slightly ascending. FLOWERS funnelform, laterally placed, 15-20 mm. long, 18 mm. wide. *Outer perianth-segments* light green base, brownish green tapering mid-stripe to reddish at tip, lanceolate, tip acuminate, margins ciliate. *Inner perianth-segments* reddish white with brownish mid-line, pink tip, lanceolate, tip acute, margins entire. *Filaments* whitish to very pale pink. *Anthers* yellow to pale rose. *Style* cream to pale rose-red. *Stigma-lobes* 5-6, yellowish to rose-red, to 3 mm. long. FRUIT red, clavate, 10-18 mm. long. SEEDS brown, small.

Distribution: Hidalgo, San Luis Potosi, Guanajuato; Mexico.

Type locality: None stated but reported from Pachuca and Real del Monte, Hgo.

Illustration: Fig. 24 is a reproduction of an illustration, Fig. 313, in Bravo (Las Cact. Mex., 685), as *Mammillaria uncinata*.

A more complete description was published by Zuccarini (Plant. Nov. Monac., 715, 1837), but he refers to the Pfeiffer publication which was published in the same year.

a. var. **biuncinata** Lemaire

Mammillaria bihamata Pfeiffer, Allg. Gartenz., 6:274, 1838.
Mammillaria uncinata biuncinata Lemaire, Cact. Gen. Nov. Sp., 96, 1839.
Cactus bihamatus Kuntze, Rev. Gen. Pl., 1:260, 1891.

BODY simple, flattened globular. TUBERCLES arranged in 13 and 21 spirals. CENTRAL SPINES 2-4, 8-11 mm. long, strong subulate, lower heavier and longer, incompletely hooked, often more of a recurve, all more or less tortuous, triangular to square in cross section. RADIAL SPINES 5-6, 3-6 mm. long, strong acicular to subulate. FLOWERS unknown.

Type locality: Mineral del Monte, Hgo.

FIG. 25. *Mammillaria mendeliana* x 0.8

20. **Mammillaria mendeliana** (Bravo) Werdermann

Neomammillaria mendeliana Bravo, Anal. Inst. Biol. Mex., 2:195, 1931.
Mammillaria mendeliana Werdermann in Backeberg, Neue Kakteen, 100, 1931.

BODY simple, globose to cylindric, 8-9 cm. wide. TUBERCLES arranged in 13 and 21 spirals, firm in texture, dark olive-green, reddish at apex, quadrangular at base, nearly terete above, somewhat keeled ventrally, with a milky sap which forms a transparent rosin, 8-9 mm. long, 6-7 mm. wide at base. AREOLES oval with scant white wool in youth. AXILS with numerous long setose white hairs and abundant white wool later becoming gray, (hairs are often lost in cultivation). CENTRAL SPINES 2-4, 15-20 mm. long, lower longest, all straight, stiff, pubescent, brownish red with nearly black tip, becoming gray, when 2: divergent dorsally and ventrally, when 4: in cross formation. RADIAL SPINES not well differentiated from short white setose hairs. FLOWER funnelform, 10 mm. long, April. *Outer perianth-segments* very pale tan mid-stripe, cream-colored margins, lanceolate, tip acuminate, margins entire. *Inner perianth-segments* very pale pink margins, darker at tip, deep pink tapering mid-stripe, lanceolate, tip acuminate, margins entire. *Filaments* pale pink above, nearly white below. *Anthers* yellow. *Style* light tan. *Stigma-lobes* 3-5, yellow, less than 1 mm. FRUIT bright purple to carmine-red, compressed clavate, 20 mm. long, with dried perianth persisting. SEEDS dull light brown, curved pyriform, faintly pitted, 1.5x0.7 mm.

Distribution: Guanjuato and Queretaro, Mexico.

Type locality: None given but reported from Tarajeas.

Illustration: Fig. 25 is from a photograph of a plant sent to us by Sr. F. Schmoll of Cadereyta, Qro.

M. maltrata without author is reported in Baltimore Cact. Journ., 2:229, 1895-96 as having 3-5 central spines, very weak or no radial spines and with wool and small spines in the axils. It might be related to the above species but the lack of important details makes it uncertain.

Fig. 26. *Mammillaria carnea* Fig. 27. *Mammillaria carnea* var. *robustispina* x 0.8

21. Mammillaria carnea Zuccarini

Mammillaria carnea Zuccarini in Pfeiffer, Enum. Cact., 19, 1837.
Mammillaria subtetragona Dietrich, Allg. Gartenz., 8:169, 1840.
Mammillaria aeruginosa Scheidweiler, Allg. Gartenz., 8:338, 1840.
Mammillaria pallescens Scheidweiler, Allg. Gartenz., 9:42, 1841.
Mamillaria villifera carnea Salm-Dyck, Cact. Hort. Dyck. 1849, 16, 1850.
Mamillaria villifera aeruginosa Salm-Dyck, Cact. Hort. Dyck. 1849, 16, 1850.
Mammillaria villifera cirrosa Salm-Dyck, Cact. Hort. Dyck. 1849, 115, 1850.
Cactus aeruginosus Kuntze, Rev. Gen. Pl., 1:260, 1891.
Cactus carneus Kuntze, Rev. Gen. Pl., 1:260, 1891.
Cactus pallescens Kuntze, Rev. Gen. Pl., 1:261, 1891.
Cactus subtetragonus Kuntze, Rev. Gen. Pl., 1:261, 1891.
Mammillaria carnea cirrosa Gürke, Blühende Kakteen, 1 under pl. 60, 1905.
Mammillaria carnea aeruginosa Gürke, Blühende Kakteen, 1, under pl. 60, 1905.
Neomammillaria carnea Britton & Rose, The Cactaceae, 4:88, 1923.

BODY simple, also lateral branching, globose to cylindric, rounded above, to 100 mm. high, 85 mm. wide. TUBERCLES arranged in 8 and 13 spirals, firm in texture, dull, light to dark green, pyramidal and angled to tip, with milky sap, 13 mm. long, 8-10 mm. wide at base. AREOLES round to quadrangular, to 4 mm. wide, with whitish wool in youth but soon becoming naked. AXILS with yellowish wool in flowering area but no bristles. CENTRAL SPINES 4, occasionally 5, unequal and variable in length, longer and heavier if grown in full sun, upper 10-20 mm. long, lateral 6-15 mm. long, lower 15-40 mm. long, all subulate, straight, upper and lower slightly recurved, all stiff, flesh-colored with black tip, in nearly cross formation, strongly ascending. RADIAL SPINES none or seldom an upper bristle or two. FLOWERS funnelform, forming a ring in the top, 15-20 mm. long, 12-15 mm. wide. *Outer perianth-segments* pale green base, flesh-color to pale rose above, brownish red mid-stripe, lanceolate, tip acute, margins ciliate or serrate. *Inner perianth-segments* pale green base (in throat), pink above to darker at tip, darker to brownish mid-stripe, darker ventrally, oblong-clavate, tip mucronate, some with ends split, margins ciliate. *Filaments* pale greenish yellow. *Anthers* yellow. *Style* pale greenish yellow. *Stigma-lobes* 4-7, green to pale yellow, with brown line ventrally, 1.5 mm. long. FRUIT red, clavate, 20-25x5 mm., with dried perianth not persisting. SEEDS light brown, curved pyriform with lateral hilum, reticulate, 1.3x0.6 mm.

Distribution: Hidalgo, Puebla (Tehuacan), Guerrero (Taxco), Oaxaca; Mexico.

Type locality: Ixmiquilpan, Hgo.

Illustration: Fig. 26 is from a photograph of a plant sent to us by Sr. F. Schmoll of Cadereyta, Qro.

The original description gave the type locality as Ixmiquilpan, Hidalgo, but this is open to question as subsequent collections have all been made farther south.

a. var. **robustispina** var. nov.

Spinae robustior.

Spines heavier, lower longer to 50 mm.

Type locality: Taxco, Guerrero.

Illustration: Fig. 27 is from a photograph of a plant sent to us by Mr. George Lindsay who collected it near Taxco.

22. Mammillaria melispina Werdermann

Mammillaria melispina Werdermann, Notizbl. Bot. Gart. Berlin, 12:226, 1934.

BODY simple, sparingly basal branching, short cylindric, 8 cm. high, 6 cm. wide, found under bushes. TUBERCLES closely set, somewhat glossy green, later becoming bluish, triangular to quadrangular, with milky sap, 10 mm. long and wide at base. AREOLES slightly sunken with white wool in youth but later becoming naked. AXILS with wool in youth but later becoming naked. CENTRAL SPINES 2 (rarely 1-4), 8-12 mm. long, acicular, nearly flexuous, smooth, base enlarged, in youth yellow, later transparent and darker, base darker, porrect. RADIAL SPINES none. FLOWERS isolated near apex, funnelform, 30 mm. long. *Outer perianth-segments* yellow to greenish margins, rust-colored mid-stripe, tip acute, margins lightly serrate. *Inner perianth-segments* pale citron-yellow, tip acute, margins serrate. *Filaments* white. *Anthers* yellow. *Style* white. *Stigma-lobes* 7, pale yellow, overtop anthers. FRUIT unknown. SEEDS brown.

Distribution: Tamaulipas, Mexico.

Type locality: Jaumave.

Illustration: None known.

We know this species only from the original description. Although we have collected near this locality we have not observed any specimens of this species.

Fig. 28. *Mammillaria orcuttii*

23. **Mammillaria orcuttii** Boedeker

Mamillaria orcuttii Boedeker in Berger, Kakteen, 323, 1929. Nomen.
Mamillaria orcuttii Boedeker, Monatsschr. Deutsch. Kakt. Ges., 2:258, 1930.

BODY simple, globular to subcylindric, with depressed apex, 6-7 cm. wide. TUBERCLES arranged in 13 and 21 spirals, glossy bluish green, quadrangular at base, short conic, almost "eggshaped", keeled ventrally, with apex bent downward quite a bit, with milky sap, 6 mm. long, 4-5 mm. wide at base. AREOLES round with abundant white wool in upper part of the body, becoming naked. AXILS with persisting white wool but no bristles. CENTRAL SPINES 4, (seldom 5), lower 20 mm. long, lateral 15 mm. long, upper 8-10 mm. long, (length of spines depends on whether the plant is grown in the sun or the shade), all acicular, straight, smooth, rigid, base not enlarged, black-brown to pitch-black in new growth, brownish at base, later becoming gray, porrect. RADIAL SPINES deciduous, only on young areoles, 6-8, very short to 2 mm. long, hair-like, white. FLOWERS funnelform, 12 mm. long and wide. *Outer perianth-segments* brownish, lanceolate, tip acute, margins serrate. *Inner perianth segments* bright carmine, with darker mid-rib, lanceolate, tip acute, margins serrate, 10x2 mm. *Filaments* bright rose. *Anthers* yellow. *Style* bright rose. *Stigma-lobes* 4, dark rose, overtop anthers. FRUIT red, clavate, 13x4 mm. SEEDS reddish brown, dull, curved pyriform with lateral white hilum near base, 1.2x0.6 mm.

Distribution: Puebla, Mexico.

Type locality: Near Esperanza.

Illustration: Fig. 28 is a reproduction of an illustration in Monatsschr. Deutsch. Kakt. Ges., 2:259, as *Mamillaria orcuttii*.

24. **Mammillaria polythele** Martius

Mammillaria polythele Martius, Nov. Act. Nat. Cur., 16:328, 1832.
Mammillaria quadrispina Martius, Nov. Act. Nat. Cur., 16:329, 1832.
Mammillaria columnaris Martius, Nov. Act. Nat. Cur., 16:330, 1832.
Mammillaria affinis DeCandolle, Mem. Cact., 11, 1834.
Mammillaria setosa Pfeiffer, Allg. Gartenz., 3:379, 1835.

M. polythele

FIG. 29. *Mammillaria polythele* x 1

Mammillaria stenocephala Scheidweiler, Allg. Gartenz., 9:43, 1841.
Mammillaria polythele quadrispina Salm-Dyck in Walpers, Repert. Bot., 2:271, 1843.
Mammillaria polythele columnaris Salm-Dyck in Walpers, Repert. Bot., 2:271, 1843.
Mamillaria polythele setosa Salm-Dyck, Hort. Dyck. 1844, 9, 1845.
Mamillaria polythele hexacantha Salm-Dyck, Cact. Hort. Dyck. 1849, 15, 1850.
Mmillaria polythele latimamma Salm-Dyck, Cact. Hort. Dyck. 1849, 112, 1850.
Cactus affinis Kuntze, Rev. Gen. Pl., 1:260, 1891.
Cactus quadrispinus Kuntze, Rev. Gen. Pl., 1:261, 1891.
Cactus setosus Kuntze, Rev. Gen. Pl., 1:261, 1891.
Cactus polythele Kuntze, Rev. Gen. Pl., 1:261, 1891.
Mamillaria rhodantha stenocephala Schumann, Gesamtb. Kakteen, 550, 1898.
Mamillaria polythele affinis Schelle, Handb. Cact., 260, 1907.
Neomammillaria polythele Britton & Rose, The Cactaceae 4:88, 1923.
Mamillaria multimamma Knuth in Backeberg & Knuth, Kaktus A.B.C. 392, 1935.

BODY simple, occasionally branching, globose to cylindric, rounded above, to 50 cm. high, to 10 cm. wide. TUBERCLES arranged in 13 and 21 spirals, dark bluish green, globular conic, nearly terete, with milky sap, 10-13 mm. long, 7-8 mm. wide. AREOLES oval to 3 mm. wide, with snow white wool. AXILS with dense white wool only in apex, soon becoming naked but no bristles. CENTRAL SPINES 1-4, seldom 5, lower longest to 25 mm., all straight to slightly bent, stout acicular, stiff, yellowish brown in youth, to reddish brown, to dark brown, widely divergent, when 2: dorso-ventrally, when 4: in cross formation. RADIAL SPINES usually entirely wanting. FLOWERS funnelform, in crown, to 20 mm. long, 10 mm. wide. *Outer perianth-segments* brownish red midstripe, pale margins, pale purple ventrally, white below, linear-lancolate, tip acuminate, margins entire to lightly serrate. *Inner perianth-segments* rose to carmine, tip acute, margins occasionally serrate to entire. *Filaments* carmine to purplish red. *Anthers* yellow, small globular. *Style* below whitish or greenish yellow, above purplish. *Stigma-lobes* 3-5, reddish purple, 1 mm. long, just overtop anthers. FRUIT red, clavate, with dried perianth persisting. SEEDS dull dark brown, slightly curved pyriform with lateral hilum, roughened, less than 1 mm.

Distribution: Hidalgo, Mexico.

Type locality: None given but reported from Ixmiquilpan, Hgo.

Illustrations: Figs. 29 and 30 are from photographs of plants sent to us by Sr. F. Schmoll of Cadereyta, Qro.

Schumann places this species in the group with the watery sap but the original description by Martius states that it is milky as were the plants that we obtained from Mexico.

M. aciculata Otto in Pfeiffer (Enum. Cact., 29, 1837) and *M. polythele aciculata* Salm-Dyck (Hort. Dyck. 1844, 9, 1845) were referred here by Schumann but it is more probable they are referrable to *M. discolor* because it was described as having 20 radial spines.

"*M. columnaris minor* Martius and *M. quadrispina major* mentioned by Foerster (Handb. Cact., 214, 215, 1846) probably belong here.

"*M. cataphracta* Martius was given by Pfeiffer (Enum. Cact., 11, 1837) as a synonym of *M. affinis* and by Salm-Dyck (Hort. Dyck. 155, 1834) as a synonym of *M. angularis*." Britton & Rose, Cact., 4:88, 1923.

M. caudata was referred to *M. affinis* by Martius ex Salm-Dyck (Hort. Dyck. 155, 1834) and Martius ex Pfeiffer (Enum. Cact. 11, 1837) according to Index Kewensis 156, 1895.

M. multimamma Knuth was erected on the strength of the radial spines as reported by Schumann in his treatment of this species. These reported spines were found on our specimens on only a few of the areoles and they were more or less deciduous or mostly lacking entirely. We do not consider them as true radial spines as they probably represent more or less of a recessive characteristic and are not typical.

Fig. 30. *Mammillaria polythele*

FIG. 31. *Mammillaria woburnensis*

25. Mammillaria woburnensis Scheer

Mamillaria voburnensis* Scheer, London Journ. Bot., 4:136, 1845.
Cactus woburnensis Kuntze, Rev. Gen. Pl., 1:261, 1891.
Mamillaria chapinensis Eichlam & Quehl, Monatsschr. Kakteenk., 19:1, 1909.
Neomammillaria woburnensis Britton & Rose, The Cactaceae, 4:100, 1923.
Mammillaria chapinensis rubescens Hort. ex Schelle, Kakteen, 336, 1926.

BODY cespitose from base and body, cylindric, with rounded apex which is very slightly sunken, 50 mm. high, 30 mm. wide. TUBERCLES closely set in 8 and 13 spirals, dark green, reddish toward apex, blunt pyramidal, many sided, somewhat angled above, with milky sap, 10 mm. long, 8 mm. wide at base. AREOLES oval, with white wool in youth, later becoming naked. AXILS with white wool nearly covering the tubercles and persisting, also white somewhat tortuous bristles 4-10 mm. long. CENTRAL SPINES 1-3, 7 mm. long, straight, acicular, stiff, dark brown in youth, later ivory colored with dark reddish brown tip, porrect. RADIAL SPINES 9, 4-5 mm. long, lower 3 longer, all straight, acicular, cream with reddish brown tip, horizontal to irregularly recurved. FLOWERS funnelform, to 20 mm. long, open for several days. *Outer perianth-segments* base pale yellow, above darker yellow, often with greenish tinge, linear-lanceolate, margins irregular to very fine ciliate. *Inner perianth-segments* yellow with reddish or brownish ventral mid-stripe, 10 mm. long, tip acuminate, margins entire (?). *Filaments* white. *Anthers* bright yellow. *Style* white, 1 mm. thick. *Stigma-lobes* 5, greenish yellow, 1.5 mm. long, overtop anthers 2 mm. FRUIT red to carmine, clavate, 25x5 mm., with dried perianth persisting. SEEDS yellowish brown, 0.5 mm.

Distribution: Guatemala.

Type locality: None given.

Illustration: Fig. 31 is a reproduction of an illustration, Fig. 98, in Britton & Rose (Cact., 4:100) as *Neomammillaria woburnensis*.

Britton & Rose reported as many as 8 central spines but the original description gives only 1-2 and our imported plants had at the most only 3.

---

*The spelling of *M. voburnensis* by Scheer might very probably have been a typographical error.

26. **Mammillaria polygona** Salm-Dyck

Mamillaria polygona Salm-Dyck, Cact. Hort. Dyck. 1849, 120, 1850.
Cactus polygonus Kuntze, Rev. Gen. Pl., 1:261, 1891.
Neomammillaria polygona Britton & Rose, The Cactaceae, 4:101, 1923.

BODY simple, clavate, to 10 cm. high. TUBERCLES pyramidal, 4-angled, keeled ventrally, with milky sap. AREOLES naked. AXILS with wool and hairs. CENTRAL SPINES 2, upper longer to 25 mm. and somewhat recurved, all subulate, carmine at base, reddish yellow above. RADIAL SPINES 8, 1-4 mm. long, upper 2-3 minute (occasionally missing), lower longest, apex brownish. FLOWERS *Inner perianth-segments* pale rose, straight, not recurved. *Stigma-lobes* 5-6, recurved. FRUIT and SEEDS unknown.

Distribution: Puebla and Morelos (Boedeker), Mexico.

Type locality: None stated.

The above description is based on the original description. Inasmuch as we have never seen any collected plants of this species, it will have to remain incompletely described until more material can be collected.

M. subpolygona is listed by Haage (Cact. Kult., 137, 1900) as a synonym without reference.

FIG. 32. *Mammillaria tenampensis*

27. **Mammillaria tenampensis** (B. & R.) Berger

Neomammillaria tenampensis Britton & Rose, The Cactaceae, 4:101, 1923.
Mamillaria tenampensis Berger, Kakteen, 325, 1929.

BODY simple, globose, 50-60 mm. wide. TUBERCLES loosely arranged in 8 and 13 spirals, firm in texture, light yellowish green, 4-angled to tip, with milky sap, 6-7 mm. long and wide at the base. AREOLES nearly round, less than 1 mm., with yellow felt in youth. AXILS with yellow wool and numerous yellow bristles in flowering area, later wool disappears and the bristles become whitish. CENTRAL SPINES 4-6, 1-6 mm. long, acicular, straight, smooth, stiff, brownish with dark tip, ascending, upper one is more central. RADIAL SPINES 8-10, only small white bristles. FLOWERS funnelform,

12 mm. long, 8 mm. wide. *Outer perianth-segments* brownish, small lanceolate, tip acuminate, margins serrate to ciliate. *Inner perianth-segments* reddish purple, lanceolate, tip apiculate, margins serrate, 8-10 mm. long. *Filaments* pale below, purplish above. *Anthers* light yellow. *Style* reddish (cream below, top pink). *Stigma-lobes* 4-5, deep purple, less than 1 mm. long, overtop anthers 1-2 mm. FRUIT unknown. SEEDS unknown but probably brown.

Distribution: State of Mexico, Mexico.

Type locality: Barranca de Tenampa.

Illustration: Fig. 32 is a reproduction of an illustration, Fig. 102, in Britton & Rose (Cact., 4:104) as *Neomammillaria tenampensis*.

FIG. 33. *Mammillaria petrophila* x 1

28. Mammillaria petrophila Brandegee

Mamillaria petrophila Brandegee, Zoe, 5:193, 1904.
Neomammillaria petrophila Britton & Rose, The Cactaceae, 4:73, 1923.

BODY simple and lateral branching, depressed globular to cylindric-globular, to 15 cm. wide. TUBERCLES arranged in 8 and 13 spirals, grayish green, 4-sided at base, irregularly faintly angled conic, keeled ventrally, with milky sap, 10 mm. long, 7 mm. wide at base. AREOLES oval with white wool, soon becoming naked. AXILS with long dense tan wool, more so in the fruiting area, some of it persisting, also 3-6 bristles 5-8 mm. long. CENTRAL SPINES 1-2, to 20 mm. long, acicular, straight, smooth, stiff, dark chestnut brown, becoming paler, porrect. RADIAL SPINES 8-10, 10-15 mm. long, slender acicular, straight, smooth, ascending. FLOWERS 18-20 mm. long. *Outer perianth-segments* bright greenish yellow mid-stripe, lighter margins, lanceolate, tip nearly acute, margins lightly serrate. *Inner perianth-segments* same color and shape as the outer. *Filaments* greenish yellow. *Anthers* yellow. *Style* greenish yellow. *Stigma-lobes* 5-6, greenish yellow. FRUIT red, round, small. SEEDS reddish brown, smooth (?), less than 1 mm.

Distribution: Baja California, Mexico.

Type locality: Sierra de la Laguna and Sierra Francisquito in cape area.

Illustration: Fig. 33 is from a photograph of a plant collected for us by Mr. Howard Gates in the type area.

Fig. 34. *Mammillaria eichlamii* x 0.8

29. **Mammillaria eichlamii** Quehl

Mamillaria eichlamii Quehl, Monatsschr. Kakteenk., 18:65, 1908.
Neomammillaria eichlamii Britton & Rose, The Cactaceae, 4:94, 1923.
Mammillaria eichlamii albida Hort. in Borg, Cacti, 349, 1937.

BODY cylindric to somewhat clavate, richly cespitose, offshoots easily broken off, rounded above, with slightly sunken apex, 15-25 cm. high, 5-6 cm. wide. TUBERCLES arranged in 8 and 13 spirals, yellowish green, conic, not angled, truncate at apex, with very milky sap, 7-8 mm. long. AREOLES round, 2 mm. wide, with scant dirty pale yellow woolly mat. AXILS with dirty yellow (sometimes whitish) wool which later disappears, also 5-6 white tortuous bristles to 10 mm. long. CENTRAL SPINES 1 to rarely 2, 10 mm. long, stout acicular, straight, base yellow, above brownish red, later gray, porrect. RADIAL SPINES 6 (7-8), 5-7 mm. long, upper 3 shorter and weaker, all acicular, straight or very slightly bent, yellowish white with darker tip, later becoming gray, uniformily distributed, somewhat ascending. FLOWERS funnelform, 20 mm. long, September. *Outer perianth-segments* cream with yellowish brownish red ventral stripe, lanceolate, tip acuminate, margins slightly ciliate. *Inner perianth-segments* cream to lemon-yellow, bright carmine red mid-line, darker ventrally, linear-lanceolate, tip acuminate, margins entire to lightly serrate. *Filaments* whitish to light yellow. *Anthers* yellow. *Style* whitish. *Stigma-lobes* 5 (4-6), greenish yellow, obtuse, overtop anthers. FRUIT clavate, with dried perianth persisting. SEEDS light brown, curved pyriform with lateral hilum near base, very lightly pitted, 1.5x0.8 mm.

Distribution: Guatemala, (Honduras).

Type locality: Near Sabantas.

Illustration: Fig. 34 is from a photograph of a plant obtained from Mr. Yale Dawson of Long Beach, California.

FIG. 35. *Mammillaria collinsii*

30. Mammillaria collinsii (B. & R.) Orcutt

Neomammillaria collinsii Britton & Rose, The Cactaceae, 4:101, 1923.
Mamillaria collinsii Orcutt, Cactography, 7, 1926.

BODY simple and cespitose, forming large clusters, individuals 40 mm. wide. TUBERCLES arranged in 8 and 13 spirals, firm in texture, green becoming bronzed to even dark purple, conic-cylindric, terete, somewhat keeled ventrally, with milky sap, 8-10 mm. long, 6-7 mm. wide at base. AREOLES oval, with white wool in youth, soon becoming naked. AXILS with white wool and several tortuous white bristles which are longer than the tubercles. CENTRAL SPINES 1, to 8 mm. long, stout acicular, straight, stiff, smooth, dark brown, porrect. RADIAL SPINES 7, 5-7 mm. long, acicular, straight, smooth, stiff, base pale yellow to dark brown to blackish at tip, somewhat ascending. FLOWERS funnelform, 12-15 mm. long. *Outer perianth-segments* very pale green below, tapering reddish mid-stripe above, yellow margins, linear-lanceolate, tip acuminate, margins ciliate. *Inner perianth-segments* rose mid-line, yellowish rose margins, lanceolate, tip acuminate, margins mostly entire but a few ciliate. *Filaments* white. *Anthers* yellow. *Style* white. *Stigma-lobes* 4, pale greenish yellow, 1-2 mm. long, overtop anthers. FRUIT deep red, clavate, 15-20x5 mm., with dried perianth persisting. SEEDS light tan, curved pyriform with lateral hilum at base, slightly roughened, 1 mm.

Distribution: Oaxaca, Mexico.

Type locality: San Jeronimo, near Isthmus of Tehuantepec.

Illustration: Fig. 35 is a reproduction of an illustration, Fig. 96, in Britton & Rose (Cact., 4:99) as *Neomammillaria collinsii*.

This species is very close to if not identical with *M. confusa centrispina* that we collected near Oaxaca City, but which is quite some distince inland from the type locality of this species.

Bravo gives the type locality as in the state of Chiapas but our maps give it as in the state of Oaxaca.

Fig. 36. *Mammillaria mystax* x 0.7

31. **Mammillaria mystax** Martius

Mammillaria mystax Martius, Hort. Reg. Monac., 127, 1829. Nomen.
Mammillaria mystax Martius, Nov. Act. Nat. Cur., **16**:332, 1832.
Mammillaria leucotricha Scheidweiler, Allg. Gartenz., **8**:338, 1840.
Mammillaria xanthotricha (1) Scheidweiler, Allg. Gartenz., **8**:338, 1840.
Mammillaria xanthotricha aculeis axillaribus robustioribus Scheidweiler, Allg. Gartenz., **8**:338, 1840.
Mammillaria mutabilis Scheidweiler, Allg. Gartenz., 9:43, 1841.
Mammillaria funkii Scheidweiler, Allg. Gartenz., 9:43, 1841.
Mammillaria senkei (2) Foerster, Handb. Cact., 227, 1846.
Mamillaria autumnalis Dietrich, Allg. Gartenz., **16**:297, 1848.
Mamillaria mutabilis xanthotricha Salm-Dyck, Cact. Hort. Dyck. 1849, 17, 120, 1850.
Mamillaria mashalacantha (3) Monville in Labouret, Monogr. Cact., 106, 1853.
Mamillaria mashalacantha leucotricha Monville in Labouret, Mongr. Cact., 106, 1853.
Mamillaria mashalacantha xanthotricha Monville in Labouret, Mongr. Cact., 106, 1853.
Cactus funkii Kuntze, Rev. Gen. Pl., 1:260, 1891.
Cactus leucotrichus Kuntze, Rev. Gen. Pl., 1:260, 1891.
Cactus mashalacanthus Kuntze, Rev. Gen. Pl., 1:260, 1891.
Cactus mutabilis Kuntze, Rev. Gen. Pl., 1:261, 1891.
Cactus mystax Kuntze, Rev. Gen. Pl., 1:261, 1891.
Cactus xanthotrichus Kuntze, Rev. Gen. Pl., 1:261, 1891.
Mamillaria mutabilis leucocarpa Schelle, Handb. Kakteenk., 273, 1907.
Neomammillaria mystax Britton & Rose, The Cactaceae, 4:92, 1923.

BODY simple to later branching from base, globose to cylindric, flattened to rounded above, with sunken apex, 15 cm. high, 7-10 cm. wide. TUBERCLES closely arranged in 13 and 21 spirals, firm in texture, dark grayish green, pyramidal 4-6 angled, sharply keeled ventrally, with milky sap, 10-15 mm. long, 8 mm. wide at base. AREOLES round to 3 mm. wide, with scant white wool, later becoming naked. AXILS with white wool and white tortuous bristles which are often heavy. CENTRAL SPINES 3-4, (frequently 1 more centrally placed), three 15-20 mm. long, more central one to 70 mm., tortuous, all strong acicular, angled, dark purplish, dark brown at tip, irregularly interlacing, overtop apex. RADIAL SPINES 5-6 later to 10, 4-8 mm. long, lower longest, all fine acicular

---

(1) This reference was given by Britton & Rose as *M. zanthotricha* which is probably a typographical error.
(2) This was originally written *M. senkii* but the species was named after F. Senke of Leipzig.
(3) Salm-Dyck (Hort. Dyck. 10, 1845) listed this name as *M. meshalacantha* without description.

to slender subulate, often tortuous, white with brown tip, ascending to later becoming nearly horizontal. FLOWERS campanulate, in upper axils, 25 mm. long, 20 mm. wide. *Outer perianth-segments* 6-8, green base, orange-brown to purplish wide tapering midstripe, narrow pink to white margins, linear-lanceolate, tip acute, margins serrate, length unequal, recurved. *Inner perianth-segments* 8-10, glossy rose-purple above, pinker below, silvery margins, linear-lanceolate, tip acuminate, margins entire to finely serrate at tip, recurved, 25 mm. long. *Filaments* white to very pale pink. *Anthers* citron-yellow. *Style* white below, rose to very pale yellow above. *Stigma lobes* 4-5, pale yellowish-green, 2 mm. long, overtop anthers 5 mm. FRUIT red, clavate, 20-25 mm. long, with dried perianth not persisting. SEEDS dull dark brown, pyriform with lateral hilum near base, roughened, 1.2x0.6 mm.

Distribution: Hidalgo to Oaxaca, Mexico.

Type locality: None given but reported from Ixmiquilpan and San Pedro Nolasco, Hgo.; Tehuacan, Puebla (Schmoll).

Illustration: Fig. 36 is from a photograph of a plant sent to us by Sr. F. Schmoll of Cadereyta, Qro.

M. hermantiana is only a name referred by Labouret (Monogr. Cact. 107, 1853) as a synonym of *M. autumnalis*. Rebut spells it *M. hermantii*.

"*M. krauseana,* a name from Gruson's Catalogue, is cited by Schumann (Gesamtb. Kakteen, 595, 1898) as a synonym of *M. mutabilis.*

"*M. mashalacantha dolichacantha* Monville was given as a doubtful synonym of *M. mashalacantha* by Labouret (Monogr. Cact., 106, 1853).

"*M. mutabilis autumnalis* (Monatsschr. Kakteenk., 30: February, 1920) is offered for sale by Grässner.

"*M. mutabilis laevior* Salm-Dyck (Cact. Hort. Dyck. 1849, 17, 120, 1850), with *M. leucocarpa* Scheidweiler as a synonym, was given as a variety of *M. mutabilis,* but it was not described. *M. xanthotricha laevior* Salm-Dyck (Hort. Dyck. 1844, 11, 1845), also undescribed, seems to be the same." Britton & Rose, Cact., 4:93-4, 1923.

Schumann referred *M. senkei* to *M. pyrrhocephala* and was likewise followed by subsequent authors but the original description would indicate that it should be referred here as was done by Salm-Dyck.

32. Mammillaria flavovirens Salm Dyck

Mamillaria flavovirens Salm-Dyck, Cact. Hort. Dyck., 1849, 117, 1850.
Mamillaria flavovirens cristata Salm-Dyck, Cact. Hort. Dyck. 1849, 118, 1850.
Neomammillaria flavovirens Britton & Rose, The Cactaceae, 4:85, 1923.

BODY simple to sometimes cespitose, flattened globose to short cylindric with sunken apex, 60-80 mm. high. TUBERCLES arranged in 8 and 13 spirals, firm in texture, pale to yellowish green, 4-sided pyramidal, angled, keeled ventrally, with milky sap, 10 mm. long, 7 mm. wide at base. AREOLES oval, sunken, with scant wool. AXILS with white wool in flowering area. CENTRAL SPINES 1-(2), 10-15 mm. long, subulate, to somewhat square in cross section, straight to slight recurve, smooth, stiff, base enlarged. RADIAL SPINES 4-5 (sometimes 2 extra very small additional deciduous ones), 5-6 mm. long, slender subulate, straight, smooth, reddish brown, somewhat ascending. FLOWERS in upper part of the body, wide funnelform, 20 mm. long, 12 mm. wide. *Outer perianth-segments* bright greenish at base, reddish brown to rose-red tapering midstripe above, cream margins, lanceolate, tip acute, margins serrate. *Inner perianth-segments* white to cream, with rose to reddish brown mid-line, lanceolate, tip obtuse, margins entire. *Filaments* white. *Anthers* golden yellow. *Style* white. *Stigma-lobes* 4-7, tan to greenish. FRUIT carmine red, clavate, 17x5 mm., with dried perianth persisting. SEEDS

Fig. 37. *Mammillaria flavovirens* x 0.8

light tan, pyriform with lateral hilum at base, slightly roughened but not pitted, 1.2x.6 mm.

Distribution: Guanajuato, Mexico. (Schmoll)

Type locality: None given but reported from Hacienda de las Barrances (Schmoll).

Illustration: Fig. 37 is from a photograph of a plant sent to us by Sr. F. Schmoll of Cadereyta, Qro.

Labouret (Monogr. Cact., 100, 1855) refers *M. daedalea viridis* Fennel in Foerster (Handb. Cacteenk., 254, 1846) as a synonym of *M. flavovirens* but without description.

M. pentacantha Pfeiffer (Allg. Gartenz., 8:406, 1840), *Cactus pentacanthus* Kuntz (Rev. Gen. Pl., 1:261, 1891) and *M. centricirrha pentacantha* were referred to *M. magnimamma* by Britton & Rose (Cact., 4:77, 1923) but the species was originally described with a central spine which would exclude it from that species. This species was described without flower, fruit or seed data, so its exact relationship is uncertain. Being the older name, it would have priority over the above species if the complete data were available and it were found to correspond.

33. Mammillaria bocensis sp. nov.

Mammillaria sp. 685 Gentry, Rio Mayo Plants, 196, 1942. (Distribution only.)

Corpus simplex, globosus, mamillis ad 8 et 13, itidem 13 et 21 seriebus ordinatis, pyramidatis, suco lacteo, areolis rotundis, axillis lanatis, albis; spinis centralibus 1, 8-12 mm., acicularibus, robustis, spadicis; spinis radialibus 6-8, 5-14 mm., rectis acicularibus, rubris; flores infundibuliformes, sepalis viridiflavis, linea media spadici, ciliatis brevibus, petalis alboflavis, stigmatibus 5-7, flavoviridis; fructus ruber; semina fusca.

BODY simple, flattened globular, becoming short cylindric, with sunken apex, to 85 mm. wide, to 90 mm. high. TUBERCLES arranged in 8 and 13 also 13 and 21 spirals, firm in texture, dark green, pyramidal, variously angled, keeled ventrally, with milky sap, 11-13 mm. long, 8-12 mm. wide at base. AREOLES nearly round, with scant dirty white wool only in youngest. AXILS with scant dirty white wool but no bristles (occa-

FIG. 38. *Mammillaria bocensis* x 0.8

sionally 1-2 very short white bristles, but not typical). CENTRAL SPINES 1, 8-12 mm. long, heavy acicular to slender subulate, straight or slight dorsal curve, smooth, stiff, with base slightly enlarged, reddish brown with darker to black tip, porrect. RADIAL SPINES 6-8, 5-14 mm. long, upper ones shorter and more slender, all stout acicular, straight, smooth, stiff, chalky white (lower ones more reddish) with brown to black tip, somewhat ascending. FLOWERS wide funnelform, 15-20 mm. long, May to June. *Outer perianth-segments* pale greenish yellow base, reddish brownish green mid-stripe above (more so ventrally), pale pink to light green margins, lanceolate, tip acuminate, margins short ciliate. *Inner perianth-segments* greenish cream, pinker at margins, light brownish green mid-line only near the apex, but darker and more pronounced ventrally, eliptical, tip acute, margins entire and ciliate, ends often split. *Filaments* white to pale pink. *Anthers* light yellow. *Style* white to very light tannish green. *Stigma-lobes* 7, pale yellowish green, 2-3 mm. long, overtop anthers. FRUIT reddish, clavate, 25x12 mm., with dried perianth persisting. SEEDS light brown, curved pyriform with lateral hilum near base, roughened. ROOTS fibrous and also large tap.

Distribution: Sonora, Mexico.

Type locality: Las Bocas.

Illustration: Fig. 38 is from a photograph of a plant collected for us by Mr. George Lindsay.

This species was sent to us by Mr. Howard S. Gentry in 1937 (671 and 688) and by Mr. George Lindsay in 1938 (904) and also in 1941. They report that it is found in the silty flats of the coastal plain, deep seated under the bushes and "adjacent to the beach."

It may be near *M. seemannii* Scheer (Bot. Herald, 288, 1856) but the latter was described as having 11-13 radial spines but inasmuch as its flower was not described, there can be no certainty of its relationship.

A very closely related series of plants was sent to us by Mr. Gentry in 1937 who collected them near Tesopaco, Sonora, which is a mountainous country. Some of them we have referred to *M. tesopacensis* (c.f.) while others shade into this species.

Mr. Lindsay brought us plants in 1938 and again in 1941 from Mazatlan, Sinaloa, that were very similar to this species except that the flower is more pinkish.

We obtained what appears to be the same plant from Sr. F. Schmoll of Cadereyta, Qro., who reports that they came from Chihuahua but with no type locality known.

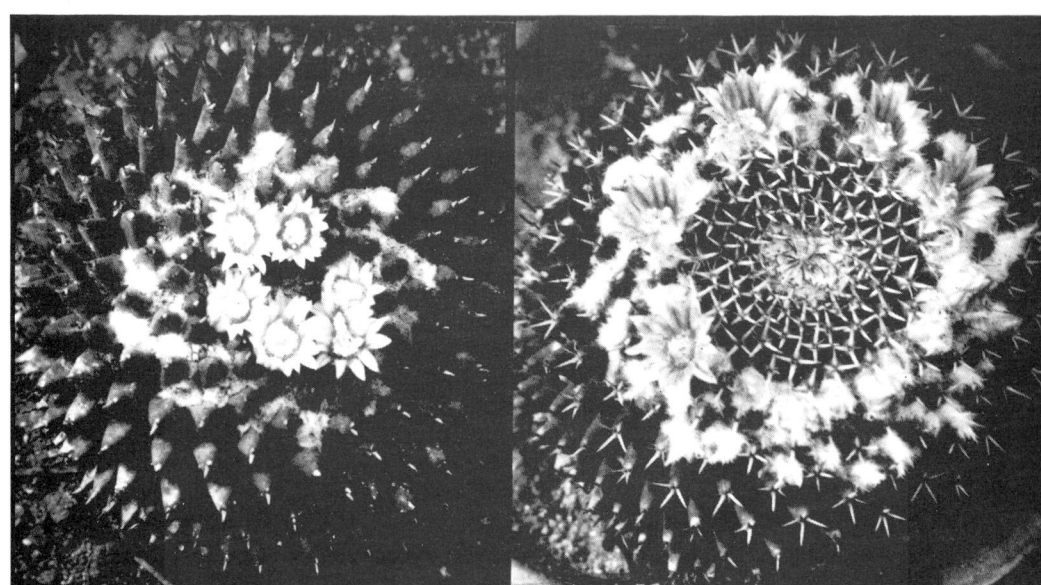

FIG. 39. *Mammillaria sempervivi*   FIG. 40. *Mammillaria sempervivi* var. *tetracantha* x 1

34. Mammillaria sempervivi DeCandolle

Mammillaria sempervivi DeCandolle, Mem. Mus. Hist. Nat. Paris, 17:114, 1828.
Mammillaria caput medusae Otto in Pfeiffer, Enum. Cact., 22, 1837.
Mammillaria diacantha Lemaire, Cact. Aliq. Nov., 2, 1838.
Mamillaria caput medusae centrispina Salm-Dyck in Labouret, Monogr. Cact., 91, 1853.
Mamillaria caput medusae crassior Salm-Dyck in Labouret, Monogr. Cact., 91, 1853.
Cactus sempervivi Kuntze, Rev. Gen. Pl., 1:261, 1891.
Neomammillaria sempervivi Britton & Rose, The Cactaceae, 4:86, 1923.

BODY simple and branching from base, depressed globular to short cylindric, rounded above, with sunken apex, to 10 cm. wide, deep seated so that more of the plant is below the ground than above. TUBERCLES arranged in 13 and 21 spirals, firm in texture, dull to grayish green, slender pyramidal, angled to tip, with milky sap, 10 mm. long, 6-7 mm. wide at base. AREOLES round to oval, 1.5 mm. wide, with white wool in youth, later becoming naked. AXILS with abundant wool, especially in the flowering area. CENTRAL SPINES 2 (some varieties 4), 4 mm. long, subulate, straight to slightly bent, smooth, pale reddish brown in youth, later becoming horn-color, divergent dorsally and ventrally. RADIAL SPINES 3-4 (sometimes to 7 or more), usually present only in youth, to 3 mm. long, hair-like, white, somewhat ascending. FLOWERS funnelform, near apex but in old growth, 10 mm. wide and long. *Outer perianth-segments* olive-green below, light brownish orange to reddish mid-stripe above, oblong, tip obtuse to acute, margins mostly entire to occasionally irregularly serrated. *Inner perianth-segments* dirty white to yellowish to rose with reddish mid-stripe, lanceolate, tip acute, margins entire. *Filaments* white below to pale pink above. *Anthers* yellow. *Style* very pale green below, rose above. *Stigma-lobes* 4-6, orange-yellow to rose, 1.5 mm. long, overtop anthers. FRUIT red, clavate, 8 mm. long, with dried perianth persisting. SEEDS light tan, dull, curved pyriform with lateral hilum near base, faintly reticulate, 1.4x0.5 mm. ROOTS tuberous.

Distribution: Hidalgo, Vera Cruz, central Mexico.

Type locality: None given but reported from Meztitlan and Ixmiquilpan, Hgo., Jalapa, V. C.

Illustration: Fig. 39 is from a photograph of a plant sent to us by Sr. F. Schmoll of Cadereyta, Qro.

M. dicantha Pfeiffer, which is referred to *M. haageana,* is obviously a different plant from that described by Lemaire.

a. var. **tetracantha** DeCandolle

Mammillaria sempervivi tetracantha DeCandolle, Mem. Mus. Hist. Nat. Paris, 17:114, 1828.
Mamillaria sempervivi laetiviridis Salm-Dyck, Cact. Hort. Dyck. 1849, 113, 1850.
Mamillaria caput medusae tetracantha Salm-Dyck in Labouret, Monogr. Cact., 91, 1853.

Central spines 4.

Illustration: Fig. 40 is from a photograph of a plant sent to us by Sr. F. Schmoll of Cadereyta, Qro.

"*M. staurotypa* Foerster (Handb. Cact., 221, 1846), credited to Scheidweiler by Schumann and referred by him as a synonym of *M. caput medusae,* seems never to have been described but may belong here.

"The two varieties of *M. caput medusae, tetracantha* and *hexacantha,* given by Salm-Dyck (Hort. Dyck. 1844, 10, 1845) are without description. The first was afterwards described by Labouret." Britton & Rose, Cact., 4:87.

We collected specimens of both the type and the variety near Ixmiquilpan and observed it at various places in the central plateau desert regions.

FIG. 41. *Mammillaria obscura* x 0.7

35. Mammillaria obscura Hildmann

Mamillaria obscura Hildmann, Monatsschr. Kakteenk., 1:52, 1891.
Neomammillaria obscura Britton & Rose, The Cactaceae, 4:87, 1923.
Mammillaria wagneriana Boedeker, Monatsschr. Deutsch. Kakt. Ges., 4:199, 1932.

BODY simple, globular, somewhat depressed, with flattened apex, to 11 cm. wide, to 8 cm. high. TUBERCLES arranged in 13 and 21 spirals, firm in texture, dark green, quadrangular at base, short pyramidal, with milky sap, 7-8 mm. long, 7 mm. wide at base. AREOLES round, 4 mm. wide in youth, soon smaller, with white wool in youth. AXILS with wool in crown, later becoming naked. CENTRAL SPINES 2-4, lower longer to 20 mm., upper 1-3 shorter, all straight, slender subulate, roughened, enlarged at base, stiff, reddish horn-colored, brown to almost back at tip, later gray, divergent, lower incurved. RADIAL SPINES 6-8, 8-14 mm. long, stiff, slender subulate, straight, whitish horn-colored with short brown tip, slightly ascending. FLOWERS small to 15 mm. wide, campanulate, near top. *Outer perianth-segments* brownish rose, yellowish margins, linear-lanceolate, tip acuminate, margins short ciliate. *Inner perianth-segments* dirty whitish yellow, rose mid-stripe, linear-lanceolate, tip acuminate, margins ciliate. *Anthers* bright yellow. *Filaments* and *Style* whitish. *Stigma-lobes* 5-6, bright yellowish green, overtop anthers. FRUIT red, clavate. SEEDS dark brown, curved pyriform with lateral hilum, hardly 1 mm.

Distribution: Queretaro, Zacatecas, Mexico.

Type locality: None given but reported from near Toliman, Qro. (Schmoll).

Illustration: Fig. 41 is from a photograph of a plant received from Schwarz & Georgi of San Luis Potosi, S.L.P.

Fig. 42. *Mammillaria obscura* var. *wagneriana tortulospina*

a. var. wagneriana tortulospina comb. nov.

Mammillaria wagneriana var. Boedeker, Monatsschr. Deutsch, Kakt. Ges., 4:199, 1932.

Spinae centrales ad 45 mm., tortuosae.

Central spines longer to 45 mm. and tortuous.

Illustration: Fig. 42 is a reproduction of an illustration in Monatsschr. Deutsch. Kakt. Ges. 4:199 as *Mammillaria wagneriana*.

Fig. 43. *Mammillaria bucareliensis* x 0.7

36. Mammillaria bucareliensis sp. nov.

Corpus subglobosus, mamillis ad 8 et 13 itidem 13 et 21 seriebus ordinatis, pyramidis, suco lacteo, axillis lanatis; spinis centralibus 2-4, 5-40 mm., rectis ad reflexis, robustis; spinis radialibus 0-3-5, 1-5 mm., tenuibus; flores campanulato-infundibuliformes, sepalis serratis, petalis purpureis, stigmatibus 5; fructus coccineus; semina fusca.

BODY simple, flattened globular, with sunken apex, 5 cm. high, 9 cm. wide. TUBERCLES arranged in 8 and 13 also 13 and 21 spirals, firm in texture, bluish grayish green, quadrangular, sharply 4-angled, somewhat flattened dorsally, with milky sap, 10 mm. long, 8-12 mm. wide at base. AREOLES depressed, pyriform, with white wool in youth, soon becoming naked. AXILS with dense white wool which persists for some time. CENTRAL SPINES 2-4, 5-40 mm. long, lower longest, all stout acicular to subulate, slight recurve, stiff, smooth, lower heavier, all light brown with black tip, becoming gray, divergent to nearly horizontal. RADIAL SPINES 3-5 or sometimes missing, less than 1 mm. to 5 mm. long, slender acicular to slightly subulate, straight, stiff, smooth, whitish with brown tips, somewhat ascending. FLOWERS campanulate-funnelform, short tube, lateral, 18 mm. wide, 15 mm. long, June. *Outer perianth-segments* purplish brown mid-stripe, more greenish at base, pinkish tan margins, lanceolate, tip acuminate, margins finely serrate. *Inner perianth-segments* deep pink to purplish, darker mid-line, lower half shading to very pale greenish white at base, linear-lanceolate, tip acute, margins entire. *Filaments* pale pink above, white below. *Anthers* yellow. *Style* deep pink above, very pale pink to greenish white below. *Stigma-lobes* 5, pink to carmine, overtop anthers. FRUIT carmine-red, clavate, 15x6 mm., with dried perianth not persisting. SEEDS light tan, dull, lightly reticulate, curved pyriform with lateral hilum, 1x0.6 mm.

Distribution: Guanajuato, Mexico.

Type locality: Bucarel.

Illustration: Fig. 43 is from a photograph of a plant sent to us by Sr. F. Schmoll of Cadereyta, Qro.

According to Neale (Cact. Other Succ., 83, 1935) the name of this species was suggested by Ernest Tiegel of Germany, but as yet we have found no description of it.

a. var. **bicornuta** Schmoll var. nov.

Corpus et spinae idem exempli; flores infundibuliformes, sepalis integris, petalis, subalbaflavis, stigmatibus 7, subflavis.

Body and spine arrangement are the same as the type. FLOWERS wide funnelform, lateral, 20 mm. long, 15 mm. wide. *Outer perianth-segments* pale green base, wide tapering greenish brownish red mid-stripe, pale green margins, lanceolate, tip acute, margins entire. *Inner perianth-segments* pale cream, outer row bright red mid-line, inner row pale green mid-line, oblong, tip obtuse, margins entire. *Filaments* white. *Anthers* yellow. *Style* white below, pale pink at very top. *Stigma-lobes* 7, pale greenish tan, ventral pink line.

Type locality: Bucarel.

This variety has some of the characteristics of the type and some of the preceeding species, *M. obscura*. The body and spine arrangement is like the type. The color of the perianth-segments and stigma-lobes are like *M. obscura* but the margins of the perianth-segments are entire and not serrate like the type nor ciliate like *M. obscura*.

This name was briefly characterized by Neale (Cact. Other Succ., 83, 1935) but not described.

Fig. 44. *Mammillaria vagaspina* x 1

37. **Mammillaria vagaspina** sp. nov.

Corpus simplex, subglobosus, mamillis ad 8 et 13 seriebus ordinatis, suco lateo, axillis lanatis; spinis centralibus 2, ad 60 mm. longis, tortuosis; spinis radialibus 3 (2-5), 1-10 mm. longis, acicularibus, tenuibus; flores infundibuliformes, sepalis integris, petalis subrubris, stigmatibus 4, fulvis; fructus coccineus; semina fusca.

BODY simple, depressed globose with sunken apex, to 85 mm. wide. TUBERCLES arranged in 8 and 13 spirals, firm in texture, dull dark grayish green, pyramidal, more or less angled, somewhat flattened dorsally, keeled ventrally, with milky sap, 13 mm. long, 8-11 mm. wide at base. AREOLES sunken, sharp edge around, somewhat diamond shaped, with white wool only in youngest. AXILS with white wool. CENTRAL SPINES 2, very irregular, 6-60 mm. long, lower longer, all heavy acicular, very tortuous, somewhat angled, smooth, flexuous in the longer ones, chalky tan, divergent dorsally and ventrally. RADIAL SPINES 3 (2-5), (upper 2 usually deciduous), 1-10 mm. long, lateral longest, all acicular, usually straight to slight recurve, smooth, stiff, chalky tannish white, ascending, when 2: directed laterally, when 3: the third one is under the lower central, when 5: the upper 2 are very short and under the upper central. FLOWERS funnelform, 10 mm. wide. *Outer perianth-segments* whitish base, pink mid-stripe, brownish near tip, pale pink margins, lanceolate, tip acute, margins entire. *Inner perianth-segments* pink margins, darker mid-stripe, linear-lanceolate, tip acuminate to split, margins entire. *Filaments* pale pink. *Anthers* brownish yellow. *Style* pink. *Stigma-lobes* 4, brownish yellow with pink mid-stripe ventrally. FRUIT reddish vermilion, curved clavate, 11x4 mm. SEEDS light brown, dull, curved pyriform with lateral hilum, 1.2x0.7 mm. ROOTS fibrous.

Distribution: Queretaro, Mexico.

Type locality: Tierra Blanca.

Illustration: Fig. 44 is from a photograph of a plant sent to us by Sr. F. Schmoll of Cadereyta, Qro., in 1936 and again in 1942 as a sp. nov.

It is somewhat similar to the previous species but it differs in several respects: spines longer and tortuous, flowers lighter in color, outer perianth-segments entire instead of serrate, stigma-lobes yellowish instead of carmine.

38. **Mammillaria craigii** Lindsay

Mammillaria craigii Lindsay, Cact. Succ. Journ., 14:107, 1942.

BODY simple and with dichotomous branching, with sunken and quite woolly apex. TUBERCLES closely set in 13 and 21 spirals, firm in texture, light yellowish gray-green, 4-sided, sharply angled to tip, with milky sap, 6-7 mm. long, 9-10 mm. wide at base. AREOLES oval, 2 mm. long, sunken, with abundant light tan wool which persists for sometime. AXILS with little white wool in flowering area but no bristles. CENTRAL SPINES 1-3, mostly 2, 10-20 mm. long, lower longest, all slender acicular, stiff, smooth, nearly straight, upper slightly recurved, all slightly enlarged at base, brownish golden, slightly divergent dorsally and ventrally from porrect. RADIAL SPINES 7-8, 4-12 mm. long, upper 3 shorter, lower longest, all fine acicular, straight, smooth, mostly stiff to semiflexuous, base slightly enlarged, brownish golden, markedly ascending. FLOWERS campanulate, somewhat lateral, 15-20 mm. long, 10-15 mm. wide, February and March, open 2-3 days. *Outer perianth-segments* 15, very pale greenish below, brownish pink above, linear, tip obtuse, margins serrate and also slightly ciliate. *Inner perianth-segments* deep pink, darker mid-line ventrally, linear-spatulate, tip obtuse to emarginate, margins mostly entire, sometimes serrate at tip. *Filaments* cream to yellow to pink above. *Anthers* sulphur-yellow. *Style* yellow to very pale pink above. *Stigma-lobes* 7, greenish yellow, 3 mm. long, slightly overtop anthers. FRUIT red, clavate, 12x8 mm., with dried perianth persisting. SEEDS light brown, curved pyriform with lateral hilum near the base, glossy, faintly reticulate, 1x0.4 mm. ROOTS fibrous.

Distribution: SW. Chihuahua and SE. Sonora, Mexico.

Type locality: Barranca del Rio Urique, few miles south of Choro in the Sierra Tarahumara.

FIG. 45. *Mammillaria craigii* x 0.6

Illustration: Fig. 45 is a reproduction of an illustration in the Cact. Succ. Journ., *14*:107 as *Mammillaria craigii*.

Habitat: Mountainous of about 6,000 feet. The plants were found in the leaf mold in the crevices of the rocks, mostly in partial shade.

39. Mammillaria melanocentra Poselger var. typica nom. nov.

Mammillaria melanocentra Poselger, Allg. Gartenz., 23:17, 1855.
Mammillaria erinacea Poselger, Allg. Gartenz., 23:18, 1855. Not Wendland.
?*Mamillaria melanacantha* Foerster, Handb. Cact., 386, 1892.
Mammillaria valida Weber in Bois, Hort. Dict., 806, 1898.
Neomammillaria melanocentra Britton & Rose, The Cactaceae, 4:81, 1923.
Mammillaria saltillensis Boedeker, Zeitsschr. Sukkulentenk., 3:270, 1928.

BODY simple, depressed globose, with somewhat sunken apex, 11 cm. wide, 16 cm. high. TUBERCLES arranged in 8 and 13 spirals, separated, firm in texture, dull bluish green, strongly 4-angled to tip, pyramidal, strongly keeled ventrally, with milky sap, 10-20 mm. long, 13-15 mm. wide at base. AREOLES nearly round, 3-4 mm. wide, with white wool, later becoming naked. AXILS with white wool in youth. CENTRAL SPINES 1, 20-55 mm. long, strong acicular to subulate, straight, stiff, smooth, black, soon becoming brownish, mostly porrect. RADIAL SPINES 7-9, unequal, upper 3 or 4: 4-6 mm. long, lower one to 25 mm. long, all straight or somewhat bent, smooth, stiff, at first black then becoming gray with black tip, ascending. FLOWERS campanulate, 18 mm. long, 25 mm. wide, April. *Outer perianth-segments* pinkish olive-green tapering mid-stripe, darker mid-line, pinkish margins, linear-lanceolate, tip acute, margins entire. *Inner perianth-segments* deep pinkish red, darker mid-line to apex, margins paler, linear-

FIG. 46. *Mammillaria melanocentra* x 0.8

spatulate, 3 mm. wide, tip acute, margins entire. *Filaments* cream to very pale yellow. *Anthers* deep yellow. *Style* very pale pink above, pale greenish white below. *Stigma-lobes* 7, bright olive-green, 2-3 mm. long. FRUIT pink to scarlet, long clavate to 30 mm., with dried perianth missing. SEEDS reddish brown, glossy, nearly globular with narrow subbasal hilum, somewhat pitted and roughened, less than 1 mm. ROOTS fibrous.

Distribution: New Mexico, Texas, U.S.; Nuevo Leon, Coahuila, Mexico.

Type locality: Monterrey and Saltillo, N.L.

Illustration: Fig. 46 is from a photograph of a plant collected for us by Mr. George Lindsay in Nuevo Leon in 1938.

M. erinacea was described as having 11-12 radial spines but inasmuch as the type locality is Saltillo, the same as the type of this species, it is being referred here as a possible variation.

M. pachethele Poselger (Allg. Gartenz., 23:17, 1855) and *M. centricirrha pachythele* Schelle (Handb. Kakt., 267, 1907) was described as having 9-12 radial spines and 1 (2) central spines of 18 mm., but no flower data. It is possibly referrable to this group of plants as a variation as it is from the same locality, rather than to *M. magnimamma* to which it has been referred by previous authors.

a. var **runyonii** (B. & R.) (Boedeker) comb. nov.

Neomammillaria runyonii Britton & Rose, The Cactaceae, 4:81, 1923.
Mammillaria runyonii Boedeker, Mammill. Vergl. Schluss., 52, 1933.

Central spines 10-14 mm. long.

Distribution: Nuevo Leon, Mexico.

Type locality: El Mirador near Monterrey.

Illustration: In Britton & Rose (Cact., 4:Pl. X) as *Neomammillaria runyonii*.

b. var. **meiacantha** (Engelmann) comb. nov.

Mamillaria meiacantha Engelmann, Pro. Amer. Acad., 3:263, 1856.
Cactus meiacanthus Kuntze, Rev. Gen. Pl., 1:260, 1891.
Neomammillaria meiacantha Britton & Rose, The Cactaceae, 4:84, 1923.

Central spines 6-7 mm. long, often deep seated with tuberous root.

Distribution: New Mexico, Texas, U.S., and Northern Mexico.

Type locality: New Mexico.

Illustration: Cact. Mex. Bound. Pl. 9, Fig. 1-3 as *Mamillaria meiacantha*.

The three varieties of this species have heretofore been classified as separate species. The body and tubercle shape, the flower, and the fruit and seed characteristics are so nearly identical that the length of the spines is the only separation factor. Our observation in the field has brought us to the opinion that this is hardly a consistent dividing characteristic because wide variations can easily be observed within a very limited area. The protection of the other flora as to shade, and the soil and moisture conditions greatly affect the plant armament.

M. meiacantha longispina Manchester (Collect. Cact., 44, 1908) may belong to one of the above varieties.

The above names have been variously misspelled as *meonacantha, melonocantha* and are probably referrable here.

FIG. 47. *Mammillaria nivosa*

40. **Mammillaria nivosa** Link

Mammillaria nivosa Link in Pfeiffer, Enum. Cact., 11, 1837.
Cactus nivosus Kuntze, Rev. Gen. Pl., 1:259, 1891.
Coryphantha nivosa Britton, Ann. Mo. Bot. Gard., 2:45, 1915.
Neomammillaria nivosa Britton & Rose, The Cactaceae, 4:71, 1923.

BODY simple and cespitose from the base, globose to nearly cylindric, 8-18 cm. wide. TUBERCLES arranged in 8 and 13, 11 and 17, also 13 and 21 spirals, firm in texture, dark green to bronze, elongate conic, laterally compressed, with milky sap, 10-15 mm. long, 10 mm. wide at base. AREOLES round, 2-3 mm. wide, with white wool in youth, becoming naked. AXILS with dense white wool which persists but no bristles. CENTRAL SPINES 1, to 20 mm. long, stout acicular, straight, smooth, bright yellow becoming dark brown, porrect. RADIAL SPINES 6-8, 10-30 mm. long, lateral longest, all acicular, straight or somewhat bent, stiff, smooth, bright yellow becoming dark brown, ascending. SUB RADIAL SPINES 5-7, 2-8 mm., upper shorter, bristle-like to acicular, lateral heavier, lighter in color than radials, horizontal. FLOWERS 15-20 mm. long. *Outer perianth-segments* pale yellow at base, very faint orange mid-line, orange at very tip, lanceolate, tip acuminate, margins finely serrate. *Inner perianth-segments* cream to citron-yellow, linear-lanceolate, tip acuminate to acute, margins serrate. *Filaments* pale tannish cream. *Anthers* yellow. *Style* very pale greenish cream. *Stigma-lobes* 6, very pale tannish yellow, 1 mm. long, just overtop anthers. FRUIT red to scarlet, clavate, 12-18x7 mm., with dried perianth not persisting. SEEDS light brown, dull, pyriform with lateral hilum, roughened, 1.1x0.8 mm.

Distribution: Southern Bahamas, Mona, Desecheo, Culebra, Buck, St. Thomas, Little St. James, Tortola, Antigua, Porto Rico, Virgin and other islands in the West Indies.

Type locality: Tortola and Virgin Islands.

Illustration: Fig. 47 is a reproduction of an illustration, Fig. 66, in Britton & Rose (Cact., 4:72) as *Neomammillaria nivosa*.

The three mature and flowering specimens sent to us by Mr. George Anton of Porto Rico had the three different spiral arrangements of the tubercles.

"*M. tortolensis* Pfeiffer, (Enum. Cact., 11, 1837) was published as a synonym of *M. nivosa*. The same or similar plant was briefly described by Forbes (Journ. Hort. Tour, 148, 1937)." Britton & Rose, Cact., 4:72.

Mammillaria flavescens (DC) Haworth

Cactus mammillaris lanuginosus DeCandolle, Plant. Succ., 111, 1799.
Cactus flavescens DeCandolle, Cat. Plant. Hort. Monosp., 83, 1813.
Mammillaria flavescens Haworth, Suppl. Pl. Succ., 71, 1819.
Mammillaria straminea Haworth, Suppl. Pl. Succ., 71, 1819.
Cactus stramineus Sprengel, Syst., 2:494, 1825. Nomen.
?*Mammillaria parmentieri* Link & Otto, Verh. Beford. Gartenb., 6:429, 1830.
Mammillaria simplex flavescens Schumann, Gesamtb. Kakteen, 573, 1898.

This species was very briefly described by DeCandolle as a variation of *M. mammillaris* from the "warm parts of America" as having long yellow spines, white wool, yellow flowers like *M. mammillaris* but with the perianth-segments ciliate. This appears to us to come fairly close to the limits of *M. nivosa*.

41. Mammillaria dawsonii (Houghton) comb. nov.

Neomammillaria dawsonii Houghton, Cact. Succ. Journ., 7:88, 1935.

BODY simple, very low growing, visible portion above the ground 15-20 mm., flattened globular, 30-50 mm. wide. TUBERCLES arranged in 5 and 8 spirals, light green, dull, somewhat triangular conic, not sharply angled, keeled ventrally, rounded dorsally, with milky sap, 6 mm. long, 4 mm. wide at base. AREOLES round, small, with scant dirty white wool in youth, very soon becoming naked. AXILS with only slight wool and no bristles. CENTRAL SPINES 1, to 6 mm., acicular, stiff, slightly curved near apex to straight, smooth, dull light brown at base, darker tip, porrect. RADIAL SPINES 6-10, 1-5 mm., upper 3 shortest, all acicular, straight, smooth, stiff, horn-colored with

Fig. 48. *Mammillaria dawsonii* x 1

dark brown tip, horizontal. FLOWERS from near center of plant but in old axils, funnelform, 12 mm. long. *Outer perianth-segments* 8, reddish brown tapering wide midstripe, pale greenish yellow margins, tan base, lanceolate, tip acute, margins serrate. *Inner perianth-segments* 16, pale greenish yellow, greener ventrally, several brownish green mid-lines that are more prominent ventrally, linear-lanceolate, tip acute, margins some serrated and some entire. *Filaments* white to cream, 5 mm. long. *Anthers* yellow, very small, over 50. *Style* greenish yellow. *Stigma-lobes* 4-5, greenish yellow. FRUIT very light pink above, whitish below, 15x5 mm., clavate, acid taste. SEEDS dull red, less than 1 mm. ROOTS 1, sometimes 2, very thick fleshy tap roots which end in smaller rootlets about 40-50 mm. below the surface.

Distribution: Baja California, Mexico.

Type locality: Punta Prieta.

Illustration: Fig. 48 is a reproduction of our illustration in Cact. Succ. Journ., 7:61 as *Neomammillaria dawsonii* of plants collected for us by Mr. George Lindsay in March, 1935, at Punta Blanca which is near the type locality.

Habitat: On the ocean front, usually growing under bushes in deep shade.

The plants flowered for us in 1936, but they are very hard to keep alive in cultivation.

42. Mammillaria gatesii Jones

Mammillaria gatesii Jones, Cact. Succ. Journ., 8:99, 1937.

BODY simple and cespitose, globular to short cylindric, deep seated, 10-15 cm. wide, 20 cm. high. TUBERCLES arranged in 8 and 13 spirals, firm in texture, conic, terete, with milky sap, 9-15 mm. long, to 15 mm. wide at base. AREOLES round to slightly oval, with white wool, becoming naked. AXILS with white wool but no bristles. CENTRAL SPINES 1, 25-30 mm. long, stout acicular to slender subulate, straight, smooth, stiff, dark reddish brown to purplish in youth, yellowish brown at base, later becoming horn-color, porrect. RADIAL SPINES 6-8, 8-13 mm. long, upper 2 shorter,

FIG. 49. *Mammillaria gatesii*

lower one longer, all heavy acicular, straight, smooth, stiff, yellowish white, brown to black tip, later becoming whitish, ascending. FLOWERS campanulate, from upper part of the body, 16 mm. long, 20 mm. wide. *Outer perianth-segments* greenish yellow, purplish mid-stripe, ovate-lanceolate, tip obtuse, margins often ciliate. *Inner perianth-segments* golden to greenish yellow, lanceolate, tip broadly acuminate, margins lacerate. *Filaments* light yellow. *Anthers* light yellow. *Style* light green. *Stigma-lobes* 5, light green. FRUIT red, short clavate, 15 mm. long, naked. SEEDS dark brown, pitted, 1 mm.

Distribution: Baja California, Mexico.

Type locality: Sea coast at the extreme end of the peninsula between Cape San Lucas and San Jose del Cabo.

Illustration: Fig. 49 is a reproduction of an illustration in Cact. Succ. Journ., 8:99 as *Mammillaria gatesii*.

43. **Mammillaria baxteriana** (Gates) Boedeker

Neomammillaria baxteriana Gates, Cact. Succ. Journ., 6:3, 1934.
Neomammillaria marshalliana Gates, Cact. Succ. Journ., 6:4, 1934.
Neomammillaria pacifica Gates, Cact. Succ. Journ., 6:5, 1934.
Mamillaria baxteriana Boedeker in Back. & Knuth, Kaktus ABC, 398, 1935.
Mamillaria pacifica Boedeker in Back. & Knuth, Kaktus ABC, 398, 1935.
Mamillaria marshalliana Boedeker in Back. & Knuth, Kaktus ABC, 398, 1935.

BODY simple, occasionally cespitose, branching by dichotomous division, flattened globular, with sunken apex, deep seated, to 15 cm. wide and high. TUBERCLES arranged in 13 and 21 spirals, firm in texture, quadrangular base but conic above, with milky sap, 10-13 mm. long, 6-9 mm. wide at base. AREOLES round, with short white wool in youth, later becoming naked. AXILS with slight wool especially in flowering area but no bristles. CENTRAL SPINES 1, 10-20 mm. long, stout acicular, straight, smooth, stiff, white with brown tip, becoming chalky, porrect. RADIAL SPINES 7-13, 10-15 mm. long, lower longer, all acicular, straight, smooth, white, occasionally with brown tip in youth, somewhat ascending. FLOWERS funnelform, 15 mm. long, 20 mm. wide. *Outer perianth-segments* wide brownish purple mid-stripe, yellow margins, ovate-lanceolate, tip acute, margins entire. *Inner perianth-segments* greenish yellow, rose-purple

Fig. 50. *Mammillaria baxteriana* x 1

mid-line, lanceolate, tip acute, margins entire to serrate at tip. *Filaments* pale yellow. *Anthers* yellow. *Style* faint yellow. *Stigma-lobes* 7-9, greenish yellow, overtop anthers. FRUIT red, clavate, 20 mm. long, with dried perianth persisting. SEEDS light to reddish brown, dull, pyriform with lateral hilum, pitted, 1x0.5 mm.

Distribution: Baja California, Mexico.

Type locality: La Paz, cape area.

Illustration: Fig. 50 is from a photograph of a plant collected for us by Mr. Howard Gates.

These three species described from the southern end of Baja California are so nearly identical that the slight variations are to be regarded as geographical variations.

44. Mammillaria brandegeei (Coulter) Brandegee

Cactus brandegeei Coulter, Contr. U.S. Nat. Herb., 3:96, 1894.
Mamillaria brandegeei K. Brandegee, Erythea, 5:116, 1897.
Neomammillaria brandegeei Britton & Rose, The Cactaceae, 4:73, 1923.

BODY simple and cespitose, globose to cylindric, with slightly sunken apex, to 90 mm. wide. TUBERCLES arranged in 13 and 21 spirals, firm in texture, dark green, sharply quadrangular at base, more conic above, with milky sap 6-10 mm. long. AREOLES round, large, with grayish white woolly felt, soon becoming naked. AXILS with dense white wool in youth (occasionally a single white bristle, but not typical). CENTRAL SPINES 2-3-4, to 20 mm. long, stout acicular, straight to slight curve, stiff, smooth, reddish brown base, darker above, divergent porrect. RADIAL SPINES* 8-10, 7-10 mm. long, slender acicular, stiff, straight, smooth, whitish with dark to yellowish brown tips, slightly ascending. FLOWERS small, to 8 mm. long. *Outer perianth-segments* bright brown mid-stripe, darker at tip, bright green margins, ovate, tip acute, margins ciliate. *Inner perianth-segments* greenish yellow, reddish mid-line, lanceolate, tip acute, margins

---

*Brandegee reports that the radial spines vary from 9-16.

FIG. 51. *Mammillaria brandegeei* x 0.8

entire. *Filaments* white. *Anthers* yellow. *Style* green. *Stigma-lobes* 4-7, green. FRUIT pink to bright red, lighter at base, clavate, 15x7 mm., with dried perianth not persisting, also with a few scales. SEEDS dull brown, curved pyriform, 1x0.8 mm.

Distribution: Baja California, Mexico.

Type locality: San Jorge and it ranges from Hamilton's Ranch to Calmalli.

Illustration: Fig. 51 is from a photograph of a plant collected for us by Mr. George Lindsay at the type locality.

a. var. **gabbii** (Coulter) (Engelmann) comb. nov.

Cactus gabbii Coulter, Contr. U. S. Nat. Herb., 3:109, 1894.
Mammillaria gabbii Engelmann in K. Brandegee, Erythea, 5:116, 1897.

This variation of the type presents less central spines 1-2, and more radial spines 11-15.

Type locality: San Ignacio to Mission San Fernando, Baja California.

45. **Mammillaria sartorii** Purpus

Mamillaria sartorii J. A. Purpus, Monatsschr. Kakteenk., 21:50, 1911.
Neomammillaria sartorii Britton & Rose, The Cactaceae, 4:82, 1923.

BODY simple and cespitose, globular to short cylindric, with apex flattened and a little sunken, 8-12 cm. wide, deep seated. TUBERCLES closely set in 8 and 13 spirals, firm in texture, dark bluish green, dull, closely white pitted, pyramidal, strongly 4-sided at base, many sided to conical above, rounded dorsally, with very milky sap, 10-12 mm. long. AREOLES round to oval, sunken, with abundant short white wool, later becoming naked. AXILS with abundant white curly wool (occasional very small bristle, but not typical). CENTRAL SPINES none to occasionally 1 or several, 1-8 mm. long, strong acicular, straight, white to brownish at tip, downward bent. RADIAL SPINES 4, rarely 6, often accessory bristles, 5-8 mm. long, upper longer, all acicular, slight recurve, stiff,

Fig. 52. *Mammillaria sartorii*

smooth, not enlarged base, dirty white base to brownish white, brown tip, strongly ascending, overtop apex. FLOWERS campanulate, to 20 mm. long. *Outer perianth-segments* bright chestnut-red mid-stripe, linear-lanceolate to oblong, tip acuminate, margins serrate. *Inner perianth-segments* yellow carmine to bright carmine with dark mid-stripe, linear-lanceolate to oblong, tip acuminate, margins serrate. *Filaments* reddish. *Anthers* light yellow. *Style* reddish. *Stigma-lobes* 4-6, reddish (carmine to purplish). FRUIT carmine red, long clavate, 16x6 mm., with dried perianth persisting. SEEDS dull reddish brown, slightly curved elongated pyriform with lateral hilum near base, roughened but not pitted, very small. ROOTS fibrous.

Distribution: Vera Cruz (Sierras), Mexico.

Type locality: Barranca de Tenampa and Atlyae near Zaenapan and also reported from near Hacienda de El Mirador in district of Huatusco.

Illustration: Fig. 52 is a reproduction of an illustration, Fig. 75, in Britton & Rose (Cact., 4:83) as *Neomammillaria sartorii*.

a. var. **brevispina** Purpus

Mamillaria sartorii brevispina Purpus, Monatsschr. Kakteenk., 21:50, 1911.

Dark green, areoles scant wool, spines 1-2 mm. long, brown.

b. var. **longispina** Purpus

Mamillaria sartorii longispina Purpus, Monatsschr. Kakteenk., 21:50, 1911.

Central spines to 80 mm. long, curved.

These varieties were not recognized by Britton & Rose or other authors although they were a part of the original description.

"Here may or may not belong *M. rebsamiana* (Cact. Journ., 2:17) advertised as a new discovery by Louis Murillo, who lived at Jalapa, Mexico." Britton & Rose, Cact., 4:82.

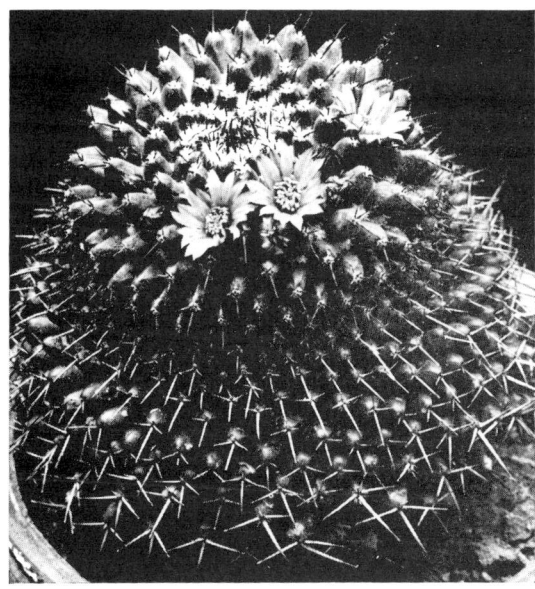

FIG. 53. *Mammillaria zuccariniana*

46. **Mammillaria zuccariniana** Martius

Mammillaria zuccariniana Martius, Nov. Act. Nat. Cur., 16:331, 1832.
Neomammillaria zucacriniana Britton & Rose, The Cactaceae, 4:89, 1923.
Chilita zuccariniana Orcutt, Cactography, 3, 1926.

BODY simple, globose to cylindric, 8-20 cm. high. TUBERCLES closely set in 16 and 26 spirals, conic, obscurely angled, flattened dorsally, with milky sap, 13-15 mm. long, 8-11 mm. wide at base. AREOLES oval, with white flocculent wool in youth. AXILS with wool only in upper axils. CENTRAL SPINES 2 (4), upper 8-13 mm. long, lower to 25 mm. long, all slender to stout acicular, straight to recurved, stiff, smooth, whitish with purplish tip, becoming ashen-horn color, divergent dorsally and ventrally. RADIAL SPINES 3-4, frequently deciduous or wanting, 2-6 mm. long, straight, stout bristle-like, white with brown tip. FLOWERS campanulate, 25 mm. long, with broad open throat, near top of plant. *Outer perianth-segments* 7-8, brownish below, purplish above, light rose margins, linear-oblong, tip acute, margins entire, 10-12x3 mm. *Inner perianth-segments* 15-16, magenta, darker mid-line and more brownish ventrally, cream throat, lanceolate, tip acute, margins entire, longer than outer. *Filaments* white to rose, subulate, 4-6 mm. long. *Anthers* pale yellow, oval globose. *Style* white below, rose above, subulate. *Stigma-lobes* 4-5 (7), yellowish pink or rose to purplish with magenta line ventrally, pyramidal, grooved ventrally. FRUIT red, clavate, 10 mm. long. SEEDS brownish.

Distribution: San Luis Potosi, Hidalgo; Mexico.

Type locality: None given but reported from Alvarez, S.L.P., and Ixmiquilpan, Hgo.

Illustration: Fig. 52 is a reproduction of an illustration, Fig. 83, in Britton & Rose (Cact., 4:90) as *Neomammillaria zuccariniana*.

Schumann refers this species to *M. centricirrha* (see *M. magnimamma*) but the color of the flower is very different.

The following species may be referrable here as closely related forms although there are some differences in the spine count.

Mammillaria monocentra Jacobi

Mamillaria monocentra Jacobi, Allg. Gartenz., 24:90, 1856.

BODY depressed globose with sunken apex, 85 mm. wide, 125 mm. high. TUBERCLES dark green, 4-sided at the base, obtuse above, pyramidal, with laterally flattened apex, with milky sap. AREOLES with white wool, later becoming naked. AXILS in youth naked, later with white wool, later becoming naked. CENTRAL SPINES 1, to 25 mm. long, straight, strong acicular, little dorsally recurved, yellowish brown with black tip, later becoming gray, porrect. RADIAL SPINES 6, 2-5 mm. long, upper shorter, lower longer and heavier, all ascending. FLOWERS near apex, hardly longer than tubercles. *Inner perianth-segments* rose, darker mid-line. *Filaments* rose. *Anthers* yellowish red. *Style* rose. *Stigma-lobes* 5, reddish yellow. FRUIT and SEEDS unknown.

Distribution: Mexico. Type locality: None given.

Mammillaria diacentra Jacobi

Mamillaria diacentra Jacobi, Allg. Gartenz., 24:91, 1856.

BODY globose, with somewhat sunken apex, 85 mm. wide. TUBERCLES dark green, rhomboidal at base, rounded above, with milky sap. AREOLES oval. AXILS with white wool, later becoming naked. CENTRAL SPINES 2, upper 12-16 mm. long, lower 25 mm. or more long, both stouter and stronger than the radials, reddish brown with black tip, upper erect, lower porrect, both recurved. RADIAL SPINES 5-6, 2-4 mm., lower three longer, all stout acicular, white with brown tip, radiating, recurved. FLOWERS small, rose with purple mid-stripe, lanceolate, short recurved. *Anthers* yellow. *Filaments* and *Style* rose. *Stigma-lobes* 6, rose.

Distribution: Unknown.

Schumann did not know them. Britton & Rose listed them under their little known species.

Fig. 54. *Mammillaria bachmannii* x 0.5

47. Mammillaria bachmannii Boedeker

Mamillaria bachmannii Berger, Kakteen, 323, 1929, as Hort. (Hesse) as belonging to *M. sempervivi*.
Mammillaria bachmannii Boedeker, Mammill. Vergl. Schluss., 59, 1933.

BODY simple, large globular to 18 cm. wide. TUBERCLES arranged in 8 and 13 spirals, firm in texture, dark green, conic to nearly globular, with blunt apex, with milky sap, 8 mm. long, 10 mm. wide at base. AREOLES diamond-shaped, with white wool. AXILS with abundant white wool so as to nearly cover the tubercles in flowering area and persisting for some time. CENTRAL SPINES 4, upper three 6 mm. long, lower one 17-20 mm. long, all very stout acicular, straight to slight recurve, stiff, smooth, black, in cross formation, spreading from porrect. RADIAL SPINES 6-10, 2-5 mm. long, hair-like, straight to slight curve, light tan with black tip, somewhat ascending only from lower part of the areole. FLOWER April, open for several days, campanulate, 17 mm. long, 20 mm. wide. *Outer perianth-segments* very pale olive-green below to white at base, brownish magenta wide tapering mid-stripe above, very narrow pale pink margins, linear, tip acuminate, margins mostly entire but occasionally ciliate. *Inner perianth-segments* very deep pink to magenta wide mid-line, darker at tip, wide pale pink margins, lanceolate, tip acuminate, margins entire. *Filaments* white, pale pink at very top. *Anthers* yellow-tan. *Style* very pale cream, heavy. *Stigma-lobes* 4-8, tannish pink, 2 mm. long, overtop anthers. FRUIT carmine red above to nearly white below, wide clavate, 18x9 mm., with dried perianth persisting. SEEDS brown, globular-pyriform with lateral hilum near the base, faintly wrinkled, 1.4x1 mm.

Distribution: Central Mexico (?)

Type locality: None given.

Illustration: Fig. 54 is from a photograph of a seedling plant in our garden.

This species was evidently described from garden specimens. Whether it is a true species or a garden hybrid, we have not been able to definitely determine as we have not been able to obtain any plants of it in Mexico. Our plant is a seedling obtained from a local nursery which in turn probably obtained the seed from Germany. It is a very striking specimen with the abundant white wool in the axils and it also flowers and fruits most satisfactorily.

48. Mammillaria phymatothele Berg

Mammillaria phymatothele Berg, Allg. Gartenz., 8:129, 1840.
Mammillaria ludwigii Ehrenberg, Linnaea, 14:376, 1840.
Mammillaria ludwigii clavata Walper, Repert. Bot. Syst., 2:299, 1843.
Cactus ludwigii Kuntze, Rev. Gen. Pl., 1:260, 1891.
Cactus phymatothele Kuntze, Rev. Gen. Pl., 1:261, 1891.
Neomammillaria phymatothele Britton & Rose, The Cactaceae, 4:76, 1923.

BODY simple, occasionally branching from the base, nearly globose, with apex sunken and white woolly, 75 mm. high, 60-90 mm. wide. TUBERCLES arranged in 8 and 13 spirals, firm in texture, dark bluish grayish green, pyramidal, keeled ventrally, rounded dorsally, terete at apex, with milky sap, 9-14 mm. long, 19 mm. wide at base. AREOLES round to 4 mm., sunken, with white wool, later becoming naked. AXILS with wool in youth only. CENTRAL SPINES 1 (2), 10-20 mm. long, strong subulate, straight to somewhat curved, smooth, stiff, dark red to reddish yellow, later becoming gray with darker tips, directed downward. RADIAL SPINES 3-7, occasionally 1-3 short accessory spines, upper 3-6 mm. long, lateral 13-17 mm. long, lowest 25-35 mm. long, all subulate, straight to recurved in lowest, stiff, smooth, grayish white with reddish brown tip, newly formed ones yellowish, ascending. FLOWERS funnelform, 10-15 mm. long. *Outer perianth-segments* dark brown, lighter margins, lanceolate. *Inner perianth-segments* fiery carmine, lanceolate, tip acute. *Filaments* white. *Anthers* chrome yellow. *Style* greenish yellow below, rose red above. *Stigma-lobes* 7, yellowish. FRUIT unknown. SEEDS unknown but probably brown.

Distribution: Hidalgo and central Mexico.

Type locality: None given but reported from San Felipe and Real del Monte, Hgo.

Illustration: The illustrations, Figs. 69 and 70, in Britton & Rose (Cact., 4:77) are referrable elsewhere as they definitely do not picture the spine arrangement called for in the original description.

We have not been able to recollect this species, so the description is compiled. The original description gave no data on the flower but we have used the incomplete description as given by Schumann and Berger. No description is known of the fruit and seed characteristics but it is assumed that the seed is brown in color.

FIG. 55. *Mammillaria phymatothele* var. *trohartii* x 0.9

a. var. **trohartii** (Hildmann) comb. nov.

Mamillaria trohartii Hildmann in Schumann, Gesamtb. Kakteen, 586, 1898.

BODY more depressed and cespitose. CENTRAL SPINES 1-2 (4). FLOWERS funnelform. *Outer perianth-segments* linear-oblong, tip acute, margins serrate. *Inner perianth-segments* deep purplish pink mid-stripe, paler margins, linear-oblong, tip obtuse, margins entire. *Filaments* deep purplish pink. *Anthers* yellowish tan. *Style* pale pink. *Stigma-lobes* 5, pale yellowish tan, fine pink line ventrally, 1 mm. long, overtop anthers 2-3 mm. FRUIT reddish, clavate, 21x7 mm., dried perianth missing. SEEDS light tan, curved pyriform with lateral hilum at base, faintly rugose, 1 mm.

Distribution: Mexico.

Type locality: None given.

Illustration: Fig. 55 is from a photograph of a plant obtained from Mexico.

Fig. 56. *Mammillaria hamiltonhoytae* x 0.7

49. **Mammillaria hamiltonhoytae** (Bravo) Werdermann

Neomammillaria hamiltonhoytae Bravo, An. Inst. Biol. Mex., 2:130,1931.
Mammillaria hamiltonhoytae Werdermann in Backeberg, Neue Kakt., 99, 1931.

BODY simple, occasionally small lateral heads from base, depressed globose with sunken apex, to 18 cm. wide. TUBERCLES arranged in 13 and 21 spirals, closely set, firm in texture, dark dull olive-green, slightly 4-sided at base, rounded at apex, flattened dorsally, keeled ventrally, with very thick milky sap, 10-14 mm. long, 9-10 mm. wide at base. AREOLES round to slightly elongate, with very short white wool, soon becoming naked. AXILS with white wool in flowering area, (occasional bristle, but not typical). CENTRAL SPINES 2-3, upper 10-20 mm. long, lower to 35 mm. long and heavier, all slender subulate, straight to sometimes tortuous, stiff, creamy white base, to pinkish orange, to reddish brown to black at tip, divergent porrect, lower more descending. RADIAL SPINES 5 (sometimes to 8), length variable, 3 uppers 3-5 mm., lower one to 17 mm. long, all heavy acicular, bent, stiff, smooth, white to creamy yellow with dark brown to black tip, somewhat ascending. FLOWERS funnelform, 20 mm. long. *Outer perianth-segments* tannish olive-green below, brownish pink wide tapering mid-stripe above, pale pink margins, lanceolate, tip acuminate, margins serrate. *Inner perianth-segments* deep purplish pink mid-stripe, wide white to pale pink margins, lanceolate, tip acute, margins entire. *Filaments* white. *Anthers* sulphur-yellow, small. *Style* greenish white, top pale pink. *Stigma-lobes* 4-5, light tannish cream, tannish pink line ventrally, 2 mm. long, overtop anthers 2-3 mm. FRUIT purple, clavate, 20 mm. long, with dried perianth not persisting. SEEDS reddish brown, globular pyriform with lateral hilum, very slightly roughened, 1.5 mm. long. ROOTS fibrous.

Distribution: Queretaro, Mexico.

Type locality: None given but reported from San Moran.

Illustration: Fig. 56 is from a photograph of a plant sent to us by Sr. F. Schmoll of Cadereyta, Qro.

The name *M. fortispina* characterized by E. Shurly (Journal British, 11, 1935), but not formally described is probably referrable here. It is compared with *M. polyedra* as well as with the above species.

a. var. **fulvaflora** var. nov.

Corpus et spinae idem, flores ad 15 mm. longibus, sepalis serratis ad ciliatis ad integris, petalis fulvis, stigmatibus 6.

Body and spine arrangement the same as the type. FLOWERS smaller, 12-15 mm. long. *Outer perianth-segments* pale tannish green below, wide tapering orange-brownish pink mid-stripe above, pinkish orange at tip, linear-lanceolate, tip acute, margins serrate to fine ciliate in scales and lower row to entire in inner row. *Inner perianth-segments* ventrally orange-tan wide mid-stripe, dorsally pale pink, yellowish tan margins, lanceolate, tip acute, margins mostly entire. *Filaments* white below, pale pink above. *Anthers* bright yellow, nearly globular. *Style* very pale greenish white. *Stigma-lobes* to 6, cream, tannish pink ventrally, 1 mm. long, just overtop anthers.

Distribution: Same as type.

FIG. 57. *Mammillaria scrippsiana*

50. **Mammillaria scrippsiana** (B. & R.) Orcutt

Neomammillaria scrippsiana Britton & Rose, The Cactaceae, 4:84, 1923.
Mamillaria scrippsiana Orcutt, Cactography, 8, 1926.

BODY simple, globose to short cylindric with apex sunken and woolly, 60 mm. high. TUBERCLES arranged in 26 (13 and 21) spirals, firm in texture, bluish green, nearly oval, terete, with milky sap, 8-10 mm. long. AREOLES nearly oval, with considerable wool in youth. AXILS with abundant wool. CENTRAL SPINES 2, 8-10 mm. long, stout acicular, straight, smooth, brown, slightly divergent dorsally and ventrally. RADIAL SPINES 8-10, 6-8 mm. long, slender acicular, straight, stiff, pale reddish at tip, somewhat ascending. FLOWERS wide funnelform, 15 mm. wide. *Inner perianth-segments* pinkish, margins paler. *Anthers* pinkish. *Stigma-lobes* 6, cream, recurved. FRUIT red, clavate. SEEDS brown.

Distribution: Jalisco, Mexico.

Type locality: Barranca de Guadalajara.

Illustration: Fig. 57 is a reproduction of an illustration, Fig. 78, in Britton & Rose (Cact., 4:85) as *Neomammillaria scrippsiana*.

We have not been able to recollect this species to complete the description.

Fig. 58. *Mammillaria esseriana*

51. Mammillaria esseriana Boedeker

Mamillaria esseriana Boedeker, Zeitschr. Sukkulentenk., 3:289, 1928.
Mamillaria brevispina Boedeker, Zeitschr. Sukkulentenk., 3:289, 1928.

BODY usually simple, may cluster from base, clavate, with slightly sunken apex which the central spines overtop, 10 cm. high, 6 cm. wide. TUBERCLES arranged in 13 and 21 spirals, light leaf-green, quadrangular at base (but not very sharp), many sided at apex, keeled and longer ventrally, apex rounded over so that the areole is ventral to the apex, with milky sap, 8 mm. long and also 8 mm. wide at base. AREOLES eliptical, 3-4 mm. wide, with considerable white wool in youth, becoming gradually naked. AXILS naked in apex, later in flowering area with wool and white tortuous bristles to 15 mm. long. CENTRAL SPINES 6, upper to 7 mm., lower to 15 mm. long, all straight, acicular to subulate, stiff, not enlarged at base, bright and transparent amber-yellow, ascending and spreading. RADIAL SPINES 10, lateral 3 mm. long, lower to 8 mm. long, all stiff bristle-like to fine acicular, pure white, slightly ascending, directed only laterally and ventrally. FLOWERS funnelform, surrounded by abundant wool, 12 mm. wide. *Outer perianth-segments* greenish, reddish brown mid-stripe, coppery margins, margins serrate. *Inner perianth-segments* first row carmine with darker mid-stripe, second row fiery carmine-red and longer, all slender lanceolate, tip acuminate, 3 mm. wide. *Anthers* white. *Filaments* and *Style* rose-carmine-red. *Stigma-lobes* 5-6, white to pale rose. FRUIT unknown. SEEDS unknown but probably brown.

Distribution: Southern Mexico to Central America.

Type locality: None given but reported from Chiapas (Schmoll).

Illustration: Fig. 58 is a reproduction of an illustration in Zeitschr. Sukkulentenk., 3:289 as *Mamillaria esseriana*.

FIG. 59. *Mammillaria geminispina* x 0.8

52. Mammillaria geminispina Haworth

Mammillaria geminispina Haworth in Till, Phil. Mag., 63:42, 1824.
Mammillaria bicolor Lehmann, Samen. Handb. Gartz., 7, 1830.
Mammillaria nivea Wendland in Pfeiffer, Enum. Cact., 27, 1837.
Mammillaria daedalea Scheidweiler, Hort. Belg., 4:16, 1837.
Mammillaria toaldoae Lehmann, Linnaea, **12**:13, 1838.
Mammillaria eburnea Miquel, Linnaea, **12**:14, 1838.
Mammillaria nivea daedalea Lemaire, Cact. Gen. Nov. Sp., 101, 1839.
Mammillaria nobilis Pfeiffer, Allg. Gartenz., **8**:282, 1840.
Mammillaria bicolor longispina Salm-Dyck, Hort. Dyck. 1844, 6, 1845.
Mammillaria bicolor cristata Salm-Dyck, Hort. Dyck. 1844, 198, 1845.
Mammillaria bicolor nobilis Foerster, Handb. Cact., 198, 1846.
Mammillaria nivea brevispina Hildmann, Verzeichnis, 4, 1888.
Cactus geminispinus Kuntze, Rev. Gen. Pl., **1**:260, 1891.
Cactus niveus Kuntze, Rev. Gen. Pl., **1**:261, 1891.
Cactus nobilis Kuntze, Rev. Gen. Pl., **1**:261, 1891.
Mamillaria bicolor nivea Schumann, Gesamtb. Kakteen, 569, 1898.
Neomammillaria geminispina Britton & Rose, The Cactaceae, 4:98, 1923.
Mammillaria geminispina nivea Borg, Cacti, 339, 1937.

BODY simple and cespitose to forming large mounds, individual heads globose to cylindric or clavate, with sunken apex, to 8 cm. wide, 9-18 cm. high. TUBERCLES arranged in 13 and 21 spirals, firm in texture, base 4-sided, not angled above, terete at apex, flattened dorsally, with milky sap, 7 mm. long, also 7 mm. wide at base. AREOLES oval, 2-3 mm. wide, with white wool in youth but soon becoming naked. AXILS with white wool and 10-20 white tortuous bristles to the length of the tubercles. CENTRAL SPINES 2-4 (6), upper one 25-40 mm. long, others 7-15 mm. long, all stout acicular, straight or slightly curved, smooth, stiff, base not enlarged, chalky-white with very tip brownish, divergent from porrect. RADIAL SPINES 16-20, 5-7 mm. long, slender acicular, straight to more or less bent, smooth, semiflexuous, chalky-white, horizontal. FLOWERS wide funnelform, 17-19 mm. long. *Outer perianth-segments* cherry-red midstripe, white margins, lanceolate, tip acute, margins serrate. *Inner perianth-segments* carmine-red mid-stripe, lighter margins, oblong-lanceolate, tip acute, margins finely serrate. *Filaments* white. *Anthers* yellow. *Style* white below, red above. *Stigma-lobes* 4-5, rose-red to tannish yellow, 1 mm. long. FRUIT carmine-red, globular to clavate, 10 mm. long, with dried perianth persisting. SEEDS brown, curved pyriform with lateral hilum, faintly roughened but not pitted.

Distribution: Hidalgo, San Luis Potosi, Vera Cruz; Mexico.

Type locality: None given but reported from Ixmiquilpan and Zimapan, Hgo., Venados, S.L.P., between Real del Monte and Tampico, V.C.

Illustration: Fig. 59 is from a photograph of a plant sent to us by Schwarz & Georgi, of San Luis Potosi. Fig. 60 is from a photograph by Mr. William Hertrich of a plant in the Huntington Gardens at San Marino, California.

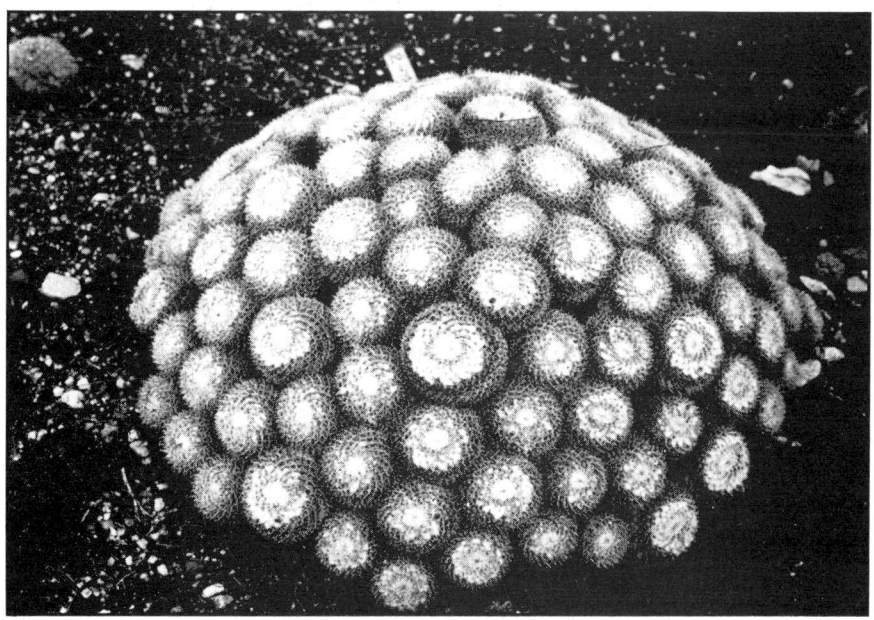

FIG. 60. *Mammillaria geminispina*

"*M. daedalea* which is referred here by Schumann, is based on an abnormal specimen which has elongated, contorted stems and looks very unlike the typical plant. Scheidweiler illustrates this species.

"*M. nivea cristata* Salm-Dyck (Walpers, Repert. Bot., 2:270, 1843) is only a name. *M. nivea wendlei* Pfeiffer (Labouret, Monogr. Cact., 57, 1853) was given as a synonym of *M. bicolor*." Britton & Rose (Cact., 4:98-99, 1923).

M. bicorem, a name used by Salm-Dyck (Cact. Hort. Dyck. 1849, 86, 1850) and attributed to Lemaire, was evidently intended for *M. bicolor*.

There is much confusion in the literature under this specific name because DeCandolle used this name for an entirely different plant after it had been previously used by Haworth. This accounts for many of the early references associating it with the plant that we now know as *M. elegans*.

Fig. 61. *Mammillaria evermanniana* x 0.5

53. **Mammillaria evermanniana** (B. & R.) Orcutt

Neomammillaria evermanniana Britton & Rose, The Cactaceae, 4:97, 1923.
Mamillaria evermannina Orcutt, Cactography, 7, 1926.

BODY simple and cespitose, globose to cylindric with rounded apex, 5-7 cm. wide. TUBERCLES closely set in 13 and 21 spirals, firm in texture, light green, conic, terete, with very small apex, with milky sap, 8-9 mm. long, 6 mm. wide at base. AREOLES oval, with dense white wool, later becoming naked. AXILS with dense white wool in youth, and also white tortuous bristles 5-8 mm. long. CENTRAL SPINES 2-4, 12-15 mm. long, heavy acicular, straight to slightly recurved, smooth, stiff, not enlarged at base, reddish brown in youth, later becoming chalky-white with brown tip, divergent. RADIAL SPINES 12-15, 5-8 mm. long, slender acicular, straight, semi-flexuous, smooth, chalky-white, lower ones reddish tan at tip, all somewhat ascending in youth, later horizontal. FLOWERS tubular, 15 mm. long, April and May. *Outer perianth-segments* brownish magenta wide mid-stripe, narrow pink margins, lanceolate, tip acute, margins serrate. *Inner perianth-segments* reddish pink to purplish mid-stripe, pale olive-green to cream margins, lanceolate, tip acute, margins entire. *Filaments* white. *Anthers* yellow. *Style* pale greenish yellow. *Stigma-lobes* 5, olive-green, 1 mm. long, just overtop anthers. FRUIT red, clavate, 10 mm. long, with dried perianth persisting. SEEDS light brown, curved pyriform with lateral hilum near base, faintly rugose, 1 mm. long.

Distribution: Baja California and adjacent islands in the Gulf of California, Mexico.

Type locality: Cerralboa Island, also reported from Nolasco Island and from the mainland near Loreto.

Illustration: Fig. 61 is from a photograph of a plant collected by Mr. Howard Gates at Loreto.

The original description is supplemented with data from material collected for us by Mr. Howard Gates and Mr. Howard Gentry.

M. nolascana was briefly characterized but not described by Radley (Cact. Succ. Journ., *12*:5, 1940) and should doubtlessly be referred here. We have the plant but as yet we have not seen the flower but from all available data it appears to come within the limits of the above species, although it has more wool in the axils.

Fig. 62. *Mammillaria saetigera*

54. **Mammillaria saetigera** Boedeker & Tiegel

Mammillaria saetigera Boedeker & Tiegel in Boedeker, Mammill. Vergl. Schluss., 49, 1933.

BODY simple, globose with somewhat sunken apex, 6-7 cm. high. TUBERCLES very loosely arranged in 13 and 21 spirals, glossy dark green, pyramidal but not angled, blunt truncate at apex, with milky sap, 12 mm. long, 5-6 mm. wide at base. AREOLES eliptical, 1-2 mm. wide, with white wool only in youngest. AXILS with white wool, also white bristles, especially in the lower part of the plant. CENTRAL SPINES 2, 7-11 mm. long, thin subulate, smooth, straight, white with brown tip, strongly spreading, often bent dorsally and ventrally. RADIAL SPINES 15-20, 7 mm. long, upper ones shorter, all straight, smooth, thin acicular, white, nearly horizontal to slightly ascending. FLOWERS lateral, 20 mm. long and wide, flattened funnelform. *Outer perianth-segments* almost white, rose mid-stripe, very slender linear-lanceolate, tip acute, margins serrate, 10 mm. long. *Inner perianth-segments* same shape and color except that they are darker. *Filaments* white below, rose above. *Anthers* yellow. *Style* rose. *Stigma-lobes* 4-5, dark yellow. FRUIT red, slender clavate, 18x2 mm., with dried perianth persisting. SEEDS yellowish brown, curved pyriform with small narrow lateral hilum, smooth (?), 1 mm.

Distribution: Queretaro, San Luis Potosi; Mexico.

Type locality: Hacienda Cenca.

Illustration: Fig. 62 is a reproduction of an illustration in Kakteenk., 191, 1934, as *Mamillaria saetigera*.

The place of the original description is usually given as Kakteenkunde 190, 1934, where it was described as a sp. nov. but the earlier publication in Boedeker's Comparison Key has priority even though it is only a brief which credits the authorship to "Bod. and Tieg. 1932" but we have not been able to find anything of that date. Index Kewensis credits the 1934 reference.

a. var. **quadricentralis** var. nov.

Corpus simplex, mamillis ad 12 et 21 seriebus ordinatis, pyramidatis, suco lacteo, axillis setis multis; spinis centralibus 4, 4-14 mm., rectis, subulatis; spinis radialibus 16-18, 2-4 mm., acicularibus; flores campanulata, sepalis ciliatis, petalis subrufis, stigmatibus 5, flavis; fructus et semina ignota.

BODY simple, flattened globose with apex sunken and woolly, 8 cm. wide, 5 cm. high. TUBERCLES arranged in 13 and 21 spirals, firm in texture, bluish grayish green, quadrangular, blunt angles, terete at apex, rounded dorsally, with milky sap, 8-9 mm. long,

9 mm. wide at base. AREOLES oval, with white wool in youth, soon becoming naked. AXILS with numerous yellow bristles and dirty white wool. CENTRAL SPINES 4, 4-14 mm. long, lower longest, lateral shortest, all subulate, straight, smooth, slightly enlarged at base, light purplish chalky, dark brown at apex, in cross formation, ascending. RADIAL SPINES 16-18, 2-4 mm. long, upper 4-5 shortest and more slender, all acicular, smooth, straight, stiff, faint yellowish chalky white, horizontal. FLOWERS campanulate, 13 mm. long, 15 mm. wide, May. *Outer perianth-segments* light pinkish tan mid-stripe above, very light tannish cream below, cream margins, linear-lanceolate, tip acute, margins ciliate. *Inner perianth-segments* pale pink with darker mid-line, much darker ventrally, spatulate, tip truncate to emarginate, margins entire with some serrations at apex. *Filaments* very pale pink. *Anthers* bright yellow. *Style* pale cream, pinkish at very top. *Stigma-lobes* 5, yellow, 1 mm. long, just overtop anthers. FRUIT and SEEDS unknown.

FIG. 63. *Mammillaria saetigera* var. *quadricentralis* x 1

Distribution: Mexico.

Type locality: Unknown.

Illustration: Fig. 63 is from a photograph of a plant sent to us by Sr. F. Schmoll of Cadereyta, Qro., as a new and unknown species.

We received this plant without any data on the type locality and we are not certain whether it is a true variety of the above species but we are referring it here because the characteristics break-down are so close to the type that there is hardly any justification for a new species on the strength of the information available at this time.

Fig. 64. *Mammillaria standleyi*

55. Mammillaria standleyi (B. & R.) Orcutt

Neomammillaria standleyi Britton & Rose, The Cactaceae, 4:97, 1923.
Mamillaria standleyi Orcutt, Cactography, 8, 1926.

BODY simple and cespitose in clumps of nearly 1 meter wide, individual heads nearly globular with somewhat sunken apex, to 15 cm. wide. TUBERCLES arranged in 13 and 21 spirals, firm in texture, glossy medium dark green, conic, terete at apex, keeled ventrally, blunt at apex where areole is placed somewhat ventrally, with milky sap, 9 mm. long, 7-8 mm. wide at base. AREOLES oval, with dense white wool in youth, soon becoming naked. AXILS with white bristles and with dense white wool in flowering and fruiting areas. CENTRAL SPINES 4, 6-8 mm. long, lower longest, all straight, stout acicular, smooth, reddish brown, divergent porrect. RADIAL SPINES 16-19, 4-8 mm. long, upper ones shortest, all fine acicular, straight, smooth, semiflexuous, white with dark tips, slightly ascending. FLOWERS funnelform, in ring near the apex, 12 mm. wide, May. *Outer perianth-segments* reddish purple margins, purple mid-line, linear, tip obtuse, margins ciliate. *Inner perianth-segments* purplish pink, darker mid-line, silvery margins, linear-lanceolate, tip obtuse, margins irregularly ciliate to serrate, in some entire. *Filaments* pale purplish pink. *Anthers* deep yellow. *Style* pale yellow. *Stigma-lobes* 5, light green, 1 mm. long, overtop anthers 2 mm. FRUIT scarlet, clavate, 12-16 mm. long, with dried perianth persisting. SEEDS brownish.

Distribution: Sonora, Mexico.

Type locality: Sierra de Alamos, near Alamos.

Illustration: Fig. 64 is a reproduction of an illustration, Fig. 93, in Britton & Rose (Cact., 4:96) as *Neomammillaria standleyi*.

The original description is supplemented with data from plants collected by Mr. and Mrs. John Hilton and the author in the canyons of the Sierra de Alamos above the town of Alamos, Sonora, in November, 1936. The plants that we collected higher in the canyons showed more of a tendency to form clusters than those that were reported by Dr. Rose and his party. He reports finding mostly solitary specimens "in the dry stony places above Alamos" while we found beautiful clusters of over a meter in width farther up in the canyons where it is often associated with *Cephalocereus leucocephalus*.

Similar to this species is a type from the Sierra Canelo, Rio Mayo, Sonora, that we have tentatively designated as *Mammillaria canelensis* (cf). It has more radial spines and longer central spines but inasmuch as we have not observed the flower, its relationship is uncertain and it may only be a variety of the above species.

FIG. 65. *Mammillaria standleyi* var. *robustispina* x 0.7

a. var. **robustispina** var. nov.

Mamillae maximae, spinis centralibus 5-6, robustis; spinis radialibus 20, 4-6 mm., acicularibus.

TUBERCLES large to 10 mm. wide at base. CENTRAL SPINES 5-6, 5-8 mm. long, subulate, dark cream with brown tip. RADIAL SPINES 20, 4-6 mm. long, slender acicular, white to cream. FLOWER, FRUIT and SEEDS unknown.

Type locality: Sierra de Alamos, Sonora, Mexico.

Illustration: Fig. 65 is a reproduction of our illustration in *Desert Plant Life* 9:28, 1937, as Mammillaria sp. No. 617.

This variety is being referred here on the strength of the information supplied by Mr. Howard Gentry who reports that he collected it in the Sierra de Alamos, the type locality of the above species. When the missing details are known, it may be referrable elsewhere.

FIG. 66. *Mammillaria lindsayi* x 0.8

56. **Mammillaria lindsayi** Craig

Mammillaria lindsayi Craig, Cact. Succ. Journ., **12**:182, 1940.

BODY simple and cespitose from base to form clumps up to 1 meter wide, individual heads globular with flattened and sunken apex, to 15 cm. wide and high. TUBERCLES arranged in 13 and 21 spirals, firm in texture, dull grayish green, conical quadrangular but not angled, keeled ventrally, with milky sap, 6-10 mm. long, 5-7 mm. wide at base. AREOLES oval, to 2 mm. wide with scant brownish felt only in youth. AXILS with dense white wool which nearly covers the tubercles in a circle in the flowering and fruiting area, also with as many as 8 tortuous white bristles which are longer than the tubercles and are persisting. CENTRAL SPINES 2-4 (mostly 4), 4-12 mm. long, upper 2-3 shortest, lower one longest and heavier, all acicular, straight, smooth, stiff, bulbous at base, light golden brown to somewhat reddish becoming grayish horn in age, spreading, lower one nearly porrect, others dorsally. RADIAL SPINES 10-14, 2-8 mm. long, upper 3-4 shortest and very slender acicular, lower little heavier, all straight, smooth, upper white, lower tan to golden yellow, all horizontal. FLOWERS form a ring in the top but not in the new growth, February to May, open during sunny hours for several days, almost tubular, 15-20 mm. long, 10 mm. wide. *Outer perianth-segments* 24, light greenish yellow with orange yellow mid-stripe, eliptical, tip obtuse, margins finely serrate, 15x2 mm. *Inner perianth-segments* 20, light greenish yellow, eliptical, tip obtuse to emarginate, margins fine ciliate to entire, 15x2.5 mm. *Filaments* very pale yellow to white. *Anthers* lemon-yellow, small, flattened. *Style* light yellow, 10-20 mm. long. *Stigma-lobes* 4-5, tannish greenish yellow, 2 mm. long, overtop anthers 1 mm. FRUIT scarlet, cylindric-clavate, 20x5 mm., appear in July of same year as flowering, with dried perianth persisting. SEEDS light brown, elongate curved pyriform with lateral hilum, reticulate, not pitted, 1x0.4 mm. ROOTS fibrous.

Distribution: SW. Chihuahua, Mexico.

Type locality: Molinas to Sierra Colorado (10-15 miles northeast of junction of the Rio Chinipas and Rio Fuerte, Lat. 27° 10′, Long. 108° 15′).

Illustration: Fig. 66 is a reproduction of an illustration in Cact. Succ. Journ., *12*:182 as *Mammillaria lindsayi*.

Habitat: In the partial shade on canyon walls and slopes, in leaf mold in crevices in rocks, mountainous at about 4000 feet.

Several variations of the type material were collected by Mr. George Lindsay and the author in April, 1939, in the immediate vicinity of the type locality and along the Rio Watchera but the salient characteristics of all of them are so nearly identical that they have all been included in the one species.

The original description gave the height of the heads as 300 mm. but it is in error as it was confused with the height of the cluster while the individual heads usually are to 15 cm. high.

a. var. **robustior** Craig

Mammillaria lindsayi robustior Craig, Cact. Succ. Journ., **12**:183, 1940.

Mamillae maximae ad 13 mm. per longitudinem, spinis centralibus ponderosioribus ad 20 mm. longis; spinis radialibus, etiam ponderosioribus ad 13 mm. longis.

Tubercles large to 13 mm. wide at base. Central spines heavier to subulate, to 20 mm. long. Radial spines heavier to strong acicular, to 13 mm. long.

Type locality: Molinas, Chihuahua, Mexico.

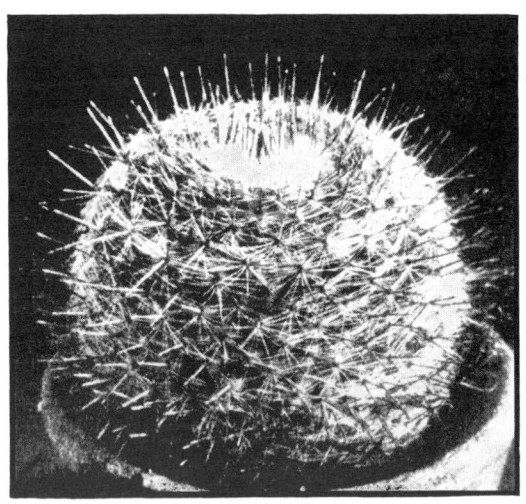

FIG. 67. *Mammillaria ritteriana*

57. **Mammillaria ritteriana** Boedeker

Mamillaria ritteriana Boedeker, Monatsschr. Deutsch. Kakt. Ges., 1:173, 1929.

BODY simple, globular with slightly sunken apex, 5-6 cm. wide. TUBERCLES arranged in 13 and 21 spirals, dull dark leaf-green, with very fine white pits, rounded 4-sided pyramidal, more or less sharp ventrally, terete above, with milky sap, 6 mm. long, 3 mm. wide at base. AREOLES round, 1-1.5 mm. wide, with white wool only in apex. AXILS with short white wool and strong white bristles. CENTRAL SPINES 1-2, 10 mm. long, thin subulate, upper slightly upward bent, all with thickened base, color variable from opaque white to yellowish brown to nearly black, porrect or divergent dorsally and ventrally. RADIAL SPINES 18-20, 5-7 mm. long, very fine hairlike, semi-stiff, somewhat recurved, pure white, horizontal, mostly lateral, nearly covering the plant. FLOWERS funnelform, 12-14 mm. long and wide, in crown but somewhat removed from the apex. *Outer perianth-segments* rose with white margins, broad-lanceolate, tip acuminate, margins serrate. *Inner perianth-segments* white, rose mid-stripe, linear-oblong, tip acute, margins serrate, 5-6x2 mm. *Filaments* white at base, rose above. *Anthers* bright yellow. *Style* yellowish white below, rose above. *Stigma-lobes* 4-5, short, reddish yellow, overtop anthers. FRUIT red, clavate, 10-12 mm. long. SEEDS pale reddish brown, pyriform with lateral hilum, 1 mm. long.

Distribution: Coahuila, Mexico.

Type locality: Higueras, between Monterrey and Saltillo.

Illustration: Fig. 67 is a reproduction of an illustration in Monatsschr. Deutsch. Kakt. Ges., *1*:173 as *Mamillaria ritteriana*.

FIG. 68. *Mammillaria ritteriana* var. *quadricentralis*

a. var. **quadricentralis** var. nov.

Spinis centralibus 4, 5-6 mm. longis, acicularibus, crassis; spinis radialibus 18-20, 3-7 mm. longis, acicularibus, tenuibus; flores infundibuliformes; sepalis serratis; petalis albis viridis; stigmatibus 5, subrubris ad viridis flavis.

BODY same as type. CENTRAL SPINES 4, 5-6 mm. long, straight, stout acicular, purplish in youth, later becoming yellowish tan with black tip, spreading porrect. RADIAL SPINES 18-20, 3-7 mm. long, upper shorter, lateral longer, all very fine acicular, white, slightly ascending. FLOWERS funnelform. *Outer perianth-segments* light green at base, reddish brown mid-stripe, light green margins, wide lanceolate, tip acute, margins lightly serrate. *Inner perianth-segments* very pale greenish white, pale green mid-line, narrow spatulate, tip obtuse, margins entire. *Filaments* white, pale pink at top. *Anthers* sulphur-yellow. *Style* white to cream, pink at top. *Stigma-lobes* 5, pink center, border and top olive-green.

Distribution: Mexico.

Type locality: Unknown.

Illustration: Fig. 68 is from a photograph of a plant sent to us by Sr. F. Schmoll of Cadereyta, Qro.

Top left, FIG. 69. *Mammillaria sonorensis* var. *brevispina* x 0.8
Bottom left, FIG. 71. *Mammillaria sonorensis* var. *hiltonii* x0.8
Top right, FIG. 70. *Mammillaria sonorensis* var. *gentryi* x 0.4
Bottom right, FIG. 72. *Mammillaria sonorensis* var. *longispina* x 0.8

58. Mammillaria sonorensis Craig

Mammillaria sonorensis Craig, Cact. Succ. Journ., 12:155, 1940.

BODY simple and cespitose from base, globose with sunken apex. TUBERCLES arranged in 5 and 8, 8 and 13, also 10 and 16 spirals (not consistent), firm in texture, dull bluish green, globular quadrangular but not sharply angled, keeled ventrally, with milky sap, 8-15 mm. long, 8-18 mm. wide at base. AREOLES oval, with white wool in youth, later becoming naked. AXILS with tufts of white wool and occasionally a few scattered hair-like bristles (but not typically characteristic as different heads of the same cluster may or may not have them). CENTRAL SPINES 1-4, length extremely variable from 5-45 mm. long, acicular to subulate, straight to sometimes recurved, stiff, smooth, enlarged at base, reddish brown, porrect to divergent, upper one subcentral. RADIAL SPINES 8-15, 1-20 mm. long, upper ones shortest, all slender acicular to acicular, straight, smooth, stiff, whitish to cream to reddish brown at tip, slightly ascending. FLOWERS campanulate, March to April, 20 mm. long. *Outer perianth-segments* brownish olive-green wide tapering mid-stripe, deep pink margins, lanceolate to ovate, tip acuminate, margins serrate to long ciliate. *Inner perianth-segments* deep pink, darker fine mid-line, especially ventrally, linear-lanceolate, 2-2.5 mm. wide, tip acuminate, margins entire or finely serrate, ends split. *Filaments* pink to reddish purple. *Anthers* yellow to orange. *Style* olive-green to pinkish above. *Stigma-lobes* 7, olive-green, 2.5 mm. long, just overtop anthers. FRUIT scarlet, clavate, 12x5 mm. SEEDS brown, pyriform with lateral hilum near base, less than 1 mm. ROOTS fibrous.

Distribution: SE. Sonora, NE. Sinaloa, SW. Chihuahua, Mexico.

Type locality: Near Guirocoba, SE. of Alamos, Sonora.

a. var. **longispina** Craig

Mammillaria sonorensis longispina Craig, Cact. Succ. Journ., 12:156, 1940.

Spinis centralibus 3-4, ad 35 mm. longis; spinis radialibus 14-15, mamillis 10 mm. latis.

Central spines 3-4, to 35 mm. long, acicular; radial spines 14-15; tubercles 10 mm. wide at base.

Type locality: Rio Mayo, Sonora.

Illustration: Fig. 72 is a reproduction of an illustration in Cact. Succ. Journ., *12*:155 as *Mammillaria sonorensis longispina*.

b. var. **brevispina** Craig

Mammillaria sonorensis brevispina Craig, Cact. Succ. Journ., 12:156, 1940.

Spinis centralibus 1-2, ad 8 mm. longis; spinis radialibus 8; mamillis 7-8 mm. latis.

Central spines 1-2, to 8 mm. long, acicular; radial spines 8; tubercles 7-8 mm. wide at base.

Type locality: Guirocoba, Sonora.

Illustration: Fig. 69 is a reproduction of an illustration in Cact. Succ. Journ., *12*:155 as *Mammillaria sonorensis brevispina*.

c. var. **gentryi** Craig

Mammillaria sonorensis gentryi Craig, Cact. Succ. Journ., 12:156, 1940.

Spinis centralibus 1, ad 45 mm. longis; spinis radialibus 9; mamillis 16 mm. latis.

Central spines 1, to 45 mm. long, subulate; radial spines 9, tubercles 16 mm. wide at base.

Type locality: Rio Mayo near Carimechi, Sonora.

Illustration: Fig. 70 is a reproduction of an illustration in Cact. Succ. Journ., *12*:155 as *Mammillaria sonorensis gentryi*.

d. var. hiltonii Craig

Mammillaria sonorensis hiltonii Craig, Cact. Succ. Journ., 12:156, 1940.

Spinis centralibus 1-3, ad 14 mm. longis; spinis radialibus 7-8; mamillis 16 mm. latis.

Central spines 1-3, to 14 mm. long, subulate; radial spines 7-8; tubercles 16 mm. wide at base.

Type locality: Guirocoba and San Bernardo, Sonora.

Illustration: Fig. 71 is a reproduction of an illustration in Cact. Succ. Journ., *12*:155 as *Mammillaria sonorensis hiltonii*.

Fig. 73. *Mammillaria sonorensis* var. *maccartyi* x 0.7

e. var. maccartyi Craig

Mammillaria sonorensis maccartyi Craig, Cact. Succ. Journ., 12:156, 1940.

Spinis centralibus 1, ad 20 mm. longis; spinis radialibus 8-10; mamillis 15 mm. latis.

Central spines 1, to 20 mm. long, stout acicular, strongly recurved dorsally; radial spines 8-10; tubercles to 15 mm. wide at base.

Type locality: Guirocoba, Sonora.

Illustration: Fig. 73 is a reproduction of an illustration in Cact. Succ. Journ., *12*:153 as *Mammillaria sonorensis maccartyi*.

59. Mammillaria hertrichiana sp. nov.

Corpus simplex et cespitosus; mamillis ad 13 et 21 seriebus ordinatis, pyramidatis, suco lacteo, 8-10 mm. per longitudinem, areolis ovatis, axillis lanatis albis, setis infrequentibus; spinis centralibus 4-5, ad 25 mm. longis, acicularibus; spinis radialibus 12-15, acicularibus, tenuibus, albis; flores infundibuliformes; sepalis ciliatis, petalis puniceis, stigmatibus 7, viridibus; fructus coccineus; semina fusca.

BODY simple to very cespitose forming clumps of nearly 1 meter wide, individual heads flattened globular with slightly sunken apex. TUBERCLES arranged in 13 and 21 spirals, firm in texture, dull medium dark green, quadrangular at base, irregularly angled

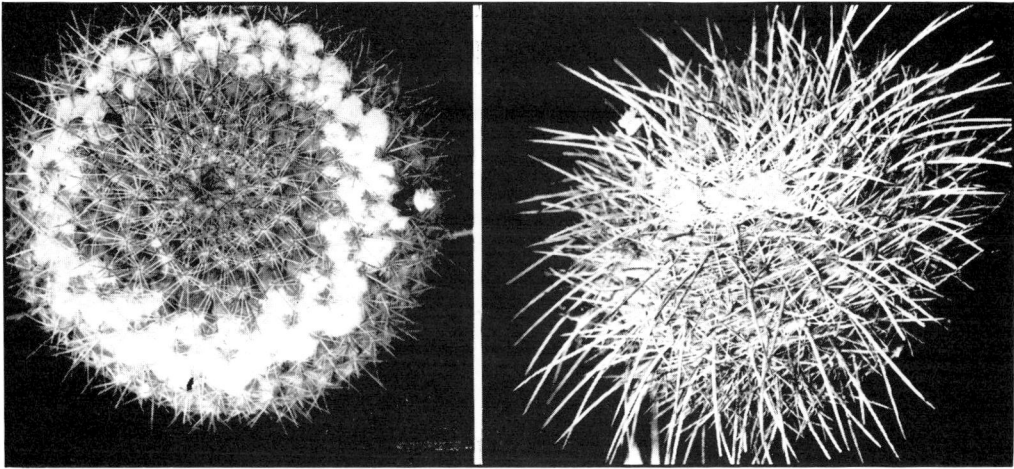

Fig. 74. *Mammillaria hertrichiana*   Fig. 75. *Mammillaria hertrichiana* var. *robustior*
x 0.5

to tip, with very milky sap, 8-10 mm. long, 8 mm. wide at base. AREOLES oval, with white wool in youngest. AXILS with dense white wool in flowering area to top of tubercles, only rarely an occasional bristle. CENTRAL SPINES 4-5, uppers 5-10 mm. long, lower to 25 mm. long and heavier, all acicular, straight, smooth, stiff, with slightly enlarged base, chestnut brown, divergent from porrect. RADIAL SPINES 12-15, 3-10 mm. long, upper shorter, all slender acicular, straight, smooth, stiff, white to pale tan with brown tip, ascending to later becoming horizontal. FLOWERS funnelform, 18 mm. wide, 10 mm. long. *Outer perianth-segments* light green at base, brownish red mid-stripe at tip, white to pale pink margins, broad-oblong to slightly lanceolate, tip acute to obtuse, margins long ciliate. *Inner perianth-segments* very deep pink to purplish pink, darker mid-stripe, linear, 2-3 mm. wide, tip obtuse and split to aristate, margins mostly entire. *Filaments* white below, pink above. *Anthers* yellow. *Style* nearly white, to pale pink at very top. *Stigma-lobes* 7, pale olive-green, 1.5 mm. long, overtop anthers 2 mm. FRUIT scarlet, clavate, 30x8 mm., with dried perianth persisting. SEEDS light brown, curved pyriform with lateral hilum, faintly roughened. ROOTS fibrous.

Distribution: Sonora, Mexico.

Type locality: Rancho El Agriminsor, east of Tesopaco.

Illustration: Fig. 74 is from a photograph of a plant collected for us by Mr. Howard S. Gentry in 1937.

This species is named in honor of Mr. William Hertrich, Curator of the Huntington Botanical Gardens at San Marino, California.

a. var. **robustior** var. nov.

Spinis centralibus robustioribus, ad 35 mm. longis.

Central spines are much heavier and longer to 35 mm. long.

Type locality: Same.

Illustration: Fig. 75 is from a photograph of a plant collected for us by Mr. Howard S. Gentry in 1937.

Fig. 76. *Mammillaria hemisphaerica* x 0.5

60. **Mammillaria hemisphaerica** Engelmann

Mamillaria hemisphaerica Engelmann, Boston Journ. Nat. Hist., 4:199, 1845.
Mamillaria heyderi hemisphaerica Engelmann, Proc. Amer. Acad., 3:263, 1856.
Cactus heyderi hemisphaericus Coulter, Contr. U. S. Nat. Herb., 3:97, 1894.
Cactus hemisphaericus Small, Fl. Southeast. U. S., 811, 1903.
Neomammillaria hemisphaerica Britton & Rose, The Cactaceae, 4:75, 1923.

BODY simple, flattened globular with apex rounded or hemispherical, 8-12 cm. wide, deep seated. TUBERCLES widely separated in 8 and 13, also 13 and 21 spirals, firm in texture, dark bluish green, dull, elongate quadrangular, slightly angled below, nearly terete at apex, keeled ventrally, with milky sap, 10-15 mm. long, 8 mm. wide at base. AREOLES nearly round, 1 mm. wide, with short white wool in youth, soon becoming naked. AXILS nearly naked and no bristles. CENTRAL SPINES 1, 3-4- (8) mm. long, slender subulate, straight to slight dorsal recurve, stiff, smooth, light brown at base, darker to black at tip, porrect to slightly erect. RADIAL SPINES 9-13, 2-8 mm. long, upper shorter, slender acicular, straight, stiff, smooth, cream to tan, darker to black at tip, slightly ascending. FLOWERS wide funnelform, 10-15 mm. long. *Outer perianth-segments* cream margins, light grayish purplish brown mid-stripe, linear-lanceolate, tip acute to obtuse, margins entire. *Inner perianth-segments* nearly white margins, pink to reddish mid-line, oblong-lanceolate, tip mucronate to acute, margins entire, tip often split. *Filaments* white below, pale pink top. *Anthers* sulphur-yellow. *Style* pale cream. *Stigma-lobes* 5-8, greenish tan, to 3 mm. long, overtop anthers 3-4 mm. FRUIT red, long clavate, 10-15 mm. long, with dried perianth not persisting. SEEDS dull reddish brown, pyriform with lateral hilum near base, roughened, 1.2x0.7 mm. ROOTS tap and fibrous.

Distribution: SE. Texas, U.S.; Tamaulipas, Nuevo Leon, Sonora (?), Mexico.

Type lociliay: Below Matamoros on the Rio Grande.

Illustration: Fig. 76 is from a photograph of a plant collected by us near Monterrey, N. L., Mexico.

This species is very closely related to *M. heyderi* and *M. applanata* but from which it differs in the number of radial spines. It is also related to *M. melanocentra* var. *meiacantha* but differs from it in the degree of angulation and size of the tubercles. Specimens that we collected along the highway north of Monterrey, N.L., had central spines that were from hardly discernable up to 3-4 mm. long. Specimens that we collected between Naco and Cananea, Sonora (650 miles north and west of the other collections) have central spines up to 8 mm. long but all the other characteristics were similar to the type so they were referred here despite the long distance from the type locality.

a. var. **waltheri** (Boedeker) comb. nov.

Mammillaria waltheri Boedeker, Zeitsschr. Sukkulentenk., 3:72, 1927.

Central spines usually 2, longer than type to 7-8 mm. long; perianth-segments midstripe more olive-green.

Fig. 77. *Mammillaria macdougalii* x 0.3

61. **Mammillaria macdougalii** Rose

Mammillaria macdougalii Rose in Bailey, Stand. Cyclo. Hort., 4:1982, 1916.
Neomammillaria macdougalii Britton & Rose, The Cactaceae, 4:74, 1923.

BODY simple and cespitose, flattened globular with sunken apex, 12-15 cm. wide. TUBERCLES arranged in 21 and 34 spirals, firm in texture, deep green, strongly angled at base, nearly terete at apex, flattened dorsally, with milky sap, to 12 mm. long, 9-12 mm. wide at base. AREOLES slightly oblong, with white wool in youth, soon becoming naked. AXILS often with long white wool but no bristles. CENTRAL SPINES 1-2, to 10 mm. long, acicular, straight, smooth, stiff, similar to radials, yellowish with brown tip, porrect. RADIAL SPINES 10-12, to 20 mm. long, lower 3 longest and stouter, all acicular, straight, stiff, smooth, white to somewhat yellowish with brown or black tip, slightly ascending. FLOWER campanulate, 35 mm. long, 40 mm. wide. *Outer perianth-segments* olive-green mid-stripe tapering to point, pale greenish cream margins, lanceolate, tip acute, margins ciliate. *Inner perianth-segments* pale greenish cream, very pale tan mid-line, linear, tip acute and often split, margins ciliate, 25x2.5 mm. *Filaments* very pale green. *Anthers* yellow. *Style* very pale green. *Stigma-lobes* 7, light green, level with anthers. FRUIT green with pink tip, clavate, with dried perianth persisting. SEEDS dull dark brown, curved pyriform to somewhat rounded with lateral hilum near base, lightly rugose. ROOTS fibrous.

Distribution: SE. Arizona, U. S.; northern Sonora, Mexico.

Type locality: Near Nogales.

Illustration: Fig. 77 is from a photograph of a plant we collected near Nogales, Arizona.

This species is fairly common in the grass lands of the upper hilly country, especially in the vicinity of Nogales. We collected specimens of it on both sides of the Arizona-Sonora border.

FIG. 78. *Mammillaria gummifera* x 0.7

62. Mammillaria gummifera Engelmann

Mamillaria gummifera Engelmann in Wislizenus, Mem. Tour North. Mex., 105, 1848.
Cactus gummifer Kuntze, Rev. Gen. Pl., 1:260, 1891.
Cactus gummiferus Coulter, Contr. U. S. Nat. Herb., 3:98, 1894.
Mamillaria buchheimeana Quehl, Monatsschr. Kakteenk., 27:97, 1907.
Neomammillaria gummifera Britton & Rose, The Cactaceae, 4:74, 1923.

BODY simple, hemispherical with sunken apex, to 12 cm. wide, to 11 cm. high. TUBERCLES arranged in 13 and 21 spirals, firm in texture, light grayish green, 4-sided pyramidal but not sharply angled, with milky sap, 13-15 mm. long, 11-13 mm. wide at base. AREOLES pyriform, with dirty white wool in youth. AXILS with scant white wool in youth but no bristles. CENTRAL SPINES 1-2, 4 mm. long, slender subulate, straight, stiff, smooth, light brown base, black tip, porrect to slightly erect. RADIAL SPINES 10-12, uppers 4-6 mm. long, lower 13-15 (20-25) mm. long, upper setaceous, lower heavy acicular, all straight to somewhat recurved, stiff, smooth, white with brownish or black tip, somewhat ascending. FLOWERS funnelform, 25-30 mm. long, 15-25 mm. wide. *Outer perianth-segments* brownish pink wide mid-stripe, pale cream to white margins, lanceolate, tip acute, margins ciliate, 3-3.5 mm. wide. *Inner perianth-segments* narrow magenta mid-stripe, nearly white margins, linear-lanceolate, tip acute, margins entire to serrate. *Filaments* pink above, nearly white below. *Anthers* yellow. *Style* very pale greenish tan above, paler below, 18 mm. long. *Stigma-lobes* 6, reddish, 2 mm. long, overtop anthers. FRUIT scarlet, elongate clavate, with dried perianth persisting. SEEDS dark brown.

Distribution: Chihuahua, Mexico.

Type locality: Cosihuiriachi.

Illustration: Fig. 78 is from a photograph of a plant sent to us by Schwarz & Georgi of San Luis Potosi, S.L.P., Mexico.

M. bucheimeana Quehl was referred with a question by Britton & Rose to *M. heyderi* but it appears to be nearer to this species.

63. Mammillaria pachyrhiza Backeberg

Mamillaria pachyrhiza Backeberg, Beit. Sukkulentenk. Pflege, 2:39, 1939.

BODY simple, deep seated, shrinks into ground in dry season, with apex a little depressed, 3-4 cm. high, 15 cm. wide. TUBERCLES leaf-green, somewhat compressed, with milky sap (?). AREOLES nearly naked. AXILS naked. CENTRAL SPINES 4-6, upper 12 mm. long, lower to 16 mm. long, all straight to a little curved, white to yellow to reddish. RADIAL SPINES 15-18, to 10 mm. long, white, horizontal, interlocking laterally. FLOWERS white. FRUIT smudgy red, clavate. SEEDS unknown.

Distribution: Vera Cruz and Puebla, Mexico.

Type locality: Near Las Derrumbadas.

Illustration: Beit. Sukkulentenk. Pflege, 2:39, as *Mamillaria pachyrhiza*. See appendix for photograph.

The original description makes no mention of the sap but we are assuming that it is milky on the strength that other species of this type with the deep seated root that occur in this same general region do have a milky sap. Further investigations will be necessary at the type locality to obtain additional material so that the correct relationship of this species can be determined.

64. Mammillaria applanata Engelmann

Mamillaria applanata Engelmann in Wislizenus, Mem. Tour North. Mex., 105, 1848.
Mamillaria declivis Dietrich, Allg. Gartenz., 18:235, 1850.
Mamillaria texensis Labouret, Monogr. Cact., 89, 1853.
Mamillaria heyderi applanata Engelmann, Proc. Amer. Acad., 3:263, 1856.
Cactus texensis Kuntze, Rev. Gen. Pl., 1:261, 1891.
Neomammillaria applanata Britton & Rose, The Cactaceae, 4:76, 1923.

BODY simple, almost level top, with depressed apex, to 11 cm. wide, 25-50 mm. high. TUBERCLES arranged in 13 and 21 spirals, firm in texture, dull green, elongate pyramidal, nearly quadrangular, with milky sap, 8-10 mm. long. AREOLES round, with white wool, soon becoming naked. AXILS naked. CENTRAL SPINES 1, 4-8 (mostly 6) mm. long, straight, stout acicular, stiff, smooth, dark brown in youth, soon becoming light yellowish brown with brown to black tips, porrect to erect. RADIAL SPINES 15-20, 5-12 mm. long, upper shorter, lower stouter and longer, all slender acicular, straight, smooth, whitish, lower light brown, all mostly horizontal. FLOWER funnelform, May, 18-25 mm. long, 25-35 mm. wide. *Outer perianth-segments* 8-13, greenish, lanceolate, tip acuminate, margins entire. *Inner perianth-segments* cream to faint reddish margins, greenish mid-stripe, linear-lanceolate, tip mucronate, margins entire to serrate at tip. *Filaments* white. *Anthers* golden yellow. *Style* white. *Stigma-lobes* 5-8 greenish. FRUIT scarlet, elongate clavate, to 35 mm. long, with dried perianth not persisting. SEEDS reddish brown, subglobose to ovate, slightly grooved rugose.

Distribution: Central and southern Texas, U. S.

Type locality: Rocky plains on the Pierdenales, also reported from Corpus Christi.

"*M. lindheimeri* Engelmann, is given by Hemsley (Biol. Centr. Amer. Bot., *1*:525, 1880) and by Index Kewensis as a synonym of *M. texensis*, belongs here." Britton & Rose, (Cact., 4:76.)

FIG. 79. *Mammillaria johnstonii* var. *typica* x 0.5

FIG. 80. *Mammillaria johnstonii* var. *scarcarlensis* x 0.5

65. **Mammillaria johnstonii** (B. & R.) Orcutt

Neomammillaria johnstonii Britton & Rose, The Cactaceae, 4:80, 1923.
Mamillaria johnstonii Orcutt, Cactography, 7, 1926.

BODY simple, occasionally cespitose from the base, globose with slightly sunken apex, 15-20 cm. high. TUBERCLES arranged in 13 and 21 spirals, firm in texture, somewhat dull bluish grayish green, 4-angled to tip, with milky sap, 10-13 mm. long, 6-9 mm. wide at base. AREOLES round to oval, with scant white wool only in youngest. AXILS with scant white wool but no bristles. CENTRAL SPINES mostly 2, occasionally 4-6 (see varieties), 10-25 mm. long, subulate, straight to occasionally tortuous, smooth, stiff, light purplish to black, divergent, upper erect. RADIAL SPINES 10-18, 6-9 mm. long, upper shorter, all acicular, stiff, straight, smooth, white to horn-color with reddish brown to black tips, slightly ascending. FLOWERS from near top of plant but in old axils, campanulate, 15-20 mm. long and wide. *Outer perianth-segments* reddish to purplish brown tapering mid-stripe, bordered with pale green, pinkish cream margins, ovate-lanceolate, tip acute to obtuse, margins ciliate to serrate. *Inner perianth-segments* pink to brownish mid-line to mid-stripe, white to cream margins, linear-lanceolate, 1.5-2.5 mm. wide, tip acuminate, margins mostly entire or with ends split. *Filaments* pink above, cream below. *Anthers* bright yellow. *Style* pale brownish pink to sometimes greenish. *Stigma-lobes* 6-7, olive-green, linear, 2 mm. long, just overtop anthers. FRUIT red to scarlet, clavate, 25x8 mm., with dried perianth persisting. SEEDS dull light brown, curved pyriform with lateral hilum, reticulate, 1.2x0.6 mm.

Distribution: Sonora, Mexico.

Type locality: San Carlos Bay near Guaymas, known as yet from only the type locality.

a. var. **typica** var. nov.

Spinis centralibus 2, 10 mm. longis, rectis; spinis radialibus 10-14.

Central spines 2, 10 mm. long, straight, stiff; radial spines 10-14; inner perianth-segments acuminate.

Type locality: Same.

Illustration: Fig. 79 is from a photograph of a plant we collected at the type locality in 1936.

b. var. **sancarlensis** var. nov.

Spinis centralibus 2, 25 mm. longis, tortuosis; spinis radialibus 15-18.

Central spines 2, to 25 mm. long, tortuous, semiflexuous; radial spines 15-18; inner perianth-segments obtuse.

Type locality: Same.

Illustration: Fig. 80 is from a photograph of a plant we collected at the type locality.

FIG. 81. *Mammillaria johnstonii* var. *guaymensis* x 0.6

c. var. **guaymensis** var. nov.

Spinis centralibus 4-6, 18 mm. longis, rectis; spinis radialibus 18.

Central spines 4-6, 18 mm. long, straight, stiff, subulate; radial spines 18; inner perianth-segments acuminate.

Type locality: Same.

Illustration: Fig. 81 is from a photograph of a plant we collected at the type locality.

This species is known to us as only from the type locality of San Carlos Bay near Guaymas on the Gulf of California where we found it quite abundantly in 1936, 8 and 9. It is quite variable in its spine arrangement, both as to length and number of the central spines. Several separate species might be erected on the different variations but inasmuch as they are all found intermixed in such a very limited area, we consider them as only variations of the same species.

It is a very strange environment in which this species grows as the soil is very salty and contains many sea shells. They are often found so close to the edge of the water that they are sprayed with the sea water.

"Believe it or not" but in this same locality we observed good sized specimens of *Pachycereus pringlei* actually growing in the sea water of the bay.

66. Mammillaria arida Rose

Mamillaria arida Rose, Monatsschr. Kakteenk., 23:181, 1913.
Neomammillaria arida Britton & Rose, The Cactaceae, 4:73, 1923.

BODY simple, globular, deep seated, to 6 cm. wide. TUBERCLES arranged in 8 and 13 (?), spirals, firm in texture, dull grayish green, quadrangular at base, nearly terete above, keeled ventrally, with very milky sap. AREOLES slightly oblong, becoming depressed, with white wool only in the very youngest. AXILS with scant white wool but no bristles. CENTRAL SPINES 4-7 (mostly 5-6), 12-16 mm. long, stout acicular, straight, stiff, smooth, dark brown, porrect. RADIAL SPINES 15, 6-10 mm. long, upper shorter, all slender acicular, straight, stiff, smooth, somewhat yellowish, darker at tip, somewhat ascending. FLOWERS funnelform, 10 mm. long, 25 mm. wide. *Outer perianth-segments* dull reddish brown to purplish mid-stripe, bright green margins, lanceolate, tip acute (?), margins entire, 6-15x2-3 mm. *Inner perianth-segments* cream to greenish light yellow, dark red mid-stripe ventrally, lanceolate, tip obtuse, margins entire. *Filaments* white. *Anthers* dirty yellow. *Style* white. *Stigma-lobes* 4, greenish to yellowish. FRUIT red, clavate, 15 mm. long. SEEDS brown.

Distribution: Baja California, Mexico.

Type locality: La Paz.

Examination of the type material in the U. S. National Museum at Washington, D.C., by W. R. Maxon and C. V. Morton at our request revealed that the central spines vary from 4-7 and are usually 5-6. Recent but separate explorations by Mr. Howard Gates and Mr. George Lindsay at the type locality of La Paz failed to relocate any of this species. From the same locality, Mr. Gates has described *M. baxteriana* in which there is only one central spine instead of the 4-7 in the above species. Whether the greater number of central spines was an unusual sport or was a species now extinct, will require more field work to clear up this uncertainty.

FIG. 82. *Mammillaria pseudocrucigera* x 1

67. Mammillaria pseudocrucigera sp. nov.

Corpus simplex, subglobosus; mamillis ad 8 et 13 seriebus ordinatis, quadrangularis, suco lacteo, axillis lanatis albis; spinis centralibus 4, 3-4 mm. longis, rectis subulatis, horizontaliter radiantes; spinis radialibus 12-13, 1-2 mm. longis, acicularibus, tenuibus, albis, horizontaliter radiantes; flores campanulatae, sepalis serratis ad integris; petalis albis cum linea media subissimmarubra; stigmatibus 3, subflavis; fructus coccineus; semina fusca.

BODY simple, deep seated, flattened globose with sunken apex, to 50 mm. wide. TUBERCLES arranged in 8 and 13 spirals, dark olive-gray-green, 4-angled pyramidal, keeled ventrally, with milky sap, 7 mm. long, 5 mm. wide at base. AREOLES nearly round, with scant white wool only in the youngest. AXILS with a little white wool but no bristles. CENTRAL SPINES 4, 3-4 mm. long, subulate, straight, stiff, smooth, base enlarged, chalky white with very tip dark brown to black, orange brown at base, horizontal. RADIAL SPINES 12-13, 1-2 mm. long, fine acicular, straight, smooth, white, horizontal. FLOWERS campanulate, 15 mm. wide, 12 mm. long, May. *Outer perianth-segments* olive-green below, pink to reddish brown tapering mid-stripe above, pale pink wide margins, lanceolate, tip acute, margins lightly serrate in lower part to entire near the top. *Inner perianth-segments* pink mid-line, darker ventrally, wide white margins, lanceolate, tip acute, margins entire. *Filaments* white, very top pale pink. *Anthers* pale greenish yellow. *Style* cream below, very pale pink at top. *Stigma-lobes* 3, pale yellow, pale pink ventral line. FRUIT bright red, slender clavate, 20x5 mm., with dried perianth persisting. SEEDS brown, curved pyriform with lateral hilum, 1.4x0.5 mm., very faintly pitted.

Distribution: Queretaro, Mexico.

Type locality: Between Cadereyta and Colon.

Illustration: Fig. 82 is from a photograph of a plant obtained from Sr. F. Schmoll of Cadereyta, Qro., in 1937 and again in 1942.

This specific name was mentioned as to name only by Neal (Cact. Other Succ., 93, 1935) and in the Cact. Succ. Journ., *10*:137, 1938.

Fig. 83. *Mammillaria gigantea* x 0.8

68. **Mammillaria gigantea** Hildmann

Mamillaria gigantea Hildmann in Schumann, Gesamtb. Kakteen, 578, 1898.
Neomammillaria gigantea Britton & Rose, The Cactaceae, 4:85, 1923.

BODY simple, depressed globose with apex strongly sunken and woolly, to 17 cm. wide, 9-10 cm. high. TUBERCLES closely set in 13 and 21 spirals, bluish grayish green, pyramidal, 4-angled, keeled ventrally, with milky sap, 10 mm. long, 10-12 mm. wide at base. AREOLES eliptical to pyriform, 6 mm. wide in youth, with abundant pure white flocculent wool in youth but later becoming naked. AXILS with white wool especially in flowering area. CENTRAL SPINES 4-6, lower longer to 20 mm. long, upper to 15 mm. long, all very robust subulate, straight to curved to recurved, smooth, yellowish reddish purple with brown to black tip in youth, later all dark yellowish grayish horn-color, nearly horizontal to a little ascending, one often more central and nearly porrect so as to appear as a true central with others as subcentrals. RADIAL SPINES 12, short to 5 mm. long, slender to heavy (lower ones) acicular, straight to slight curve, smooth, nearly white, horizontal. FLOWERS funnelform, April, 15 mm. long and wide. *Outer perianth-segments* light green below, wide tapering reddish brown mid-stripe above, lanceolate, tip acute, margins serrate. *Inner perianth-segments* wide reddish straight midline, bordered with light green, wide light greenish yellow margins, spatulate, tip obtuse, margins entire. *Filaments* very pale yellow. *Anthers* pale golden yellow. *Style* very pale greenish white. *Stigma-lobes* 5, very pale greenish yellow, 2 mm. long, just overtop anthers. FRUIT purplish pink, clavate, 30x8 mm., juicy, with dried perianth persisting. SEEDS light tan, curved pyriform with small darker lateral hilum at base.

Distribution: Guanajuato, Queretaro; Mexico.

Type locality: None given but reported from San Moran.

Illustration: Fig. 83 is from a photograph of a plant sent to us by Schwarz & Georgi of San Luis Potosi, S. L. P.

"*M. macdowellii* Heese and *M. guanajuatensis* Rünge are two names referred here by Schumann (Gesamtb. Kakteen, 578, 1898), but they were never published." Britton & Rose (Cact., 4:85).

FIG. 84. *Mammillaria gaumeri* x 1

69. Mammillaria gaumeri (B. & R.) Orcutt

Neomammillaria gaumeri Britton & Rose, The Cactaceae, 4:72, 1923.
Mamillaria gaumeri Orcutt, Cactography, 7, 1926.

BODY simple and cespitose, globose to short cylindric, to 15 cm. high, grows half hidden in sand dunes. TUBERCLES arranged in 8 and 13 spirals, dark green, 4-sided at base, nearly terete above, obtuse, with very milky sap, 5-7 mm. long. AREOLES with conspicuous white woolly mat in youth, soon becoming naked. AXILS naked or with scant wool but no bristles. CENTRAL SPINES 1, to 10 mm. long, subulate, straight, smooth, purplish brown with black tip, porrect. RADIAL SPINES 12-14, 5-7 mm. long, acicular, straight, smooth, white with brown tips, lower ones darker, all ascending. FLOWERS 10-14 mm. long. *Outer perianth-segments* greenish with brown tip, broad lanceolate, margins serrate. *Inner perianth-segments* creamy white. FRUIT crimson, clavate, 18-20 mm. long, with dried perianth persisting. SEEDS brown.

Distribution: Yucatan, Mexico.

Type locality: Progresso.

Illustration: Fig. 84 is from a photograph of a seedling plant in our garden.

Our plant has as yet not bloomed for us, so the flower's description is incomplete.

Fig. 85. *Mammillaria xanthina*

70. Mammillaria xanthina (B. & R.) Boedeker

Neomammillaria xanthina Britton & Rose, The Cactaceae, 4:164, 1923.
Chilita xanthina Orcutt, Cactography, 2, 1926.
Mammillaria xanthina Boedeker, Mammill. Vergl. Schluss., 47, 1933.

BODY depressed globose, 8-9 cm. wide. TUBERCLES arranged in 13 and 21 spirals, dull bluish green, short wide conic, broader than tall, flattened dorsally, with milky sap, 5 mm. long. AREOLES circular, small. AXILS with dense white wool in youth, later becoming naked. CENTRAL SPINES 2, longer than radials, 4-6 mm. long, stout acicular, lower porrect, upper nearly erect. RADIAL SPINES 10-12, to 4 mm. long, acicular, straight, white, spreading. FLOWERS rotate, from top of plant but in old axils, tube not exserted, limb appressed against adjacent tubercles. *Outer perianth-segments* lemon-yellow, oblong, tip obtuse, margins ciliate. *Inner perianth-segments* lemon-yellow, oblong, tip retuse to sometimes apiculate, margins entire. *Filaments* pale lemon-yellow. *Anthers* yellow. *Style* pale lemon-yellow. *Stigma-lobes* not given. FRUIT unknown. SEEDS unknown but probably brown.

Distribution: Durango, Mexico.

Type locality: Near Monte Mercado.

Illustration: Fig. 85 is a reproduction of an illustration, Fig. 184, in Britton & Rose (Cact., 4:164) as *Neomammillaria xanthina*.

We have not been able to recollect this species to complete the description.

71. Mammillaria tesopacensis sp. nov.

Corpus simplex, globosus; mamillis ad 12 et 21 seriebus ordinatis, conicis, suco lacteo, 10-12 mm. per longitudinem; areolis rotundis; axillis nudis; spinis centralibus 1, 10-12 mm. longis, acicularibus crassis, luteis, porrectis; spinis radialibus 10-15, 4-7 mm. longis, acicularibus, tenuibus, luteis; flores infundibuliformis; sepalis ciliatis, petalis alboflavis, stigmatibus 5, viridibus; fructus coccineus; semina fusca.

BODY simple, globose to cylindric, to 18 cm. high, to 13 mm. wide. TUBERCLES arranged in 13 and 21 spirals, firm in texture, bluish grayish green, pyramidal conic, very

FIG. 86. *Mammillaria tesopacensis* x 1

faintly angled below, terete at apex, with milky sap, 10-12 mm. long, 7-8 mm. wide at base. AREOLES round, with very scant white wool in youngest. AXILS naked or with very scant wool in flowering area but no bristles. CENTRAL SPINES 1, occasionally 2, 10-12 mm. long, stout acicular, straight, smooth, stiff, in youth reddish tan below, black at top, becoming all ashen-brown, porrect. RADIAL SPINES 10-15, 4-7 mm. long, slender acicular, straight, smooth, stiff, same color as centrals, strongly ascending in youth, later nearly horizontal. FLOWERS funnelform, 20 mm. long and wide, May. *Outer perianth-segments* wide brownish red mid-stripe, narrow greenish cream margins, wide lanceolate, tip obtuse, margins ciliate. *Inner perianth-segments* cream, pink mid-line in upper half, ventrally mid-line increases into a tapering mid-stripe, linear, 2-3 mm. wide, tip acute, margins some ciliate, some with fine serrations especially at the tip, others entire. *Filaments* pink. *Anthers* yellow. *Style* tannish pink above, lighter below. *Stigma-lobes* 5, light green, 3 mm. long, overtop anthers 2 mm. FRUIT scarlet, short clavate, 18x10 mm., dried perianth persisting. SEEDS light brown, pyriform with lateral hilum, faintly rugose, 1x0.6 mm. ROOTS fibrous.

Distribution: Sonora, Mexico.

Type locality: Tesopaco.

Illustration: Fig. 86 is from a photograph of a plant collected for us by Mr. Howard S. Gentry while he was making a botanical survey of the Rio Mayo.

a. var. **rubraflora** var. nov.

Corpus et spinae sunt eadem et typica; flores petalis puniceis, stigmatibus 7, flavoviridis.

Body and spine arrangement are the same as the type. FLOWER *Inner perianth-seg-*

ments deep purplish pink, darker mid-line, margins entire. *Stigma-lobes* 7, bright yellowish green.

Type locality: Movas, Sonora, about 15 miles north of Tesopaco.

A series of closely related plants were collected in near-by areas by Mr. Gentry but these showed considerable variation:

From Botania: Tubercles angled, 12 mm. long. Central spines 1. Radial spines 7-9-12.

From Movas: (a) Tubercles obscurely 4-sided, 12 mm. long. Central spines 1. Radial spines 11. (b) Tubercles obscurely 4-sided, 8 mm. long. Central spines 1. Radial spines 9.

From Rio del Media: Tubercles terete, 12 mm. long. Central spines 1. Radial spines 11-12.

FIG. 87. *Mammillaria mammillaris*

72. Mammillaria mammillaris (Morison) Karsten

Melocarduus mamillaris Morison, Plant. Univer. Oxon., 171, 1715.
Melocactus mammillaris Plukenet, Opera Omnia Bot., 4:148, 1720.
Cactus mammillaris Linnaeus, Species Plant., 1:466, 1753.
Cactus mammillaris glaber DeCandolle, Pl. Succ., 2:111, 1799.
Cactus mammillaris lanuginosis DeCandolle, Pl. Succ., 2:111, 1799.
Mammillaria simplex Haworth, Syn. Pl. Succ., 117, 1812.
Mammillaria parvimamma Haworth, Suppl. Pl. Succ., 72, 1819.
Cactus microthele Sprengel, Syst. Veg., 2:494, 1825.
Mammillaria microthele Martius, Hort. Reg. Monac., 127, 1829. Nomen.
Mammillaria simplex parvimamma Lemaire, Cact. Gen. Nov. Sp., 98, 1839.
Mammillaria caracassana (*) Otto in Salm-Dyck, Cact. Hort. Dyck. 1849, 107, 1850.
Mamillaria karstenii Poselger, Allg. Gartenz., 21:95, 1853.
Mammillaria mammillaris Karsten, Deutsch. Fl., 888, 1882.
Cactus parvimammus Kuntze, Rev. Gen. Pl., 1:259, 1891.
Neomammillaria mammillaris Britton & Rose, The Cactaceae, 4:70, 1923.
Mammillaria ekmanni Werdermann in Fedde, Repertorum, 242, 1931.

(*) Schumann spelled it *M. caracasana*.

BODY simple and cespitose, globose to short cylindric with sunken apex, to 20 cm. high. TUBERCLES arranged in 8 and 13 spirals, glossy light to dark green, conic, nearly terete, with milky sap, 7-12 mm. long, 4-8 mm. wide at base. AREOLES round, 2-3 mm. wide, with dense light brown wool in youth. AXILS with scant wool. CENTRAL SPINES 3-5, 7-8 mm. long, upper ones longer, all subulate, straight or upper occasionally slightly bent, all smooth, reddish brown with darker tip, becoming gray, spreading porrect. RADIAL SPINES 10-16, 5-8 mm. long, acicular, straight, smooth, reddish, later gray, nearly horizontal to somewhat ascending. FLOWERS funnelform, 10-12 mm. long, July-September. *Outer perianth-segments* brownish green, top row dark brown, eliptical, tip long mucronate, margins entire (?) *Inner perianth-segments* whitish cream to pale yellow, lanceolate, tip acute, margins entire (?) *Anthers* yellowish. *Style* yellowish. *Stigma-lobes* 5-6, greenish yellow. FRUIT carmine red, clavate, 10-20x6-8 mm., with dried perianth not persisting. SEEDS brown, minute, roughened.

Distribution: Northern Venezuela and adjacent islands.

Type locality: None given but reported from northern Venezuela and Paos, Curacao and Margarita Islands.

Illustration: Fig. 87 is a reproduction of an illustration, Fig. 64, in Britton & Rose (Cact., 4:70) as *Neomammillaria mammillaris*.

"*M. microthele* Monville and *M. micrantha* Hortus are names which Rümpler (Förster, Handb. Cact., ed. 2, 335, 1885) refers to *M. caracassana;* Salm-Dyck (Hort. Dyck. 1844, 9, 1845) also referred to it *M. micracantha* Monville.

"*M. simplex affinis* Otto is mentioned by Förster (Handb. Cact., 217, 1846), but is not described.

"*M. karstenii* Poselger (Allg. Gartenz., 21:95, 1853) is listed by Schumann among his little known species. The Index Kewensis states that it comes from Argentina, which is doubtless a mistake. The type locality is given as 'La Canada,' a common Spanish locality name. If collected by Karsten, it probably was obtained in Venezuela, in which case it would probably be referrable *Neomammillaria mammillaris.*" Britton & Rose (Cact., 4:71). A footnote to original description states that it is named in honor of Prof. Karsten of Venezuela which verifies the Britton & Rose treatment.

M. micracantha (Linnaea 16, 1838) is not referrable here but possibly to *M. polyedra.*

M. microthele Mühlenpfordt (Allg. Gartenz., 11, 1848) is not referrable here as it was described as having 2 central spines and 22-24 radial spines.

M. microthele brongniartii Salm-Dyck (Cact. Hort. Dyck. 1849, 9, 1850. Ref. Mühlenpf.) is likewise not referrable here.

73. Mammillaria zeyeriana Haage

Mamillaria zeyeriana Haage, Jr., in Schumann, Gesamtb. Kakteen, 574, 1898.
Neomammillaria zeyeriana Fosberg, Bull. So. Calif. Acad. Sci., 30:56, 1930.

BODY simple, globular to pyramidal with sunken apex, to 10 cm. wide. TUBERCLES arranged in 13 and 21 spirals, pale bluish green, egg-shaped, faintly angled, oblique truncate, with milky sap, 10-12 mm. long. AREOLES oval, 3 mm. wide, with white wool, later becoming naked. AXILS naked. CENTRAL SPINES 4, 15-25 mm. long, acicular, semi-flexuous, straight except the upper one has a strong recurve, all ruby-red, later becoming chestnut-brown, upper one porrect in center of areole, lower 3 descending. RADIAL SPINES 10, 3-10 mm. long, upper 4-5 shorter, all fine acicular, straight, smooth, white, lower ones intertwining with lower centrals. FLOWERS funnelform. *Outer perianth-segments* wide reddish brown mid-stripe, yellowish tan margins, lanceolate, tip acute, margins lacerate. *Inner perianth-segments* reddish orange mid-stripe, yellow margins, lanceolate, tip acuminate, margins slightly lacerate. *Anthers* yellow.

Fig. 88. *Mammillaria zeyeriana* x 0.9

Stigma-lobes 5-7, yellowish brown. FRUIT unknown. SEEDS dull brown, globular pyriform with lateral hilum, roughened, 1.3x0.8 mm.

Distribution: Coahuila and Durango, Mexico.

Type locality: None given but reported from Viesca, Coahuila.

Illustration: Fig. 88 is from a photograph of a seedling plant in our garden.

Britton and Rose did not recognize this species but referred it to the little-known group. Although we have not seen any collected plants of it, the seedling plants grown here, do appear to be distinct.

74. **Mammillaria petterssonii** Hildmann

Mammillaria petterssonii Hildmann, Deut. Garten-Zeitung, 185, 1886.
Mamillaria heeseana McDowell, Monatsschr. Kakteenk., 6:125, 1896.
Mamillaria heeseana brevispina Schelle, Handb. Kakteenk., 266, 1907.
Mamillaria heeseana longispina Schelle, Handb. Kakteenk., 266, 1907.
Neomammillaria petterssonii Britton & Rose, The Cactaceae, 4:94, 1923.

BODY simple and cespitose from body, wide globose, rounded above, with sunken apex, to 12 cm. wide. TUBERCLES arranged in 13 and 21 spirals, firm in texture, dull light green, obscurely 4-sided, angles rounded, flattened dorsally, keeled ventrally, with milky sap, 10 mm. in length and also in width at base. AREOLES oval, 2-3 mm. wide, with abundant white curly wool in youth, later becoming naked. AXILS with curly white wool but no bristles. CENTRAL SPINES 1-4, upper 10-20 mm. long, lower one to 45 mm. long, all acicular, lower heavier, all straight, smooth, stiff, enlarged at base, tan with darker tips, becoming dirty white, strongly divergent. RADIAL SPINES 10-12, three uppers 3 mm. long, lowers 10-15 mm. long, slender acicular, upper weaker, lower stronger, all straight, smooth, stiff, upper white, lower dirty white with darker tips, all horizontal. FLOWERS funnelform, 24 mm. long. *Outer perianth-segments* pale yellowish green base, reddish brown tapering mid-stripe above, lanceolate, tip acuminate, margins ciliate. *Inner perianth-segments* deep purplish pink mid-line, pale pink margins, lanceolate, tip caudate-acuminate, margins mostly entire with some serrations. *Filaments*

Fig. 89. *Mammillaria petterssonii* x 0.7

whitish below, very pale pink above. *Anthers* sulphur-yellow. *Style* whitish below, very pale pink at top. *Stigma-lobes* 5, brownish olive-green, 2-3 mm. long, overtop anthers 2 mm. FRUIT unknown. SEEDS unknown but probably brown.

Distribution: Guanajuato, Mexico.

Type locality: None given but reported from San Moran.

Illustration: Fig. 89 is from a photograph of a plant sent to us by Schwarz & Georgi of San Luis Potosi, S.L.P., in 1940.

75. Mammillaria brauneana Boedeker

Mammillaria brauneana Boedeker, Kakteenkunde, 113, 1933.

BODY mostly solitary, rarely divided, branching from base, thick clavate with sunken apex, 8 cm. wide. TUBERCLES loosely arranged in 21 and 34 spirals, grayish green, short broad pyramidal, with milky sap, 3-4 mm. wide at base. AREOLES round, small, with some white wool present only occasionally in youth, otherwise naked. AXILS with dense white wool and moderately long white tortuous thin hair-like bristles. CENTRAL SPINES 2-4, 5-7 mm. long, slender to strong subulate, reddish with blackish tips, becoming horn-colored, lighter at base, divergent porrect. RADIAL SPINES 25-30, about 5 mm. long, upper somewhat shorter, all nearly straight, hair-like, white, horizontal. FLOWER isolated, funnelform, to 13 mm. wide. *Outer perianth-segments* bright green below, violet-red above, darker ventral-stripe, linear-lanceolate, tip acute, margins serrate, to 7 mm. wide. *Inner perianth-segments* same color as outer, somewhat longer and wider than the outer, blunter tip, margins entire. *Filaments* bright green below, violet-red above. *Anthers* whitish yellow. *Style* (?) *Stigma-lobes* 4, very small, violet-red. FRUIT unknown. SEEDS yellow to dark brown, dull.

Fig. 90. *Mammillaria brauneana*

Distribution: Tamaulipas, Mexico.

Type locality: None given but reported from Jaumave.

Illustration: Fig. 90 is a reproduction of an illustration in Kakteenk. 113, as *Mammillaria brauneana*.

76. Mammillaria hahniana Werdermann

Mamillaria hahniana Werdermann, Monatsschr. Deutsch. Kakt. Ges., 77, 1929.

BODY mostly cespitose, slightly flattened globose with slightly sunken apex, heads to 9 cm. high, 10 cm. wide. TUBERCLES closely set in 13 and 21 spirals, light green, conical, almost terete to slightly triangular at apex, rounded dorsally, keeled ventrally, with very milky sap, 5-6 mm. long, 2-4 mm. wide at base. AREOLES oval, hardly 1 mm. wide, with white wool in apex, later with brownish mat, later becoming naked. AXILS with short white wool and a tuft of 20 or more long white hair-like tortuous bristles of uneven length to 35-40 mm. long. CENTRAL SPINES present only in new growth, older ones missing as they are easily dislodged, 1 (2-5), 4 (5-8) mm. long, straight, acicular, smooth, somewhat thickened at base, whitish transparent below, reddish brown at very tip, porrect or divergent dorsally and ventrally. RADIAL SPINES 20-30, 5-15 mm. long, very fine hair-like, soft, tortuous, white, extending mostly laterally, interlacing with approximating ones and axillary hairs, horizontal to somewhat ascending. FLOWERS funnelform, 15-20 mm. wide, July, open for several days. *Outer perianth-segments* greenish yellow below, purplish above, greenish white margins, dark brown ventral mid-stripe, linear-lanceolate, tip acute, margins coarsely ciliate to almost lacerate, 6-8 mm. long. *Inner perianth-segments* purplish red, margins transparent, linear-lanceolate, tip mucronate to almost acuminate, margins mostly entire. *Filaments* white below, purple above. *Anthers* cream. *Style* white. *Stigma-lobes* 3-5, yellowish white. FRUIT reddish purple, oval, 7x4 mm., with dried perianth persisting. SEEDS dull dirty brown, pyriform with lateral hilum, pitted, 1.5x1 mm. ROOTS fibrous.

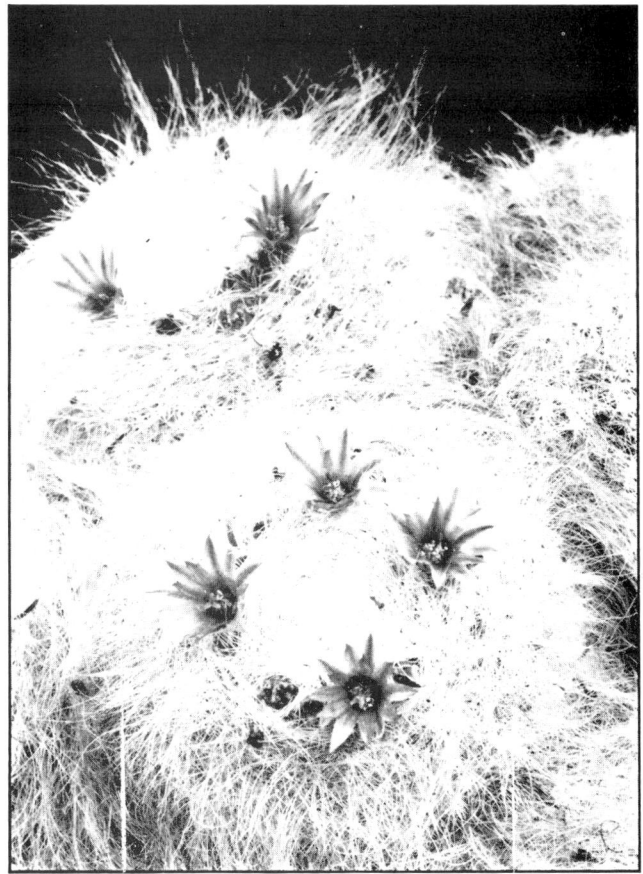

FIG. 91. *Mammillaria hahniana* x 0.8

Distribution: Guanajuato and Queretaro, Mexico.

Type locality: None given but reported from Sierra de Jalapa and Ocotitlan.

Illustration: Fig. 91 is a reproduction of an illustration in Cact. Succ. Journ., 6:101 as *Neomammillaria hahniana*.

This species is quite variable in the amount and length of the axillary hairy bristles. It also shades into the forms which we have designated as *M. bravoae* and *M. woodsii*. The latter two species are very close to the type form in some respects but on the other hand they differ in other respects to such an extent that we are presenting them as separate species.

a. var. **giseliana** Neale

Mammillaria hahniana giseliana Neale, Cact. Other Succ., 88, 1935.

Axillary hairs fewer to 15 mm. long; radial spines to 40.

Distribution: Guanajuato.

Type locality: Tarajeas.

Illustration: Fig. 92 is from a photograph of a plant sent to us by Sr. F. Schmoll of Cadereyta, Qro.

Here is referrable the plants offered by Sr. F. Schmoll as *M. giseliana, M. tarajensis* and *M. hahniana tarajensis*.

FIG. 92. *Mammillaria hahniana* var. *giseliana* x 0.6

FIG. 93. *Mammillaria hahniana* var. *werdermanniana* x 0.6

b. var. **werdermanniana** Schmoll var. nov.

Axillis setis ad 25 mm. longis; stigmatibus puniceis fuscis.

Axillary hairs to 25 mm. long; stigma-lobes brownish pink, tubercles more angled, outer perianth-segments short ciliate.

Distribution: Guanajuato, Mexico.

Type locality: Not known.

Illustration: Fig. 93 is from a photograph of a plant sent to us by Sr. F. Schmoll of Cadereyta, Qro.

77. **Mammillaria bravoae** sp. nov.

Corpus simplex et cespitosus, globosus; mamillis ad 13 et 21 seriebus ordinatis, conoidis, suco lacteo, axillis lanatis albis et setis; spinis centralibus 2, 6-8 mm., subulatis tenuibus, rectis; spinis radialibus 28-30, 4-7 mm., acicularibus, rectis, albis; flores infundibuliformes, sepalis ciliatis, petalis magentis, integris, stigmatibus 4; semina fulva.

BODY simple and cespitose from base, globose, with strongly sunken apex, 65 mm. wide and high. TUBERCLES arranged in 13 and 21 spirals, almost hidden by spines and wool, glossy bright grass-green, conic, nearly terete, sharply keeled ventrally, rounded apex, with milky sap, 8 mm. long, 4-7 mm. wide at base, wider laterally. AREOLES oblong, with white wool in youth, soon becoming naked. AXILS with dense white wool, covering tubercles in upper part of plant and also white hair-like bristles extending beyond tubercles. CENTRAL SPINES 2, 6-8 mm. long, slender subulate, straight, stiff, smooth, cream at base to faint pinkish tan, black at tip, ascending divergent dorsally and ventrally. RADIAL SPINES 28-30, 4-7 mm. long, longer laterally, all acicular, straight, smooth, white to the tip, horizontal, interlacing laterally. FLOWERS funnelform, 10 mm. wide and long. *Outer perianth-segments* pinkish tan mid-stripe, outer end yellowish, black tip, pale pink margins, linear-lanceolate, tip acuminate, margins ciliate, 1.5 mm. wide. *Inner perianth-segments* very deep pink, darker narrow mid-stripe, black at very tip, linear-lanceolate, tip acute, margins entire. *Filaments* pale pink above, white below. *Anthers* bright sulphur-yellow. *Style* pink at very top, white below. *Stigma-lobes* 4-5, pinkish tan, 0.8 mm. long. FRUIT carmine, clavate, 15x4 mm. with dried perianth

FIG. 94. *Mammillaria bravoae* x 1

persisting. SEEDS light tan, curved pyriform with lateral hilum near base, 1.5x0.8 mm.

Distribution: Guanajuato, Mexico.

Type locality: Rio Blanco.

Illustration: Fig. 94 is from a photograph of a plant sent to us by Sr. F. Schmoll of Cadereyta, Qro., Mexico.

This species is named in honor of Helia Bravo H., formerly with the Instituto de Biologia de Mexico.

This name was mentioned in Cact. & Succ. Journ., *12*:145, 1940, but was not described.

Mammillaria quevedoi, an unpublished name in Schmoll's Catalogue, is only a variation of this species in which the central spines are of a darker color.

78. **Mammillaria woodsii** Craig

Mammillaria woodsii Craig, Cact. Succ. Journ., 15:33, 1943.

BODY simple, flattened globular to clavate with strongly sunken apex, 8 cm. wide, 5 cm. high. TUBERCLES arranged in 13 and 21 spirals, dull grass-green, angled below, nearly rounded at apex, strongly keeled ventrally, with milky sap, 7 mm. long, 6-7 mm. wide at base. AREOLES oval, with white wool, soon becoming naked. AXILS in flowering and fruiting area with dense white wool almost covering the apex of the tubercles, and also numerous white hair-like bristles to 25 mm. long. CENTRAL SPINES mostly 2, occasionally 4, lower one to 16 mm. long, upper ones 4-5 mm. long, lower slender subulate, uppers heavy acicular, all smooth, mostly straight or with very slight recurve, dull chalky purplish pink with black tip, lower directed ventrally, uppers directed fan-like dorsally. RADIAL SPINES 25-30, 4-8 mm. long, fine hair-like, smooth, tortuous, flexuous, white, somewhat ascending. FLOWERS funnelform, March and April, 10-12 mm. long, 12-15 mm. wide. *Outer perianth segments* brownish dark pink mid-stripe, pale pink margins, lanceolate, tip acute, margins finely serrate, 2 mm. wide. *Inner perianth-*

Fig. 95. *Mammillaria woodsii* x 1

segments dark pink tapering mid-stripe to tip, pink margins, eliptical, tip obtuse, margins mostly entire. *Filaments* white below, pink above. *Anthers* yellow. *Style* pale pink. *Stigma-lobes* 3-6, pale brownish pink, 1 mm. long. FRUIT deep pink, clavate, 15x6 mm., with dried perianth persisting. SEEDS dull brown, curved pyriform with lateral hilum, faintly pitted.

Distribution: Guanajuato, Mexico.

Type locality: Hacienda de Tarajeas.

Illustration: Fig. 95 is a reproduction of an illustration in Cact. Succ. Journ., *15*:33 as *Mammillaria woodsii*.

This species is named in honor of Mr. Robert S. Woods of Azusa, California, for his enthusiastic efforts in the culture and research in the Cactaceae.

Sr. Ferdinand Schmoll of Cadereyta, Queretaro, Mexico, has sent to us a whole series of plants that could be placed in a group with the three preceeding species. They are all very similar in many respects but still they show just a little difference but it is hardly enough to set up specific outstanding distinctions so as to justify specific rank. They are all from the same general area of Queretaro and Guanajuato and doubtlessly they are all closely related. This is a typical illustration of the fact that this genus contains many species that do not have definite dividing lines but they merge from one into another.

Fig. 96. *Mammillaria chionocephala*

79. **Mammillaria chionocephala** Purpus

Mamillaria chionocephala Purpus, Monatsschr. Kakteenk., 16:41, 1916.
Neomammillaria chionocephala Britton & Rose, The Cactaceae, 4:101, 1923.

    BODY usually simple, occasionally branching, conic to globose, rounded above with apex slightly sunken, to 12 cm. wide, 8-12 cm. high. TUBERCLES arranged in 13 and 21 spirals, firm in texture, bluish green, finely white pitted, blunt 4-angled, pyramidal, keeled ventrally, rounded dorsally, with very milky sap, 6-8 mm. long. AREOLES eliptical, 2x1 mm., with short white wool turning brown, later becoming naked. AXILS with dense white wool in apex and persisting, also numerous white hairs, 2 mm. long. CENTRAL SPINES 2-4 (6 is exceptional), 4-6 mm. long, subulate, stiff, straight or slightly bent, thickened base, dirty white to brownish with blackish brown tip, when 2: divergent dorsally and ventrally, when 4: in cross formation, overtop apex. RADIAL SPINES 22-24, to 8 mm. long, stiff, setaceous, white, horizontal to somewhat ascending. FLOWERS in crown near top, 18-22 mm. long and wide. *Outer perianth-segments* white or flesh-color, tapering red or brownish red mid-stripe, browner at point, lanceolate, tip acuminate, margins serrate also often entire. *Inner perianth-segments* white or flesh-color with bright red mid-stripe and darker brownish red point, eliptical-lanceolate, tip acute, margins mostly entire. *Filaments* white below, pink above. *Anthers* bright yellow. *Style* white below, rose above. *Stigma-lobes* 6 (5-7), yellowish, reddish or bright rose-red. FRUIT dark carmine, clavate, to 22 mm. long, with dried perianth persisting. SEEDS bright brown, curved pyriform with lateral hilum, reticulate, 1.5x0.8 mm.

    Distribution: Coahuila, Durango, Mexico.

    Type locality: Sierra de Parras, also reported from Canyon de Arteaga near Monterrey, Coa.; and also Nazas, D., Sierra del Rosario, D. (Patoni).

    Illustration: Fig. 96 is a reproduction of an illustration, Fig. 99, in Britton & Rose, (Cact. 4:100) as *Neomammillaria chionocephala*.

    "It was distributed by Pringle in 1890 as *Mammillaria acanthophelgma*. It resembles very much a large plant of *Mammillaria elegans,* but the tubercles are milky and bear setae in their axils." Britton & Rose, (Cact. 4:101.)

    The difference in the spelling by Britton & Rose was very probably a typographical error.

Fig. 97. *Mammillaria mayensis* x 0.3

80. **Mammillaria mayensis** sp. nov.

Corpus simplex, columnus, 19 cm. altus, mamillis ad 13 et 21 seriebus ordinatis, quadrangularis, autem non cum angulis, suco lacto; axillis setis; spinis centralibus 6, 5 mm., rectis; spinis radialibus 25-30, 2-6 mm., flores infundibuliformes, sepalis serratis; petalis puniceis, serratis, stigmatibus 5, subviridis; semina probabilis flavidofusca.

BODY simple, flat top cylindric with apex sunken and very woolly, 19 cm. high, 15 cm. wide. TUBERCLES arranged in 13 and 21 spirals, firm in texture, light olive-gray-green, quadrangular, but not angled, keeled ventrally, rounded dorsally, with milky sap, 8 mm. long, 8-12 mm. wide at base, wider laterally. AREOLES oval with considerable white wool in youth, later becoming naked. AXILS with abundant white wool, and long white somewhat tortuous bristles, not extending beyond tubercles but interlacing over the wool. CENTRAL SPINES 6, 5 mm. long, slender subulate, straight, stiff, smooth, base not enlarged, brownish tan in youth, later horn-color, spreading divergent. RADIAL SPINES 25-30, 2-6 mm. long, upper 4-6 shorter, all slender acicular, straight, stiff, smooth, chalky white, some with very tip yellowish to light brown, slightly ascending. FLOWERS funnelform, 12 mm. long, 15 mm. wide. Flowering in April. *Outer perianth-segments* pink tapering mid-stripe, paler margins, lanceolate, tip acute, margins serrate. *Inner perianth-segments* deep pink, darker mid-stripe, lanceolate, tip acute, margins serrate irregularly. *Filaments* very pale pink. *Anthers* yellow. *Style* tannish yellow. *Stigma-lobes* 5, light tannish green. FRUIT unknown. SEEDS unknown but probably brown.

Distribution: SW. Chihuahua and SE. Sonora, Mexico.

Type locality: Sierra Cajurichic and Sierra Canelo, Rio Mayo.

Illustration: Fig. 97 is from a photograph of a plant collected by Mr. Howard S. Gentry in 1936, while on a botanical survey of the Rio Mayo. He reports that it is very rare.

This plant is named in honor of the Mayo Indian people who have been very kind to us in our explorations of their country.

FIG. 98. *Mammillaria neopotosina* x 0.8

81. Mammillaria neopotosina nom. nov.

Corpus simplex, globosus, mamillis ad 13 et 21 seriebus, ordinatis, pyramidatis, suco lacteo, 8 mm. per longitudinem; areolis ovatis; axillis lanatis albis et setis; spinis centralibus 4, 4-35 mm. longis, rectis, acicularibus, fulvis; spinis radialibus 40-50, 2-6 mm. longis, setiformis; flores infundibuliformes, sepalis ciliatis, petalis puniceis, stigmatibus 5-6, rubris; fructus coccineus; semina fusca.

BODY simple, globose to cylindric with strongly sunken apex, 10 cm. high, 8 cm. wide. TUBERCLES arranged in 13 and 21 spirals, firm in texture, dark green, finely pitted with white dots, pyramidal, strongly angled to tip, flattened dorsally, with milky sap, 8 mm. long, 5-6 mm. wide at base. AREOLES round to slightly oval, with white wool, later becoming naked. AXILS with white wool in youth, also bristles longer than the tubercles. CENTRAL SPINES 4 (occasionally to 6, see variety), length variable, upper 4-12 mm. long, lower 5-35 mm. long, all acicular, straight to curved, smooth, stiff, yellow with brown tip, spreading from porrect. RADIAL SPINES 40-50, 2-6 mm. long, upper shorter, all setaceous to fine acicular, straight to tortuous, semiflexuous, smooth, white, horizontal to somewhat ascending. FLOWERS lateral near top, funnelform, 15 mm. long, 10 mm. wide, May-July. *Outer perianth-segments* pale green at base, reddish brown mid-stripe above, pale tannish pink narrow margins, lanceolate, tip acute, margins long ciliate. *Inner perianth-segments* purplish red, darker mid-line, lanceolate, tip acute, margins entire. *Filaments* white below, pale pink above. *Anthers* greenish tannish yellow. *Style* very pale green below, pale pink above. *Stigma-lobes* 5-6, reddish, 1 mm. long, just overtop anthers. FRUIT red, slender clavate, to 20 mm. long, with dried perianth persisting. SEEDS light brown, dull, curved pyriform with lateral hilum near base, smooth to faintly rugose, 1x0.8 mm.

Distribution: Queretaro, Guanajuato; Mexico.

Type locality: Boundry between the two states (Schmoll).

Illustrations: Fig. 98 and 99 are from photographs of plants sent to us by Sr. F. Schmoll of Cadereyta, Qro.

a. var. **brevispina** comb. nov.

>Spinis centralibus 4, 5-6 mm. longis.
>Central spines 4, 5-6 mm. long.

b. var. **longispina** comb. nov.

>Spinis centralibus 4, ad 25 mm. longis.
>Central spines 4, up to 25 mm. long.

c. var. **hexispina** Schmoll var. nov.

>Spinis centralibus 6, ad 35 mm. longis.
>Central spines 6, to 35 mm. long; flowers smaller and more tubular, to 9 mm. long.

This species and its varieties have been known in the trade under the name of *M. potosina* for some time but for the most part they have been avoided by most authors in their consideration of this genus. The name *M. potosiana*, as proposed by Jacobi (Allg. Gartenz., 24:92, 1856), is for some other plant in which the radial spine count is less, and which possesses glands in the axils and has yellow flowers. Schumann, followed by Berger, Borg and others, refers it to *M. elegans* but inasmuch as we consider the latter as belonging to the non-milky group (tubercles do not produce a milky sap when wounded), the above species can not be referred there because the tubercles have definitely a milky sap. Britton and Rose referred it to *M. geminispina* but the radial spine count of the above species is so much greater that it is placed in an entirely different group.

It is quite variable in the number and length of the central spines but inasmuch as the other salient characteristics are so nearly uniform, the variations have been grouped under this one species as they all come from the same general territory according to Sr. F. Schmoll.

M. nealeana is only a name proposed by E. Tiegel of Germany but as far as we know it was never published but is very probably referrable here. It is briefly mentioned by Neale (Cact. Other Succ., 91, 1935) and Day (Flow. Desert, 65, 1938).

Although this species was originally named in honor of the state of San Luis Potosi, it is not actually found in that state according to Sra. Carolin Schmoll.

FIG. 99. *Mammillaria neopotosina*

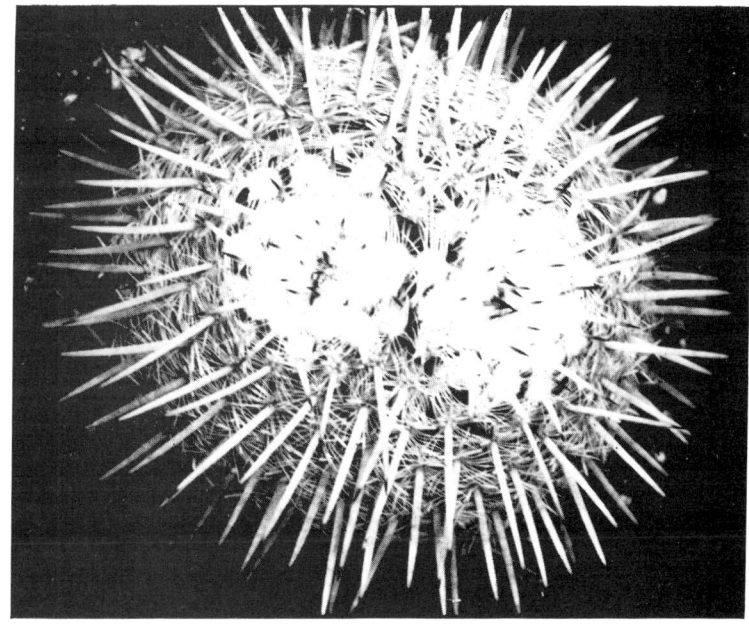

Fig. 100. *Mammillaria parkinsonii* x 0.9

82. Mammillaria parkinsonii Ehrenberg

Mammillaria parkinsonii Ehrenberg, Linnaea, 14:375, 1840.
Mamillaria dealbata Dietrich, Allg. Gartenz., 14:309, 1846.
Cactus dealbatus Kuntze, Rev. Gen. Pl., 1:260, 1891.
Cactus parkinsonii Kuntze, Rev. Gen. Pl., 1:261, 1891.
Neomammillaria parkinsonii Britton & Rose, The Cactaceae, 4:98, 1923.
Mammillaria dietrichiae Tiegel in Moeller, Deut. Gart. Zeitung, 48:413, 1933.

BODY simple and cespitose, forming large mounds, individual heads flattened globular, with apex sunken and with snow-white wool, 15 cm. high, 8 cm. wide. TUBERCLES arranged in 8 and 13 spirals, bluish green, finely white pitted, pyramidal, angles rounded, above terete, with milky sap, 8-10 mm. long, 4-6 mm. wide at base. AREOLES round, to 3 mm. wide, somewhat ventral to tip of tubercle, with white wool, later becoming naked. AXILS with thick wool especially in flowering area, also with white tortuous bristles. CENTRAL SPINES 2-4 (seldom 5), upper 6-8 mm. long, lower to 35 mm. long, all subulate, angular, straight to slight recurve, often somewhat tortuous, smooth, stiff, milky white to faint reddish with dark brown tip, not enlarged at base, upper nearly porrect, lower strongly ventral to nearly horizontal. RADIAL SPINES 30-35, 3-7 mm. long, fine acicular, straight to recurved, smooth, semi-stiff, white, horizontal, more laterally. FLOWERS funnelform, 12-15 mm. long. *Outer perianth-segments* brownish pink mid-stripe, very pale pink margins, lanceolate, tip acute, margins fine ciliate. *Inner perianth-segments* wide cream margins, tannish pink mid-line, lanceolate, tip acute to apiculate, margins entire. *Filaments* very pale pink above, white below. *Anthers* very pale cream. *Style* pale pink. *Stigma-lobes* 5, pale pink to yellowish tan, 1 mm. long, overtop anthers 1-2 mm. FRUIT scarlet, clavate, 10 mm. long, with dried perianth persisting. SEEDS light brown, pyriform with lateral exserted hilum, smooth, 1.6x0.8 mm.

Distribution: Hidalgo, Queretaro and central Mexico.

Type locality: San Onofre in Mineral del Doctor, Hgo., and also reported from San Juan del Rio, Qro.

Fig. 101. *Mammillaria parkinsonii* x 0.9

Illustrations: Figs. 100 and 101 are from photographs of plants sent to us by Sr. F. Schmoll of Cadereyta, Qro.

M. dealbata was placed in the *M. elegans* group by Schumann and was subsequently

Fig. 102. *Mammillaria parkinsonii* var. *brevispina* x 1

followed by other authors. In the original description it definitely states that it is similar to *M. parkinsonii* and *M. leucocentra* but differs from the latter in the fewer number of central spines.

M. waltonii Walton (Cact. Journ., 1:29, 1898) was only briefly characterized but not described and is probably to be referred here.

M. parkinsi Salm-Dyck (Gartenflora 66, 1883) may belong here.

a. var. **brevispina** var. nov.

Mamillae ad 13 et 21 seriebus ordinatis, spinis centralibus 4 (5), 3-5 mm. longis, subulatissimis; spinis radialibus 35-40, 1-4 mm. longis, setaeformis; sepalis integris, petalis albis subflavis, stigmatibus 5, viridis pallidis flavis.

BODY globose. TUBERCLES arranged in 13 and 21 spirals. CENTRAL SPINES 4 (occasionally 5), 3-5 mm. long, strong subulate. RADIAL SPINES 35-40, 1-4 mm. long, fine acicular. FLOWERS narrow funnelform. *Outer perianth-segments* very pale green below, brownish red mid-stripe above, light green border in outer ⅓, lanceolate, tip acute, margins entire. *Inner perianth-segments* creamy white, pale pink mid-line, spatulate, tip acute to obtuse, margins entire. *Filaments* white below, very pale pink at top. *Anthers* sulphur yellow. *Style* same as filaments. *Stigma-lobes* 5, pale yellowish green.

Illustration: Fig. 102 is from a photograph of a plant sent to us by Sr. F. Schmoll.

FIG. 103. *Mammillaria klissingiana* x 1

83. **Mammillaria klissingiana** Boedeker

Mammillaria klissingiana Boedeker, Zeitschr. Sukkulentenk., 3:123, 1927.

BODY simple and cespitose from the base, globular to cylindric with rounded apex, 9 cm. wide, 16 cm. high. TUBERCLES arranged in 13 and 21 spirals, firm in texture, rounded pyramidal to conic, apex nearly terete, with milky sap, 5 mm. long, 2 mm. wide at base. AREOLES oval, 1 mm. wide, with somewhat yellowish wool in youth, later becoming naked. AXILS with thick white wool in flowering and fruiting areas, also with many tortuous bristles to 10 mm. long. CENTRAL SPINES 2-4 (5), 2 mm. long, acicular to fine subulate, straight, smooth, not enlarged at base, white with dark brown tip, faint yellowish at base, divergent porrect. RADIAL SPINES 30-35, mostly 5 (3-7) mm. long, fine acicular, straight to slightly tortuous, smooth, white, faint yellow at base, horizontal. FLOWERS campanulate, 10 mm. long, 8 mm. wide, lateral. *Outer perianth-segments* greenish below, tannish rose mid-stripe, lighter margins, lanceolate, tip acuminate, margins ciliate. *Inner perianth-segments* delicate rose, throat greenish, margins nearly white, lanceolate, tip acuminate, margins very slightly ciliate to entire. *Filaments* delicate rose. *Anthers* yellow. *Style* delicate rose. *Stigma-lobes* 3, bright yellow. FRUIT bright carmine, clavate, 10-12 mm. long, with dried perianth persisting. SEEDS dark reddish brown, curved pyriform with lateral hilum, faintly glossy, 0.5 mm. long.

Distribution: Tamaulipas, Mexico.

Type locality: Ciudad Victoria near Calabazos.

Illustration: Fig. 103 is from a photograph of a plant sent to us by Sr. F. Schmoll of Cadereyta, Qro.

84. Mammillaria heyderi Muhlenpfordt

Mamillaria heyderi Muhlenpfordt, Allg. Gartenz., 16:20, 1848.
Cactus heyderi Kuntze, Rev. Gen. Pl., 1:260, 1891.
Neomammillaria heyderi Britton & Rose, The Cactaceae, 4:75, 1923.

BODY simple, depressed globose, with sunken apex, 80-120 mm. wide, 25-50 mm. high. TUBERCLES arranged in 13 and 21 spirals, bright to dark green, elongate pyramidal, quadrangular at base, terete at apex, with milky sap, 9-13 mm. long, 5-6 mm. wide at base. AREOLES round, 2 mm., with scant white wool in youth. AXILS with wool in youth. CENTRAL SPINES 1, 4-8 mm. long, stout acicular, straight, smooth, stiff, light yellowish gray to brown, darker at base, to reddish brown at tip, porrect. RADIAL SPINES 20-22, upper 4-6 mm. long, lower 7-8 (12) mm. long, all bristle-like, lower stouter, all white to ivory-white with brownish tips, horizontal. FLOWERS funnelform, lateral, 20-25 mm. long, April and May. *Outer perianth-segments* green at base, shading into pinkish brown mid-stripe at point, margins whitish, lanceolate, tip acute, margins entire. *Inner perianth-segments* brownish pink mid-stripe, cream margins, ventrally tinged with green, linear-lanceolate, tip notched, margins entire to slightly serrate at tip. *Filaments* white to pale pink. *Anthers* straw-yellow. *Style* white below to rose-red above. *Stigma-lobes* 6, green. FRUIT carmine red, elongate-clavate, 37x10 mm., dried perianth not persisting. SEEDS reddish brown, curved pyriform with lateral hilum, faintly rugose, 1 mm.

Distribution: Texas and New Mexico, U.S.; Tamaulipas, Nuevo Leon and Chihuahua, Mexico.

Type locality: None given but reported from San Antonio and New Braunfels, west to El Paso in Texas, north to Carlsbad in New Mexico, U.S.; south to Monterrey, N.L., east to Matamores, Tamaulipas, Mexico.

Engelmann regarded *Mammillaria meiacantha* as possibly a variety of *M. heyderi* but the spine count places it nearer to *M. hemispherica* but inasmuch as they are from the same general area, there are many intergrading forms. The type from Durango City, Durango referred to *M. heyderi* by Ochoterena should be referred to *M. applanata* because of the number of radial spines.

Fig. 104. *Mammillaria infernillensis*

85. Mammillaria infernillensis sp. nov.

Corpus simplex, mammillis ad 12 et 21 seriebus ordinatis, suco lacteo; axillis lanatis albis; spinis centralibus 1-2 (4), 4-10 mm. longis, acicularibus crassis, rectis; spinis radialibus 30, 6-10 mm. longis, acicularibus tenuis; sepalis ciliatis, petalis subpuniceis, stigmatibus 4, subflava; semina subflava, pyriformis.

BODY simple, somewhat flattened globose with sunken apex, 9 cm. wide. TUBERCLES arranged in 13 and 21 spirals, firm in texture, dull grayish green, white spotted, 4-sided but not sharply angled, with milky sap, 6-8 mm. long, 5-7 mm. wide at base. AREOLES round to oval, with scant tannish white wool only in youngest. AXILS with white wool in flowering area but no bristles. CENTRAL SPINES 1-2 (4), 4-10 mm. long, strong acicular, straight to slight recurve, smooth, stiff, chalky white to chalky lavender with dark brown to black tip, divergent porrect. RADIAL SPINES 25-30, 2-10 mm. long, upper shorter, all setaceous to slender acicular, straight to tortuous, smooth, semiflexuous, white, slightly ascending. FLOWERS funnelform, April-June. *Outer perianth-segments* pale green below, pinkish olive-green tapering mid-stripe above, very pale pink margins, linear lanceolate, tip acuminate, margins finely serrate. *Inner perianth-segments* deep purplish pink, darker mid-line, pale pink to whitish very narrow margins, linear-lanceolate to oblong, tip acute to obtuse, margins entire. *Filaments* white below, pale purplish pink above. *Anthers* sulphur-yellow. *Style* white to cream, pink above. *Stigma-lobes* 4, yellowish olive-green to tan, pink back stripe, 1 mm. long, overtop anthers 3 mm. FRUIT light pink, clavate, 12x4 mm., with dried perianth persisting. SEEDS light brown, curved pyriform with lateral hilum near base, faintly pitted, slightly glossy, 1.1x0.7 mm.

Distribution: Queretaro and Guanajuato; Mexico.

Type locality: Infernillo, Qro.

Illustration: Fig. 104 is from a photograph of a plant obtained from Sr. F. Schmoll of Cadereyta, Qro.

This species varies somewhat in the robustness and the length of the spines.

Fig. 105. *Mammillaria formosa*

86. **Mammillaria formosa** Galeotti

Mammillaria formosa Galeotti, Bull. Acad. Sci. Brux., 5:497, 1838.
Mammillaria formosa microthele Salm-Dyck, Cact. Hort. Dyck. 1849, 87, 1850.
Mammillaria formosa dispicula Monville in Labouret, Monogr. Cact., 60, 1853.
Mammillaria formosa gracilispina Monville in Labouret, Monogr. Cact., 60, 1853.
Mammillaria formosa laevior Monville in Labouret, Monogr, Cact., 60, 1853.
Cactus formosus Kuntze, Rev. Gen. Pl., 1:260, 1891.
Neomammillaria formosa Britton & Rose, The Cactaceae, 4:90, 1923.

BODY simple, later basal branching, globose to cylindric, rounded above with apex sunken, 70-80 mm. wide. TUBERCLES arranged in 13 and 21 spirals, light green, white pitted, slender pyramidal, obscurely 4-angled, with milky sap, 7-9 mm. long, 4 mm. wide at base. AREOLES oval, 2 mm. wide, with scant wool, soon becoming naked. AXILS with white wool, especially in the flowering area. CENTRAL SPINES 4-6, (seldom a 7th one more central), to 8 mm. long, stout acicular to subulate, straight, smooth, stiff, base enlarged, flesh-colored with black tip, later chalky-white, nearly horizontal. RADIAL SPINES 20-25, 3-6 mm. long, thin acicular, straight or slightly bent, stiff, smooth, chalky-white, horizontal interlacing. FLOWERS lateral, funnelform, 10-15 mm. long and wide. *Outer perianth-segments* nearly white to cream below, mid-stripe pink in middle to pale green at top, margins pale pink to greenish white, linear to linear-lanceolate, tip acuminate, margins serrate. *Inner perianth-segments* pink margins, deep purplish pink mid-stripe, especially at the tip, lanceolate, tip acuminate, margins entire. *Filaments* white. *Anthers* light yellow. *Style* white. *Stigma-lobes* 5, light greenish yellow, overtops anthers. FRUIT bright carmine-red, dry pulp, clavate, 15x5 mm., with dried perianth persisting. SEEDS light tan, globular pyriform with lateral hilum near base, nearly smooth to faintly rugose, 1.3x0.6 mm.

Distribution: San Luis Potosi, Mexico.

Type locality: Near San Felipe.

Illustration: Fig. 105 is a reproduction of an illustration in Kakteenk., 146, 1934, as *Mammillaria formosa*.

Haage, (Cact. Cult., 133, 1900), lists the variety *nigrispina* without author or description.

Fig. 106. *Mammillaria morganiana*

87. Mammillaria morganiana Tiegel

Mammillaria (*Neomammillaria*) *morganiana* Tiegel in Moeller, Deutsch. Gartner Zeit., 48:397, 1933.

BODY simple and cespitose by dichotomous branching, globose to clavate with sunken apex, to 8 cm. wide. TUBERCLES arranged in 13 and 21 spirals, bright greenish blue, pyramidal, lightly angled, with milky sap, 10 mm. long, 5 mm. wide at base. AREOLES round, with pale yellow short woolly mat. AXILS in youth with dense white hair to 20 mm. long, extending beyond spines. CENTRAL SPINES 4-6, 10 mm. long, stout acicular to slender subulate, straight, stiff, smooth, white with brown tip, somewhat thickened base, divergent porrect. RADIAL SPINES 40-50, to 12 mm. long, fine acicular to hair-like, tortuous, smooth, flexuous, white, nearly horizontal, interlacing. FLOWERS, FRUIT and SEEDS unknown.

Distribution: Guanajuato and Queretaro, Mexico.

Type locality: None given but reported from Bucareli, Guan.

Illustration: Fig. 106 is a reproduction of an illustration in Moeller, (Deutsch. Gartner Zeit., 48:397), as *Mammillaria* (*Neomammillaria*) *morganiana*, also the same in Cact. Succ. Journ., 7:21.

88. Mammillaria crucigera Martius

Mammillaria crucigera Martius, Nov. Act. Nat. Cur., 16:340, 1832.
Cactus cruciger Kuntze, Rev. Gen. Pl., 1:260, 1891.

BODY simple, dichotomous branching and also crestate form, flattened globose to cylindric with sunken apex, deep seated, to 15 cm. high. TUBERCLES arranged in 8 and 13 spirals, firm in texture, bright to olive-gray-green, quadrangular at base, not sharply angled, terete above, keeled ventrally, with watery sap which becomes milky in growing season, 4-7 mm. long, 5 mm. wide at base. AREOLES round, with scant white wool only in the youngest. AXILS with thick white wool to top of tubercles in youth, later becoming naked. CENTRAL SPINES (2) usually 4, occasionally 5, 2-3 mm. long, strong subulate, straight, stiff, smooth, yellowish to chalky white, dark brown to black at very tip, orange brown at base, horizontal in cross formation on top of the radials. RADIAL SPINES 24 or more, 1.5-2 mm. long, fine acicular to bristle-like, straight, smooth, semi-stiff, white, horizontal. FLOWERS funnelform, May and June. *Outer perianth-segments* brownish purple mid-stripe, linear-lanceolate, tip acute, margins serrate. *Inner perianth-*

FIG. 107. *Mammillaria crucigera*

segments purplish red, lanceolate, tip acute, margins serrate. *Filaments* purplish. *Anthers* yellow. *Style* purplish above. *Stigma-lobes* 4-5, crimson-red to violet, overtop anthers. FRUIT red, clavate, 10x5 mm., with dried perianth persisting. SEEDS yellowish to dark brown, curved pyriform with lateral hilum at base, flattened on both sides, 1.2x0.5 mm.

Distribution: Oaxaca, Mexico.

Type locality: None given but reported from San Antonia Las Calla (between Tehuacan and Oaxaca City).

Illustration: Fig. 107 is a reproduction of an illustration in Cact. Succ. Journ., 6:14 as *Mammillaria crucigera*. Fig. 108 is from a photograph of the crestate form that we obtained from Sr. F. Schmoll of Cadereyta, Qro., in 1942.

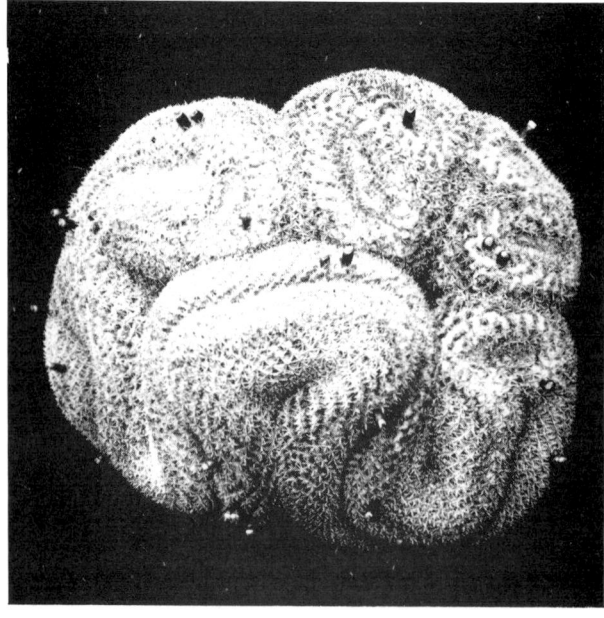

FIG. 108. *Mammillaria crucigera* x 0.2

There has been much controversy over the type of sap in this species. The original description called for a sap of "unnatural color" which is quite indefinite. Schumann referred this species to *M. formosa* in the Section Galactochylus (milky sap). Quehl (Monatsschr. Kakteenk., 190, 1907) comments to some extent on this species and states that the sap is milky. Tiegel (Cact. Succ. Journ., 6:14, 1934) adds additional information to the description but states that the sap is watery. We obtained plants from Sr. F. Schmoll, the same source as did Tiegel, and they displayed, while in a dormant condition, no milkiness to the sap but when they filled out and started to grow, the sap was milky.

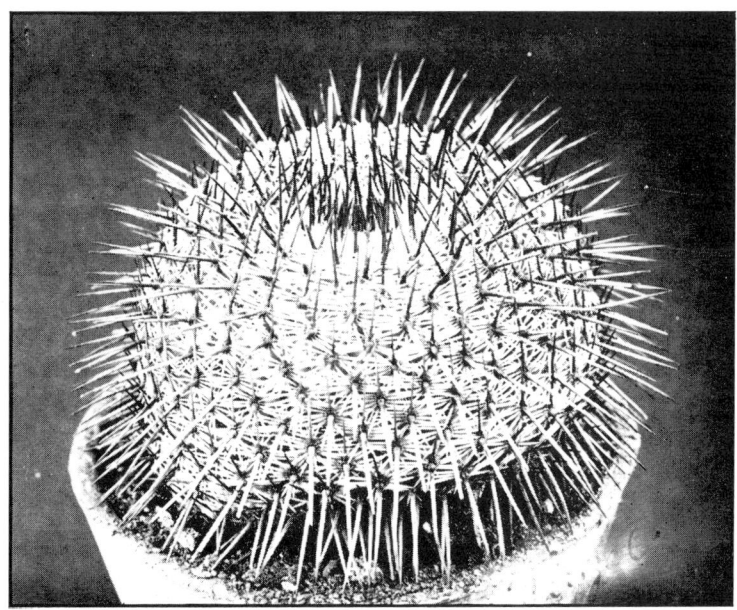

FIG. 109. *Mammillaria vaupelii* x 0.8

89. Mammillaria vaupelii Tiegel

Mammillaria (*Neomammillaria*) *vaupelii* Tiegel in Moeller, Deutsch. Gartner Zeit., 48:412, 1933.

BODY simple, flattened globular with deeply sunken and woolly apex, 6 cm. wide, 4 cm. high. TUBERCLES arranged in 13 and 21 spirals, firm in texture, bright bluish green, covered with small white dots, 4-sided at base, ovate above, with watery sap (occasionally slightly milky), 6 mm. long, 5-6 mm. wide at base. AREOLES round, 2 mm. wide, with thin curly wool in youth, soon becoming naked. AXILS with curly white wool and also grayish white bristles. CENTRAL SPINES 2 (4 see variety), lower 15 mm. long, upper 10 mm. long, all subulate, straight, stiff, smooth, orange-brown in youth, becoming creamy at base, pinkish above with brownish tip, later becoming grayish, divergent dorsally and ventrally and overtoping the apex. RADIAL SPINES 16-21, 5-6 mm. long, acicular, straight, stiff, smooth, glossy white, yellowish at base, ascending in youth, becoming horizontal. FLOWERS tubular campanulate, 17 mm. long, 10 mm. wide. *Outer perianth-segments* nearly white, light purplish pink at tip, linear lanceolate, tip acute, margins entire. *Inner perianth-segments* nearly white in lower $1/3$, light purplish pink in upper $2/3$, linear-lanceolate, tip obtuse, margins entire. *Filaments* white. *Anthers* sulphur-yellow. *Style* white. *Stigma-lobes* 3, pale yellow, 1 mm. long, overtop anthers 2-3 mm. FRUIT carmine red, clavate, 15 mm. long, with dried perianth persisting. SEEDS yellowish orange-brown, curved pyriform, smooth, 1 mm. ROOTS fibrous.

Distribution: Oaxaca, Mexico.

Type locality: None given but reported from Mixteca (Schmoll).

Illustration: Fig. 109 is a reproduction of an illustration by Tiegel in Cact. Succ. Journ., 7:21, 1935, as *Mammillaria vaupelii*.

FIG. 110. *Mammillaria vaupelii* var. *flavispina* x 0.8

a. var. **flavispina** Neal

Mammillaria vaupelii flavispina Neale, Cact. Other Succ., 94, 1935.

Central spines 4-5, mostly 4, yellowish.

Illustration: Fig. 110 is from a photograph of a plant sent to us by Sr. F. Schmoll of Cadereyta, Qro.

90. **Mammillaria rekoiana** nom. nov.

Neomammillaria rekoi Britton & Rose, The Cactaceae, 4:142, 1923.

BODY simple, clavate, 12 cm. high. TUBERCLES arranged in 8 and 13 spirals, conic, terete, with semi-milky sap, 8-10 mm. long. AXILS with tuft of white wool, also 1-8 bristles. CENTRAL SPINES 4-5, 10-15 mm. long, stout acicular, straight, brown, spreading porrect. RADIAL SPINES 20, 4-6 mm. long, slender acicular, straight, white, slightly ascending. FLOWERS campanulate, 15 mm. long. *Inner perianth-segments* deep purple, linear-oblong, tip apiculate. *Filaments* purplish. *Anthers* yellow. *Style* purplish. *Stigma-lobes* greenish. FRUIT red, clavate. SEEDS brown.

Distribution: Oaxaca, Mexico.

Type locality: None given.

Illustration: Fig. 111 is a reproduction of an illustration, Fig. 155a, in Britton & Rose (Cact., 4:142) as *Neomammillaria rekoi*.

Fig. 111. *Mammillaria rekoiana*

This species was confused by Britton & Rose with *N. rekoi* in their treatment of that species. They combined the two species on the strength that they were both sent to them by the same party from Oaxaca. They are doubtlessly very closely related but we believe that they are distinct enough to classify them as separate species.

91. **Mammillaria guerreronis** (Bravo) Backeberg & Knuth

Neomammillaria guerreronis Bravo, Inst. Biol. Mex., 3:395, 1932.
Mamillaria guerreronis Backeberg & Knuth, Kaktus A.B.C., 391, 1935.

BODY simple and cespitose from body and base, very tall columnar, to 70 cm. high, 6 cm. wide, with rounded apex, hidden by spines. TUBERCLES arranged in 8 and 13 spirals, bright green becoming gray-green, conic, with semi-milky sap, 8 mm. long, 5 mm. wide at base. AREOLES oval, with short white wool only in the youngest. AXILS with short white wool which persists for some time, also 15-20 white hair-like bristles. CENTRAL SPINES 2-4 (5), to 15 mm. long, lower longer, all acicular, straight to occasionally hooked (see variety), stiff, smooth, light orange-tan in youth, later becoming creamy white, orange yellow at very base, rose at tip, nearly horizontal with lateral ones more ventral than in cross formation. RADIAL SPINES 20-30, 5-10 mm. long, setaceous to fine acicular, straight, smooth, semiflexuous, white, orange-yellow at very base, horizontal. FLOWERS unknown. FRUIT greenish white becoming pink, paler below, clavate, 20x8 mm., with dried perianth persisting. SEEDS brown, curved globular-pyriform with lateral hilum near the base, glossy, pitted, 1.5x1 mm.

Distribution: Guerrero, Michoacan; Mexico.

Type locality: Canyon del Zapilote near Rio Balsas (Rio Mescala), Gro.

Illustration: Fig. 1 is from a photograph of a plant as we found it growing near the

Rio Balsas. Fig. 112 is from a photograph of a plant we collected at the type locality in 1941 and again in 1942.

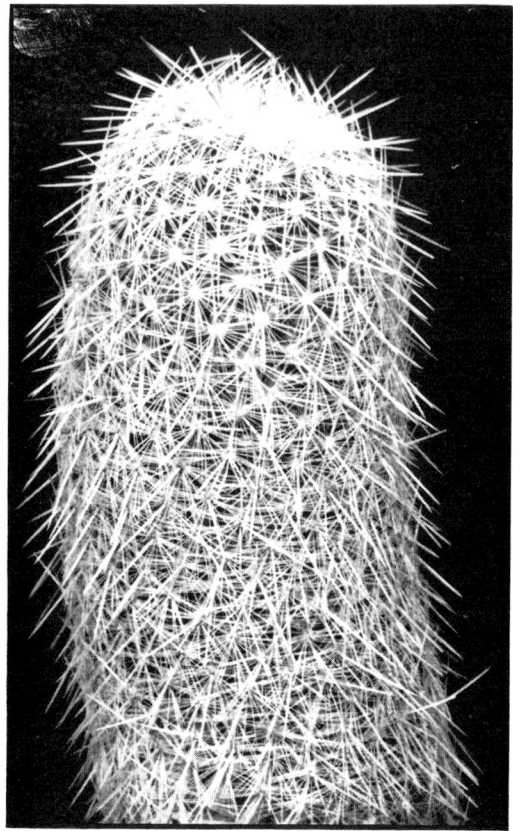

FIG. 112. *Mammillaria guerreronis* x 0.8

a. var. **recta** var. nov.

>Spinis centralibus rectis.

>All of the central spines straight.

b. var. **subhamatam** var. nov.

Neomammillaria guerreronis Bravo, Inst. Biol. Mex., 3:395, 1932.

>Una spinorum centralum est sub hamatorum.

>One of the central spines is occasionally slightly hooked.

This species, as originally described by Bravo as having one of the central spines sometimes slightly hooked, represents a possible hybrid form. Our investigation of the plants at the type locality disclosed several variations: (a) Those with all of the central spines straight—var. *recta*; (b) Those with the lower central spine occasionally slightly hooked—var. *subhamatam*; (c) Those with the lower central spine strongly hooked—*Mammillaria zapilotensis*.

This species is one of the most unusual ones in the manner of its growth and its size. We collected it along the highway that connects Taxco with Alcopulco, especially in the vicinity of the Rio Balsas where it grows quite abundantly in the cliffs of the steep canyon walls along the stream that parallels the highway particularly in the vicinity of K. 290. The roots of the plant are often fastened into the nearly vertical walls and the plants hang down sometimes for nearly a meter with the apex upturned like a big fish-hook. In all our investigations, we have never seen a plant of this genus with a normal growth of such a length.

Dr. Iwerson of Mexico City reported to us that he found this species also at San Juan de Peura, Michoacan.

FIG. 113. *Mammillaria rekoi*

92. Mammillaria rekoi (B. & R.) Vaupel

Neomammillaria rekoi Britton & Rose, The Cactaceae, 4:141, 1923.
Mammillaria rekoi Vaupel in Engler & Prantl, Nat. Pflanzenfam., ed. 2, 21:633, 1925.

BODY simple, globular to short cylindric with rounded apex, 12 cm. high, 5-6 cm. wide. TUBERCLES arranged in 8 and 13 spirals, green, conic, terete, with watery to semi-milky sap, 8-10 mm. long. AREOLES oval, with scant wool only in the youngest. AXILS with a tuft of white wool, also 1-8 long white bristles. CENTRAL SPINES 4, 10-15 mm. long, lower longer, all stout acicular, lower curved and hooked, uppers straight, all brown, upper ascending, lower nearly porrect. RADIAL SPINES 20, 4-6 mm. long, delicate acicular, straight, white, horizontal. FLOWERS near apex, 15 mm. long. *Inner perianth-segments* deep purple, linear-oblong, tip apiculate. *Filaments* purplish. *Anthers* yellow. *Style* purplish. *Stigma-lobes* greenish. FRUIT red, clavate, 12 mm. long. SEEDS brown, curved pyriform.

Distribution: Oaxaca, Mexico.
Type locality: None given.
Illustration: Fig. 113 is a reproduction of an illustration, Fig. 149, in Britton & Rose (Cact., 4:136) as *Neomammillaria rekoi*.

a. var. pseudorekoi (Boed.) comb. nov.

Mammillaria pseudorekoi Boedeker, Mammill. Vergl. Schlus., 34, 1933.

Central spines 4 7, lower hooked, upper straight, brown to black. Radial spines 20, white. Flowers red.

We have not been able to recollect either the type or the variety so as to complete the description.

Britton and Rose confused the type by including in it a closely related species in which all of the central spines are straight. We have referred the latter to *M. rekoiana*.

Fig. 114. *Mammillaria zapilotensis* x 0.8

93. **Mammillaria zapilotensis** sp. nov.

Corpus cylindricus ad 60 cm. longis, cespitosus, mamillis ad 8 et 13 seriebus ordinatis, subpyramidalibus, suco semilacteo; axillis lanatis et setis; spinis centralibus 2-4, 15-25 mm. longis, acicularibus crassis, hamatis; spinis radialibus 30, 6-8 mm. longis, acicularibus tenuis, albis; flores ignota, fructus puniceus, semina spadicia.

BODY cespitose from base and occasionally from body, cylindric to 60 cm. high, 5-6 cm. wide, with sunken apex, spines interlacing over the top. TUBERCLES arranged in 8 and 13 spirals, firm in texture, medium gray-green, cylindric, obscurely 4-sided, with semi-milky sap, 8-10 mm. long, 6 mm. wide at base. AREOLES broad oval to round, with practically no wool. AXILS with scant white wool, also 15-20 white bristles. CENTRAL SPINES 2-4, upper 15 mm. long, lateral 10 mm. long, lower 20-25 mm. long, all stout acicular, upper 1-3 straight, lower one usually strongly hooked, all smooth, stiff, orange-tan in youth, later becoming cream-colored, darker tip, in cross formation, nearly horizontal. RADIAL SPINES 30, 6-8 mm. long, upper shorter, slender acicular, straight, smooth, stiff, white, horizontal. FLOWERS (linear-lanceolate, tip acute, margins serrate, from dried perianth on fruit). FRUIT purplish pink, clavate, 25x9 mm., juicy, with dried perianth persisting. SEEDS dark reddish brown, curved pyriform with lateral hilum at base, glossy, pitted, 1.5x0.8 mm. ROOTS fibrous.

Distribution: Guerrero, Mexico.

Type locality: Zapilote Canyon near Rio Balsas.

Illustration: Fig. 114 is from a photograph of a plant collected by us at the type locality in June, 1942.

This species was found growing in the cliffs of the canyon walls that parallels the highway from Taxco to Alcopulco, especially between the Rio Balsas and Chilpancingo. It hybridizes with *M. guerreronis* (cf.) which is also found in the same territory. It is somewhat similar to *M. rekoi* but it has more radial spines. Its unusual length makes it rather conspicuous on the cliffs.

Although we have had the plant for a couple of years, it has as yet not bloomed, so the flower description is incomplete.

Fig. 115. *Mammillaria camptotricha* x 1

94. Mammillaria camptotricha Dams

Mamillaria camptotricha Dams, Gartenwelt, 10:14, 1906.
Neomammillaria camptotricha Britton & Rose, The Cactaceae, 4:126, 1923.
Chilita eschauzieri Orcutt, Cactography, 2, 1926.

BODY cespitose from base, globose with flattened and somewhat sunken top, 4-7 cm. wide. TUBERCLES arranged in 8 and 13 spirals, deep green to very light green at base, to sometimes reddish bronze at tip when grown in full sun, long slender conic, sometimes bent, terete, with watery sap, 20 mm. long, 7 mm. wide at base. AREOLES round, with short white wool, later becoming naked. AXILS with little wool, also 2-5 yellowish bristles to 15 mm. long. CENTRAL SPINES none. RADIAL SPINES 4-6 (8), upper to 30 mm. long, lower to 15 mm. long, all very slender acicular, tortuous, smooth, flexuous, in youth greenish white, becoming yellow, later gray, ascending in youth, later becoming horizontal. FLOWERS funnelform, 13 mm. long. *Outer perianth-segments* pale greenish white mid-stripe, very pale greenish white margins, linear-lanceolate, tip acuminate to acute, margins finely serrate, 1 mm. wide. *Inner perianth-segments* white, several pale green mid-lines, linear lanceolate, tip acute, margins mostly entire, 2 mm. wide. *Filaments* white, 5-6 mm. long. *Anthers* bright sulphur-yellow, very small, eliptical. *Style* very pale greenish white. *Stigma-lobes* 4-6, yellowish to pale greenish white. FRUIT very pale pink to pale greenish at top, slender clavate, 20x4 mm., with dried perianth not persisting. SEEDS light brown, curved pyriform with lateral hilum near base, faintly pitted, 0.8x0.5 mm.

Distribution: Queretaro, Mexico.

Type locality: None given but reported from eastern deserts between Higuerillas and San Pablo.

Illustration: Fig. 115 is from a photograph of a plant sent to us by Sr. F. Schmoll of Cadereyta, Qro.

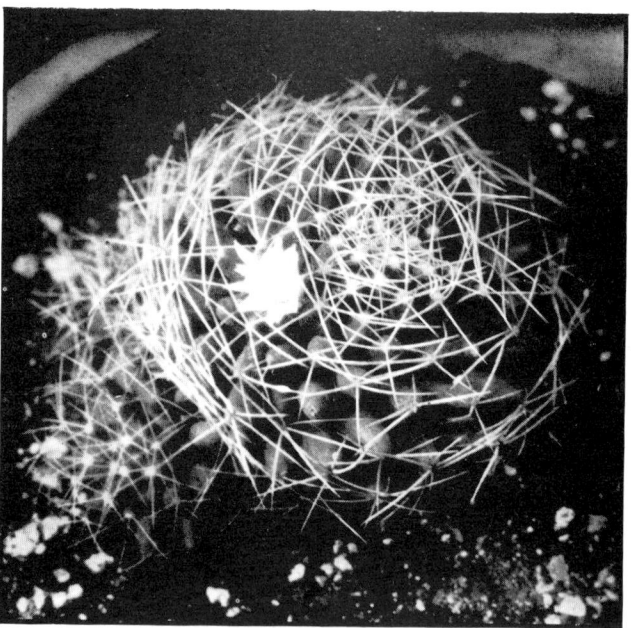

FIG. 116. *Mammillaria albescens* x 1

95. **Mammillaria albescens** Tiegel

Mammillaria albescens Tiegel in Moeller, Deutsch. Gartner Zeit., 48:260, 1933.

BODY cespitose from base, globular with sunken apex, to 8 cm. high, 5 cm. wide. TUBERCLES arranged in 5 and 8 spirals, flabby in texture, more or less leaf green, slender conic, with watery sap, 15-20 mm. long, 5-7 mm. wide. AREOLES round, very small to only 0.5 mm. wide, with short yellowish white wool only in the youngest. AXILS 2-4 strong yellowish white bristles, but no wool. CENTRAL SPINES none to occasionally 1 (when present to 15 mm. long, straight, acicular, stiff, white, porrect). RADIAL SPINES 5 (4-6), 8-15 mm. long, acicular, nearly straight, semi-stiff, smooth, pale yellow to greenish white in youth, later becoming chalky white, somewhat irregularly ascending, later becoming horizontal. FLOWERS funnelform, to 20 mm. long, 15-18 mm. wide. *Outer perianth-segments* very pale greenish white, light green to rose-brown mid-lines which are more pronounced ventrally, lanceolate, tip obtuse, margins entire. *Inner perianth-segments* very pale greenish white to nearly white, linear lanceolate, 3 mm. wide, tip acute, margins entire, tip often split. *Filaments* white, somewhat bent. *Anthers* pale yellow. *Style* yellow below, rose above. *Stigma-lobes* 5, white to yellow. FRUIT green to reddish, slender clavate, to 25 mm. long. SEEDS bright to dark brown, curved pyriform, finely pitted.

Distribution: Queretaro, Mexico.

Type locality: None given, but reported from La Templadora, Qro. (Schmoll).

Illustration: Fig. 116 is from a photograph of a plant obtained from Sr. F. Schmoll of Cadereyta, Qro.

This species probably represents an intermediate form between *M. camptotricha* and *M. decipiens* as it has some of the characteristics of both.

FIG. 117. *Mammillaria kewensis* x 0.8

96. Mammillaria kewensis Salm-Dyck

Mamillaria kewensis Salm-Dyck, Cact. Hort. Dyck. 1849, 112, 1850.
Mamillaria kewensis albispina Salm-Dyck, Cact. Hort. Dyck. 1849, 15, 1850.
Neomammillaria kewensis Britton & Rose, The Cactaceae, 4:104, 1923.

BODY simple and cespitose, clavate with rounded apex, 9 cm. wide, 12 cm. high. TUBERCLES widely separated in 8 and 13 spirals, firm in texture, bluish grayish green, wide conic, with watery sap, 8 mm. long, 9-15 mm. wide at base. AREOLES oval, with white wool in youth. AXILS with white wool which persists, but no bristles. CENTRAL SPINES none. RADIAL SPINES 6, uppers 6-12 mm. long, lower one to 30 mm. long, all stout acicular, slight recurve, smooth, stiff, not enlarged at base, reddish brown to black in youth, later the lower ones becoming lighter, all ascending and recurved slightly. FLOWERS campanulate, June-September, 20 mm. long, 12-15 mm. wide. *Outer perianth-segments* greenish at base, brownish magenta tapering mid-stripe above, paler margins, lanceolate, tip acute, margins usually entire but occasionally lightly serrate at tip. *Inner perianth-segments* deep purplish pink, lanceolate, tip acute, margins entire. *Filaments* deep purplish pink, hooked. *Anthers* yellowish tan. *Style* cream to pale pink above. *Stigma-lobes* 5, pale pink, 1.5 mm. long. FRUIT greenish to pale pink to rose red, broad clavate, 13-20x5-9 mm., with dried perianth persisting. SEEDS light tan, curved pyriform with lateral hilum at base, roughened, 1.2x0.6 mm. ROOTS fibrous.

Distribution: Hidalgo and central Mexico.

Type locality: None given but reported from Ixmiquilpan.

Illustration: Fig. 117 is from a photograph of a plant we collected near Ixmiquilpan.

The original description is supplemented with data from plants that we collected in 1941 and 1942 near Ixmiquilpan, Hgo., on the canyon walls in partial shade. *M. hidalgensis* was collected by us a few miles distant but in that species the tubercles are more slender conic, the spines are all central, and the outer perianth-segments are serrate instead of mostly entire as with this species.

Britton & Rose (Cact., 4:104, 1923), report "crisp hairs" in the axils but the original description makes no mention of their presence nor did we find any in the plants that we collected.

M. spectabilis Mühlenpfordt (Allg. Gartenz., *13*:346, 1845), *Cactus spectabilis* Kuntze (Rev. Gen. Pl., *1*:261, 1891) has been doubtfully associated with the above species

Fig. 118. *Mammillaria kewensis* var. *craigiana* x 0.8

but the description calls for tubercles that are more angled and a different spine arrangement.

M. radianti is only a name reported by Walper (Repert., 5:810, 1845) as near *M. spectabilis*.

a. var. **craigiana** Schmoll var. nov.

Mamillis tenuioris.

Tubercles more slender, 4-5 mm. wide at base, spines shorter and strongly ascending.

Distribution: Guanajuato and Queretaro; Mexico.

Type locality: Hacienda "Ojo de Agua" near San Juan del Rio, Qro.

Illustration: Fig. 118 is from a photograph of a plant sent to us by Sr. F. Schmoll of Cadereyta, Qro.

This is the plant that Sr. Schmoll has been exporting under the name of *M. craigiana*.

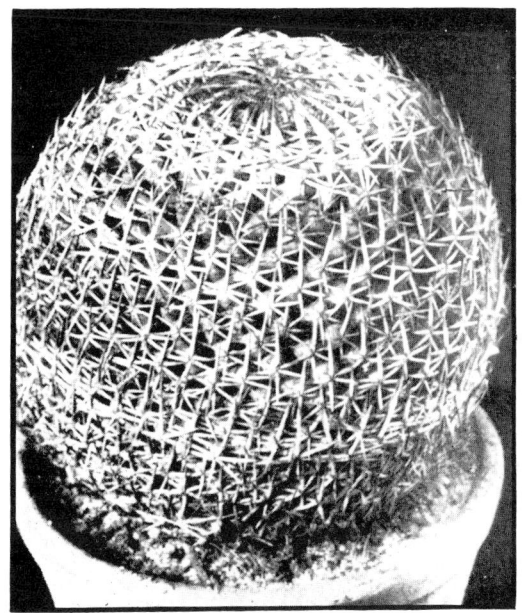

FIG. 119. *Mammillaria durispina* x 0.7

97. **Mammillaria durispina** Boedeker

Mammillaria durispina Boedeker, Zeitschr. Sukkulentenk., 3:342, 1928.

BODY simple, globular to cylindric with only slightly sunken apex, 20 cm. high, 11 cm. wide. TUBERCLES arranged in 8 and 13, also 21 and 34 spirals, firm in texture, dull dark green, conic, truncate at apex, with watery sap, 8-11 mm. long, 8-9 mm. wide at base. AREOLES oval, 5 mm. wide in youth, later becoming only 1.5 mm. wide, with white wool only in the youngest. AXILS with coarse white wool only in the youngest and no bristles. CENTRAL SPINES usually none (occasionally a few scattered areoles with 1-2 central spines that are similar to the radial spines, sometimes porrect to nearly in the same plane as the radials). RADIAL SPINES 6-8, upper 15 mm. long, lateral and lower 7-8 mm. long, subulate, straight to slight recurve, especially in upper, all smooth, not enlarged at base, color alternating in more or less distinct bands of faint grayish white and with dark reddish brown or black, horizontal to somewhat ascending, especially in youth. FLOWERS campanulate, 15 mm. long and wide. *Outer perianth-segments* cream to bright green in lower part, deep carmine red above, brownish red tapering mid-stripe, linear-lanceolate, tip acuminate, margins lightly serrate.* *Inner perianth-segments* pale green to nearly white at base, purplish pink above, darker mid-line, margins lighter to nearly transparent, linear-lanceolate, to 3 mm. wide, tip acute, margins entire. *Filaments* light green below, purplish pink above. *Anthers* whitish to tannish yellow. *Style* whitish to cream below, rose to deep pink above. *Stigma-lobes* 5, deep purplish pink, short globular clavate, 1 mm. long, level with anthers. FRUIT carmine red to green at top, clavate, 20 mm. long, with dried perianth persisting. SEEDS faint yellow, curved pyriform with lateral hilum at base, 1 mm., finely pitted.

Distribution: Guanajuato, Queretaro; Mexico.

Type locality: None given, but reported from San Moran, Qro. (Schmoll).

Illustration: Fig. 119 is a reproduction of an illustration in Zeitschr. Sukkuleentenk., 3:343 as *Mammillaria durispina*.

FIG. 120. *Mammillaria napina* x 0.8

98. Mammillaria napina Purpus

Mamillaria napina Purpus, Monatsschr. Kakteenk., **22**:161, 1912.
Neomammillaria napina Britton & Rose, The Cactaceae, 4:104, 1923.

BODY simple, semi-globose, hemispherical, flattened above with slightly sunken apex, 4-6 cm. wide. TUBERCLES arranged in 8 and 13 spirals, firm in texture, conic, laterally compressed at base, with watery sap, 8-10 mm. long, 7-8 mm. wide at base. AREOLES oval, 3x2 mm., with yellow wool in youth, later becoming naked. AXILS with scant wool or even naked. CENTRAL SPINES usually none (see variety). RADIAL SPINES 10-12, 8-9 mm. long, thin subulate, somewhat bent, very stiff, smooth, light yellow, pectinate, interlacing. FLOWERS campanulate, lateral near top, 40 mm. wide. *Outer perianth-segments* rose, brownish rose mid-stripe, narrow lighter margins, linear-lanceolate, tip obtuse, margins entire, 25x2 mm. *Inner perianth-segments* rose, upper surface dark rose to bright violet-rose, very slender linear, tip obtuse, margins entire, 35x3 mm. *Filaments*

---

*The original description gives the margins as entire but our imported plants were found to be lightly serrate.

white below, more yellowish above. *Anthers* yellow. *Style* yellowish. *Stigma-lobes* 6, whitish, slender, overtop anthers. FRUIT and SEEDS unknown.

Distribution: Puebla, Mexico.

Type locality: Tehuacan.

Illustration: Fig. 120 is a reproduction on an illustration in Schwarz & Georgi's Catalogue: Kakteensammier, 12 as *Mamillaria napina*.

a. var. **centrispina** (Britton & Rose) comb. nov.

Neomammillaria napina Britton & Rose, The Cactaceae, 4:104, 1923.

"These differ from the type plant chiefly in having usuallly one porrect central spine, 5-8 mm. long."

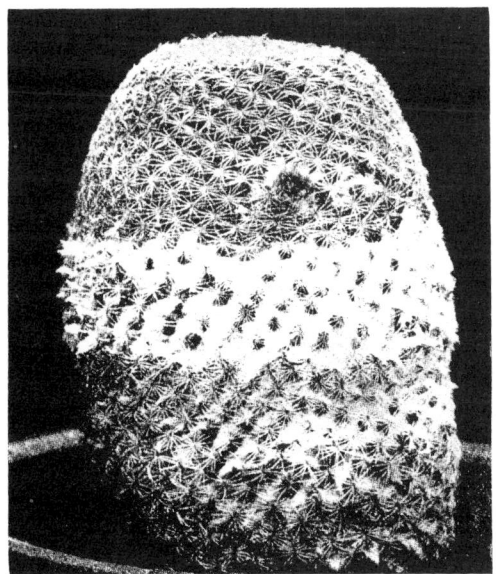

FIG. 121. *Mammillaria lanata*

99. **Mammillaria lanata** (B. & R.) Orcutt

Neomammillaria lanata Britton & Rose, The Cactaceae, 4:104, 1923.
Mammillaria lanata Orcutt, Cactography, 7, 1926.

BODY simple and cespitose from base, short cylindric, with slightly sunken apex, clumps to 25 cm. wide, individual heads 25-30 mm. wide. TUBERCLES arranged in 13 and 21, also in 21 and 34 spirals, firm in texture, light green, conic, with watery sap, 2-4 mm. long, 2 mm. wide at base. AREOLES short oval, with scant white wool only in youth. AXILS with considerable wool in flowering area. CENTRAL SPINES only very rudimentary when seen with magnifying glass. RADIAL SPINES 12-20, 1-2 mm. long, very fine acicular, straight, smooth, semi-flexuous, pale brownish white at base, horizontal. FLOWERS funnelform, lateral, 10 mm. long, 6-10 mm. wide. *Outer perianth-segments* wide pinkish tan mid-stripe, pale pink margins, lanceolate, tip acute, margins somewhat serrate. *Inner perianth-segments* bright pink tapering narrow mid-stripe, whitish to very pale pink wide margins, linear, tip acute, margins entire. *Filaments* pink. *Anthers* yellow. *Style* cream below, deep pink above. *Stigma-lobes* 3 purplish pink, less than 1 mm., overtop anthers 1-2 mm., not wide spreading. FRUIT scarlet, elongate globular, 5x2 mm., dried perianth persisting, very few seeds (5-6) per pod. SEEDS olive-greenish brown, semi-glossy, globular pyriform with lateral hilum at base, finely pitted. ROOTS heavy tap.

Distribution: Puebla, Oaxaca; Mexico.

Type locality: Rio de Santa Luisa, Puebla (?), reported from Tomellin Canyon, Oaxaca.

Illustration: Fig. 121 is a reproduction of an illustration, Fig. 105, in Britton & Rose, (Cact. 4:105) as *Neomammillaria lanata.*

This species merges into *M. elegans* through the latter's variety *supertexta* (cf.).

The original description is supplemented with data from plants collected by Mr. George Lindsay near the railroad station in Tomelin Canyon, 50 miles south of Tehucan, Puebla, toward Oaxaca, on the north side of a steep canyon.

While we were traveling by train between Tehucan and Oaxaca, we observed this species growing in clusters along the canyon walls close to the railroad. The train was going so slowly that we could tantalizingly see them just beyond our reach, still it was going too fast to enable us to get off to do any collecting.

FIG. 122. *Mammillaria fragilis* x 1

100. **Mammillaria fragilis** Salm-Dyck

Mamillaria fragilis Salm-Dyck, Cact. Hort. Dyck. 1849, 103, 1850.
Neomammillaria fragilis Britton & Rose, The Cactaceae, 4:133, 1923.
Chilita fragilis Orcutt, Cactography, 2, 1926.

BODY very cespitose from base and body, branches easily dislodged, cylindric to clavate. TUBERCLES loosely arranged in 5 and 8 spirals, bright light green, blunt conic, terete, with watery sap, 4-5 mm. long, 3 mm. wide at base. AREOLES round to slightly oval with scant white wool in youth. AXILS with slight tufts of white wool. CENTRAL SPINES 0-1 (present only on old stems, seldom on branches, to 10 mm. long, acicular, straight, smooth, brownish at tip, lighter below, porrect). RADIAL SPINES 12-16, 4-5 mm. long, very slender acicular, straight or slightly recurved, smooth, white, upper ones sometimes with brownish tip, slightly ascending in youth, later horizontal. FLOWERS funnelform, 12 mm. long, near but not in apex, lasting several days. *Outer perianth-segments* with light yellow margins, pale red tapering mid-stripe, wide-lanceolate, tip obtuse, margins mostly entire. *Inner perianth-segments* light yellow, pale brownish mid-line, oblong, tip obtuse, margins mostly entire. *Filaments* pale yellow. *Anthers* yellow. *Style* pale yellow. *Stigma-lobes* 4, greenish yellow, 2 mm. long, overtop anthers 2 mm. FRUIT red, juicy, clavate, very few seeds, 12x5 mm., dried perianth not persisting. SEEDS glossy black, curved pyriform with small lateral hilum, minutely pitted, 2x1.4 mm.

Distribution: Hidalgo, Mexico.

Type locality: None given, but reported from Metztitlan and Zimapan at Puente de Dios.

Illustration: Fig. 122 is a reproduction of an illustration, Fig. 145, in Britton & Rose (Cact., 4:133) as *Neomammillaria fragilis*.

The original description is supplemented with data from plants sent to us by Sr. F. Schmoll of Cadereyta, Quo., Mexico, and from flower data from plants in the collection of Mr. R. W. Kelly of Temple City, California.

Mammillaria gracilis Pfeiffer (Allg. Gartenz., 6:275, 1838) to which the above species is often referred, should be referred to possibly *M. echinaria* because it calls for 2 strong central spines of 10-12 mm.

Torrey (Coll. Rocky Mt. Pl., 202, 1826) mentioned *M. fragilis* but confused it with *Coryphantha vivipara*. *M. gracilis pusilla* (Monat. Kakteenk. 62,, 1893) is only a name, but may be referrable here.

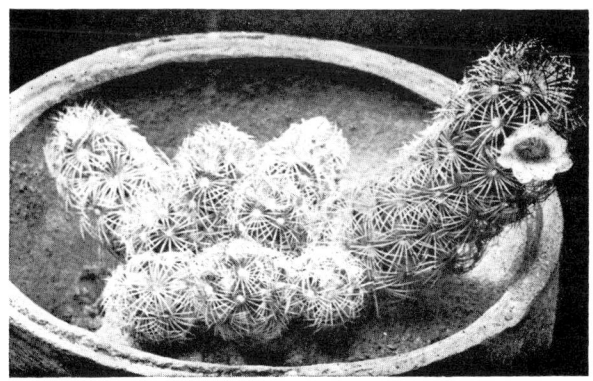

FIG. 123. *Mammillaria elongata* x 1

101. Mammillaria elongata DeCandolle

Mammillaria elongata DeCandolle, Mem. Mus. Hist. Nat. Paris, **17**:109, 1828.
Mammillaria subcrocea DeCandolle, Mem. Mus. Hist. Nat. Paris, **17**:110, 1828.
Mammillaria intertexta DeCandolle, Mem. Mus. Hist. Nat. Paris, **17**:110, 1828.
Mammillaria tenuis media DeCandolle, Mem. Mus. Hist. Nat. Paris, **17**:110, 1828.
Mammillaria stella aurata Martius in Zuccarini, Plant. Nov. Plant., 701, 1837.
Mamillaria subcrocea intertexta Salm-Dyck, Hort. Dyck. 1844, 13, 1845.
Mammillaria elongata subcrocea Salm-Dyck, Cact. Hort. Dyck. 1849, 12, 1850.
Mamillaria elongata intertexta Salm-Dyck, Cact. Hort. Dyck. 1849, 12, 1850.
Mamillaria elongata rufescens Salm-Dyck, Cact. Hort. Dyck. 1849, 100, 1850.
Mamillaria stella aurata gracilispina Salm-Dyck, Cact. Hort. Dyck. 1849, 101, 1850.
Cactus elongatus Kuntze, Rev. Gen. Pl., **1**:260, 1891. (Not Willdenow 1813.)
Cactus intertextus Kuntze, Rev. Gen. Pl., **1**:260, 1891.
Cactus stella auratus Kuntze, Rev. Gen. Pl., **1**:261, 1891.
Cactus subcroceus Kuntze, Rev. Gen. Pl., **1**:261, 1891.
Mamillaria elongata stella aurata Schumann, Gesamtb. Kakteen, 520, 1898.
Neomammillaria elongata Britton & Rose, The Cactaceae, 4:134, 1923.
Chilita elongata Orcutt, Cactography, 2, 1926.
Mamillaria gracilis pulchella Berger, Kakteen, 304, 1929.
Mammillaria elongata schmollii Borg, Cacti, 312, 1937.

BODY cespitose from the base, elongate, with rounded apex, 3-10 cm. high, 2-3 cm. wide. TUBERCLES arranged in 3 and 5, also 5 and 8 spirals, light green in youth, later becoming darker, conic, with watery sap, 2-4 mm. long. AREOLES round to oval, with slight wool in youth. AXILS with very slight wool to naked. CENTRAL SPINES none to very rarely 1 (6-8 mm. long, straight, stiff, smooth, yellow, porrect). RADIAL SPINES

15-20, 5-12 mm. long, acicular, straight to recurved, smooth, stiff, yellow, slightly ascending and recurved to body. FLOWERS funnelform, 6-15 mm. long, 10-12 mm. wide. *Outer perianth-segments* very pale greenish yellow, faint pink mid-stripe, not to tip, linear-lanceolate, tip acute, margins entire. *Inner perianth-segments* white to yellowish with red mid-stripe, lanceolate, tip obtuse to acute, margins entire. *Filaments* white to yellow. *Anthers* pale yellow. *Style* white to yellow. *Stigma-lobes* 4, greenish yellow. FRUIT red, curved clavate, with dried perianth persisting. SEEDS light brown, curved pyriform with lateral hilum near the base, faintly roughened.

Distribution: Hidalgo, Mexico.

Type locality: None given but reported from Ixmiquilpan and Zimapan.

Illustration: Fig. 123 is a reproduction of an illustration, Fig. 146, in Britton & Rose (Cact., 4:133) as *Neomammillaria elongata*.

This species is close to and often intergrades into *M. echinaria* but the major difference between them is the presence of central spines in the latter species. In clusters of the above species a few scattered central spines may often be present.

M. supertexta rufa and *M. supertexta rosea* are reported to be referrable here but we have not seen the description.

M. stella aurata Martius was given by Britton & Rose as having been published in Zuccarini (Abh. Bayer Akad. München, 2:101, 1837) but examination of that reference failed to reveal any such description of this species but it did appear in another work by Zuccarini (Plant. Nov. Plant., 701, 1837). This specific name was first listed by Martius (Hort. Reg. Monac., 128, 1829) but it was not described until 1837.

M. caespitosa was first listed by Link & Otto (Uber Gatt. Melo. Echinoc., 21, 1827) but it was referred to *M. densa* by Martius (Hort. Reg. Monac., 127, 1829) and also by Index of Berlin Bot. Gard., 429, 1829. Later Martius (Besch. Neuen Nopal., 338, 1832) refers it to *M. glochidiata* but Pfeiffer (Enum. Cact., 5, 1837) considered it as a synonym of *M. echinata* while Labouret (Des Cactees, 25, 1853) refers it to *M. pusilla*. Gray (Intro. Struct. & Syst. Bot., 421, 1862) offers it only as a named illustration of an entirely different plant, *M. missouriensis* which was later referred to *Echinocereus reichenbachii*.

a. var. **tenuis** (DC) Schumann

Mammillaria tenuis DeCandolle, Mem. Mus. Hist. Nat. Paris, **17**:110, 1828.
Mammillaria minima Reichenbach in Terscheck, Suppl. Cact. Verz., 1.
Mammillaria tenuis minima Salm-Dyck in Walpers, Repert. Bot., **2**:272, 1843.
Cactus minimus Kuntze, Rev. Gen. Pl., **1**:260, 1891.
Cactus tenuis Kuntze, Rev. Gen. Pl., **1**:261, 1891.
Mamillaria elongata tenuis Schumann, Gesamtb. Kakteen, 520, 1898.
Mamillaria elongata minima Schelle, Kakteen, 304, 1926.

BODY slender clavate, 10 mm. wide. TUBERCLES arranged in 3 and 5 spirals. CENTRAL SPINES none. RADIAL SPINES 20-25, setaceous, yellow, orange at base.

102. **Mammillaria humboldtii** Ehrenberg

Mammillaria humboldtii Ehrenberg, Linnaea, **14**:378, 1840.
Cactus humboldtii Kuntze, Rev. Gen. Pl., **1**:260, 1891.

BODY simple and cespitose from base, flattened globose to slender columnar. TUBERCLES arranged in 13 and 21 spirals, light green, cylindric, rounded above, with watery sap, 12 mm. long, 2-3 mm. wide at base. AREOLES round to slightly oval, at first with scant wool, later becoming naked and yellowish. AXILS with scant white wool and 7-8 white straight bristles of various lengths. CENTRAL SPINES none. RADIAL SPINES 80 or more, several series, unequal lengths of 2-8 mm. long, very slender acicular,

FIG. 124. *Mammillaria humboldtii* x 1

straight, smooth, semi-flexuous, snow white, very base slightly yellowish, horizontal. FLOWERS 15 mm. wide. *Outer perianth-segments* unknown. *Inner perianth-segments* bright red, linear, tip obtuse, margins (?) *Filaments* red. *Anthers* orange. *Style* yellow. *Stigma-lobes* 3, green. FRUIT reddish, dried perianth not persisting, clavate. SEEDS black, glossy, obovate, constricted above base, with basal hilum, pitted, 1.4x1 mm.

Distribution: Hidalgo, Mexico.

Type locality: Ixmiquilpan and Metztitlan.

Illustrations: Fig. 124 is from a photograph of a plant obtained from Sr. F. Schmoll in 1942. Fig. 125 is from a photograph of a hybrid plant obtained from Sr. F. Schmoll in 1942.

This species is often found growing with *M. schiedeana* and hybrids of the two have been collected as is shown in Fig. 125 which was collected by Sra. Schmoll.

This species has been referred by some authors to *M. candida* but inasmuch as the original description calls for no central spines and the collected plants were found to conform to it, it is here considered a distinct species.

FIG. 125. *Mammillaria humboldtii* (hybrid)

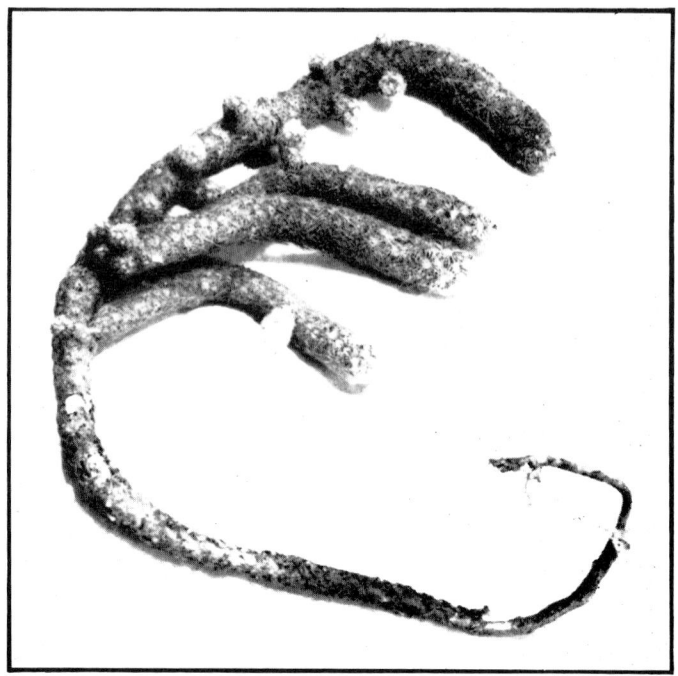

Fig. 126. *Mammillaria viperina* x 0.5

103. **Mammillaria viperina** Purpus

Mamillaria viperina Purpus, Monatsschr. Kakteenk., 148, 1912.
Neomammillaria viperina Bravo, Anal. Inst. Biol. Mex., 1:124, 1930.
Mammillaria elongata viperina Marshall & Bock, Cactaceae, 180, 1941.

BODY simple and cespitose from body, grows upright at first, later falls over and creeps along the ground with the apex somewhat erect, new growth may send out new roots as the old stems become corky and dry up, cylindric with apex rounded, 10-15 mm. wide. TUBERCLES arranged in 3 and 5 spirals, light green, firm in texture, sometimes nearly globular to short cylindric, with watery sap, 5 mm. long, 3 mm. wide at base. AREOLES oval with short white wool in youth. AXILS sometimes with white bristles or hair to length of tubercles (have been reported but not found in collected plants). CENTRAL SPINES none. RADIAL SPINES 25-30, 3-5 mm. long, acicular, smooth, stiff, recurved, variable white to yellow to sometimes brownish black, evenly distributed, ascending but tips recurved back to body. FLOWERS funnelform, 15 mm. long, 20 mm. wide. *Outer perianth-segments* olive-green with purplish brown mid-stripe, blunt-lanceolate to spatulate, tip obtuse, margins slightly serrate. *Inner perianth-segments* clear carmine-red, throat little lighter, lanceolate, ends split, margins mostly entire. *Filaments* greenish below, carmine red above. *Anthers* orange-yellow. *Style* short. *Stigma-lobes* 4, greenish white. FRUIT carmine-red, cylindric clavate, 8x3 mm., with dried perianth persisting. SEEDS light tan, curved pyriform with lateral hilum at base, finely pitted, less than 1 mm.

Distribution: Puebla, Mexico.

Type locality: Rio de Zapotitlan, also Sierra de Mixteca, on road from Tehuacan to San Luis Teutitlanpa.

Illustration: Fig. 126 is from a photograph of a plant obtained from Sr. F. Schmoll of Cadereyta, Qro., in 1942.

This species has been referred to as related to *Mammillaria sphacelata* which is also found in the same type locality but it is definitely a distinct species.

The nature of the characteristic growth of this species makes pot culture in cultivation unsatisfactory as it does not have the opportunity to spread out and as a result the new roots do not have an opportunity to reach the food supply.

FIG. 127. *Mammillaria lenta*

104. **Mammillaria lenta** Brandegee

Mamillaria lenta K. Brandegee, Zoe, 5:194, 1904.
Neomammillaria lenta Britton & Rose, The Cactaceae, 4:129, 1923.
Chilita lenta Orcutt, Cactography, 2, 1926.

BODY simple and cespitose from base and by dichotomous branching, globose to slightly cylindric with flattened top, 30-50 mm. wide, to 60 mm. high. TUBERCLES arranged in 13 and 21 spirals, semi-firm in texture, bright to yellowish green, slender conic, with watery sap, 8-10 mm. long, 2 mm. wide at base. AREOLES round, nearly naked. AXILS with short persistent white wool also an occasional bristle. CENTRAL SPINES none. RADIAL SPINES 30-40 in several series, to 5 mm. long, bristle-like, straight, smooth, soft, yellow to transparent white, some ascending, others horizontal and interlacing. FLOWERS 20 mm. long, 25 mm. wide. *Outer perianth-segments* white with hardly noticeable tint of rose, red mid-stripe, lanceolate, tip obtuse, margins finely serrate. *Inner perianth-segments* whitish, lanceolate, tip acute, margins mostly entire. *Filaments* light rose below, dark rose above. *Anthers* chrome-yellow. *Style* white to faint rose. *Stigma-lobes* 4, bright olive-green. FRUIT red, clavate, 10 mm. long, with only a few seeds per pod. SEEDS black, dull, strongly constricted above basal hilum, upper part globose, tuberculate, not pitted. ROOTS large knarly tuberous.

Distribution: Coahuila, Mexico.

Type locality: Viesca, also reported from Torreon.

Illustration: Fig. 127 is a reproduction of an illustration, Fig. 137, in Britton & Rose (Cact. 4:128) as *Neomammillaria lenta* (a grafted plant). Fig. 128 is from a photograph of a plant obtained from Sr. F. Schmoll in 1942.

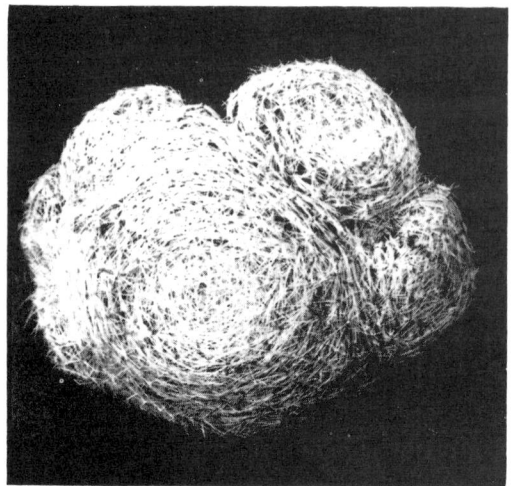

FIG. 128. *Mammillaria lenta* x 1

FIG. 129. *Mammilaria plumosa* x 1

105. Mammillaria plumosa Weber

Mamillaria plumosa Weber in Bois, Dict. Hort., 2:804, 1893-9.
Neomammillaria plumosa Britton & Rose, The Cactaceae, 4:123, 1923.
Chilita plumosa Orcutt, Cactography, 2, 1926.

BODY cespitose from body and base, globular, 60-70 mm. high and wide. TUBERCLES arranged irregularly in 8 and 13 spirals, very soft in texture, light green, cylindric, with watery sap, 12 mm. long, 2-3 mm. wide at base. AREOLES round, with scant very short wool. AXILS with long white wool. CENTRAL SPINES none. RADIAL SPINES to 40, 3-7 mm. long, plumose, soft silky featherlike, tortuous, white, ascending. FLOWERS campanulate, 15 mm. long, 14 mm. wide. *Outer perianth-segments* pale yellowish green, margins nearly white, lanceolate to clavate, tip obtuse, margins serrate above, entire below. *Inner perianth-segments* pale green throat, very pale greenish white margins, pale greenish reddish brown mid-line ventrally at tip, nearly clavate, tip obtuse, margins entire. *Filaments* and *Style* very pale green. *Anthers* sulphur-yellow. *Stigma-lobes* 3-5, pale greenish yellow, 1.5 mm. long, overtop anthers 3 mm. FRUIT unknown. SEEDS black, pitted.

Distribution: Coahuila and northern Mexico.

Type locality: None given but reported from Monterrey to Saltillo.

Illustration: Fig. 129 is from a photograph by Mr. J. R. Brown.

This most interesting species is so different from all of the other members of this genus that the spines hardly appear as such because of their soft feathery plumose nature. It forms small whitish rounded mounds in the more or less protected locations in the crevices of rocks.

M. scheideana plumosa Rebut, ined. (Cat., 7, 1896) may be referrable here.

M. lasiacantha plumosa Watson (Cact. Cult., 82, 1889) which has longer spines distinctly feathered, may be referrable here or possibly a hybrid with *M. lasiacantha*.

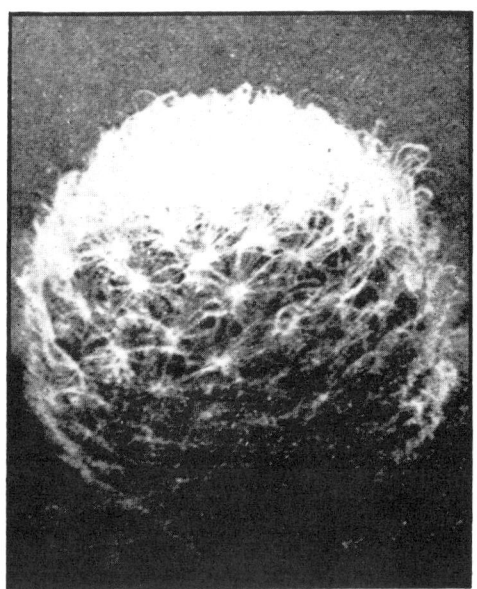

Fig. 130. *Mammillaria aureilanata*

106. Mammillaria aureilanata Backeberg

Mamillaria cephalophora Quehl, Monatsschr. Kakteenk., 24:158, 1914.
Chilita eschauzieri Orcutt, Cactography, 2, 1926.
Mamillaria aureilanata Backeberg, Breitr. Sukkul. Pfl., 13, 1938.

BODY simple, globose, with slightly sunken apex, 75 mm. high. TUBERCLES widely separated in 8 and 13 spirals, pitted, glossy dark green, cylindric, sloping, with apex smaller, turned upward, with watery sap, 10 mm. long, 7 mm. wide at base. AREOLES round, naked. AXILS naked. CENTRAL SPINES none. RADIAL SPINES 25-30 in two series, to 15 mm. long, thin setaceous, smooth, tortuous, transparent white to golden yellow, darker in age, somewhat ascending, interlacing. FLOWERS campanulate, 30 mm. long, 15-18 mm. wide. *Outer perianth-segments* green at base, purplish green wide mid-stripe, pink tip, very pale margins, lanceolate, tip acute, margins entire. *Inner perianth-segments* white to pale pink, bright rose-red mid-stripe, lanceolate to oblong, tip aristate, margins entire. *Filaments* white to pale pink. *Anthers* yellow to orange. *Style* greenish white to pale yellowish pink. *Stigma-lobes* 5, deep yellow. FRUIT pinkish white, clavate. SEEDS dull black. ROOTS heavy tuberous.

Distribution: San Luis Potosi, Mexico.

Type locality: None given but reported from Villar.

Illustration: Fig. 130 is a reproduction of an illustration in Blätter Kakteenf. (Bull. Cact. Research) 131, 5, 1934-9, as *Mammillaria cephalaphora*.

The name *M. cephalaphora* was used previously by Salm-Dyck (Cact. Hort. Dyck. 1849, 137, 1850) for a different plant which is now referred to *Coryphantha pycnacantha*.

FIG. 131. *Mammillaria herrerae* x 1

107. Mammillaria herrerae Werdermann

Mammillaria herrerae Werdermann, Notizbl. Bot. Gart. Mus. Berlin, 11:276, 1931.

BODY simple and cespitose from base, globose, with apex flattened and a little sunken, 35 mm. wide. TUBERCLES arranged in 8 and 13 spirals, closely set, cylindric, terete, with tip nearly truncate, with watery sap, 5-6 mm. long, 2 mm. wide at base, becoming spineless and corky in age. AREOLES round to oval, 1.5 mm., with scant short white wool in youth below the spines, later becoming naked. AXILS naked. CENTRAL SPINES none. RADIAL SPINES 100, several series superimposed, 1-5 mm. long, setaceous, straight, base not enlarged, pure white to whitish ashen-gray, slightly yellowish at very base, horizontal to little appressed, interlacing so as to completely cover the body. FLOWERS funnelform, lateral, 20-25 mm. long. *Outer perianth-segments* olive-green mid-stripe, rose margins, lanceolate, tip obtuse, margins entire, 10x2 mm. *Inner perianth-segments* pale rose to violet, darker mid-stripe, lanceolate, tip acute and often split, margins entire, 10 mm. *Filaments* above pale rose-violet. *Anthers* orange-yellow. *Style* white. *Stigma-lobes* 6, greenish. FRUIT carmine, nearly globose, fleshy, retained behind the spines, 6 mm. wide. SEEDS dull black, irregular pyriform, tuberculate, with two small projections at lateral hilum, small.

Distribution: Queretaro, Mexico.

Type locality: Cadereyta.

Illustration: Fig. 131 is a reproduction of an illustration in Monatsschr. Deut. Kakteenk., 4:248, 1932, as *Mammillaria herrerae*.

Orcutt (Cactography 2, 1926) mentioned this specific name and its variety *intertexta* (Fric) without reference. He considered it as synonymous with *Chilita eschauzieri*.

a. var. **albiflora** Werdermann

Mammillaria herrerae v. *albiflora* Werdermann, Notizb. Bot. Gard. Mus. Berlin, 11:277, 1931.

BODY more cylindric. RADIAL SPINES 60-80. FLOWERS longer to 35 mm. and snow white.

Distribution: San Luis Potosi and Queretaro.

Type locality: None given.

FIG. 132. *Mammillaria schiedeana* x 3

108. **Mammillaria schiedeana** Ehrenberg

Mammillaria schiedeana Ehrenberg, Allg. Gartenz., 6:275, 1838.
Mammillaria sericata Lemaire, Cact. Gen. Nov. Sp., 44, 1839.
Cactus shiedeanus Kuntze, Rev. Gen. Pl., 1:261, 1891.
Mamillaria dumetorum J. A. Purpus, Monatsschr. Kakteenk., 22:149, 1912.
Neomammillaria schiedeana Britton & Rose, The Cactaceae, 4:128, 1923.
Chilita scheideana Orcutt, Cactography, 2, 1926.

BODY simple and cespitose, globose to elongate, with somewhat sunken apex, 25-100 mm. high, 25-40 mm. wide. TUBERCLES arranged in 13 and 21 spirals, soft in texture, dark green, slender conic to cylindric, terete, with watery sap, 6-10 mm. long, 3-4 mm. wide at base. AXILS with white wool longer than the tubercles. CENTRAL SPINES none. RADIAL SPINES very numerous to about 75, in more than 1 series, 2-5 mm. long, very slender acicular, straight, pubescent, white at base, to yellow, to golden yellow at tip, horizontal to depressed. FLOWERS wide funnelform, to 20 mm. long, 15 mm. wide. *Outer perianth-segments* greenish white, darker ventral-stripe, lanceolate, tip acute, margins serrate. *Inner perianth-segments* white to yellowish white, clavate, tip acute, margins entire with split ends. *Filaments* white. *Anthers* bright canary-yellow. *Style* white to cream. *Stigma-lobes* 4, bright yellow, short obtuse. FRUIT bright carmine, elongate globular, 23x12 mm., with dried perianth persisting. SEEDS glossy black, elongate globular, slightly curved, constricted and truncate at base, with hilum basal and extending slightly onto lateral, 1.3x0.8 mm.

Distribution: Hidalgo and central Mexico.

Type locality: None given, but reported from Puente de Dios, Metztitlan.

Illustration: Fig. 132 is a reproduction of an illustration in Borg (Cacti, pl. XLI) as *Mammillaria schiedeana*.

Britton & Rose (Cact., 4:128) report the presence of 6-10 central spines but they must have confused their description with some other species as the original description calls for only radial spines and the plants that we obtained in Mexico coincide with it.

The specific name is being retained under *nomen conservanda* because the plant is so well known under this name even though this principle is not usually applied to species. The name was first only listed as a Hort. synonym by Pfeiffer (Enum. Cact., 14, 1837) of *M. magnimamma* and was subsequently listed by various authors under the synonymy of the later species even though it was not described.

FIG. 133. *Mammillaria lasiacantha* x 1

109. **Mammillaria lasiacantha** Engelmann

Mamillaria lasiacantha Engelmann, Pro. Amer. Acad., 3:261, 1846.
Mamillaria lasiacantha minor Engelmann, Cact. Mex. Bound., 5, 1859.
Cactus lasiacanthus Kuntze, Rev. Gen. Pl., 1:259, 1891.
Neomammillaria lasiacantha Britton & Rose, The Cactaceae, 4:128, 1923.
Chilita lasiacantha Orcutt, Cactography, 2, 1926.

BODY simple and cespitose, globose to oval globose, rounded above, 18-25 mm. wide. TUBERCLES arranged in 8 and 13 spirals, medium gray-green, cylindric, terete, with watery sap, 4 mm. long, 2 mm. wide at base. AREOLES round to oval, with white wool in youth. AXILS naked. CENTRAL SPINES none. RADIAL SPINES 40-60, in more than 1 series, 3-4 mm. long, upper ones longer, inner series shorter, all setaceous, straight or slightly recurved, pubescent, white, horizontal. FLOWERS funnelform, lateral, 12 mm. long and wide. April and May. *Outer perianth-segments* white, brownish lavender mid-stripe, linear-lanceolate, tip obtuse to acute, margins serrate. *Inner perianth-segments* white, reddish purple mid-stripe extending to base where it forms a red circle in the throat, oblong, tip obtuse to emarginate, margins entire. *Filaments* yellow to pale lavender. *Anthers* yellow. *Style* pale greenish yellow. *Stigma-lobes* 4-5, yellowish green, short, oval, erect. FRUIT scarlet, 10-20 mm. long, clavate, with dried perianth persisting, and dried fruit remaining sunken in axils for some time, 10-25 seeds per pod. SEEDS black, oval to globose with oval basal hilum, slightly pitted, 1x0.8 mm.

Distribution: W. Texas, SE. New Mexico, U.S.; N. Chihuahua, Mexico.

Type locality: Leon Spring and Camanche Spring, west of Pecos, also reported from Carlsbad, N. M.

Illustration: Fig. 133 is from a photograph of a plant we collected near Carlsbad, New Mexico.

The credit for the original description is given as "Proceedings of the American Academy of Arts & Sciences" but this contained only a short synopsis and referred to the full description which appeared in the "Cactaceae of the Mexican Boundary" which was published at a later date. The latter publication was in reality prepared first but the publication was delayed because of the work involved in the preparation of the illustration plates.

The original description is supplemented with data from Engelmann's other works and from plants that we collected in the limestone hills of south-eastern New Mexico in the vicinity of Carlsbad and in western Texas. At one time it was very abundant but it is becoming harder to find because of indiscriminate collecting in the past. It does not do well in cultivation but the addition of lime rock to the soil helps remind the plant of its native habitat.

a var. **denudata** Engelmann

Mamillaria lasiacantha denuata Engelmann, Cact. Mex. Bound., 2, Pt. 1, 5, 1859.
Cactus lasiacanthus denudatus Coulter, Contr. U.S. Nat. Herb., 3:100, 1894.
Mamillaria lasiandra denudata Quehl, Monatsschr. Kakteenk., 19:79, 1909.
Neomammillaria denudata Britton & Rose, The Cactaceae, 4:129, 1923.
Chilita denudata Orcutt, Cactography, 2, 1926.
Mamillaria denudata Berger, Kakteen, 288, 1929.
Mammillaria lengdobleriana Boedeker, Jahrbuch Deut. Kakt. Ges., 1:68, 1935-6.

BODY larger. TUBERCLES in 13 and 21 spirals. RADIAL SPINES 50-80, longer, not pubescent.

Distribution and type locality: Same as type.

This variety is treated here as a variety of *M. lasiacantha* as originally described, because the body size is such a variable characteristic. The pubescence of the spines is not a constant characteristic as there are varying degrees of it on the same plant.

"*Mammillaria rungei* Schumann, (Gesamtb. Kakteen, 522, 1898) an unpublished garden name, was supposed by Schumann to be referrable to *M. lasiacantha denudata*." Britton & Rose (Cact. 4:129, 1923).

Mammillaria scheideana v. *denudata* Haage, (Cact. Catl., 138, 1900) is only a name without description, but may be referrable here.

110. Mammillaria criniformis DeCandolle

Mammillaria criniformis DeCandolle, Mem. Cact., 8, 1834.
Mammillaria criniformis rosea DeCandolle, Mem. Cact., 8, 1834.
Mammillaria criniformis albida DeCandolle, Mem. Cact., 8, 1834.
?*Mammillaria crinita rubra* Forbes, Journ. Hort. Tours, 151, 1837.

BODY cespitose from base. TUBERCLES soft in texture, glossy light green, ovate to oblong, with watery sap, 6-8 mm. long, 3 mm. wide at base. AREOLES with short white wool. AXILS nearly naked. CENTRAL SPINES 1, hooked, rigid, yellow to rose, porrect. RADIAL SPINES 8-10, 8 mm. long, setaceous, pubescent, white, horizontal. FLOWERS tubular, 14-16 mm. long. *Outer perianth-segments* greenish margins, linear-lanceolate. *Filaments* rose-red. *Anthers* whitish. *Style* light rose-red. *Stigma-lobes* 4, whitish, open. FRUIT unknown. SEEDS unknown but probably black.

Distribution: Hidalgo, Mexico.

Type locality: None given but reported from near Zimapan (Schmoll).

Illustration: DeCandolle, Mem. Cact. 8, 1834.

This species has usually been referred to *M. glochidiata* but it should be excluded from there because of the fewer number of both central and radial spines. This species was collected in Mexico by Coulter, but he reported no definite locality but Sr. Schmoll supplied us with the above distribution data.

M. cuneiformis (Loudon, Gard. Mag. 314, 1841) may probably be a misprint of the above specific name although it has been referred to *M. glochidiata*.

111. Mammillaria wildii Dietrich

Mammillaria wildii Allg. Gartenz., 4:137, 1836.
Mammillaria wildiana Otto in Pfeiffer, Enum. Cact., 37, 1837.
Mamillaria wildiana major Salm-Dyck, Hort. Dyck. 1844, 5, 1845.
Mamillaria wildiana rosea Salm-Dyck, Cact. Hort. Dyck. 1849, 7, 1850.
Mammillaria wildiana compacta Hort. in Förster, Handb. Cact. ed. 2, 258, 1885.
Mammillaria wildiana cristata Hort. in Förster, Handb. Cact. ed. 2, 258, 1885.
Cactus wildianus Kuntze, Rev. Gen. Pl., 1:261, 1891.
Neomammillaria wildii Britton & Rose, The Cactaceae, 4:143, 1923.
Chilita wildii Orcutt, Cactography, 2, 1926.
Mammillaria wildii cristata Hort. ex Schelle, Kakteen. 305, 1926.
Mammillaria wildii monstrosa Hort. ex Schelle, Kakteen. 305, 1926.
Mammillaria wildii compacta Hort. ex Schelle, Kakteen. 305, 1926.

BODY cespitose from base and body, cylindric, with rounded apex, 80-150 mm., long and 40-60 mm. wide. TUBERCLES widely separated in 13 and 21 spirals, dark bluish green, occasionally rose in axils, cylindric to clavate, terete, with watery sap, 6-10 mm. long, 3-6 mm. wide at base. AREOLES round, 1.5 mm. wide, with scant wool only in youngest. AXILS with 1 to several long hairs, more or less tortuous. CENTRAL SPINES 3-4, 8-10 mm. long, slender acicular, lower hooked, 3 upper ones straight, all pubescent, stronger than radials, pale yellow, nearly transparent, becoming muddy brown, lower porrect, uppers nearly horizontal. RADIAL SPINES 8-10, 6-8 mm. long, straight, pubescent, bristle-like, white, horizontal. FLOWERS funnelform, 12 mm. long and wide. *Outer perianth-segments* 6, pale green at base, whitish above, transparent margins, wide pale brownish mid-stripe, linear-lanceolate, tip acuminate, margins entire. *Inner perianth-segments* 6, white, transparent with faint greenish or brownish luster, linear-lanceolate, tip acute to acuminate, margins entire to slightly serrate. *Filaments* greenish below, white above. *Anthers* very pale yellow, globular. *Style* pale greenish white. *Stigma-lobes* 4-5, yellowish green. FRUIT brownish red, clavate with slightly sunken apex, with dried perianth persisting. SEEDS dull black, globular with exserted ventral hilum, pitted, 0.8 mm.

FIG. 134. *Mammillaria wildii* x 1

Distribution: Hidalgo and Queretaro, Mexico.

Type locality: None stated but reported from Venados, Hgo.

Illustration: Fig. 134 is from a photograph of a plant sent to us by Sr. F. Schmoll of Cadereyta, Qro.

Mammillaria crinita v. *pauciseta* DeCandolle (Mem. Mus. Hist. Nat. Paris, 112, 1828) may be referrable here.

Mammillaria wildiana v. *aurea* is only a name used in Monatsschr. Kakteenk., 157, 1891, which attributes the author as Salm-Dyck but it most likely refers to the variety *rosea* Salm-Dyck.

Mammillaria wildii rosiflora is only a hyb. hort. Steinick (Kakteenk., 6:106, 1936).

112. **Mammillaria aurihamata** Boedeker

Mamillaria aurihamata Boedeker, Zeitschr. Sukkulentenk., 3:341, 1928.

BODY simple and cespitose from body, globular to ovate, 60 mm. high, 40 mm. wide. TUBERCLES arranged in 8 and 13 spirals, leaf-green, short cylindric, keeled ventrally, terete at apex, with watery sap, 6 mm. long, 3 mm. wide. AREOLES roundish, 1 mm. wide, with considerable white wool in youth. AXILS with no wool, but 8 white fine tortuous hair-like bristles to length of tubercles. CENTRAL SPINES 4, upper three 10 mm. long, lower one 15-25 mm. long, all slender acicular, smooth to somewhat pubescent, lower ones hooked, 3 uppers nearly straight, all with base slightly enlarged, whitish yellow in youth, becoming golden yellow to brownish yellow, lower one porrect, upper ones nearly horizontal. RADIAL SPINES 15-20, 8 mm. long, nearly hair-like, semi-stiff, smooth, yellowish white, in upper part of the body strongly ascending, later horizontal. FLOWERS funnelform, 15 mm. long, 12 mm. wide. *Outer perianth-segments* bright olive green, darker mid-stripe, linear-lanceolate, tip acute, margins serrate, 10x2 mm. *Inner perianth-segments* bright sulphur-yellow, linear-lanceolate, tip acute, margins serrate

Fig. 135. *Mammillaria aurihamata*

but tips less so than outer segments, 13 mm. long. *Filaments* and *Style* whitish green. *Anthers* bright yellow. *Stigma-lobes* 3-4, short bright green. FRUIT red, clavate, small. SEEDS dark brown, to nearly black, oval with lateral hilum, pitted, 1 mm.

Distribution: San Luis Potosi and central Mexico.

Type locality: None given, but reported from Monte Gordo, Guanajuato.

Illustration: Fig. 135 is a reproduction of an illustration in Zeitsschr. Sukkulentenk. 3:340 as *Mamillaria aurihamata*.

113. **Mammillaria leucantha** Boedeker

Mammillaria leucantha Boedeker, Kakteenk., 233, 1933.

BODY simple or rarely branching, globular cylindric, with slightly sunken apex, 35 mm. wide. TUBERCLES widely separated in 8 and 13 spirals, dull dark leaf-green, cylindric, with watery sap, 7mm. long, 2 mm. wide at base. AREOLES round with wool in youth, later becoming naked. AXILS with short white wool, becoming naked, also white bristles 5 mm. long. CENTRAL SPINES 3-4, 5-6 mm. long, 3 hooked, the fourth when present is straight, all smooth, heavier than radials, dark amber-yellow, spreading porrect. RADIAL SPINES 18, 5-6 mm. long, straight, very slender, very weakly pubescent, white with hardly perceptible light brownish yellow tip, horizontal. FLOWERS slender funnelform, 15 mm. long, copious bloomer. *Outer perianth-segments* white, light olive-green mid-stripe, linear-lanceolate, margins serrate, 10x1 mm. *Inner perianth-segments* white, faint olive-green mid-line, green at base, linear-lanceolate, margins serrate, 10x2 mm. *Filaments* and *Style* bright green below, white above. *Anthers* light yellow. *Stigma-lobes* 4, yellowish white. FRUIT unknown. SEEDS dark grayish brown, dull, pyriform, pitted, 1 mm. ROOTS tuberous.

Distribution: San Luis Potosi, Mexico.

Type locality: Soledad Diez Gutierrez.

Fig. 136. *Mammillaria leucantha*

Illustration: Fig. 136 is a reproduction of an illustration in Kakteenk., 233, 1933, as *Mammillaria leucantha*.

Habitat: In fissures in almost perpendicular walls in the hills.

The name *leucacantha*, a *Coryphantha*, has been misspelled in the literature as *leucantha*. Britton & Rose (Cact. 4:31, 1923) refers both names under *Coryphantha octocantha*. The use of the name by Boedeker adds to the confusion.

Fig. 137. *Mammillaria capensis* x 0.5

114. Mammillaria capensis (Gates) comb. nov.

Neomammillaria capensis Gates, Cact. Succ. Journ., 4:372, 1933.

BODY cespitose mostly from base, cylindric, to 25 cm. high, 3-5 cm. wide. TUBERCLES closely set in 5 and 8 spirals, olive-green, ovoid cylindric, with watery sap, 4-5 mm. long, 5 mm. wide at base. AREOLES oval with scant cream wool. AXILS naked or with 1-3 short bristles. CENTRAL SPINES 1, 15-20 mm. long, acicular, stiff, strongly hooked, usually turned upward, white at base, shading through red-brown to black at tip, porrect. RADIAL SPINES 13, 8-15 mm. long, acicular, stiff, white at base, shading through red-brown to black tips, horizontal. FLOWERS lateral but near top, funnelform. *Outer perianth-segments* greenish pink mid-stripe, pale pink margins, oblong, tip obtuse, margins (?) *Inner perianth-segments* pale pink, darker mid-line, oblong, tip obtuse, margins entire. *Anthers* yellow. *Stigma-lobes* 6, elongate, greenish yellow. FRUIT orange to scarlet, 20 mm. long, clavate. SEEDS black, glossy, globular with extended ventral hilum, constricted above base, pitted, 1.3x1 mm. ROOTS fibrous.

Distribution: Baja California, Mexico.

Type locality: Puerto de Bahia de los Muertos (Cape district).

Illustration: Fig. 137 is from a photograph by Mr. Howard Gates.

Mammillaria cochemoides mentioned in Cactus Journal (British) 18, 1936-37, and also listed in Backeberg's catalogue is referrable here according to Mr. E. Shurly.

FIG. 138. *Mammillaria erythrosperma*

115. Mammillaria erythrosperma Boedeker

Mamillaria erythrosperma Boedeker, Monatsschr. Kakteenk., 28:101, 1918.
Chilita eschauzieri Orcutt, Cactography, 2, 1926.

BODY simple and cespitose from base and body, with rounded apex, globose, 50 mm. high, 40 mm. wide. TUBERCLES closely set in 8 and 13 spirals, glossy dark green, white spotted, cylindric, rounded tip, bent ventrally, with watery sap, 10 mm. long, 5 mm. wide at base. AREOLES round, 1 mm. wide, with yellow mat but not woolly. AXILS with no wool, but thin tortuous hair-like bristles. CENTRAL SPINES 1-3, seldom 4, 10 mm. long, pubescent, stiff, lower hooked, upper ones (when present) straight, all acicular, base thickened, yellow, tip reddish brown, lower one porrect, upper ones nearly horizontal. RADIAL SPINES 15-20, 8-10 mm. long, very thin, in youth pubescent but later smooth, straight, base not enlarged, white with yellowish base, horizontal to appressed. FLOWERS funnelform, 15 mm. long and wide. *Outer perianth-segments* green below, brownish carmine-red above, darker mid-stripe, linear-lanceolate, tip acute, margins (?), 10x2 mm. *Inner perianth-segments* carmine-red, lighter margins, linear-lanceolate, tip acuminate, margins (?), 15x2 mm. *Filaments* light carmine-red, darker above. *Anthers* bright yellow. *Style* light carmine-red. *Stigma-lobes* 3-4, carmine-red. FRUIT carmine-red, clavate, 20 mm. long. SEEDS dark red to almost black, glossy, curved pyriform, pitted.

Distribution: San Luis Potosi, Mexico.

Type locality: None given but reported from Alvarez, S.L.P.

Illustration: Fig. 138 is a reproduction of an illustration in Monatsschr. Deutsch. Kakt. Ges., 195, 1931, as *Mamillaria erythrosperma*.

Britton & Rose (Cact., 4:151) refer this species to *N. painteri* but we believe that it deserves specific rank: 1—This species has bristles in axils vs. none in *M. painteri*, 2—Stigma-lobes carmine-red vs. pale greenish cream, 3—Inner perianth-segments carmine-red vs. greenish white, 4—Central spines 1-3 vs. 4-5, 5—Radial spines 15-20 vs. over 20.

a. var. **similis** DeLaet.

Mamillaria erythrosperma var. *similis* DeLaet in Boedeker, Monatsschr. Kakteenk., 28:102, 1918.

BODY smaller. CENTRAL SPINES often yellow. RADIAL SPINES shorter, pubescent. FLOWERS lighter, stigma-lobes yellow.

This variety has been referred to *M. multiformis* but the latter has 30 radial spines. Further investigations at the type locality will be necessary to clearify this matter.

Possibly referrable to this variety is the following uncertain species:

Mammillaria monancistra Berg

Mamillaria monancistra Hort. ex Salm-Dyck, Cat. Hort. Dyck., 1849, 7, 1850. Nomen as syn. of
  M. wildiana rosea.
Mamillaria monancistra Berg in Schumann, Gesamtb. Kakteen, 533, 1898.

BODY cespitose, globose to cylindric. TUBERCLES arranged in 8 and 13 spirals, light green, later becoming darker, cylindric. AREOLES with scant yellowish wool. AXILS with bristles. CENTRAL SPINES 1-4, pubescent, 1-2 hooked. RADIAL SPINES 9-11, 10 mm. long, setaceous, white. FLOWERS funnelform, 12-13 mm. long. *Outer perianth-segments* rose, darker mid-stripe. *Inner perianth-segments* lighter rose, tip acuminate, margins serrate. *Filaments* white below, carmine-red above. *Anthers* yellow. *Style* white below, bright rose-red above. *Stigma-lobes* 3, yellow.

Distribution: Mexico.

Type locality: None stated.

Because of the incomplete data, it is uncertain as to the exact relationship of this species but it appears to be associated here.

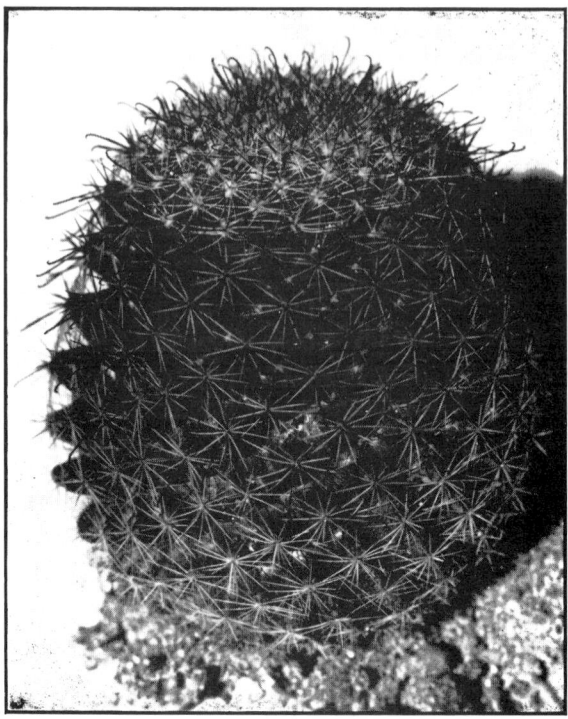

Fig. 139. *Mammillaria balsasoides* x 1

116. **Mammillaria balsasoides** sp. nov.

Corpus simplex et cespitosus, mamillis textis mollis, viribibus fuscis, conoidis, suco aquario, axillis 2-6 setibus; spinis centralibus 4, 4-9 mm., 3 superioribus rectis, 1 inferioribus hamatis aliquando; spinis radialibus 10-11, 4-6 mm., acicularibus tenuis, albis; flores campanulata, 40 mm. latitudo, sepalis viridibus splendidis, integris, petalis aureis, 20x3 mm., stigmatibus 3, aureis, 3 mm., fructus et semina ignota.

BODY simple and cespitose from base, heads to 70 mm. wide. TUBERCLES arranged in 11 and 18 spirals, soft in texture, dark grayish green, becoming pale purplish red in direct sunlight, dull, lighter in axils, conic, blunt apex, with watery sap, 8 mm. long, 4 mm. wide at base. AREOLES round with practically no wool. AXILS with 2-6 white bristles to length of tubercles, but no wool. CENTRAL SPINES 4, 4-9 mm. long, lower ones longer and hooked, 3 upper ones straight, all acicular, smooth, semi-stiff, base slightly enlarged, dark reddish brown, top nearly black, spreading, upper ones nearly horizontal, lower ones nearly porrect. RADIAL SPINES 10-11, 4-6 mm. long, slender acicular, straight, smooth, white, horizontal. FLOWERS June, campanulate, 40 mm. wide. *Outer perianth-segments* bright green, lanceolate, tip obtuse, margins entire, to 3 mm. wide. *Inner perianth-segments* orange, broad lanceolate, tip acute, margins entire, 20x3 mm. *Filaments* nearly white, very short as compared to other species, orange pink at top. *Anthers* deep yellow, each lobe partially divided lengthwise. *Style* pale cream. *Stigma-lobes* 3, orange, to 3 mm. long, each lobe strongly grooved ventrally. FRUIT and SEEDS unknown.

Distribution: Guerrero, Mexico.

Type locality: Near Rio Balsas along the highway between Taxco and Alcopulco.

Illustration: Fig. 139 is from a photograph of a plant from the type locality.

This species was collected by Dr. and Mrs. Craig on June 11, 1942, in the shade of

the trees on the mountain sides along the highway between the Balsas River and Chilpancingo. It is certainly one of the most striking species of this genus with its large orange flowers. It resembles *Mammillaria beneckei* Ehrenberg but differs from it:
Bristles in axils.
Radial spines 10-11 vs. 13-15.
Perianth-segments entire vs. serrate.
Outer perianth-segments bright green vs. violet-rose.
Stigma-lobes 3 vs. 5.

Type material has been deposited in the Dudley Herbarium at Stanford University, California.

Fig. 140. *Mammillaria fraileana* x 0.6

117. Mammillaria fraileana (B. & R.) Boedeker

Neomammillaria fraileana Britton & Rose, The Cactaceae, 4:157, 1923.
Chilita fraileana Orcutt, Cactography, 2, 1926.
Mammillaria fraileana Boedeker, Mammill. Vergl. Schluss., 30, 1933.

BODY simple and cespitose from base, cylindric. TUBERCLES arranged in 5 and 8 spirals, firm in texture, short conic, with watery sap, 5 mm. long, 3-4 mm. wide at base. AREOLES oval, with short dirty white wool, soon becoming naked. AXILS with 1 to several bristles. CENTRAL SPINES 3-4, 10 mm. long, upper shorter, thinner and straight, lower hooked, all acicular, stiff, dark brown, upper nearly horizontal, lower porrect. RADIAL SPINES 11-12, 8-10 mm. long, thin acicular, straight, stiff, white becoming chalky, in new growth reddish brown, nearly horizontal. FLOWERS funnelform, 20 mm. wide. *Outer perianth-segments* unknown. *Inner perianth-segments* pinkish, tip acuminate and often lacerate, 20-25 mm. long. *Filaments* pinkish. *Anthers* yellow. *Style* pinkish. *Stigma-lobes* 6-7, long slender. FRUIT lilac-pink, clavate, with dried perianth persisting. SEEDS black, globular with extending ventral hilum, constricted above the base.

Distribution: Pichilinque, Cerralboa, Catalina Islands, Baja California; Mexico.
Type locality: Pichilinque Island.
Illustration: Fig. 140 is a reproduction of an illustration in Blatter Kakteen., Backeberg, 4, 1934, as *Mamillaria fraileana*.

FIG. 141. *Mammillaria verhaertiana* x 1

118. Mammillaria verhaertiana Boedeker

Mamillaria verhaertiana Boedeker, Monatsschr. Kakteenk., 22:152, 1912.
Neomammillaria verhaertiana Britton & Rose, The Cactaceae, 4:164, 1923.
Chilita verhaertiana Orcutt, Cactography, 2, 1926.

BODY simple, short cylindric, with apex rounded and little sunken, 90 mm. high, 40 mm. wide. TUBERCLES arranged in 8 and 13 spirals, firm in texture, bright leaf-green, short conic, keeled ventrally, with watery sap, 7-10 mm. long, 6-8 mm. wide at base. AREOLES round to oval to 3 mm. wide, with abundant short whitish wool, later becoming naked. AXILS with white wool and 15-20 snow white bristles. CENTRAL SPINES 4-6, to 12 mm. long, upper ones straight, lower one hooked, all heavy acicular, yellowish white, tip yellowish brown, wide spreading. RADIAL SPINES 15-20, upper ones 8 mm. long, lower ones to 16 mm. long, all straight, thin acicular, dark yellowish brown in new growth, soon yellowish white, horizontal, directed more laterally. FLOWERS funnelform, lateral, 20 mm. long and wide. *Outer perianth-segments* yellowish white margins, olive-green mid-stripe, darker ventrally, wide lanceolate to oblong, tip acute to obtuse, margins serrate. *Inner perianth-segments* wide whiter margins, pale pink mid-stripe, wide lanceolate to clavate, tip obtuse, margins entire. *Filaments* rose to whitish. *Anthers* golden yellow. *Style* pale pink above. *Stigma-lobes* 8-9, orange-yellow. FRUIT orange-red, clavate with dried perianth persisting. SEEDS black, pitted.

Distribution: Baja California, Mexico.

Type locality: None stated but reported from Los Angeles Bay.

Illustration: Fig. 141 is from a photograph of a plant collected for us by Mr. George Lindsay at Los Angeles Bay.

Fig. 142. *Mammillaria dioica* x 0.7

119. **Mammillaria dioica** K. Brandegee

Mamillaria dioica K. Brandegee, Erythea, 5:115, 1897.
Mamillaria fordii Orcutt, West. Amer. Sci., 13:49, 1902.
Neomammillaria dioica Britton & Rose, The Cactaceae, 4:158, 1923.
Chilita fordii Orcutt, Cactography, 2, 1926.
Mammillaria incerta Parish in Jepson, Flora Calif., 2:549, 1936.

BODY simple and cespitose from base, cylindric, with rounded apex, to as much as 33 cm. high, to 10 cm. wide but mostly smaller. TUBERCLES arranged in 8 and 13 spirals, firm in texture, bluish green, short cylindric, somewhat 4-sided at base, with watery sap, 4-6 mm. long, 3-4 mm. wide at base. AREOLES round with wool in youth, soon becoming naked. AXILS with sparse wool and 5-15 bristles as long as the tubercles. CENTRAL SPINES 1-4, 8-15 mm. long, lower longest and hooked, all stout acicular, smooth, stiff, brownish to brownish black, uppers nearly like radials, lowers more porrect. RADIAL SPINES 11-22 (usually nearer 15), 5-7 mm. long, acicular, straight, smooth, stiff, usually white, sometimes rose, often brownish to black at tip, horizontal. FLOWERS funnelform, incompletely dioecious, 10-30 mm. long. July. *Outer perianth-segments* reddish to purple mid-stripe, very narrow yellow margins, lanceolate, tip acute to acuminate, margins entire, much longer and spreading in the male flower. *Inner perianth-segments* cream, fine purple mid-stripe, which is darker and wider ventrally, lanceolate to oblong, tip acute and sometimes split, margins entire. *Filaments* white. *Anthers* deep yellow. *Style* white to pale yellow. *Stigma-lobes* 5-6, light green (yellowish to brownish green). FRUIT scarlet, oval to clavate, 10-25 mm. long, with dried perianth persisting. SEEDS black, pyriform with lateral hilum, minutely pitted.

Distribution: Southern California, U.S.; Baja California, Mexico.

Type locality: None given but common around San Diego, California, eastward to the desert, and southward along the coast to Cape San Lucas at the southern tip of Baja California.

Illustration: Fig. 142 is from a photograph of a plant we collected near San Diego, California.

This species is extremely variable in body size and coloration of the spines. The plants from the coastal area around San Diego are mostly small globose and often cespitose. Those we collected in the Borego Canyon on the desert side of the mountains were more prone to be simple and very columnar, often to over 30 cm. high. As one proceeds down the peninsula of Baja California, many apparently different forms are observed but when the salient characters are all checked, it is found that they are only slight variations of differences in spine coloration but they are all readily referrable to the type species.

Mrs. K. Brandegee points out the interesting feature of this species is that there are two types of flowers as the name implies, they are dioecious or unisexual (but incompletely so or not always). The male flowers are brighter colored and larger, while the female ones have narrower petals.

The form described by Parish as *M. incerta* is based on those plants from the desert side of the mountains that have whiter flowers with a reddish brown mid-stripe.

FIG. 143. *Mammillaria armillata* x 0.8

120. **Mammillaria armillata** K. Brandegee

Mamillaria armillata K. Brandegee, Zoe, 5:7, 1900.
Neomammillaria armillata Britton & Rose, The Cactaceae, 4:157, 1923.
Chilita armillata Orcutt, Cactography, 2, 1926.
Neomammillaria lapacena Gates, Desert Plant Life, 3:117, 1932.

BODY simple and cespitose from base and body, cylindric, to 30 cm. high, 4-5 cm. wide. TUBERCLES arranged in 5 and 8 spirals, leathery in texture, bluish green, 4-sided at base, conic to cylindric above, with watery sap, 5 mm. long, 3-4 mm. wide at base. AREOLES oval, with short dirty white wool, soon becoming naked. AXILS with scant wool and 1-3 bristles. CENTRAL SPINES 1-4, 10-20 mm. long, lower ones longer and hooked, upper ones straight, all stout acicular, smooth, yellowish gray to dark brown, upper ones ascending to horizontal, lower ones porrect. RADIAL SPINES 9-15, 7-12 mm. long, slender acicular, straight, smooth, stiff, white to yellow below, dark brown at tip, somewhat ascending. FLOWERS funnelform, 10-20 mm. long, 20 mm. wide. *Outer perianth-segments* greenish, lanceolate, tip acute, margins serrate. *Inner perianth-segments* white to pink, lanceolate, tip acute, margins serrate. *Filaments* pinkish. *Anthers* yellow. *Style* pinkish. *Stigma-lobes* 6-7, rose. FRUIT red, clavate, 15-30 mm. long, with dried perianth persisting. SEEDS black, dull, oblique obovate, constricted above slender basal hilum, very finely pitted and wrinkled.

Distribution: Baja California, Mexico.

Type locality: San Jose del Cabo.

Illustration: Fig. 143 is a reproduction of an illustration in Monatsschr. Deutsch. Kakt. Ges., 120, 1932, as *Mammillaria armillata*.

Circling the plant are alternating bands of lighter and darker colored spines.

We are referring the name *Neomammillaria lapacena* here upon the advice of Mr. H. Gates who collected and briefly described the plant.

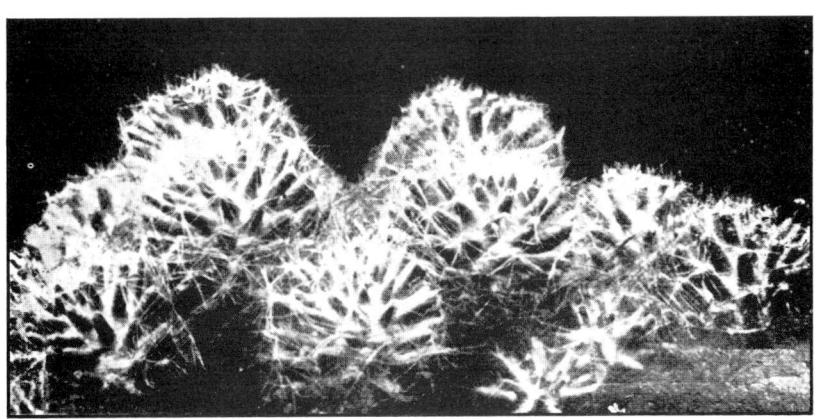

Fig. 144. *Mammillaria glochidiata* x 1

121. Mammillaria glochidiata Martius

Mammillaria glochidiata Martius, Nov. Act. Nat. Cur., 16:337, 1832.
Cactus glochidiatus Kuntze, Rev. Gen. Pl., 1:260, 1891.
Mamillaria glochidiata prolifera Schumann, Gesamtb. Kakteen, 532, 1898.
Neomammillaria glochidiata Britton & Rose, The Cactaceae, 4:149, 1923.
Chilita glochidiata Orcutt, Cactography, 2, 1926.

BODY cespitose from base, cylindric, rounded above, 20-35 mm. wide. TUBERCLES widely separated in 5 and 8 spirals, flabby in texture, glossy bright green, cylindric to slender conic or sometimes clavate, terete, with watery sap, 12-16 mm. long, 4-6 mm. wide at base. AREOLES nearly round, with scant white wool only in youth. AXILS with 1-5 fine bristles. CENTRAL SPINES 3-4, 6-12 mm. long, 3 upper ones straight, lower one hooked, all lightly pubescent, acicular, dark yellow to reddish brown, upper ones nearly horizontal, lower one porrect. RADIAL SPINES 12-15, to 12 mm. long, setaceous, smooth, flexuous, white, horizontal. FLOWERS campanulate, 15 mm. long, 12 mm. wide. *Outer perianth-segments* green at base, greenish red to yellow mid-stripe above, white margins, linear, tip acute and sometimes split, margins entire. *Inner perianth-segments* tan to rose-red mid-stripe, white margins, lanceolate, tip acuminate, margins entire. *Filaments* white to pale yellow. *Anthers* yellow, nearly globose. *Style* white to pale green. *Stigma-lobes* 4-5, pale yellowish green. FRUIT scarlet, clavate, 16 mm. long, with dried perianth persisting. SEEDS black.

Distribution: Hidalgo, Mexico.

Type locality: None stated but reported from San Pedro Nolasco, Zimapan and Ixmiquilpan.

Illustration: Fig. 144 is a reproduction of an illustration in Berger (Kakteen 294, 1929) as *Mamillaria glochidiata*.

Mammillaria cuneiformis is an unpublished name referred to this species by Loudon Gardeners Mag., 314, 1841, or it may be a misprint of *M. criniformis*.

Fig. 145. *Mammillaria pygmaea* x 1

122. **Mammillaria pygmaea** (B. & R.) Berger

Neomammillaria pygmaea Britton & Rose, The Cactaceae, 4:142, 1923.
Chilita eschauzieri Orcutt, Cactography, 2, 1926.
Mammillaria pygmaea Berger, Kakteen, 296, 1929.

BODY simple and cespitose from base, globose to cylindric, 2-3 cm. wide. TUBERCLES arranged in 13 and 21 spirals, dark purplish dull green, soft in texture, cylindric, tip obtuse, terete, with watery sap, 6-10 mm. long, 2-4 mm. wide at the base. AREOLES round, with scant wool only in youngest. AXILS with long tortuous very fine white bristles. CENTRAL SPINES 4, 5-8 mm. long, fine acicular, lower one hooked, others straight, all smooth, golden yellow to somewhat brownish yellow, lower nearly porrect, uppers ascending. RADIAL SPINES to 15, to 12 mm. long, fine hair-like, hardly pubescent, straight, white, horizontal. FLOWERS 10-12 mm. long. *Outer perianth-segments* tan, reddish tan mid-stripe, linear-lanceolate, tip acute, margins entire. *Inner perianth-segments* cream, tan mid-stripe, linear, tip acute, margins entire. *Filaments* pale greenish yellow. *Stigma-lobes* 4, whitish green to greenish yellow. FRUIT reddish, curved clavate, with dried perianth persisting. SEEDS dull black, curved pyriform with oblong sloping hilum, finely pitted in grooves, 0.5 mm. long.

Distribution: Queretaro, Mexico.

Type locality: Cadereyta.

Illustration: Fig. 145 is from a photograph of a plant we obtained near Cadereyta in 1942.

FIG. 146. *Mammillaria angelensis* x 1

123. **Mammillaria angelensis** sp. nov.

Corpus simplex, cylindratus; mamillis ad 8 et 13 seriebus ordinatis, conoidis, suco aquario; areolis orbicularibus, axillis lanatis albis et setis; spinis centralibus 3-5, 8-14 mm., inferioribus hamatis et porrectis, superioribus rectis, acicularibus; spinis radialibus 16, 5-10 mm. acicularibus; flores infundibuliformes, sepalis alboflavis, ciliatis brevis, petalis, albis, stigmatibus 5, flavaviridis 3 mm. longis.

BODY simple, columnar, 15 cm. high, 6 cm. wide. TUBERCLES arranged in 8 and 13 spirals, light yellowish olive-green, conical, not angled, keeled ventrally, with watery sap, 9 mm. long, 7 mm. wide at base. AREOLES oval, with white wool persisting. AXILS with abundant white wool in flowering and fruiting area, later becoming less, also 15 or more stiff and somewhat tortuous white bristles, 10 mm. long. CENTRAL SPINES 3-4, 8-14 mm. long, lower hooked and longer, all acicular, smooth, base not enlarged, dull purplish brown, lighter to tannish cream at base, hooked porrect, straight dorsally spreading and nearly in the same plain as radials. RADIAL SPINES 16, 5-10 mm. long, lower ones longer, all acicular, straight, stiff, smooth, white, in some tan at very tip, in youth ascending, later nearly horizontal and mostly lateral. FLOWERS campanulate, 20 mm. long, 30 mm. wide. *Outer perianth-segments* creamish white, pink mid-stripe at apex, linear, tip obtuse, margins short ciliate. *Inner perianth-segments* widely separated, white, linear, 3 mm. wide, tip acute, margins entire. *Filaments* white. *Anthers* very pale yellow. *Style* white. *Stigma-lobes* 4-5 (mostly 5), yellowish olive-green, 3 mm. long, overtop anthers 4 mm. FRUIT reddish, clavate. SEEDS black, pitted.

Distribution: Baja California.

Type locality: Angel de la Guardia Island in Los Angeles Bay.

Illustration: Fig. 146 is from a photograph of a plant collected for us by Mr. George Lindsay at the type locality in 1936.

Type material has been deposited in the Dudley Herbarium at Stanford University, California.

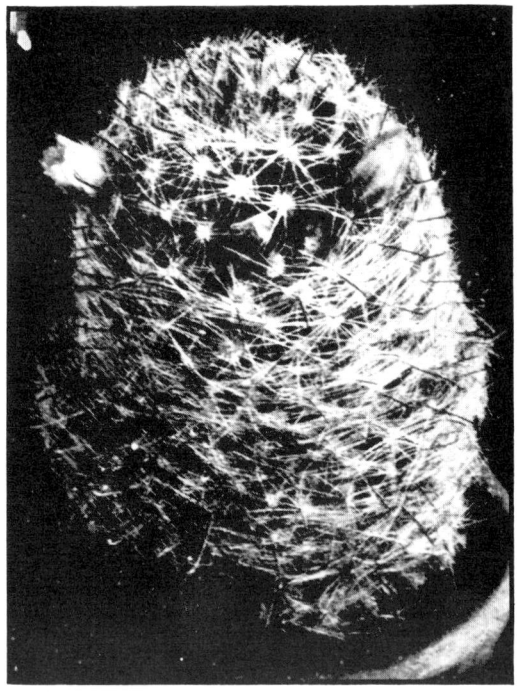

FIG. 147. *Mammillaria icamolensis*

124. **Mammillaria icamolensis** Boedeker

Mammillaria icamolensis Boedeker, Kakteenk., 168, 1933.

BODY simple, short cylindric, with rounded apex, 6 cm. high, 4 cm. wide. TUBERCLES moderately closely set in 13 and 21 spirals, leaf-green, cylindric, rounded apex, with watery sap, 7-8 mm. long, 2 mm. wide at base. AREOLES round, 1 mm. wide, with some white wool in youth, later becoming naked. AXILS with few short hairs but no wool. CENTRAL SPINES 4, 7 mm. long, lower hooked, upper straight, all smooth, acicular, uppers reddish brown with lighter base, lower more brownish, uppers fan-like horizontal, lower porrect. RADIAL SPINES 16-20, 5-7 mm. long, thin acicular to hair-like, smooth, white, horizontal. FLOWERS wide funnelform, 12 mm. wide, near top. *Outer perianth-segments* pale rose with lighter margins, linear-lanceolate, tip acute, margins entire. *Inner perianth-segments* pale rose-white, pale rose mid-stripe, slender oblong, tip obtuse, margins entire. *Filaments* green below, rose above. *Anthers* bright yellow. *Style* bright green. *Stigma-lobes* 5, white, overtop anthers. FRUIT clavate, small. SEEDS olive-gray (?), 1 mm. long, pyriform with white lateral hilum, pitted.

Distribution: Nuevo Leon, Mexico.

Type locality: Icamol and Monterrey.

Illustration: Fig. 147 is a reproduction of an illustration in Kakteenk., 168, 1933, as *Mammillaria icamolensis*.

The color of the seeds as reported by Boedeker is rather unusual and probably open to question. He might have had immature seeds or sometimes very old seeds will appear grayish until they are washed and the dried pulp removed.

FIG. 148. *Mammillaria pubispina* x 1

125. Mammillaria pubispina Boedeker

Mamillaria pubispina Boedeker, Monatsschr. Deutsch. Kakt. Ges., 61, 1930.

BODY simple, globular, with very slightly depressed apex, to 40 mm. wide. TUBERCLES arranged in 13 and 21 spirals, soft in texture, dull leaf-green to rose in axils, cylindric, with watery sap, 8 mm. long, 2 mm. wide at base. AREOLES roundish, slightly depressed, 1 mm. wide, with white wool in youth, soon becoming naked. AXILS with slight wool and few slender hair-like tortuous bristles. CENTRAL SPINES 4, seldom 3, 9-10 mm. long, lower one hooked, upper ones slightly longer and straight, all very thin acicular, lightly pubescent, red to blackish brown, becoming lighter, lower porrect, uppers spreading ascending, do not overlock apex. RADIAL SPINES 15, 8-12 mm. long, thin hair-like, pubescent, more or less bent and tortuous, white, horizontal to slightly ascending, interlacing. FLOWERS wide funnelform, 18 mm. long, 15 mm. wide. *Outer perianth-segments* dirty rose, margins whitish, linear-lanceolate, tip acute, margins serrate, 8x1.5 mm. *Inner perianth-segments* pure white to cream, delicate rose mid-stripe, linear-lanceolate, tip short acuminate, margins serrate. *Filaments* deep rose. *Anthers* bright yellow. *Style* white. *Stigma-lobes* 5, whitish. FRUIT unknown. SEEDS unknown but probably black.

Distribution: Hidalgo, Mexico.

Type locality: Ixmiquilpan in lava rocks.

Illustration: Fig. 148 is a reproduction of an illustration in Monatsschr. Deutsch. Kakt. Ges. 62, 1930, as *Mamillaria pubispina*.

FIG. 149. *Mammillaria occidentalis*

126. **Mammillaria occidentalis** (B. & R.) Boedeker

Neomammillaria occidentalis Britton & Rose, The Cactaceae, 4:161, 1923.
Chilita occidentalis Orcutt, Cactography, 2, 1926.
Mammillaria occidentalis Boedeker, Mammill. Vergl. Schluss., 36, 1933.

BODY cespitose from base, slender cylindric, to 15 cm. high, 2-3 cm. wide. TUBERCLES arranged in 5 and 8 spirals, firm in texture, conic, with watery sap, 5 mm. long, 3-4 mm. wide at base. AREOLES oval, with dirty white wool in youth, soon becoming naked. AXILS with no wool, but occasional scattered white bristles. CENTRAL SPINES 4-5, 5-12 mm. long, acicular, upper ones straight, lower one hooked or sometimes straight, all smooth, stiff, reddish brown, divergent porrect. RADIAL SPINES 12-18, 3-8 mm. long, slender acicular, straight, smooth, stiff, white to yellowish, at tip brown, slightly ascending. FLOWERS tubular, 10 mm. long. *Outer perianth-segments* brownish pink mid-stripe, pink margins, lanceolate, tip acuminate, margins serrate. *Inner perianth-segments* pink to deep pink, darker mid-line, linear-lancolate, tip acuminate, margins mostly entire. *Filaments* reddish. *Anthers* yellow. *Stigma-lobes* 8-9, olive-green. FRUIT reddish, clavate. SEEDS black, glossy.

Distribution: Colima, Nayarit, Sinaloa; Mexico.

Type locality: Manzanillo, Colima.

Illustration: Fig. 149 is a reproduction of an illustration, Fig. 179, in Britton & Rose, (Cact. 4:160) as *Neomammillaria occidentalis*. Fig. 150 is from a photograph of a plant collected by Mr. E. Baxter near Mazatlan, Sinaloa, in the spring of 1940.

a. var. **patonii** (Bravo) (Boedeker) comb. nov.

Neomammillaria patonii Bravo, Anal. Inst. Biol. Mex., 2:129, 1931.
Mammillaria patonii Boedeker, Mammill. Vergl. Schluss., 33, 1933.

BODY little more robust, CENTRAL SPINES 4, 12 mm. long, 1 lightly hooked. RADIAL SPINES 13-15, 5-8 mm. long. FLOWERS purple.

Distribution: Nayarit, Mexico.

Type locality: Tres Marias Islands.

Illustration: Anal. Inst. Biol, Mex., 2:129 as *Neomammillaria patonii*.

b. var. **sinalensis** var. nov.

Spinis centralibus 1, 8-10 mm., hamatis; spinis radialibus 10-12, 5-8 mm.; flores ignotes.

CENTRAL SPINES 1, 8-10 mm. long, strongly hooked, porrect. RADIAL SPINES 10-12, 5-7 mm. long. FLOWERS unknown.

Distribution: Sinaloa, Mexico.

Type locality: Arroyo de Ibarra, near Rosario.

Illustration: Fig. 151 is from a photograph of a plant collected by Mr. E. Baxter in the spring of 1940.

This may be a distinct species but we are referring it here until more data can be obtained on the flower.

FIG. 150. *Mammillaria occidentalis*   FIG. 151. *Mammillaria occidentalis* var. *sinalensis* x 1

Fig. 152. *Mammillaria swinglei* x 1

127. **Mammillaria swinglei** (B. & R.) Boedeker

Neomammillaria swinglei Britton & Rose, The Cactaceae, 4:158, 1923.
Chilita swinglei Orcutt, Cactography, 2, 1926.
Mammillaria swinglei Boedeker, Mammill. Vergl. Schluss., 33, 1933.

BODY simple and cespitose from base, cylindric, 10-20 cm. high, 3-5 cm. wide. TUBERCLES arranged in 8 and 13 spirals, firm in texture, dark green, cylindric to conic, with watery sap, 8-10 mm. long, 5-7 mm. wide at base. AREOLES oval, with scant wool only in youngest. AXILS with more or less bristles, irregular or sometimes not present. CENTRAL SPINES 1-4, 8-15 mm. long, lower longest, all stout acicular to subulate, smooth, stiff, lower hooked or sometimes straight, uppers straight, all dark brown to black, lower porrect, uppers subcentral in same plain as the radials. RADIAL SPINES 11-18, 7-14 mm. long (mostly 8-9 mm.), rather stout acicular, straight, stiff, smooth, dull white with dark tips, horizontal. FLOWERS campanulate, to 3 cm. wide, near the top. *Outer perianth-segments* purplish brownish green mid-stripe, margins white to greenish, linear-lanceolate, tip obtuse, margins mostly serrate. *Inner perianth-segments* pink mid-stripe (sometimes greenish to tannish), white to greenish cream margins, linear-lanceolate to eliptical, tip acute, margins entire. *Filaments* pink. *Anthers* orange. *Style* yellow to sometimes pink at top. *Stigma-lobes* 7-9, yellowish green, overtop anthers. FRUIT dark red, clavate, 14-18 mm. long, with dried perianth persisting. SEEDS black, glossy, globular with basal hilum, constricted above the base, pitted, 1 mm. ROOTS fibrous.

Distribution: Sonora, Mexico.

Type locality: Guaymas and along the coast.

Illustration: Fig. 152 is from a photograph of a plant we collected at Guaymas in 1935.

The original description is supplemented with data from plants that we collected at the type locality. This species, which is somewhat common along the coast in the vicinity of Guaymas, is quite variable in some of the characteristics as given in the original description. In the 14 specimens that we collected, only 5 showed bristles in the axils, central spines 1-4, B. & R. give 4. Of the 5 specimens that have bloomed so far, the outer perianth-segments were 1 entire, 1 ciliate, 3 serrate; the inner perianth-segments were all entire, color varies from 2 pink, 1 white (B. & R.), 1 greenish, 1 cream. *Stigma lobes* were: 3 had 7, 1 had 8 (B. & R.), 1 had 9. Style was: 2 pink (B. & R.), 3 yellow.

The name *Neomammillaria swingleri* is just a typographical error.

FIG. 153. *Mammillaria trichacantha* x 1

128. Mammillaria trichacantha Schumann

Mamillaria trichacantha Schumann, Gesamtb. Kakteen Nachtr., 133, 1903.
Neomammillaria trichacantha Britton & Rose, The Cactaceae, 4:151, 1923.
Chilita eschauzieri Orcutt, Cactography, 2, 1926.

BODY simple, globose to short cylindric, with sunken apex, to 5 cm. wide. TUBERCLES arranged in 13 and 21 spirals, bluish green, in axils occasionally reddish, clavate, terete, with watery sap, 8-10 mm. long, 4-5 mm. wide at base. AREOLES round, 1 mm. wide, with scant white wool. AXILS with few scattered hair-like bristles. CENTRAL SPINES 2- (3), lower to 12 mm. long and hooked, upper little shorter and straight, all pubescent, acicular, chestnut-brown, in youth reddish, later grayish with brown tip, lower porrect, upper horizontal. RADIAL SPINES 15-18, lower longer to 8 mm. long, all fine acicular, finely pubescent, white at base, transparent yellow above, interlacing, nearly horizontal. FLOWERS wide funnelform, 10-15 mm. long and wide. *Outer perianth-segments* light green at base, orange-brown to pale rose-red mid-stripe, cream margins, linear-lanceolate, tip acute, margins somewhat serrate. *Inner perianth-segments* greenish cream to yellowish, light green mid-line, lanceolate to spatulate, tip acute, margins entire. *Filaments* greenish. *Anthers* bright yellow. *Style* pale green. *Stigma-lobes* 3, white to light tannish green, 0.5 mm. long. FRUIT red, slender clavate, 10 mm. long. SEEDS black, curved pyriform, strongly pitted.

Distribution: San Luis Potosi, Queretaro; Mexico.

Type locality: None given.

Illustration: Fig. 153 is from a photograph of a plant obtained from Sr. F. Schmoll of Cadereyta, Qro.

This species was somewhat confused by Britton & Rose in that they stated that Schumann described the flowers as red, while Quehl reported them as pale yellow. In Schumann's original description the outer perianth-segments underside is pale rose-red and the inner perianth-segments are yellow. This accounts for the discrepency in the B. & R. treatment of this species which was caused by their failure to take into account all of the data in the original description.

M. hamauligera (*M. lamuligera*) is an undescribed name but it has often been referred to the above species. Boedeker earlier considered it as a distinct species but he did not consider it in his book (Mammill. Vergl. Schluss., 1933).

Fig. 154. *Mammillaria mainae* x 0.8

129. Mammillaria mainae K. Brandegee

Mamillaria mainae K. Brandegee, Zoe, 5:31, 1900.
Neomammillaria mainae Britton & Rose, The Cactaceae, 4:154, 1923.
Chilita mainae Orcutt, Cactography, 2, 1926.

BODY simple and branching from base and body, hemispherical to conic globular, to 10 cm. high. TUBERCLES arranged in 8 and 13 spirals, more or less firm in texture, somewhat incurved, varies from pale to bluish gray-green, sometimes reddish in axils, cylindric becoming conic, terete, with watery sap, 10-15 mm. long, 8-10 mm. wide at base. AREOLES round to slightly oblong, with very little wool, very soon becoming naked. AXILS naked. CENTRAL SPINES 1-2, rarely 3, 15-20 mm. long, hooked, somewhat twisted, stout acicular, smooth or sometimes pubescent in youth, yellowish with dark tip, to all black (in type from Sinaloa), porrect, hook often turned to side. RADIAL SPINES 10-15, 6-10 mm. long, upper ones shorter, all straight, stiff, slender acicular, sometimes pubescent in youth (not constant in Sinaloa type), yellowish with brown tips becoming chalky, horizontal. FLOWERS funnelform, broad open throat, in crown in upper part of plant, 10-20 mm. long, to 25 mm. wide. *Outer perianth-segments* pinkish cream margins, brownish green mid-stripe becoming reddish at point, greenish ventrally, linear-lanceolate, tip obtuse, margins ciliate, 2.5 mm. wide. *Inner perianth-segments* broad nearly white margins, pink to reddish tapering mid-stripe, lanceolate, tip acute, margins entire. *Filaments* purplish rose to red. *Anthers* yellow to orange. *Style* white to light pink. *Stigma-lobes* 5-6, rose-red to purplish, 7 mm. long. *Ovary* green, smooth, globular. FRUIT red, globular to obovate, 8x5 mm., not protruding beyond tubercles but retained under spines. SEEDS dull black, obovate with narrow basal hilum, pitted, little more than 1 mm. ROOTS fibrous.

Distribution: Arizona, U.S.; Sonora and Sinaloa, Mexico.

Type locality: South of Nogales, Sonora; also reported at Sells, Arizona; Hermosillo and Banamichi, Sonora; Fuerte, Sinaloa.

Illustration: Fig. 154 is a reproducion of an illustration, Fig. 172, in Britton & Rose (Cact. 4:153) as *Neomammillaria mainae*.

This species is quite widely distributed but still it is quite stable in its characteristics. We collected specimens in Arizona in the grassy hillsides around the Indian agency town of Sells and eastward to Nogales. In our memorable, but none too advisable, trip down the Rio Sonora in the eastern part of the Mexican state of Sonora, we collected it at various places between Naco and Ures. Around Hermosillo we collected it in the dry rocky hills but the specimens were not very abundant nor as well developed. Specimens collected for us by Mr. George Lindsay near Fuerte in the northern part of Sinaloa did show a much darker spine coloration. We have not seen nor know of any collection between Hermosillo in Sonora and Fuerte in Sinaloa, although we have crossed this area by two different routes.

FIG. 155. *Mammillaria fasciculata* x 1

130. **Mammillaria fasciculata** Engelmann

Mamillaria fasciculata Engelmann in Emory, Mil. Reconn., 157, 1848.
Cactus fasciculatus Kuntze, Rev. Gen. Pl., 1:259, 1891.
Mamillaria thornberi Orcutt, West. Amer. Sci., 12:161, 1902.
Neomammillaria fasciculata Britton & Rose, The Cactaceae, 4:162, 1923.
Chilita thornberi Orcutt, Cactography, 2, 1926.

BODY cespitose from base and sometimes from body, cylindric, clumps to 30 cm. wide, heads 5-8 cm. wide. TUBERCLES arranged in 5 and 8 spirals, somewhat flabby in texture, dull purplish green, conic to cylindric, with watery sap, 5 mm. long, 5 mm. wide at base. AREOLES round, with no wool. AXILS naked. CENTRAL SPINES 1 (occasionally 2-3), to 18 mm. long, slender acicular, stiff, smooth, hooked, brownish or black, porrect. RADIAL SPINES 13-20, 5-7 mm. long, fine acicular, smooth, straight, semi-flexuous, white, dark brown or black at tip, horizontal. FLOWERS broad funnelform, 3 cm. long, 2 cm. wide. August to September. *Outer perianth-segments* greenish at base, brownish purple above, broad lanceolate, tip acute to obtuse, margins serrate. *Inner perianth-segments* carmine-red to purplish mid-stripe, whitish margins, broad lanceolate, tip acute, margins entire. *Filaments* rose below, purplish above. *Anthers* yellow. *Style* golden yellow to reddish brown. *Stigma-lobes* 5-6, dark red, 2 mm. long, overtop anthers 2 mm. FRUIT scarlet, short clavate, very juicy, 8 mm. long, with dried perianth deciduous. SEEDS black, glossy, globular with basal hilum, constricted above the base, pitted. ROOTS fibrous.

Distribution: Southern Arizona, U.S.; northern Sonora, Mexico.

Type locality: Along the Gila River, Arizona.

Illustration: Fig. 155 is from a photograph of a plant we collected near Sells, Arizona.

This small, strongly hooked species is quite widely distributed but still it is not very plentiful throughout south central Arizona and it has been reported from northern Sonora. It is usually found along the dry washes under the thorny bushes and the various *Opuntias*. It likes its desert home but does not do well in cultivation.

We collected a closely related form in the Yaqui River valley in southern Sonora, which we have tentatively referred to as *Mammillaria yaquensis* but inasmuch as we have not as yet obtained the flower on it, we are not certain as to its true relationship.

Fig. 156. *Mammillaria beneckei*

131. **Mammillaria beneckei** Ehrenberg

Mammillaria beneckei Ehrenberg, Allg. Gartenz., 12:401, 1844.
Cactus beneckei Kuntze, Rev. Gen. Pl., 1:260, 1891.
Neomammillaria nelsonii Britton & Rose, The Cactaceae, 4:163, 1923.
Chilita nelsonii Orcutt, Cactography, 2, 1926.
Mamillaria balsasensis Boedeker, Monatsschr. Deutsch. Kakt. Ges., 121, 1931.
Mammillaria nelsonii Boedeker, Mammill. Vergl. Schluss., 37, 1933.

BODY simple and cespitose from body and base, forming clumps to 35 cm. across, flattened globular to elongate, with somewhat sunken apex, 6 cm. wide. TUBERCLES loosely arranged in 8 and 13 spirals, glossy leaf-green, often becoming deep reddish and dull, short wide conic, with watery sap, 6-7 mm. long, 8-9 mm. wide at base. AREOLES round, 1 mm. wide, with white wool in youth, soon becoming naked. AXILS with scant white wool but no bristles. CENTRAL SPINES 4 (2-6), three uppers 7 mm. long and straight, lower one 10 mm. long and hooked, all stout acicular, smooth, with base slightly enlarged, dark brown to nearly black, uppers ascending, lower depressed porrect. RADIAL SPINES 13-15, to 8 mm. long, uppers shorter, all fine acicular, straight, stiff, smooth, white with brownish tip, horizontal to slightly ascending above. FLOWERS wide funnelform, 20 mm. long, to 30 mm. wide. *Outer perianth-segments* violet-rose, wide lanceolate, tip acuminate, margins serrate, 15x4 mm. *Inner perianth-segments* gold to orange-yellow, darker at tip, light green in throat, linear-oblong, tip acute, margins serrate, 25x2 mm. *Filaments* red below, yellow above. *Anthers* chrome yellow. *Style* yellow below, orange above. *Stigma-lobes* 5, dark orange, slender, overtop anthers. FRUIT red, very slender clavate, 30 mm. long. SEEDS black, glossy, clavate to slightly curved pyriform with basal hilum which is depressed triangular and white, rugose and warty, 2x3 mm. (very large for this genus).

Distribution: Guerrero and Michoacan, Mexico.

Type locality: Balsas Gro., La Salada, Micho.

Illustration: Fig. 156 is a reproduction of an illustration in Monatsschr. Deutsch. Kakt. Ges., 121, 1931, as *Mammillaria balsasensis*. Fig. 157 is a reproduction of an illustration drawing, Fig. 182, in Britton & Rose (Cact., 4:163) as *Neomammillaria nelsonii* in which: a—fruit, b—spine clusters, c—seed.

Salm-Dyck (Cact. Hort. Dyck., 1849, 10, 1850) says that the above species is near *M. goodrichii*. Seeman (Bot. Herald, 86, 1852) in comment on this reference says, "does not appear to me to be the case." Labouret (Des Cactees, 32, 1853) gives it as a synonym of *M. goodrichii*. Schumann (Gesamtb. Kakteen, 556, 1898) refers to it as being like *M. coronaria*. Britton & Rose (Cact., 4:171, 1923) lists it as known by name only and cites the original description in error as Foerster (Handb. Cact., 210, 1846).

This species was originally described without any data on its distribution other than Mexico which made its identification rather uncertain but it did list the flower as yellow and the large seed but made no mention of the shape of the seed. Britton & Rose probably was not aware of the description and particularly the subsequent article by Ehrenberg in which he emphasized the extra large seed. Britton & Rose described the plant from Michoacan as *N. nelsonii* with seed data but without any flower description. Boedeker described the plant from Guerrero, the neighboring state, as *M. balsasensis* with a good flower description but with no seed data. The body characteristics and spine count of all three of the species are nearly identical. We were listing them as three separate species until we obtained seed of *M. balsasensis* which was so exceptionally large and its very unusual shape indicated that it coincided with *M. nelsonii*. The large seed also suggested *M. beneckei* which has yellow flowers like *M. balsasensis* thus tying the three together as one species. This illustrates how incomplete descriptions can be very misleading.

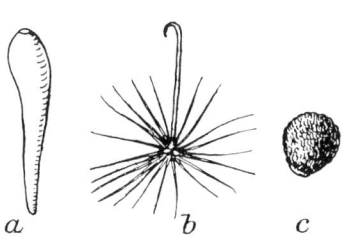

Fig. 157. *Mammillaria beneckei*

Fig. 158. *Mammillaria carretii* x 1

132. **Mammillaria carretii** Rebut

Mamillaria carretii Rebut in Schumann, Gesamtb. Kakteen, 542, 1898.
Neomammillaria saffordii Britton & Rose, The Cactaceae, 4:149, 1923.
Neomammillaria carretii Britton & Rose, The Cactaceae, 4:160, 1923.
Chilita saffordi Orcutt, Cactography, 2, 1926.
Chilita carretii Orcutt, Cactography, 2, 1926.
Mammillaria saffordii Bravo, Las Cact. Mex., 613, 1937.
Mammillaria unihamata Boedeker, Kakteenk., 3:40, 1937.

BODY simple and cespitose from base, globose to cylindric, 5-6 cm. wide. TUBERCLES arranged in 8 and 13 spirals, semi-flabby in texture, slightly conic, terete, with watery sap, 7-9 mm. long, 4-5 mm. wide at base. AREOLES round, 2 mm. wide, with very scant and very light tan wool in youth, soon becoming naked. AXILS with only scant white wool. CENTRAL SPINES 1, 14-18 mm. long, slender acicular, hooked, smooth, semi-flexuous, chestnut brown, yellowish below, porrect. RADIAL SPINES 14-15, 10-13 mm. long, very slender acicular, lightly pubescent, somewhat curved, semi-flexuous, yellow to the middle, brownish above, ascending and recurved. FLOWERS funnelform, lateral, 25 mm. long, 15 mm. wide. *Outer perianth-segments* flesh-red, tip long very thin acuminate, margins entire. *Inner perianth-segments* white, rose-red midstripe, lanceolate, tip acute, margins entire. *Filaments* white. *Anthers* chrome-yellow. *Style* olive-green. *Stigma-lobes* 5-6, greenish, moderately long. FRUIT unknown. SEEDS black, glossy, globular with exserted oval white basal hilum, finely pitted.

Distribution: Neuvo Leon, Coahuila, Mexico.

Type locality: None stated but reported from Ascension and Icamole, N.L., and to the north of Saltillo, Coahuila.

Illustration: Fig. 158 is a reproduction of an illustration, Fig. 176, in Britton & Rose, (Cact., 4:157) as *Neomammillaria carretii*.

We are referring *M. unihamata* Boedeker here because it comes from the same general locality and the body and spine description is so close to the above species that until the flower detail is known, they will be considered as one and the same.

Britton & Rose separated *N. saffordii* and *N. carretii* in their Key on the stoutness of the spines of *N. carretii* but in the description of each, the central spines of *N. saffordii* are also given as stout. Notes under *N. carretii* says, "It is related to *N. saffordii* but radial spines are yellow." In the description, the radial spines of both species are given as yellowish. For these reasons it is evident that there is not enough differences between them to justify separate species, so the older name is here used.

FIG. 159. *Mammillaria surculosa* x 1

133. Mammillaria surculosa Boedeker

Mamillaria surculosa Boedeker, Monatsschr. Deutsch. Kakt. Ges., 3:78, 1931.

BODY cespitose from base and body, 3 cm. wide, 4 cm. high, globular, rounded at apex. TUBERCLES arranged in 5 and 8 spirals, flabby in texture, dark green, lighter at base, finely white dotted, tapering cylindric, terete, with rounded apex, with watery sap, 8 mm. long, 4 mm. wide at base. AREOLES round, 1 mm. wide, with pale yellow to white wool, soon becoming naked. AXILS naked or with slight tuft of wool. CENTRAL SPINES 1, 18-20 mm. long, fine acicular, smooth, strongly hooked, reddish brown, amber yellow at base, porrect. RADIAL SPINES to 15, 8-10 mm. long, fine acicular, straight or slightly bent, smooth, glossy white to pale yellow, horizontal. FLOWERS funnelform, near top, to 18 mm. wide. *Outer perianth-segments* chrome-yellow, rose ventral mid-stripe, darker at tip, linear-lanceolate, tip acuminate, margins entire. *Inner perianth-segments* sulphur-yellow, with small red tip, orange-yellow mid-stripe dorsally, brownish orange ventrally, lanceolate to oblong, tip acute to obtuse, margins entire, 3 mm. wide. *Filaments* bright to greenish yellow. *Anthers* orange. *Style* greenish yellow. *Stigma-lobes* 5-6, yellowish green, 2 mm. long, overtop anthers. FRUIT unknown. SEEDS yellowish brown, curved pyriform with lateral hilum near base, 1 mm. long, finely pitted.

Distribution: Tamaulipas, Mexico.

Type locality: Miquihuana.

Illustration: Fig. 159 is from a photograph of a plant sent to us by Sr. F. Schmoll of Cadereyta, Qro.

Fig. 160. *Mammillaria zephyranthoides* x 1

134. Mammillaria zephyranthoides Scheidweiler

Mammillaria zephyranthoides Scheidweiler, Allg. Gartenz., 9:41, 1841.
Mammillaria fennelii Hopffer, Allg. Gartenz., 11:3, 1843.
Cactus zephyranthoides Kuntze, Rev. Gen. Pl., 1:261, 1891.
Neomammillaria zephyranthoides Britton & Rose, The Cactaceae, 4:159, 1923.
Chilita zephyranthoides Orcutt, Cactography, 2, 1926.

BODY simple, depressed globose to cylindric, 8 cm. high, to 10 cm. wide. TUBERCLES arranged in 5 and 8 spirals, soft in texture, dark bluish green, conic, flattened dorsally and ventrally, with rounded apex, with watery sap, 20-25 mm. long, 10-12 mm. wide laterally at base, 7 mm. wide dorso-ventrally at base. AREOLES small, round to oval, with white then becoming deep yellow wool, later becoming naked. AXILS naked. CENTRAL SPINES 1, occasionally 2, from very minute to 8-14 mm. long, lower ones longer, acicular and hooked, upper ones (when present) straight, all pubescent, yellowish to red-brown, newly formed ones purplish, porrect or divergent. RADIAL SPINES 12-18, 8-12 mm. long, lateral longer, slender hair-like, straight, pubescent, white, horizontal. FLOWERS funnelform, near apex, 40 mm. long and wide, in August. *Outer perianth-segments* greenish to brown, lanceolate, tip acute, margins serrate. *Inner perianth-segments* white to yellow, carmine-red mid-stripe, lanceolate, tip acute, margins serrate. *Filaments* green base, rose above. *Anthers* golden. *Style* green at base, rose above. *Stigma-lobes* 8-10, yellowish green. FRUIT red, oval. SEEDS black, thick, finely pitted.

Distribution: Oaxaca, Queretaro (Schmoll), Mexico.

Type locality: None stated but reported from El Laus, Qro. (Schmoll).

Illustration: Fig. 160 is from a photograph of a plant obtained from Johnson's Water Gardens of Hynes, California.

Mammillaria zephyranthiflora is an unpublished name used as a caption by Pfeiffer (Abbild. Beschr. Cact. 2: pl. 8) for an illustration and it may be referrable here.

Fig. 161. *Mammillaria hutchisoniana* x 1

135. Mammillaria hutchisoniana (Gates) Boedeker

Neomammillaria hutchisoniana Gates, Cact. Succ. Journ., 6:4, 1934.
Neomammillaria bullardiana Gates, Cact. Succ. Journ., 6:4, 1934.
Mammillaria hutchisoniana Boedeker in Back. & Knuth, Kaktus ABC, 387, 1935.
Mammillaria bullardiana Boedeker in Back. &. Knuth, Kaktus ABC, 387, 1935.

BODY cespitose from base, cylindric, to 15 cm. high, 4-6 cm. wide. TUBERCLES arranged in 8 and 13 spirals, firm in texture, with watery sap, 4-5 mm. long, 5 mm. wide at base. AREOLES oval, with white wool in youth, soon becoming naked. AXILS naked. CENTRAL SPINES 3, 8 mm. long, stout acicular, lower longer, stronger and hooked, upper straight, all smooth, stiff, white, purplish at tip, lower slightly depressed from porrect, uppers ascending. RADIAL SPINES 15-20, 5-8 mm. long, slender acicular, straight, smooth, purplish to black in youth, soon becoming whitish, nearly horizontal. FLOWERS recurved campanulate, 20 mm. long, 30 mm. wide. *Outer perianth-segments* light purple to purplish pink wide mid-stripe, pale greenish white wide margins, wide lanceolate, tip obtuse, margins lightly serrate. *Inner perianth-segments* very pale pinkish cream, pinker in throat, with mid-line darker pink to purplish pink to purplish, obolanceolate, tip acute, margins mostly entire, 7 mm. wide. *Filaments* white, tannish above. *Stigma-lobes* 9, golden brown, 7 mm. long, just overtop anthers. FRUIT scarlet, oval, juicy, 10x20 mm., with dried perianth not persisting. SEEDS black, globular with extending basal hilum, pitted, less than 1 mm. long.

Distribution: Baja California, Mexico.

Type locality: Vizcaino Desert, 8 miles west of Calamalli.

Illustration: Fig. 161 is a reproduction of an illustration in Cact. Succ. Journ., 6:3 as *Neomammillaria hutchisoniana*.

The original description calls for 25-35 radial spines but an examination of the type material in the Dudley Herbarium disclosed only 15-20, which number is also shown in the illustration accompanying the original description, and also on the plants that we obtained from Mr. Gates.

Fig. 162. *Mammillaria sheldonii* x 0.8

136. **Mammillaria sheldonii** (B. & R.) Boedeker

Neomammillaria sheldonii Britton & Rose, The Cactaceae, 4:156, 1923.
Chilita sheldoni Orcutt, Cactography, 2, 1926.
Mammillaria sheldonii Boedeker, Mammill. Vergl. Schluss., 30, 1933.

BODY simple and cespitose from base, cylindric, 25 cm. long, 6 cm. wide. TUBERCLES arranged in 8 and 13 spirals, semi-firm in texture, medium to dark dull green, conical cylindric, 4-sided at base, terete above, blunt apex, keeled ventrally, with watery sap, 9 mm. long, 6 mm. wide at base. AREOLES round to oval, with very scant tan wool in youth, soon becoming naked. AXILS naked. CENTRAL SPINES 1-3, 9-12 mm. long, uppers straight and shorter, lower one hooked and longer, all stout acicular, smooth, dark reddish brown, straight upper ones nearly horizontal, hooked lower one porrect. RADIAL SPINES 10-15, 6-9 mm. long, acicular, smooth, straight, upper 3-4 darker, reddish brown, whitish at base, darker at tip, mostly horizontal. FLOWERS wide funnelform, lateral near top, 20 mm. long, 30 mm. wide, open several days, April to August. *Outer perianth-segments* 10-13, light olive-green at base, wide light brownish green midstripe, green at apex, narrow very pale pinkish greenish tan margins, linear-lanceolate, tip acute, margins short ciliate (more so near base, upper part often entire), 2.5 mm. wide. *Inner perianth-segments* bright pink to purplish mid-stripe, wide white margins, brownish green ventrally, wide lanceolate, tip acute, margins entire, to 7 mm. wide. *Filaments* light purple. *Anthers* orange-yellow. *Style* yellow to light pink. *Stigma-lobes* 6-8, light olive-green, 3-5 mm. long, overtop anthers 4 mm. FRUIT pale scarlet, clavate, 25-30 mm. long, with dried perianth easily detached. SEEDS black, glossy, globular with basal hilum, constricted above base, slightly pitted, 1x0.8 mm. ROOTS fibrous.

Distribution: Southern Sonora, Mexico.

Type locality: Hermosillo, also reported from Guaymas and Yaqui Valley.

Illustration: Fig. 162 is from a photograph of a plant we collected near Hermosillo in 1935.

The original description is supplemented with data from plants that we collected at Hermosillo, near Guaymas and from the Yaqui Valley.

The illustration in Britton & Rose (Cact. 4:157), Fig. 175, does not coincide with their description as there are more radial spines in the illustration than is called for in the text. The illustration used may have been confused with *M. microcarpa*, a closely related species. Our observations indicate that the upturned hook of the central spine is not a constant characteristic as stated by Britton & Rose.

137. Mammillaria neocoronaria (Schuman) Knuth

Mamillaria coronaria Schumann, Gesamtb. Kakteen, 555, 1898.
*Mammillaria coronaria beneckei** Schelle, Kakteen, 319, 1926.
Mammillaria coronaria nigra Schelle, Kakteen, 319, 1926.
Mammillaria coronaria nigra euchlora Schelle, Kakteen, 319, 1926.
Mammillaria coronaria nigra euchlora cristata Schelle, Kakteen, 319, 1926.
Mammillaria coronaria eugenia Schelle, Kakteen, 319, 1926.
Mamillaria neocoronaria Knuth in Back. & Knuth, Kaktus ABC, 392, 1935.

BODY simple and cespitose from base, globular to cylindric, with sunken apex with white wool, 6-7 cm. wide, 7-15 cm. high. TUBERCLES arranged in 8 and 13 spirals, green to grayish green, conic, with watery sap (?), 8 mm. long. AREOLES oval, to 3 mm. wide, with white wool. AXILS naked. CENTRAL SPINES (4)-6, upper longest to 15 mm. long, lower hooked, all dark ruby-red, later brown, then yellowish to gray, spreading. RADIAL SPINES 16-18, 8-10 mm. long, lateral longer, transparent, later white, spreading. FLOWERS wide funnelform, 16-17 mm. long, 10-12 mm. wide. *Outer perianth-segments* brownish carmine. *Inner perianth-segments* light carmine, darker fiery mid-stripe, lanceolate, tip acute, margins fine ciliate. *Filaments* white. *Anthers* sulphur-yellow. *Style* white. *Stigma-lobes* 6, yellowish green. FRUIT and SEEDS unknown.

Distribution: Hidalgo, Mexico.

Type locality: Real del Monte.

Illustration: Berger, Kakteen, Fig. 92, 306 as *M. coronaria*.

We have not seen any collected plants of this species nor do the seedling plants in the trade follow the above description, so our description is compiled from previous descriptions.

See also *M. coronaria* Haworth.

138. Mammillaria goodridgei Scheer

Mamillaria goodrichii Scheer in Salm-Dyck, Cact. Hort. Dyck. 1849, 91, 1850.
Mamillaria goodridgii Scheer, Bot. Herald, 286, 1852-57.
Cactus goodridgii Kuntze, Rev. Gen. Pl., 1:260, 1891.
Neomammillaria goodridgei Britton & Rose, The Cactaceae, 4:258, 1923.
Chilita goodridgei Orcutt, Cactography, 2, 1926.

BODY simple and cespitose from base, globose to cylindric, to 10 cm. long, 4 cm. wide. TUBERCLES covered by spines, closely set in 8 and 13 spirals, firm in texture, becoming corky in age, cylindric, rounded above, faintly 4-sided below, with watery sap, 3-5 mm. long. AREOLES round, with scant short white wool, later becoming naked. AXILS with very scant wool or naked. CENTRAL SPINES 3-4, 8-10 mm. long, lower one strongly hooked, upper 2-3 straight, all acicular, white, brownish above, hooked one nearly porrect, others erect and nearly horizontal. RADIAL SPINES 11-15, 4-7 mm. long, 8 lateral, 4 lower and shorter, all rigid acicular, straight, transparent white, with

---

*This variety may be referrable to *M. beneckei* but because of the absence of definite data it is uncertain.

M. goodridgei

FIG. 163. *Mammillaria goodridgei* x 1

sometimes darker tips, horizontal. FLOWERS broad funnelform, 15 mm. long. 25 mm. wide. *Outer perianth-segments* deep pink to brownish red with light pink margins, greenish mid-stripe ventrally, oblong, tip obtuse, margins mostly entire. *Inner perianth-segments* deep pink to rose-red mid-stripe, white to cream margins, oblong, tip acute to obtuse, margins entire, 20x6 mm. *Filaments* cream. *Anthers* deep yellow, small. *Style* rose-red. *Stigma-lobes* 4-5, green, slender, 3-4 mm. long. FRUIT scarlet, clavate, 15-20 mm. long, with dried perianth persisting. SEEDS black, globular with narrow basal hilum, pitted.

Distribution: Cedros and other islands and adjacent mainland of Baja California, Mexico.

Type locality: Cedros Island.

Illustration: Fig. 163 is from a photograph by Mr. George Lindsay of a plant he collected on Guadalupe Island on July 4, 1942.

This species, as described by Scheer in Salm-Dyck, was without any data on the flower, fruit or seeds and as a result this led to much confusion by subsequent authors. Material from San Diego County in California and also from Baja California, which is close to this species but still quite distinct, has been referred to this previously described species by Engelmann (Pro. Amer. Acad. Sci., 3:263, 1856, and also his subsequent publications), Coulter (Contr. U. S. Nat. Herb., 3:2, 1894), and others. Mrs. Katherine Brandegee (Erythea, 5:111, 1897), was the first to correct the error and distinguished the two species after examining much material from the islands and reported: "The plant is of more slender growth than the San Diego species, with naked or sparsely woolly axils, and much more or less rose-colored flowers, and long slender subulate style divisions. In the two score plants I have seen in flower they were all hermaphrodite." For the San Diego type she erected the species *M. dioica* (Erythea, 5:115, 1897), which is basically a yellowish flower and has bristles in the axils. Britton & Rose (Cact., 4:69 and 158, 1923), in their Key to Species placed it with those with bristles in the axils but in the description said that there were none. Material collected by Mr. George Lindsay on Guadalupe Island and which flowered for him has proven to belong here.

FIG. 164. *Mammillaria blossfeldiana* x 0.8

139. **Mammillaria blossfeldiana** Boedeker

Mamillaria blossfeldiana Boedeker, Monatsschr. Deutsch. Kakt. Ges., 3:209, 1931.
Neomammillaria blossfeldiana Gates, Cact. Succ. Journ., 4:371, 1933.

BODY simple, occasionally cespitose from base, globular to cylindric, somewhat sunken at apex, 4-5 cm. wide. TUBERCLES arranged in 5 and 8 spirals, grayish green with sometimes a purplish tinge, brownish below, short conic, terete, with watery sap, 5-7 mm. long, 5-6 mm. wide at base. AREOLES round to slightly oval, 2 mm. wide, with some white wool, soon becoming naked. AXILS with scant white wool only. CENTRAL SPINES 4, 7-10 mm. long, lower one longest, stout acicular and hooked, 3 uppers acicular and straight, all grayish white at base, shading through purple to black at tip, uppers ascending, lower porrect. RADIAL SPINES 15-20, 5-7 mm. long, straight, slender acicular, smooth, stiff, grayish white with dark brown to black tip, horizontal. FLOWERS funnelform, near apex, 20 mm. long, 30-35 mm. wide, lasts for several days. *Outer perianth-segments* brownish red mid-stripe, pale greenish white margins, brownish green at base, linear-lanceolate, tip obtuse, margins entire, 15x2 mm. *Inner perianth-segments* wide rose-carmine mid-stripe, darker mid-line, white margins, wide lanceolate, tip obtuse, margins entire, 15x4 mm. *Filaments* whitish to pale pink. *Anthers* orange-yellow. *Style* rose-pink. *Stigma-lobes* 5-9, olive-green, later becoming golden yellow, elongate to 10 mm. long. FRUIT orange-red, clavate, 20x5 mm., with dried perianth not persisting. SEEDS black, glossy, pyriform with small lateral hilum, slightly pitted, 0.9x0.6 mm. ROOTS usually tuberous.

Distribution: Baja California, Mexico.

Type locality: Santa Rosalita Bay, Lat. 29°.

Illustrations: Fig. 164 is from a photograph by Mr. George Lindsay of a plant he collected at the type locality. Fig. 165 is a reproduction of an illustration in Cact. Succ. Journ., 8:31 by Mr. R. S. Woods of a seedling plant.

Fig. 165. *Mammillaria blossfeldiana*

a. var. shurliana Gates

Mammillaria blossfeldiana shurliana Gates, Cact. Succ. Journ., 13:78, 1941.

Roots more fibrous, 1 less central spine, filaments reddish purple instead of white, seeds smoother.

Type locality: Mesquital Ranch.

Illustration: Cact. Succ. Journ., *13*:78, 1941.

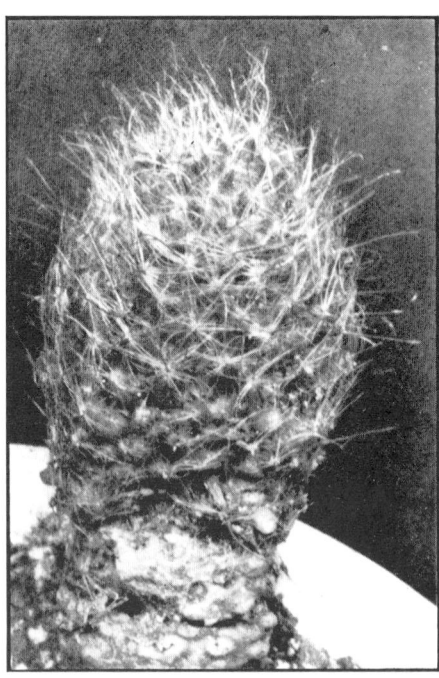

Fig. 166. *Mammillaria crinita* x 1

140. **Mammillaria crinita** DeCandolle

Mammillaria crinita DeCandolle, Mem. Mus. Hist. Nat. Paris, 17:112, 1828.
?*Mammillaria criniformis albida* DeCandolle, Mem. Cact., 8, 1834.
Cactus crinitus Kuntze, Rev. Gen. Pl., 1:260, 1891.
Mamillaria glochidiata crinita Schumann, Gesamtb. Kakteen, 532, 1898.

BODY simple and cespitose from base, globose, 4 cm. wide. TUBERCLES arranged in 8 and 13 spirals, soft in texture, dull grayish green, ovate-oblong, terete, with watery sap, 16-18 mm. long, 3-4 mm. wide at base. AREOLES round, with very scant wool only in youngest. AXILS naked. CENTRAL SPINES 4-5, 12-15 mm. long, very slender acicular, pubescent, one or more hooked, all yellow, spreading porrect. RADIAL SPINES 15-20, 16-18 mm. long, setaceous, straight, pubescent, white, horizontal. FLOWERS campanulate, 16 mm. long. *Outer perianth-segments* reddish mid-stripe ventrally, pinkish yellow margins, lanceolate, tip acute, margins serrate. *Inner perianth-segments* white to yellowish (sometimes pinkish cream), pink mid-line but paler dorsally, linear-lanceolate, tip acuminate, margins entire. *Filaments* very pale yellow. *Anthers* pale yellow. *Style* very pale greenish yellow. *Stigma-lobes* 4-5, pale yellowish green. FRUIT reddish, clavate, 15x3 mm., with dried perianth persisting. SEEDS black, glossy, globular with exserted basal hilum, pitted.

Distribution: Hidalgo, Mexico.

Type locality: None given but reported from Zimapan and Rancho de San Antonio.

Illustration: Fig. 166 is from a photograph of a plant sent to us by Sr. F. Schmoll of Cadereyta, Qro.

This species has been referred to *M. glochidiata* but it can not be placed under that species because the axils are naked.

FIG. 167. *Mammillaria wrightii* x 0.8

141. Mammillaria wrightii Engelmann

Mamillaria wrightii* Engelmann, Pro. Amer. Acad. 3:262, 1856.
Cactus wrightii Kuntze, Rev. Gen. Pl., 1:261, 1891.
Neomammillaria wrightii Britton & Rose, The Cactaceae, 4:152, 1923.
Chilita wrightii Orcutt, Cactography, 2, 1926.

BODY simple, globose to depressed globose, with sunken apex, oboconic, deep-seated, 4-8 cm. wide. TUBERCLES arranged in 8 and 13 spirals, soft in texture, dark green, cylindric, terete, with watery sap, 10-15 mm. long, 3-4 mm. wide at base. AREOLES round, to 2 mm. wide, with short white wool in youth, later becoming naked.

---

*The original description was written by Engelmann in the "Cactaceae of the Boundary," 2:Part I, 7, 1869, which was delayed in publication due to the engraving of the illustration plates, so the synopsis which appeared in the first reference takes priority.

AXILS naked. CENTRAL SPINES usually 2, occasionally 4, 10-12 mm. long, lower one hooked, upper one straight, all stout acicular, pubescent, dark brown to often nearly black, divergent laterally, rarely dorso-ventrally except when 4. RADIAL SPINES 12-14, 8-12 mm. long, lateral longer, acicular, upper 3-5 more robust, lower ones more slender or entirely wanting, all pubescent, white with darker apex, lateral and upper ones darker, horizontal, one or more of the upper ones subcentral. FLOWERS recurved funnelform, 25 mm. long and wide, June and July, open several days. *Outer perianth-segments* (13), purplish olive-green, lighter margins, triangular lanceolate, tip obtuse to acute, margins ciliate, 10x4 mm. *Inner perianth-segments* (12), purplish, linear-lanceolate, tip acuminate aristate, margins mostly entire, 20x3.5 mm. *Filaments* greenish below, cerise above. *Anthers* orange-yellow. *Style* greenish below, pink above. *Stigma-lobes* 11, lemon-yellow, linear. FRUIT purplish, large ovate globose, 25 mm. long, juicy, with dried perianth persisting. SEEDS black, obovate with very small narrow ventral hilum, pitted. ROOTS fibrous.

Distribution: New Mexico and Texas, U. S.: Chihuahua, Mexico.

Type locality: Anton Chico on Pecos River, near copper mines near El Paso, Texas, and from near Lake Santa Maria, Chihuahua.

Illustration: Fig. 167 is a reproduction of an illustration in Texas Cacti, 161 as *Neomammillaria wrightii*.

Habitat: Under juniper trees at about 6000 feet, sun and partial shade.

FIG. 168. *Mammillaria wilcoxii* x 1

142. **Mammillaria wilcoxii** Toumey

Mamillaria wilcoxii Toumey in Schumann, Gesamtb. Kakteen, 545, 1898.
Neomammillaria wilcoxii Britton & Rose, The Cactaceae, 4:153, 1923.
Chilita wilcoxii Orcutt, Cactography, 2, 1926.

BODY simple, occasionally branching from base, depressed globose to short cylindric, rounded above, to 10 cm. high. TUBERCLES widely separated in 5 and 8, also 8 and 13 spirals, flabby in texture, bright to dark glossy green, conical cylindric, terete, with watery sap, variable in length to 20 mm., 5 mm. wide at base. AREOLES round to oval, 2 mm. wide, with scant dirty white wool, very soon naked. AXILS naked. CENTRAL SPINES 1-3 (variable, given as many as 5-6), 20-30 mm. long, stout acicular, stiff, smooth to pubescent in varying degrees from entirely smooth to feathery pubescent, 1 or more hooked, dark amber to reddish brown, paler below, divergent porrect. RADIAL SPINES 14-22 (30?), 10-15 mm. long, lateral longer, bristle-like to thin acicular, white with brown tip, pubescence variable like centrals, horizontal. FLOWERS narrowly campanulate, to 35 mm. in length, 40 mm. wide. *Outer perianth-segments* 20, light cream at base, brownish green mid-stripe, pale green to cream margins, linear-lanceolate, tip acuminate, margins long ciliate, to 20 mm. long, 2-3 mm. wide. *Inner perianth-segments* cream, tannish pink to purplish mid-stripe, linear-lanceolate, tip long tapering acuminate, margins entire, 30x2 mm. *Filaments* white. *Anthers* light orange. *Style* pale green. *Stigma-lobes* 7, light green, 2 mm. long, overtop anthers 4-5 mm. FRUIT pink base, greenish pink to purplish at top, large ovate, 25x13 mm., with dried perianth persisting. SEEDS black, glossy, curved globular pyriform with white lateral hilum, pitted, 1.7x1 mm.

Distribution: Arizona, U. S.; Sonora, Mexico.

Type locality: None given but reported from near Nogales and Benson, Arizona; south of Sonoyta, Sonora.

Illustration: Fig. 168 is from a photograph of a plant we collected near Nogales, Arizona.

This species is very closely related to, if not just a geographical variation of *M. wrightii* although the radial spines count is slightly more and the stigma-lobes count is less (7 vs. 11).

Britton and Rose suggests that inasmuch as this species is "often associated with *M. grahamii* and *Coryphantha aggregata* has led to the suggestion that it might be a

FIG. 169. *Mammillaria wilcoxii* var. *viridiflora* x 1

hybrid between these species." That supposition might be possible if its distribution was limited to a very restricted area but inasmuch as it has been found in Sonora, northward through central Arizona at Miami in the varietal form of *viridiflora* and eastward where it merges into *M. wrightii*, we are of the opinion that it is a valid species and not a result of hybridization of two different genera. Likewise Britton & Rose were prone to associate with the Coryphantha genus any plant that possessed large flowers. There are several good species of the genus *Mammillaria* that possesses large flowers.

a. var. **viridiflora** (B. & R.) Marshall & Bock

Neomammillaria viridiflora Britton & Rose, The Cactaceae, 4:153, 1923.
Chilita viridiflora Orcutt, Cactography, 2, 1926.
Mammillaria viridiflora Boedeker, Mammill. Vergl. Schluss., 36, 1933.
Mammillaria wilcoxii viridiflora Marshall & Bock, Cactaceae, 182, 1941.

Inner perianth-segments greenish instead of pink to purplish.

Distribution: Pinal and Gila counties of Arizona.

Type locality: Boundary monument on Superior-Miami highway.

Illustration: Fig. 169 is a reproduction of an illustration in Cact. Succ. Journ., 9:29 as *Mammillaria viridiflora*.

Our field collection show that this variety shades from the typical greenish into the pinkish flower of the type form even when collected at the type locality of the variety, so the green flowered form is only a purely local sport variation which condition is often observed in some of the other species.

Fig. 170. *Mammillaria zeilmanniana* x 1

143. **Mammillaria zeilmanniana** Boedeker

Mamillaria zeilmanniana Boedeker, Monatsschr. Deutsch. Kakt. Ges., 3:227, 1931.

BODY simple and occasionally branching from body and base, cylindric, little sunken at apex, to 6 cm. high, 5 cm. wide. TUBERCLES arranged in 13 and 21 spirals, soft in texture, glossy dark leaf-green, ovate to short cylindric, rounded at apex, terete, with watery sap, 6 mm. long, 3-4 mm. wide at base. AREOLES round, 1 mm. wide, with scant short white wool, soon becoming naked. AXILS naked. CENTRAL SPINES 4, 8 mm. long, acicular, 3 uppers straight, lower one hooked, all enlarged at base, reddish brown, lighter to yellowish at base, lower porrect, uppers horizontal to ascending. RADIAL SPINES 15-18, 10 mm. long, very thin acicular to hair-like, straight or slightly bent, stiff, pubescent, not enlarged at base, horizontal to slightly ascending. FLOWERS campanulate, 20 mm. wide. *Outer perianth-segments* greenish at base, to purplish brown above, lanceolate, tip acute, margins entire, 4-9x1-2.5 mm. *Inner perianth-segments* violet-red to purplish, paler below, lanceolate, tip acute, margins entire, 11x2 mm. *Filaments* violet-red. *Anthers* bright yellow. *Style* greenish below, violet-red above. *Stigma-lobes* 4, yellow, small. FRUIT whitish green (?), small. SEEDS black, roundish with basal oblong hilum, finely pitted, hardly 1 mm.

Distribution: Guanajuato, Mexico.

Type locality: Near San Miquel Allando.

Illustration: Fig. 170 is from a photograph of a plant obtained from Johnson's Water Gardens of Hynes, California.

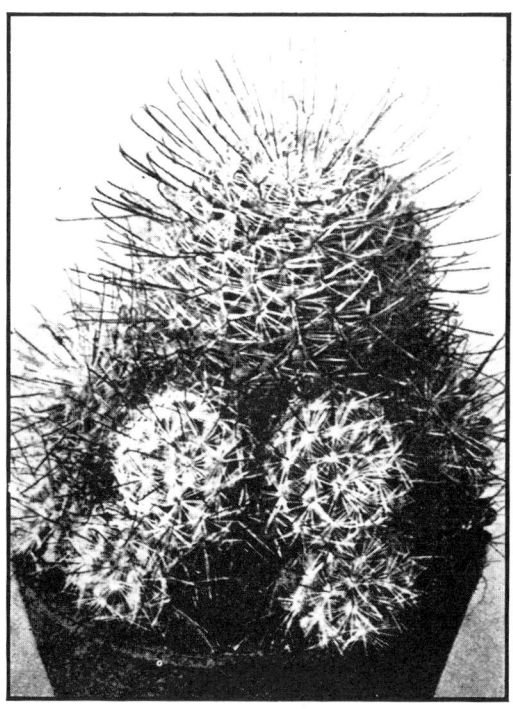

FIG. 171. *Mammillaria schelhasii*

144. **Mammillaria schelhasii** Pfeiffer

Mammillaria schelhasii Pfeiffer, Allg. Gartenz., 6:274, 1838.
Mammillaria glochidiata purpurea Scheidweiler, Bull. Acad. Sci. Brux., 5:495, 1838.
Cactus schelhasii Kuntze, Rev. Gen. Pl., 1:261, 1891.
Mamillaria schelhasei lanuginosior Hildmann ex Schumann, Gesamtb. Kakteen, 531, 1898.
Neomammillaria schelhasei Britton & Rose, The Cactaceae, 4:149, 1923.
Chilita schelhasei Orcutt, Cactography, 2, 1926.

BODY cespitose from base, globose to short cylindric. TUBERCLES arranged in 8 and 13 spirals, flabby in texture, pale green at base to darker green above, 4-sided at base, cylindric, rounded at apex, with watery sap, 20 mm. long, 8 mm. wide at base. AREOLES round, 1.5 mm. wide, sunken, with a little woolly felt to naked. AXILS with scant white wool, but no bristles. CENTRAL SPINES 3, 12-18 mm. long, lower longer heavier and hooked, uppers setaceous and straight, all reddish brown, lower darker and porrect, uppers erect. RADIAL SPINES 15-20, 8-10 mm. long, lower longer, all setaceous, straight, white, horizontal. FLOWERS funnelform, 22-25 mm. long, 12 mm. wide. *Outer perianth-segments* white, rose-red mid-stripe, linear-lanceolate, tip acute, margins entire (?). *Inner perianth-segments* white, red or rose mid-line, lanceolate, tip acute, margins entire (?). *Filaments* rose-red. *Anthers* yellowish, round. *Style* pale red. *Stigma-lobes* 4-5, whitish to yellowish. FRUIT red, clavate. SEEDS black.

Distribution: Hidalgo, Mexico.

Type locality: None given but reported from Mineral de Real del Monte (Bravo), Actopan and Ixmiquilpan (Schumann).

Illustration: Fig. 171 is a reproduction of an illustration in Monatsschr. Kakteen., 30:163 as *Mamillaria schelhasei*.

"Salm-Dyck (Cact. Hort. Dyck. 1849, 7 and 81, 1850), describes the three following varieties: *sericata, rosea* and *triuncinata,* some of which may belong elsewhere. Of these Schumann recognizes only the last. The first, Lemaire has referred to a different species, *Mammillaria glochidiata sericata* Lemaire (Cact. Gen. Nov. Sp., 40, 1839)." Britton & Rose Cact., 4:149.

FIG. 172. *Mammillaria rettigiana*

145. Mammillaria rettigiana Boedeker

Mamillaria rettigiana Boedeker, Monatsschr. Deutsch. Kakt. Ges., 2:98, 1930.

BODY simple to somewhat branched, depressed to oblong-globular, with slightly sunken apex, to 4 cm. wide. TUBERCLES loosely arranged in 8 and 13 spirals, glossy dark leaf-green, tall cylindric, conic at tip, with watery sap, 8-10 mm. long, 3 mm. wide at base. AREOLES round, 2-3 mm. wide, with white wool, later becoming naked. AXILS naked or with very scant wool. CENTRAL SPINES 3-4, 2-3 upper ones 12 mm. long, lower one to 15 mm. long, all acicular, smooth, little thickened at base, lower hooked, upper ones straight, all red to dark brown, lighter at base, uppers erect and fan-like, lower porrect. RADIAL SPINES 18-20, to 10 mm., very slender acicular, smooth, straight, white to yellow, gray in age, horizontal. FLOWERS near apex, short funnelform, 15 mm. wide. *Outer perianth-segments* greenish ventrally, yellowish green above, darker mid-rib and tip, oblong, tip acute, margins entire. *Inner perianth-segments* delicate rose mid-stripe, lighter margins, silky luster, oblong-lanceolate, tip acuminate, margins entire. *Filaments* whitish below, bright carmine-red above. *Anthers* bright yellow. *Style* creamy white. *Stigma-lobes* 4-5, whitish, fairly short, spreading. FRUIT red, clavate, small. SEEDS brownish black, glossy, ovate with oblong basal hilum, pitted, 1 mm.

Distribution: Hidalgo and Guanajuato, Mexico.

Type locality: SW. Hidalgo.

Illustration: Fig. 172 is a reproduction of an illustration in Monatsschr. Deutsch. Kakt. Ges., 2:99 as *Mamillaria rettigiana*.

146. Mammillaria umbrina Ehrenberg

Mamillaria umbrina Ehrenberg, Allg. Gartenz., 17:287, 1849.
Cactus umbrinus Kuntze, Rev. Gen. Pl., 1:261, 1891.
Mamillaria umbrina v. *roessingii* (Mathsson) Quehl, Monatsschr. Kakteenk., 177, 1905.
Neomammillaria umbrina Britton & Rose, The Cactaceae, 4:164, 1923.
Chilita umbrina Orcutt, Cactography, 2, 1926.

BODY simple and cespitose, globular to cylindric, 5-10 cm. wide, 12-15 cm. high. TUBERCLES closely set in 8 and 13 spirals, dark bluish green, 4-sided at base, conic, truncate above, with watery sap, 8 mm. long, 5 mm. wide at base. AREOLES round to oval, 2-3 mm. wide, with gray-white wool, later becoming naked. AXILS with wool and white bristles. CENTRAL SPINES 1-4 (mostly 2), upper ones 8-10 mm. long, lower ones 20-24 mm. long, stronger and hooked, all dark ruby-red in youth, later reddish tan, brown tip, spreading, porrect. RADIAL SPINES 22-24, upper ones 4 mm., lower ones to 12 mm., all setaceous, transparent to snow white, nearly horizontal, interlacing. FLOWERS 18 mm. long, 12 mm. wide. *Outer perianth-segments* greenish red, becoming dark red. *Inner perianth-segments* dark rose, paler below, linear-lanceolate, tip acute, 2 mm. wide. *Filaments* and *Anthers* yellowish white. *Stigma-lobes* 4, green. FRUIT and SEEDS unknown.

Distribution: Mexico.

Type locality: None given.

Salm-Dyck in error reported the axils as naked and was followed by most of the subsequent authors.

The plants in the trade now under this name do not coincide with the original description as they differ in several respects. Inasmuch as the type locality was not recorded, the recollection of any additional type material is problematical and uncertain. We have had to rely upon the original description and various other authors for our description.

Britton & Rose reports that it resembles *Mammillaria zephyranthoides* "but is undoubtedly distinct."

Fig. 173. *Mammillaria solisii* x 0.8

147. Mammillaria solisii (B. & R.) Boedeker

Neomammillaria solisii Britton & Rose, The Cactaceae, 4:142, 1923.
Mammillaria solisii Boedeker, Mammill. Vergl. Schluss., 35, 1933.

BODY simple and cespitose from body, globose to cylindric to 10 cm. wide, to 20 cm. high. TUBERCLES closely set in 13 and 21 spirals, greenish becoming purplish, conic, rounded at apex, with watery sap, 8-10 mm. long, 6-10 mm. wide at base. AREOLES oval, 2 mm. wide, with scant dirty white wool only in the youngest. AXILS with several white tortuous bristles but no wool. CENTRAL SPINES 3-6 (mostly 4), to 20 mm. long, uppers straight, lower one hooked and longest, all slender subulate, smooth, stiff, yellowish brown, lighter at base, nearly black at tip, spreading porrect. RADIAL SPINES 20-25, 8-10 mm. long, acicular, straight, smooth, stiff, white, nearly horizontal. FLOWERS funnelform, 15 mm. long. *Outer perianth-segments* brownish red, little darker midstripe, linear-lanceolate, tip acute, margins finely serrate. *Inner perianth-segments* deep rose to bright magenta-red, darker mid-stripe, linear-lanceolate, tip acute, margins entire. *Filaments* white below, rose above. *Anthers* very pale dirty cream. *Style* very pale pinkish cream. *Stigma-lobes* 5, yellowish green, 1 mm. long, overtop anthers 3 mm. FRUIT whitish green to rose, clavate, 20 mm. long, with dried perianth persisting. SEEDS light reddish brown, glossy, curved pyriform with lateral hilum at base, finely pitted, 1.4x0.6 mm.

Distribution: Guerrero, Mexico.

Type locality: Cerro de Buenavista de Cuellar and also reported from Taxco.

Illustration: Fig. 173 is from a photograph of a plant we collected near Taxco in 1942.

This species, along with *M. nunezii* and *M. spinosissima* and the various hybrids of these three species, were found growing in the cliffs of the canyons near Taxco by Dr. and Mrs. Craig in June of 1942. Inasmuch as Taxco is only a few miles from the type locality, our collections are here considered as nearly equivalent to the type locality material.

Plants collected by Mr. George Lindsay at Taxco in 1937 and our collections of 1941 and 1942 were used to supplement the original description.

148. Mammillaria multiformis (B. & R.) Boedeker

Neomammillaria multiformis Britton & Rose, The Cactaceae, 4:148, 1923.
Chilita multiformis Orcutt, Cactography, 2, 1926.
Mammillaria multiformis Boedeker, Mammill. Vergl. Schluss., 26, 1933.

BODY cespitose from base and body, globose to much elongate, 3-6 times as long as thick. TUBERCLES soft in texture, grayish green, conic to cylindric, terete, with watery sap, 6-8 mm. long. AREOLES oval, small. AXILS with white wool and long white bristles. CENTRAL SPINES 4, 8-10 mm. long, stout acicular, lower one hooked, upper ones straight, all reddish in upper part, upper ones nearly erect, lower one porrect. RADIAL SPINES 30 or more, 8 mm. long, slender acicular, straight, yellow, ascending. FLOWERS 8-10 mm. *Inner perianth-segments* deep purplish red, oblong, tip acute. *Filaments* red. FRUIT nearly globose. SEEDS black.

Distribution: San Luis Potosi, Mexico.

Type locality: Alvarez.

Illustration: Fig. 174 is a reproduction of an illustration, Fig. 164, in Britton & Rose (Cact., 4:148) as *Neomammillaria multiformis*.

We have not been able to recollect this species to complete the description, but we did observe it in the garden of Dr. Iwerson in Mexico City in 1942.

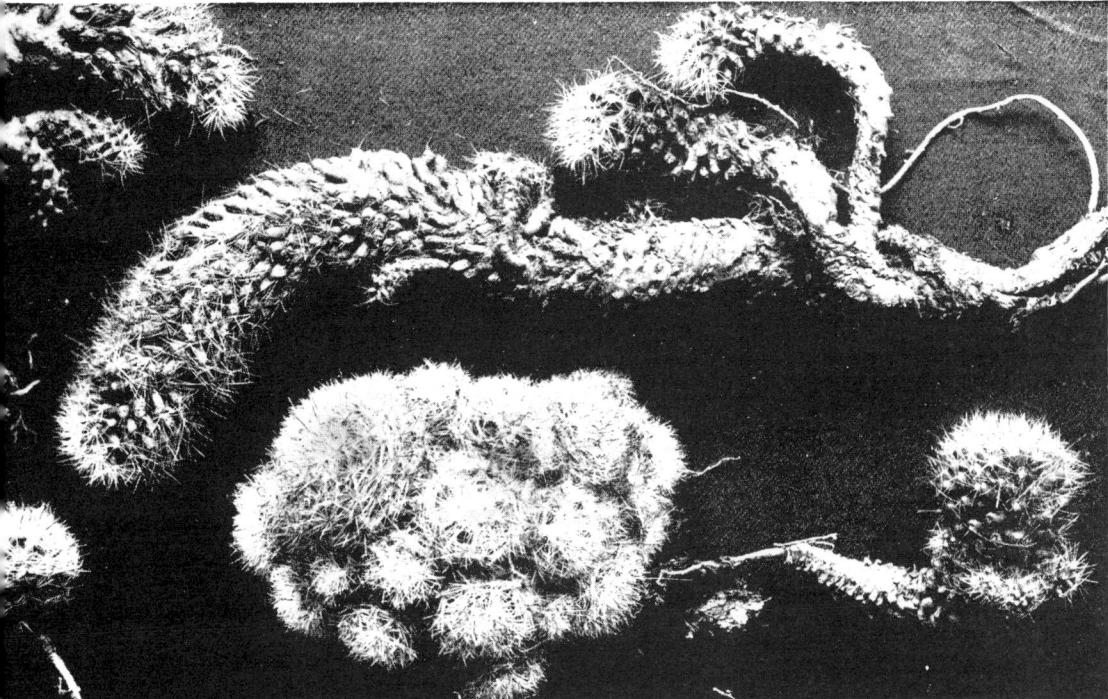

FIG. 174. *Mammillaria multiformis*

Fig. 175. *Mammillaria erectohamata*

149. Mammillaria erectohamata Boedeker

Mamillaria erectohamata Boedeker, Monatsschr. Deutsch. Kakt. Ges., 2:189, 1930.

BODY simple and cespitose from base, conic, rounded to flat apex, to 60 mm. wide. TUBERCLES arranged in 13 and 21 spirals, bright leaf-green, cylindric, terete, with apex turned ventrally, with watery sap, 8 mm. long, 5 mm. wide at base. AREOLES with very scant wool only in youngest. AXILS with no wool but few fine white hair-like tortuous bristles. CENTRAL SPINES 2 to seldom 3, to 17 mm. long, lower one longer and hooked, all strong acicular, smooth, hardly any enlarged at base, chestnut-red to black-brown, later yellowish, golden yellow at base, ascending. RADIAL SPINES to 25, 7 mm. long, very fine acicular, straight, stiff, smooth, white, horizontal. FLOWERS campanulate, lateral in crown, 18 mm. wide. *Outer perianth-segments* green below, brownish rose above, margins cream, linear-lanceolate, tip acute, margins serrate. *Inner perianth-segments* outer row cream to nearly pure white, fine faint rose mid-line, inner row pure white, linear-oblong, tip acuminate, margins entire. *Filaments* and *Style* light green below, whitish above. *Anthers* light yellow. *Stigma-lobes* 5-6, whitish, spreading. FRUIT red, clavate, 12 mm. long. SEEDS dark brown, curved pyriform, half flat, finely pitted, 0.5 mm.

Distribution: San Luis Potosi, Mexico.

Type locality: Southern part of state in the ravines.

Illustration: Fig. 175 is a reproduction of an illustration in Monatsschr. Deutsch. Kakt. Ges., 2:189 as *Mamillaria erectohamata*.

Fig. 176. *Mammillaria knebeliana*

150. **Mammillaria knebeliana** Boedeker

Mamillaria knebeliana Boedeker, Monatsschr. Deutsch. Kakt. Ges., 4:52, 1932.

BODY simple to branching from body and base, short cylindric, with flat but not sunken apex, 6 cm. high, 5 cm. wide. TUBERCLES arranged in 13 and 21 spirals, glossy leaf-green, cylindric, obtuse at apex, with watery sap, 6-7 mm. long, 3-4 mm. wide at base. AREOLES round, 1.5 mm. wide, slightly sunken, with sparce white wool in youth, soon becoming naked. AXILS with 5-8 white tortuous bristles. CENTRAL SPINES 4 (later to 5-7), upper one and laterals 8 mm. long and straight, lower one 12-15 mm. long, and hooked, all heavy acicular, smooth, enlarged at base, reddish brown, ochre-yellow at base, lower one porrect, others somewhat ascending. RADIAL SPINES 20-25, 5-7 mm. long, setaceous, almost hair-like, smooth, straight to slightly bent, glossy white, horizontal to irregularly spreading. FLOWERS 15 mm. wide. *Outer perianth-segments* yellow, pale brownish rose mid-stripe, lanceolate, tip acute, margins entire. *Inner perianth-segments* clear yellow, lanceolate, tip acute, margins entire. *Filaments* yellowish green. *Anthers* yellow. *Style* yellowish green. *Stigma-lobes* 4-5, cream to pale greenish cream. FRUIT red, slender clavate. SEEDS reddish brown, curved pyriform with lateral hilum, very finely pitted, hardly 1 mm.

Distribution: Guanajuato, Mexico.

Type locality: Sierra de San Luis Potosi.

Illustration: Fig. 176 is a reproduction of an illustration in Monatsschr. Deutsch. Kakt. Ges., 4:52, as *Mamillaria knebeliana*.

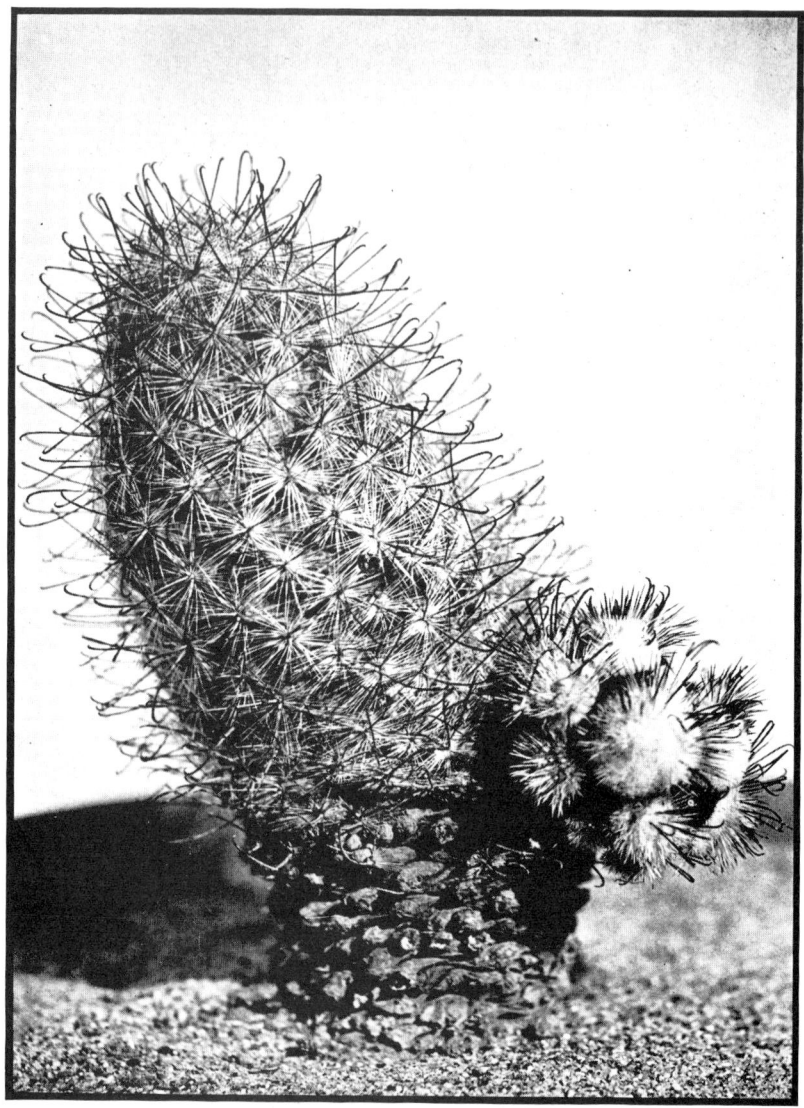

Fig. 177. *Mammillaria phellosperma* x 1

151. Mammillaria phellosperma Engelmann

Mamillaria tetrancistra Engelmann, Amer. Journ. Sci., II, 14:337, 1852.
Mamillaria phellosperma Engelmann, Pro. Amer. Acad., 3:262, 1856.
Cactus phellospermus Kuntze, Rev. Gen. Pl., 1:261, 1891.
Cactus tetrancistrus Coulter, Contr. U. S. Nat. Herb., 3:104, 1894.
Pehellosperma tetrancistra Britton & Rose, The Cactaceae, 4:60, 1923.
Neomammillaria tetrancistra Fosberg, Bull. So. Calif. Acad. Sci., 30:57, 1931.

BODY simple and cespitose, globose to cylindric, 10-15 cm. high, to 7 cm. wide. TUBERCLES loosely arranged in 8 and 13 spirals, soft in texture becoming corky in age, pale green to grayish, ovate to cylindric, with watery sap, 4-7 mm. long. AREOLES very small, round, with slight wool in youth. AXILS with some wool in youth and a few long spine-like bristles. CENTRAL SPINES 3-4, uppers 3-7 mm. long, lower 6-9 mm. long, 2-3 uppers straight or sometimes hooked, lower one hooked and stouter, all acicular, somewhat flexuous, pale brown at base, becoming darker to black at tip, spreading from porrect. RADIAL SPINES 40-60, arranged in two rows, 5-10 mm. long, slender acicular to bristle-like, straight, smooth, white, often with brown tip, nearly horizontal. FLOWERS lateral, funnelform, to 35 mm. wide. *Outer perianth-segments* ovate to oblong-linear, tip obtuse, margins ciliate. *Inner perianth-segments* white border, rose to lavender midstripe. *Stigma-lobes* 5, cream. FRUIT scarlet, obclavate, warty from protrusion of large seeds, 6-12x4-6 mm., with dried perianth not persisting but leaving a wide umbilicus. SEEDS upper seed proper dull black, rugose, 0.6-0.7 mm. wide, lower appendage-like base dusky and 1.2-1.5 mm. long. "The curious spongy or corky appendage is larger than the seed itself and buries its lower part, as it were, in a bluntly lobed cap." ROOTS large, fleshy, branched, tuberous.

Distribution: Arizona, California, and Utah, U. S.; Baja California, and Sonora, Mexico.

Type locality: San Felipe, California.

Illustration: Fig. 177 is a reproduction of an illustration in Cact. Succ. Journ., 5:451, 1933, as *Phellosperma tetrancistra*.

Britton & Rose erected a new genus for this species on the strength of the corky base of the seeds. This characteristic is also found in some of the other members of this genus but to a lesser extent, so we are returning it to Engelmann's modified classification.

FIG. 178. *Mammillaria multihamata*

152. **Mammillaria multihamata** Boedeker

Mamillaria multihamata Boedeker, Monatsschr. Kakteenk., 25:76, 1915.
Neomammillaria multihamata Britton & Rose, The Cactaceae, 4:146, 1923.
Chilita multihamata Orcutt, Cactography, 2, 1926.

BODY simple and cespitose from base, globular to short cylindric, sunken at apex, 5 cm. wide. TUBERCLES arranged in 8 and 13 spirals, glossy dark leaf-green, cylindric, with watery sap, 10 mm. long, 4 mm. wide at base. AREOLES round, 3 mm. wide, with white wool, soon becoming naked. AXILS with no wool but long hair-like bristles.

CENTRAL SPINES 7-9, 10 mm. long, slender acicular, pubescent, 2-3 upper ones straight and thinner, others hooked, all with enlarged base, spreading porrect. RADIAL SPINES 25, 8 mm. long, setaceous, mostly straight, pubescent, white, somewhat ascending. FLOWERS wide funnelform, near apex, 15 mm. long, 12 mm. wide. *Outer perianth-segments* dirty white, wide mid-stripe brownish ventrally and rose dorsally, lanceolate, tip acute, margins ciliate. *Inner perianth-segments* pure white, very fine rose mid-line only at tip, lanceolate, tip acute, margins ciliate. *Filaments* pale rose. *Anthers* bright green (?). *Style* green below, bright brownish rose above. *Stigma-lobes* 4, bright green. FRUIT unknown. SEEDS dark brown, nearly black, globular with exserted ventral hilum.

Distribution: Guanajuato, Mexico.

Type locality: None given.

Illustration: Fig. 178 is a reproduction of an illustration in Monatsschr. Kakteenk., 25:77, as *Mamillaria multihamata*.

153. Mammillaria gilensis Boedeker

Mammillaria gilensis Boedeker, Jahrb. Deutsch. Kakt. Ges., 1:60, 1936.

BODY simple, globose to short cylindric, rounded at apex, but not woolly, 4 cm. wide. TUBERCLES arranged in 8 and 13 spirals, glossy light yellowish leaf-green, thick cylindric, conic, terete at apex, with watery sap, 8-9 mm. long, 3-4 mm. wide at base. AREOLES round, 1 mm. wide, with little wool in youth, soon becoming naked. CENTRAL SPINES 3, seldom 4, upper ones 7 mm. long and straight, lower ones 10 mm. long and hooked, all acicular, smooth, thickened at base, glossy brownish yellow, brighter in new growth, upper ones horizontal, lower ones porrect. RADIAL SPINES 20-25, 5-6 mm. long, straight, nearly hair-like, smooth, whitish to often pale yellow, horizontal, regularly spreading. FLOWERS funnelform, in crown near apex. *Outer perianth-segments* green below, cream above, with wide rose mid-stripe, linear-lanceolate, tip acute, margins entire. *Inner perianth-segments* pure cream, faint rose mid-line at tip, pale green below, linear oblong, margins entire (?). *Filaments* white. *Anthers* bright yellow. *Stigma-lobes* 4-5, greenish white, small, overtop anthers. FRUIT clavate, small. SEEDS dark brownish gray (black?), globular with long white hilum, constricted above base, pitted, 1 mm.

Distribution: Aguas Calientes, Mexico.

Type locality: San Gil.

Illustration: Jahrb. Deutsch. Kakt. Ges., 1:61, as *Mammillaria gilensis*.

154. Mammillaria hirsuta Boedeker

Mamillaria hirsuta Boedeker, Monatsschr. Kakteenk., 29:130, 1919.
Neomammillaria hirsuta Britton & Rose, The Cactaceae, 4:146, 1923.
Chilita hirsuta Orcutt, Cactography, 2, 1926.
Mammillaria aureoviridis Heinrich, Kakteen Andere Sukk., 56, 1937.

BODY simple and cespitose from base, flat to rounded cylindric, rounded at apex, 6 cm. wide. TUBERCLES arranged in 8 and 13 spirals, flabby in texture, pale gray-green, spotted, cylindric, rounded at apex, with watery sap, 10 mm. long, 5 mm. wide at base. AREOLES round, 1 mm. wide, not sunken, with yellowish mat but not woolly. AXILS with no wool but few thin hair-like bristles. CENTRAL SPINES 3-4, upper ones 10-15 mm. long, lower ones 15-20 mm. long, all slender acicular, pubescent, enlarged at base, upper ones straight, lower one hooked and heavier, upper ones glossy white with brown tip, lower dark reddish brown, base ochre-yellow, upper ones ascending, lower ones porrect. RADIAL SPINES 20 or more, 10-15 mm. long, hair-like, smooth, bent, glossy

FIG. 179. *Mammillaria hirsuta* x 1

white, horizontal, interlacing. FLOWERS campanulate, near top, 10-17 mm. long. *Onter perianth-segments* pale olive-greenish brown base, dark rose to yellowish mid-stripe with rose mid-line, whitish margins, eliptical, tip acute, margins entire. *Inner perianth-segments* greenish base, light to bright yellow, rose luster, lanceolate, tip obtuse to acute, margins entire to slightly serrate at tip. *Filaments* whitish. *Anthers* bright golden. *Style* bright yellowish green below, faint rose above. *Stigma-lobes* 3-5, very pale greenish white, small. FRUIT green with slight tint of brown. SEEDS glossy black, globular with exserted ventral and more or less corky hilum, pitted, 1.3x0.7 mm.

Distribution: Guerrero, Mexico.

Type locality: None given but reported from Zapilote Canyon between Taxco and Alcopulco (Bravo).

Illustration: Fig. 179 is from a photograph of a plant obtained in Mexico.

155. Mammillaria longicoma (B. & R.) Berger

Neomammillaria longicoma Britton & Rose, The Cactaceae, 4:146, 1923.
Chilita longicoma Orcutt, Cactography, 2, 1926.
Mammillaria longicoma Berger, Kakteen, 292, 1929.

BODY simple and cespitose from base, globose to cylindric, 3-5 cm. wide. TUBERCLES arranged in 13 and 21 spirals, soft in texture, dark green, conic, obtuse, with watery sap, 4-5 mm. long, 3-4 mm. wide at base. AREOLES oval, small, with very scant wool only in the youngest. AXILS with long white hairs. CENTRAL SPINES 4, 10-12 mm. long, very slender acicular, 1-2 hooked, all pubescent, brown above, lighter below, spreading porrect. RADIAL SPINES 25 or more, 6-8 mm. long, weak hair-like, pubescent, tortous, white, nearly horizontal, interlacing. FLOWERS funnelform, from upper axils.

Fig. 180. *Mammillaria longicoma* x 1

Outer perianth-segments light green at base, pinkish olive-green mid-stripe above, cream at margins, linear-lanceolate, tip acuminate, margins entire. *Inner perianth-segments* nearly white to very pale pinkish cream, very pale green ventral mid-stripe, linear-lanceolate, tip acute to acuminate, margins entire. *Filaments* very pale pink. *Anthers* yellow. *Style* pale yellow. *Stigma-lobes* 3-5, cream to greenish yellow. FRUIT red, clavate, 15x3 mm., with dried perianth persisting. SEEDS black, glossy, globular with exserted hilum, pitted, 1x0.6 mm.

Distribution: San Luis Potosi, Mexico.

Type locality: San Luis Potosi, S. L. P.

Illustration: Fig. 180 is a reproduction of an illustration, Fig. 162, in Britton & Rose (Cact., 4:147), as *Neomammillaria longicoma*.

156. **Mammillaria kunzeana** Boedeker & Quehl

Mamillaria kunzeana Boedeker & Quhel, Monatsschr. Kakteenk. 17:117, 1912.
Mamillaria bocasana kunzeana Quehl, Monatsschr. Kakteenk., 26:46, 1916.
Neomammillaria kunzeana Britton & Rose, The Cactaceae, 4:145, 1923.
Chilita kunzeana Orcutt, Cactography, 2, 1926.
Mamillaria kunzeana flavispina Hort. ex Schelle, Kakteen, 306, 1926.
Mamillaria kunzeana longispina Hort. ex Schelle, Kakteen, 306, 1926.
Mamillaria kunzeana rubrispina Hort. ex Schelle, Kakteen, 306, 1926.

BODY simple and cespitose from body, globose to cylindric, sunken at apex, to 9 cm. high, 5-6 cm. wide. TUBERCLES arranged in 8 and 13 spirals, flabby in texture, glossy bright green, cylindric to slightly clavate, rounded above, with watery sap, 8-15 mm. long, 3-5 mm. wide at base. AREOLES oval, small, sunken, ventral to tip of tubercle, with very slight wool, soon becoming naked. AXILS with no wool but with numerous tortuous hair-like bristles. CENTRAL SPINES 3-4, 10-20 mm. long, lower longer and hooked, upper straight, all slender acicular, pubescent, semi-flexuous, white to amber-yellow at base, orange-brown to purplish brown above, with black tip, upper ascending, lower porrect. RADIAL SPINES 20-25, 4-10 mm. long, setaceous, straight, pubescent in youth, flexuous, snow white, ascending in youth, later horizontal. FLOWERS funnel-form, 15-20 mm. long, 15 mm. wide. *Outer perianth-segments* rose-red wide mid-stripe, very pale to whitish margins, linear-lanceolate, tip acute, margins entire. *Inner perianth-segments* pale cream to white, pale pink to nearly absent mid-line, lanceolate, tip acute,

Fig. 181. *Mammillaria kunzeana* x 1

margins entire. *Filaments* very pale greenish white. *Anthers* pale yellow. *Style* white below, very pale pink at very top. *Stigma-lobes* 3-5, light yellow, 0.7 mm. long, overtop anthers. FRUIT red, clavate, 15x8 mm., with dried perianth persisting. SEEDS black, glossy, globular with exserted ventral hilum, pitted, 0.9x0.7 mm.

Distribution: Zacatecas, Queretaro; Mexico.

Type locality: None given, but reported from Ocotillo in Sierra de San Moran.

Illustration: Fig. 181 is from a photograph of a plant obtained in Mexico.

The three varieties listed by Schelle are only names without any description, except that the last one might be referrable to *M. glochidiata.*

157. **Mammillaria bombycina** Quehl

Mamillaria bombycina Quehl, Monatsschr. Kakteenk., 20:149, 1910.
Neomammillaria bombycina Britton & Rose, The Cactaceae, 4:161, 1923.
Chilita bombycina Orcutt, Cactography, 2, 1926.

BODY simple and cespitose from base, globose to clavate, sunken at apex, with considerable white wool, to 20 cm. high, 6 cm. wide. TUBERCLES arranged in 11 and 18 spirals, firm in texture, bright green, conic to cylindric, rounded at apex, with watery sap, 15 mm. long, 10 mm. wide at base. AREOLES round in youth, later elongate, with a little white wool in youth, later becoming naked. AXILS with abundant white wool covering tubercles in apex, (occasional bristle but not typical). CENTRAL SPINES 2-4, upper one 7 mm. long, laterals 10 mm. long, lower one 20 mm., upper and lateral straight, lower one hooked, all slender acicular, smooth, white becoming amber-yellow, brownish red at tip, uppers dorsally spreading and ascending, lower porrect. RADIAL SPINES 30-40, 2-10 mm. long, lateral longest, all very thin acicular, stiff, straight, somewhat

Fig. 182. *Mammillaria bombycina* x 1

pubescent, glossy white, mostly lateral and horizontal, upper and lower more ascending. FLOWERS funnelform, near apex, 15 mm. long and wide. *Outer perianth-segments* greenish red at base, reddish white above, lanceolate, tip acuminate, margins ciliate. *Inner perianth-segments* bright carmine-red, darker mid-stripe, long eliptical, tip split, margins ciliate. *Filaments* white below, bright carmine-red above. *Anthers* yellowish. *Style* white below, deep red above. *Stigma-lobes* 4-6, reddish purple, linear, overtop anthers. FRUIT whitish, clavate, 15x2 mm. SEEDS black, very small.

Distribution: Coahuila, San Luis Potosi; Mexico.

Type locality: None given, but reported from Santa Maria, S. L. P.

Illustration: Fig. 182 is from a photograph of a plant we obtained from a local nursery.

"*Mammillaria cordigera* Hesse resembles this species very much in its spines and form but it is described as with grooved tubercles, which would exclude it from this genus." Britton & Rose (Cact. 4:161).

158. Mammillaria seideliana Quehl

Mamillaria seideliana Quehl, Monatsschr. Kakteenk., 21:154, 1911.
Neomammillaria seideliana Britton & Rose, The Cactaceae, 4:144, 1923.
Chilita seideliana Orcutt, Cactography, 2, 1926.

BODY simple and cespitose, globose to short cylindric, rounded above, with apex sunken, but with no wool. TUBERCLES arranged in 8 and 13 spirals, flabby in texture, dark green or purplish, slender cylindric, terete, with watery sap, 10 mm. long, 5 mm. wide at base. AREOLES oval, with white wool only in youngest. AXILS with no wool, and only occasional bristle. CENTRAL SPINES 3-4, upper ones 5-8 mm. long, straight and setaceous, lower one 10-15 mm. long, hooked and slender acicular, all pubescent in youth, white with brown tip, upper nearly radial, lower porrect. RADIAL SPINES 20 or more (18-25), 5-8 mm. long, setaceous, straight to slightly bent, pubescent in youth, white, horizontal. FLOWERS funnelform, near apex, 15-20 mm. long, 15 mm. wide. *Outer perianth-segments* bright yellow, pale rose-red back stripe with a tint of brown, oblong, tip acute, margins entire. *Inner perianth-segments* clearer yellow to nearly white,

FIG. 183. *Mammillaria seideliana*

faint brownish pink mid-line, lanceolate, tip acute, margins entire. *Filaments* white. *Anthers* yellow. *Style* white to cream. *Stigma-lobes* 3-6, pale greenish yellow, overtop anthers. FRUIT scarlet-red, slender curved clavate, 30x4 mm., persisting in axils apparently for a number of years, with dried perianth persisting. SEEDS black, glossy, oval with basal hilum, constricted above base, pitted, 1 mm.

Distribution: Zacatecas and Queretaro, Mexico.

Type locality: None given.

Illustration: Fig. 183 is a reproduction of an illustration, Fig. 157a, in Britton & Rose (Cact., 4:144) as *Neomammillaria seideliana*.

"Although the flowers appear to come from near the top of the plant, they are all from axils of old tubercles. In the single specimen examined, the flowers appeared before the plant began to form new tubercles. In *Mammillaria barbata*, a closely related species, the flowers occur at both the old and new tubercles, but so far as known, no other species possess that character, although there is no good reason for not finding it in closely related species." Britton & Rose (Cact., 4:144).

159. Mammillaria scheidweileriana Otto

Mammillaria glochidiata sericata Lemaire, Cact. Gen. Nov. Sp., 40, 1839.
Mammillaria scheidweileriana Otto in Dietrich, Allg. Gartenz., 9:179, 1841.
Mamillaria wildiana rosea Salm-Dyck, Cact. Hort. Dyck. 1849, 81, 1850.
Cactus scheidweilerianus Kuntze, Rev. Gen. Pl., 1:261, 1891.
Neomammillaria scheidweileriana Britton & Rose, The Cactaceae, 4:148, 1923.
Chilita scheidweileriana Orcutt, Cactography, 2, 1926.

BODY cespitose from base, globose to cylindric, rounded at apex. TUBERCLES arranged in 8 and 13 spirals, dark green, cylindric, terete, sunken at apex, with watery sap, 6 mm. long. AREOLES pyriform, with very scant wool only in very youngest. AXILS with little wool, but no bristles. CENTRAL SPINES 1-4, upper 3 shorter and straight, lower one hooked, all pubescent, rigid, acicular, upper reddish, lower dark purple at tip, yellowish white below, uppers nearly horizontal, lower porrect. RADIAL SPINES 25-30, 8-12 mm. long, pubescent, setaceous to very fine bristles, flexuous, silvery white, somewhat ascending. FLOWERS wide funnelform, 12-15 mm. wide. *Outer perianth-segments* (6) brownish rose, whitish margins, linear-lanceolate, tip acuminate, margins entire. *Inner perianth-segments* (9) rose, brownish point and mid-line, whitish margins, linear-lanceolate, tip acuminate or somewhat split, margins usually entire. *Filaments* deep pink. *Anthers* orange-yellow. *Style* deep pink. *Stigma-lobes* 4, pale pink. FRUIT unknown. SEEDS black.

Distribution: San Luis Potosi (?) (Ehrenb.) Hidalgo (Schmoll).

Fig. 184. *Mammillaria scheidweileriana* x 1

Type locality: None given, but reported from Zimapan, Hgo. (Schmoll).

Illustration: Fig. 184 is from a photograph of a plant sent to us by Schwarz & Georgi of San Luis Potosi, S.L.P.

Mammillaria monancistra Berg in Schumann, Gesamtb. Kakteen, 533, 1898, which is referred here by Britton & Rose should possibly be referred to *M. erythrosperma* because its description calls for bristles in the axils.

Fig. 185. *Mammillaria mercadensis* x 1

160. Mammillaria mercadensis Patoni

Mamillaria mercadensis Patoni, Alianza Cientifica Universal, 1:54, 1910.
Mamillaria ocamponis Ochoterena, Bol. Direc. Estud. Biol., 2:355, 1918.
Neomammillaria mercadensis Britton & Rose, The Cactaceae, 4:145, 1923.
Chilita mercadensis Orcutt, Cactography, 2, 1926.

BODY simple and cespitose from base, globose, sunken at apex, 5 cm. wide. TUBERCLES arranged in 8 and 13 spirals, olive to dark green, conic to short cylindric, terete, with watery sap, 10 mm. long. AREOLES small oval, 2 mm. long, with white wool only in youth. AXILS naked. CENTRAL SPINES 4-7, 15-25 mm. long, lower longer and hooked, uppers straight, all stout acicular, white at base, chestnut-brown to red in upper half, spreading from porrect. RADIAL SPINES 25-30, 5-8 mm. long, slender acicular, pubescent in youth, white, slightly ascending. FLOWERS 30 mm. wide. *Inner perianth-segments* pale rose, tip obtuse. FRUIT unknown. SEEDS unknown but probably black.

Distribution: Durango, Mexico.

Type locality: Cerro de Mercado.

Illustration: Fig. 185 is from a photograph of a plant sent to us by F. Schmoll of Cadereyta, Qro.

We had this plant but it died before it flowered, so the description will have to remain incomplete until further collection of material can be made. It has been reported that the cactaceae of this region is becoming very scarce due to the mining operations.

FIG. 186. *Mammillaria insularis* x 1

161. Mammillaria insularis Gates

Mammillaria insularis Gates, Cact. Succ. Journ., 10:25, 1938.

BODY cespitose from body and base, flattened globular, deep-seated, 6 cm. high, 5 cm. wide. TUBERCLES arranged in 5 and 8 spirals, bluish green, conical truncate, base irregularly sided, with watery viscid sap, 7 mm. long and wide at base. AREOLES round, 2 mm. wide with white wool, soon becoming naked. AXILS naked or with slight wool. CENTRAL SPINES 1, 10 mm. long, acicular, small hook turned laterally, slightly enlarged at base, black, shading through brown to yellow, porrect. RADIAL SPINES 20-30, 5 mm. long, slender acicular, stiff, white, horizontal. FLOWERS wide funnelform, 15-25 mm. long. *Outer perianth-segments* (10) light green mid-stripe, lanceolate, tip acute, margins entire. *Inner perianth-segments* (12) light pink mid-stripe, white margins, wide lanceolate, tip acute, margins entire. *Anthers* yellow. *Stigma-lobes* 4-5, greenish yellow, 3-4 mm. FRUIT orange-red, clavate, smooth, 10x2 mm., with dried perianth persisting. SEEDS black, dull, irregular globular, with hilum basal, flat, reddish brown, slightly pitted, less than 1 mm. ROOTS very heavy fleshy.

Distribution: Baja California, Mexico.

Type locality: Eastern most islet of Smith Island in Los Angeles, Bay, Lat. 29°5'N., Long. 113° 30' W.

Illustration: Fig. 186 is from a photograph by Mr. George Lindsay of plants that he collected at the type locality.

FIG. 187. *Mammillaria longiflora* x 1

162. Mammillaria longiflora (B. & R.) Berger

Neomammillaria longiflora Britton & Rose, The Cactaceae, 4:163, 1923.
Chilita longiflora Orcutt, Cactography, 2, 1926.
Mammillaria longiflora Berger, Kakteen, 296, 1929.
Krainzia longiflora Backeberg, Blatt. Kakteenforschung, 6, 1938.

BODY simple and cespitose, 3 cm. wide. TUBERCLES arranged in 5 and 8 spirals, dark green, cylindric, terete, with watery sap, 5-10 mm. long. AREOLES oval, with a little white woolly felt in youth, later becoming naked. AXILS naked or with a little wool. CENTRAL SPINES 4, lower one hooked, upper 3 straight, all 10-13 mm. long, stout acicular, pubescent, reddish brown, lower porrect, uppers ascending. RADIAL SPINES 25-30, 10-13 mm. long, setaceous, straight, pubescent, light yellow to straw-colored, somewhat ascending. FLOWERS funnelform, distinct long narrow tube to 15 mm. long, near top of plant, 25 mm. wide, 20 mm. long exclusive of tube. *Outer perianth-segments* pale brownish green mid-stripe, darker at tip, dirty cream margins, oblong, tip acute, margins entire. *Inner perianth-segments* pale pink, darker mid-line and point, oblong to spatulate, tip acute to acuminate, margins entire. *Filaments* very pale pink to violet-rose. *Anthers* yellow. *Style* very pale pink below, darker above. *Stigma-lobes* 4, light yellow, overtop anthers 4 mm. Ovary very small, ovoid, more or less sunken in axils, thin above and perhaps opening by operculum, lower part with seeds persisting for years. FRUIT unknown. SEEDS black, nearly globose with prominent white ventral hilum, minutely pitted, 1-1.5 mm.

Distribution: Durango, Mexico.

Type locality: Santiago Papasuiaro.

Illustration: Fig. 187 is from a photograph of a plant we obtained from Johnson's Water Gardens of Hynes, California.

FIG. 188. *Mammillaria posseltiana*

163. **Mammillaria posseltiana** Boedeker

Mammillaria posseltiana Boedeker, Monatsschr. Deutsch. Kakt. Ges., 4:99, 1932.

BODY simple, globose to ovate, rounded at apex, to 5 cm. wide. TUBERCLES arranged in 8 and 13 spirals, dark leaf-green, in youth in axils rose colored, conical, with watery sap, 10 mm. long, 4-5 mm. wide at base. AREOLES round, 3 mm. wide, with wool in youth. AXILS with white wool, but no bristles. CENTRAL SPINES 4, 9 mm. long, stout acicular, lower bent to hooked, uppers nearly horizontal. RADIAL SPINES 20, 9 mm. long, thin acicular, straight, smooth, white, horizontal and nearly equally spreading. FLOWERS 25 mm. long, 15 mm. wide. *Outer perianth-segments* white, bright rose to darker mid-stripe, slender-lanceolate, tip acuminate, margins entire, 14x1.5 mm. *Inner perianth-segments* white, dark rose mid-stripe, lanceolate, tip acute, margins somewhat serrate at tip, 16x2 mm. *Filaments* white below, dark red above. *Anthers* yellow. *Style* greenish below, rose above. *Stigma-lobes* 4-5, green, overtop anthers. FRUIT pale green, clavate, 15 mm. long. SEEDS black, faintly glossy, curved pyriform with long lateral hilum near the base, 1 mm.

Distribution: Guanajuato, Mexico.

Type locality: Sierra de Guanajuato.

Illustration: Fig. 188 is a reproduction of an illustration in Monatsschr. Deutsch. Kakt. Ges., 4:99, as *Mammillaria posseltiana*.

FIG. 189. *Mammillaria microcarpa* x 1

164. Mammillaria microcarpa Engelmann

Mamillaria microcarpa Engelmann in Emory, Mil. Reconn., 157, 1848.
Mamillaria grahamii Engelmann, Pro. Amer. Acad., 3:262, 1856.
Cactus grahami Kuntze, Rev. Gen. Pl., 1:260, 1891.
Mamillaria grahamii arizonica Quehl, Monatsschr. Kakteenk., 6:44, 1896.
Coryphantha grahami Rydberg, Fl. Rocky Mts., 582, 1917.
Neomammillaria microcarpa Britton & Rose, The Cactaceae, 4:155, 1923.
Neomammillaria milleri Britton & Rose, The Cactaceae, 4:156, 1923.
Chilita grahami Orcutt, Cactography, 2, 1926.
Chilita milleri Orcutt, Cactography, 2, 1926.
Mammillaria milleri Boedeker, Mammill. Vergl. Schluss., 30, 1933.

BODY simple and cespitose from base and body, globose to cylindric, 5-6 cm. wide, to 16 cm. high, rounded at apex. TUBERCLES arranged in 13 and 21 spirals, firm in texture, becoming corky in age, dark grayish green, conic, terete at apex, blunt tip, with watery sap, 7 mm. long, 6 mm. wide at base. AREOLES oval, with very scant wool only in the youngest. AXILS naked. CENTRAL SPINES 1-3, to 18 mm. long, stout acicular, smooth, stiff, 1 strongly hooked and stouter, others subcentral and straight, all with base not enlarged, extremely variable in color from light tan to black, hooked one porrect, straight upper ones horizontal. RADIAL SPINES 20-30, 6-12 mm. long, upper shorter, all acicular, straight, smooth, stiff, white to dark yellow with brown tip, horizontal. FLOWERS broad funnelform, lateral near apex, 40 mm. wide, to 25 mm. long. *Outer perianth-segments* greenish tan mid-stripe, darker at tip, pale green margins with a tinge of pink, lanceolate, tip acute, margins ciliate. *Inner perianth-segments* pink, darker mid-stripe, very pale margins, linear-oblong, tip acute, margins entire, 20x4-5 mm. *Filaments* deep pink, paler at base. *Anthers* yellow to orange. *Style* pinkish above, pale yellowish green below. *Stigma-lobes* 6-8, light green with tan tinge, 5 mm. long, overtop anthers. FRUIT dimorphic (a) scarlet, clavate, 20-25 mm. long, with dried perianth persisting, protruding beyond tubercles; (b) green, small oval, with dried perianth missing, with pod remaining inconspicuously covered by the spines until disintegration, develops from late season flowers. SEEDS black, glossy, oblique obovate with small narrow hilum, slightly pitted, 1 mm. ROOTS fibrous.

Distribution: California, Arizona to Texas, U. S.; Sonora and Chihuahua, Mexico.

Type locality: Along the Gila River, Arizona, 3000-4000 above sea level.

Illustration: Fig. 189 is from a photograph of a plant we collected in northern Sonora.

The original description is supplemented by Engelmann's other works and from plants that we collected at various places from its western limits near Parker Dam in California to various points in western, central and across the southern part of Arizona to eastern New Mexico and western Texas and southward to the southern part of Sonora, Mexico, especially along the coast near Guaymas and inland along the Sonora River. Mr. R. H. Peebles of Sacaton, Arizona, generously supplied the data on the fruiting habits.

This species has such a very wide distribution that geographical variations inevitably show up due to the differences of soil, rain fall, temperature, etc. The color of the spines shows such a wide variation that no definite coloration can be taken as definitely characteristic and likewise the length and robustness of the spines.

Engelmann's description of *M. grahamii* calls for 20-30 radial spines and "in some Sonoran specimens only 15-18." The latter form we are referring to *M. sheldonii*.

M. grahamii was described as sp. nov. in *Cactaceae of the Boundary* which was not published until 1858, due to the delay in the preparation of the illustrations. In the meantime a short synopsis was published in the *Proceedings of the American Academy of Arts and Sciences* in 1856, which is given priority of publication over the full description originally described by Engelmann.

Fig. 190. *Mammillaria weingartiana*

165. Mammillaria weingartiana Boedeker

Mammillaria weingartiana Boedeker, Monatsschr. Deutsch. Kakt. Ges., 4:219, 1932.

BODY simple and cespitose from base, globose, sunken at apex, 4-5 cm. wide. TUBERCLES loosely arranged in 13 and 21 spirals, dark leaf-green, faintly glossy, slender conic, sunken at apex, with watery sap, 8 mm. long, 3 mm. wide at base. AREOLES round, 2 mm. wide, becoming smaller, with some white wool, soon becoming naked. AXILS naked. CENTRAL SPINES 1-4, to 12 mm. long, stout acicular, 1-2 hooked, all smooth, red becoming brownish black, hooked ones porrect, others ascending. RADIAL SPINES 20-25, 6-8 mm. long, uniform, very slender, straight, pure white, horizontal. FLOWERS funnelform, 10 mm. long, 12 mm. wide. *Outer perianth-segments* pale green, pale olive-brown ventral-stripe, pale pink tips, lanceolate, tip acuminate, margins serrate, 10x1.5 mm. *Inner perianth-segments* pale greenish yellow, pale brownish rose midstripe, darker tip, green base, lanceolate, margins ciliate in middle. *Filaments* green below, rose above. *Anthers* bright yellow. *Style* green. *Stigma-lobes* 3-5, white to pale rose, overtop anthers. FRUIT red, clavate, small. SEEDS black, faintly glossy, pyriform with lateral whitish hilum, 1 mm.

Distribution: Nuevo Leon, Mexico.

Type locality: Near Ascension.

Illustration: Fig. 190 is a reproduction of an illustration in Monatsschr. Deutsch. Kakt. Ges., 4:219, and also in Cact. Succ. Journ., 8:60, as *Mammillaria weingartiana*.

FIG. 191. *Mammillaria bocasana* x 1

166. Mammillaria bocasana Poselger

Mamillaria bocasana Poselger, Allg. Gartenz., **21**:94, 1853.
Cactus bocasanus Coulter, Contr. U. S. Nat. Herb., 3:104, 1894.
Mamillaria bocasana splendens Rebut, Catalogue, 7, 1896. Nomen.
Mamillaria bocasana cristata Schelle, Handb. Kak., 250, 1907. Nomen.
Mamillaria bocasana flavispina Schelle, Handb. Kak., 250, 1907. Nomen.
Neomammillaria bocasana Britton & Rose, The Cactaceae, 4:147, 1923.
Chilita bocasana Orcutt, Cactography, 2, 1926.
Mammillaria bocasana multihamata Schelle, Kakteen, 304, 1926. Nomen.

BODY simple and cespitose, forming large mounds, globose, becoming cylindric, 4-5 cm. wide. TUBERCLES arranged in 8 and 13 spirals, flabby in texture, light to dark bluish green, slender cylindric to somewhat conic, terete, with watery sap, 6-10 mm. long, 2-4 mm. wide at base. AREOLES oval to round, with scant yellowish woolly felt. AXILS naked. CENTRAL SPINES 1 (2-3), 5-8 mm. long, thin acicular, pubescent, lower hooked, upper straight, all yellowish brown, darker at apex, lower porrect, upper nearly radial. RADIAL SPINES 25-30, 8-20 mm. long, very fine hair-like, tortuous, smooth, at first ascending, later nearly horizontal, interlacing. FLOWERS funnelform, tube as long as tubercles, 16 mm. long, 12 mm. wide. *Outer perianth-segments* green base, light reddish brown ventral mid-stripe, greenish yellow margins, lanceolate, tip acute, margins entire. *Inner perianth-segments* yellow, light reddish brown mid-line and point, linear-lanceolate, tip acute to obtuse, margins entire. *Filaments* white to pale yellow or to pale rose above. *Anthers* yellow. *Style* pale yellow. *Stigma-lobes* 4, light green, level with anthers. FRUIT red, slender clavate. SEEDS black, broad obovate with narrow basal hilum, small.

Distribution: San Luis Potosi, Mexico.

Type locality: Sierra de Bocas.

Illustration: Fig. 191 is from a photograph of a plant obtained in Mexico.

Coulter was the first to describe the seeds as "cinnamon brown" and was followed by subsequent authors. His description of the plant differs in several respects from the original, but specimens that we obtained in Mexico and which did check in other respects with the original description, were found to have black seeds as do other closely associated species.

The original description calls for naked axils but Schumann reports 3-6 bristles and he is followed by subsequent authors. So, inasmuch as his description differs in other respects from the original, we are discounting his notes on this species.

The two varieties *M. bocasana cristata* and *flavispina* are listed by Schelle (Handb. Kakteenk., 250, 1907), but not described. The former is offered by Grassner in his Kakteen for 1914. We do not find that *M. bocasana splendens* Liebner and *M. bocasana sericata* Lemaire, mentioned by Quehl (Monatsschr., 19:46, 1909), have even been described.

"*Mammillaria schelhasei lanuginosior* Hildmann in Schumann (Gesamtb. Kakteen, 531, 1898), we have not seen, but it may belong here.

"*Mammillaria bocasana splendens* credited to Schlechtendal, is offered for sale by Haage & Schmidt in their 1922 Catalogue." Britton & Rose (Cact. 4:14). This variety has been also attributed to Hildmann and also to Liebner.

Mammillaria bocasana inermis Hort. is briefly characterized but not described by Hummel (Cact. Succ. Journ., 8:102, 1936), but well illustrated.

167. Mammillaria painteri Rose

Mamillaria painteri Rose in Quehl, Monatsschr. Kakteenk., 27:22, 1917.
Neomammillaria painteri Britton & Rose, The Cactaceae, 4:151, 1923.
Chilita eschauzieri Orcutt, Cactography, 2, 1926.

BODY simple, globose to cylindric, somewhat sunken at apex, 20 mm. wide. TUBERCLES arranged in 8 and 13 spirals, flabby in texture, cylindric, terete, with watery sap, 7 mm. long, 2 mm. wide at base. AREOLES nearly round, very small, with hardly any wool. AXILS with little wool but no bristles. CENTRAL SPINES 4-5, 10 mm. long, slender acicular, uppers straight, lower one hooked, all pubescent, dark brown, darker at tip, ascending. RADIAL SPINES 20 or more, 5 mm. long, setaceous, straight, pubescent, white, somewhat ascending, later horizontal. FLOWERS funnelform, 10-15 mm. long and wide, May. *Outer perianth-segments* reddish tan mid-stripe, darker ventrally, light

FIG. 192. *Mammillaria painteri*

olive-green base, pale cream margins, linear-lanceolate, tip acute, margins entire. *Inner perianth-segments* greenish white, very pale yellow mid-line, lanceolate, tip acute, margins entire. *Filaments* white. *Anthers* pale yellow. *Style* pale green. *Stigma-lobes* 4-5, pale green to cream, 1.5 mm. long, level with anthers. FRUIT red, cylindric, 10x2 mm., with dried perianth persisting. SEEDS black, glossy, short blunt clavate with extended ventral hilum, faintly pitted, constricted above base.

Distribution: Queretaro, Mexico.

Type locality: San Juan del Rio.

Illustration: Fig. 192 is a reproduction of an illustration, Fig. 169, in Britton & Rose (Cact., 4:152), as *Neomammillaria painteri*.

168. Mammillaria boedekeriana Quehl

Mamillaria boedekeriana Quehl, Monatsschr. Kakteenk., 20:108, 1910.
Neomammillaria boedekeriana Britton & Rose, The Cactaceae, 4:154, 1923.
Chilita boedekeriana Orcutt, Cactography, 2, 1926.

BODY simple, globose to short cylindric, 6-7 cm. high, 4 cm. wide. TUBERCLES arranged in 8 and 13 spirals, flabby in texture, dark green, very finely pitted, cylindric to clavate, with watery sap, 10 mm. long, 5 mm. wide at base. AREOLES round, 2-3 mm. wide, with scant white woolly mat at first, soon becoming naked. AXILS naked. CENTRAL SPINES 3-4, 8-13 mm. long, lower longer, heavier and hooked, upper straight, all acicular, smooth, bright brown, black at tip, some coppery brown, hooked one porrect, uppers nearly radial. RADIAL SPINES 20 or more, 10 mm. long, upper shorter and thinner, all acicular, lightly pubescent, glossy white to white with yellow luster, ascending in youth, later becoming horizontal. FLOWERS funnelform, near apex, 25-30 mm. long. *Outer perianth-segments* light reddish brown mid-stripe, yellow margins, linear-lanceolate, tip acuminate, margins lightly serrate. *Inner perianth-segments* pinkish tan mid-stripe, wide tannish yellow margins, linear-lanceolate, tip acuminate, margins entire. *Filaments* pale green to white below, pale pink above. *Anthers* yellowish orange. *Style* white to pale green below, olive-green above. *Stigma-lobes* 3-4, yellow. FRUIT red, clavate. SEEDS black. ROOTS tuberous.

Distribution: Guanajuato and San Luis Potosi, Mexico.

Fig. 193. *Mammillaria boedekeriana*

Type locality: None given but reported from near La Maria, S.L.P.

Illustration: Fig. 193 is a reproduction of an illustration, Fig. 172a, in Britton & Rose (Cact., 4:154), as *Neomammillaria boedekeriana*.

Fig. 194. *Mammillaria sinistrohamata* x 1

169. Mammillaria sinistrohamata Boedeker

Mamillaria sinistrohamata Berger, Kakteen, 296, 1929. Nomen.
Mammillaria sinistrohamata Boedeker, Monatsschr. Deutsch. Kakt. Ges., 4:162, 1932.

BODY simple, globose, slightly sunken at apex, but not woolly, 4-5 cm. wide.
TUBERCLES loosely arranged in 13 and 21 spirals, glossy leaf-green, short cylindric,

somewhat bent downward, rounded dorsally, with watery sap, 8 mm. long, 4 mm. wide at base. AREOLES round, hardly 2 mm. wide, with white wool only in the youngest, soon becoming naked. AXILS naked. CENTRAL SPINES 4, three uppers 8-10 mm. long, straight and thinner, lower one 14 mm. long, hooked and turned to the left, all strong acicular, smooth, enlarged at base, translucent amber-yellow, upper ascending fan-like, lower depressed somewhat from porrect. RADIAL SPINES 20 or more, 8-10 mm. long, very thin acicular, straight, smooth, slightly enlarged at base, white to pale yellow, horizontal. FLOWERS funnelform, near apex, 15 mm. long, 12 mm. wide. *Outer perianth-segments* greenish cream, greener mid-stripe ventrally, reddish at apex, lanceolate, tip acute, margins serrate, 8x2 mm. *Inner perianth-segments* greenish cream to ivory, delicate cream mid-line, lanceolate, tip acute, margins serrate, longer and wider. *Filaments* whitish. *Anthers* bright yellow. *Style* whitish. *Stigma-lobes* 5, yellowish white, overtop anthers. FRUIT red, small. SEEDS black, somewhat glossy, curved pyriform with small elongate hilum, finely pitted, hardly 1 mm.

Distribution: Zacatecas, Durango, Coahuila; Mexico.

Type locality: Zacatecas.

Illustration: Fig. 194 is a reproduction of an illustration in Monatsschr. Deutsch. Kakt. Ges., 4:162, as *Mammillaria sinistrohamata*.

FIG. 195. *Mammillaria moelleriana*

170. **Mammillaria moelleriana** Boedeker

Mamillaria mölleriana Boedeker, Zeitschr. Sukkulentenk., 1:213, 1924.
Neomammillaria moelleriana Britton & Rose, Contr. U.S. Nat. Herb., 23:1678, 1924.

BODY simple, globose, rounded at apex but not sunken, 6 cm. wide. TUBERCLES arranged in 8 and 13 spirals, firm in texture, glossy leaf-green, cylindric to ovate, with watery sap, 8 mm. long, 8 mm. wide at base. AREOLES oval, 3-4 mm. wide, with very white wool only in youngest. AXILS naked. CENTRAL SPINES 8-9, seldom 10, 4 lower ones 20 mm. long and hooked, upper ones shorter and straight, all stout acicular, smooth, enlarged at base, bright honey-yellow above, dark reddish brown below, strongly divergent. RADIAL SPINES 35-40, 7-9 mm. long, acicular, straight, smooth stiff, snow white, faintly yellow at base, slightly ascending. FLOWERS funnelform, near top, 15 mm. long and wide. *Outer perianth-segments* dirty white to pale rose with brownish mid-stripe, lanceolate, tip acute, margins entire. *Inner perianth-segments* cream to yellow, pink mid-stripe, linear, tip obtuse, margins entire. *Filaments* white below, light rose above. *Anthers* yellow. *Style* white. *Stigma-lobes* 5-6, yellowish rose. FRUIT greenish white, clavate, very juicy, 15 mm. long. SEEDS black, glossy, oblong round with basal white hilum, smooth, 1 mm. ROOTS fibrous.

Distribution: Durango, Mexico.

Type locality: Sierra de Santa Maria.

Illustration: Fig. 195 is a reproduction of an illustration in Zeitschr. Sukkulentenk. 1:214, as *Mamillaria mölleriana.*

Fig. 196. *Mammillaria gasseriana*

171. Mammillaria gasseriana Boedeker

Mamillaria gasseriana Boedeker, Zeitschr. Sukkulentenk., 3:75, 1927.

BODY simple and cespitose from base, globular to short ovoid, sunken at apex, 3-4 cm. wide. TUBERCLES arranged in 8 and 13 spirals, faint gray-green, reddish at base, glossy, short ovoid, terete at apex, pitted, with watery sap, 6 mm. long, 5-6 mm. wide at base. AREOLES slender oval, 2 mm. long, with very white wool, soon becoming naked. AXILS naked. CENTRAL SPINES 1-2, 8 mm. long, stout acicular, hooked, white below, becoming dark brown, strongly spreading from porrect. RADIAL SPINES 40-50, in 2-3 rows in confusion, 5-8 mm. long, fine acicular, pubescent, often little bent, white, occasionally in youth with reddish tip, horizontal to little ascending. FLOWERS wide funnelform, near top, 7-8 mm. wide. *Outer perianth-segments* whitish cream, fine faint brownish mid-line, opaque glossy, wide lanceolate, tip acute, margins serrate. *Inner perianth-segments* same as outer but with throat greenish. *Filaments* whitish. *Anthers* bright yellow. *Style* light greenish yellow. *Stigma-lobes* 4, light greenish, very small, inclined together. FRUIT brownish red, clavate, very small. SEEDS blackish gray (?), glossy, long obovate with exserted ventral hilum, finely pitted, 0.5 mm. ROOTS somewhat tuberous but branching.

Distribution: Coahuila, Mexico.

Type locality: Near Torreon.

Illustration: Fig. 196 is a reproduction of an illustration in Zeitschr. Sukkulentenk., 3:76, as *Mamillaria gasseriana.*

Fig. 197. *Mammillaria barbata*

172. **Mammillaria barbata** Engelmann

Mamillaria barbata Engelmann in Wislizenus, Mem. Tour. North. Mex., 105, 1848.
Cactus barbatus Kuntze, Rev. Gen. Pl., 1:260, 1891.
Neomammillaria barbata Britton & Rose, The Cactaceae, 4:144, 1923.
Chilita barbata Orcutt, Cactography, 2, 1926.

BODY simple to cespitose from base, depressed globose, woolly at apex, 3-5 cm. wide. TUBERCLES arranged in 8 and 13 spirals, globose to cylindric, with watery sap, 8 mm. long, 3 mm. wide at base. AREOLES oval, small with scant white wool in youth, soon becoming naked. AXILS naked. CENTRAL SPINES 1 (2), 12-15 mm., stout acicular, hooked downward, pubescent in youth, brownish, porrect. RADIAL SPINES 50-60, several rows (outer to 40, inner 10-15), 6-8 mm. long, acicular, straight, outer white, inner yellowish brown, darker tip, horizontal. FLOWERS funnelform, length and width 18-20 mm., almost centrally placed. *Outer perianth-segments* (8 outer scales, 12 inner) outer ones green, inner ones reddish, lanceolate to linear-lanceolate, tip cuspidate, margins lightly ciliate. *Inner perianth-segments* rose-red, darker mid-stripe, linear-lanceolate, shorter and narrower than outer, margins entire. *Filaments* rose. *Anthers* yellow. *Stigmalobes* 5-6, greenish yellow. FRUIT oblong, 10-12 mm. long, with dried perianth remaining. SEEDS black, glossy, obovate with extended white basal hilum, pitted, minute.

Distribution: Chihuahua, Mexico.

Type locality: Cosihuiriachi.

Illustration: Fig. 197 is a reproduction of an illustration, Fig. 159, in Britton & Rose (Cact. 4:145), as *Neomammillaria barbata*.

The first flowers of the season appear in the axils of the last tubercles of the preceeding year (so in the center of the apex) while the later ones develop from the axils of the first tubercles of the same season.

Britton & Rose in error changed the description of the number of radial spines from the original of 50-60 to 20.

Schumann referred this species to *M. grahamii* (*M. microcarpa*) but in doing so he failed to take into account the number of radial spines and the position of the flowers.

Engelmann originally described the seeds as dark brown but this was questioned because other closely related species all possessed black ones. We obtained plants from Chihuahua while in Mexico in 1942 from Sr. F. Schmoll. There were some seeds in a dried fruit retained between the older tubercles that appeared dark brown but when they were washed they were found to be black.

Fig. 198. *Mammillaria guelzowiana* x 2

173. **Mammillaria guelzowiana** Werdermann

Mamillaria gülzowiana Werdermann, Zeitschr. Sukkulentenk., 3:356, 1928.
Mamillaria gülzowiana splendens Neal, Cact. other Succ. 87, 1935.

BODY simple and cespitose from base, nearly globular, flattened at top and somewhat sunken, 4-6 cm. high, 7 cm. wide. TUBERCLES arranged in 8 and 13 spirals, flabby in texture, becoming corky and dried in age, glossy leaf-green, conic, terete, slightly flattened dorsally, with watery sap, 12-13 mm. long, 4-5 mm. wide at base. AREOLES round to somewhat oval, 1 mm. wide, short yellowish white wool in youth, very soon becoming naked. AXILS naked. CENTRAL SPINES 1-2, 8-10 mm. long, slender acicular, smooth, somewhat curved, apex sharply hooked, yellowish to reddish brown, porrect. RADIAL SPINES 60-80, 15-20 mm. long, setaceous, smooth, tortuous, pure white, horizontal to ascending, 2 or more rows. FLOWERS campanulate, long tube, near top, 50 mm. long, 60 mm. wide, May to September. *Outer perianth-segments* greenish bronze brown, rose margins, lanceolate, tip acuminate, margins white ciliate, 15 mm. long. *Inner perianth-segments* intense purplish red mid-stripe, lighter in throat and at margins, oblong-lanceolate, tip acute, margins entire, 25x5 mm. *Filaments* white. *Anthers* golden yellow. *Style* white below, pale brownish green above, shorter than filaments. *Stigma-lobes* 3, indistinct, very small, greenish. FRUIT yellow, nearly globular, 8x7 mm. SEEDS black, dull with irregular corky mantle in lower half, 1.5x1 mm. ROOTS fibrous.

Distribution: Durango, Mexico.

Type locality: None given but reported from Minas de Fude.

Illustration: Fig. 198 is a reproduction of an illustration in Borg (Cacti, Pl. XLII) as *Mammillaria gülzowiana*.

FIG. 199. *Mammillaria jaliscana*

174. Mammillaria jaliscana (B. & R.) Boedeker

Neomammillaria jaliscana Britton & Rose, The Cactaceae, 4:160, 1923.
Chilita jaliscana Orcutt, Cactography, 2, 1926.
Mammillaria jaliscana Boedeker, Mammill. Vergl. Schluss., 35, 1933.

BODY simple and cespitose from base, globose, 5 cm. wide. TUBERCLES arranged in 13 and 21 spirals, bright green, cylindric, terete, with watery sap, 4-5 mm. long, 3 mm. wide at base. AREOLES oval, 2-3 mm. wide, with white wool in youth, later becoming naked. CENTRAL SPINES 4-8, one more centrally placed, uppers 7-9 mm. long and straight, lower to 12 mm. long and hooked, all thin subulate, reddish brown, darker at tip, lighter at base, lower porrect, others ascending. RADIAL SPINES 30 or more, to 8 mm. long, uppers little shorter, all straight, smooth, acicular, stiff, white, horizontal. FLOWERS lateral, 10 mm. wide, fragrant. *Outer perianth-segments* purplish, ovate-oblong, tip acute to obtuse, margins more or less serrate. *Inner perianth-segments* purplish pink, oblong, tip obtuse, margins mostly entire. *Filaments* pinkish. *Anthers* yellow. *Stigma-lobes* 3-4, white. FRUIT whitish to pink, blunt clavate, 8 mm. long, with dried perianth persisting. SEEDS black, glossy, curved pyriform with white lateral oblong hilum near the base.

Distribution: Jalisco, Mexico.

Type locality: Rio Blanco, near Guadalajara.

Illustration: Fig. 199 is a reproduction of an illustration in Kakteenk., 75, 1934, as *Mammillaria jaliscana*.

Mammillaria ancistroides Lemaire

Mammillaria ancistroides Lemaire, Cact. Gen. Nov. Spec. Mon., 38, 1839. Non Lehmann.
Mammillaria ancistrina Schelhase in Walper, Repert. Bot. Syst., 2:271, 1843. Nomen.
Cactus ancistroides Kuntze, Rev. Gen. Pl., 1:261, 1891.

BODY globose, sunken at apex, later branching from base. TUBERCLES yellowish green, somewhat cylindric, obtuse, with watery sap, 10-12 mm. long, 4-6 mm. wide at base. AREOLES oval. AXILS naked or with very scant wool. CENTRAL SPINES 5, 4 uppers 4-6 mm. long and straight, lower one 12-14 mm. long, hooked and stronger, all rigid, reddish yellow at base, dark violet at tip. RADIAL SPINES 30-40, 6 mm. long, slender, flexuous, transparent white. FLOWERS (by Salm-Dyck). *Outer perianth-segments* acute. *Inner perianth-segments* pale rose mid-line, emarginate to obtuse. *Stigma-lobes* 3, white, short.

Distribution: Unknown.

This species has usually been referred to *M. glochidiata* but the spine count is greater than the latter and it may belong here, but inasmuch as the type locality is unknown and many of the details are lacking in the description, it is uncertain.

M. ancistrata has been referred to Pfeiffer by Deitrich (Syn. Plant., 93, 1843), and to Schelhase by Walper (Repert. Bot. Syst., 2:271, 1843).

FIG. 200. *Mammillaria colonensis* x 1

175. Mammillaria colonensis sp. nov.

Corpus simplex et caespitosus, mamillis ad 8 et 13 seriebus ordinatis, conoideis, suco aquario, axillis 8-10, setibus, albis; spinis centralibus 1-4, 1-3 rectis, 1-3 mm. longis, raro 1 hamato, ad 7 mm. longo; spinis radialibus 15, 6-7 mm. longis, tenuis; sepalis ciliatis, petalis alboflavis, stigmatibus 5, alboviridibus.

BODY simple and cespitose from base, flattened globular, 6-7 cm. wide, 2-3 cm. above the ground, 5 cm. overall height. TUBERCLES arranged in 8 and 13 spirals, soft in texture, dull grayish green, conical, nearly terete, with watery sap, 10 mm. long, 5-6 mm. wide at base. AREOLES round, with scant tan wool in only newly formed ones.

AXILS with 8-10 white bristles nearly to length of the tubercles, but no wool. CENTRAL SPINES 1-4, older areoles show more spines, newer ones only 1 (but when brought into cultivation they revert to 4 and with more evidence of presence of a hooked one), 1-3 upper ones straight, 1-3 mm. long, lower sometimes hooked to 7 mm. long, all acicular, smooth, stiff, orange at base, becoming cream, reddish brown at tip, straight ones nearly porrect, hooked one strongly deflected ventrally. RADIAL SPINES 15, 6-7 mm. long, slender acicular, straight, smooth, semi-flexuous, orange at base, snow white to tip, horizontal. FLOWERS June, 23 mm. long, in top but not in apex. *Outer perianth-segments* olive-green at base, becoming somewhat brownish olive-green tapering mid-stripe, creamish margins, linear-lanceolate, tip obtuse, margins finely ciliate. *Inner perianth-segments* creamish, narrow magenta mid-stripe not tapering, oblong, tip obtuse, margins entire, 3 mm. wide. *Filaments* very pale greenish white, hooked at top. *Anthers* orange. *Style* pale green below, cream above. *Stigma-lobes* 5, light green, less than 1 mm. long, overtop anthers 7 mm. FRUIT and SEEDS unknown. ROOTS short, broad, blunt, tuberous, then fibrous.

Distribution: Guerrero, Mexico.

Type locality: Near Colonia on the Rio Balsas along the highway between Taxco and Alcopulco.

Illustration: Fig. 200 is from a photograph of a plant collected at the type locality by us in 1942.

Habitat: Partial shade on hillsides, in leaf mold, at approx. 2000 feet.

This species belongs to the transition group between the straight and the hooked central spined types, as the hook on the central spines is not a constant characteristic and is present only under varying conditions. A similar condition is exhibited by several of the following species. This characteristic may be due to a recessive factor in the evolution of the species or to a possible hybridization in the past between the two types.

Discovery of this species is credited to Mrs. Ina Craig who first observed it while trying to avoid a snake when she was exploring a steep wooded hillside during the summer of 1942.

Type material has been deposited in the Dudley Herbarium at Stanford University, California.

176. Mammillaria guirocobensis sp. nov.

Mammillaria sp. No. 645 Gentry, Rio Mayo Plants, 196, 1942. (Mention of distribution, no description.)

Corpus simplex et caespitosus, cylindratus, mamillis ad 8 et 13 seriebus ordinatis, sub pyramidalis, suco aquario, 7 mm. per longitudinem; areolis ovatis; axillis nudis; spinis centralibus 1-3 (fere 3), 2-10 mm. longis, forma "a" rectis, subradialibus, forma "b" rectis et brevis, ad 2 mm. longis, porrectis, aut forma "c" hamatis et longis ad 10 mm., porrectis, totis acicularibus, spadicibus; spinis radialibus 18-20, 5-8 mm. longis, acicularibus, rectis, albis; flores infundibuliformes, petalis ciliatis, sepalis albis cum linea media spadici, stigmatibus 8, flavoviridibus; fructus coccineus; semina nigra.

BODY simple and cespitose from body and base, cylindric, to 11 cm. high, to 5 cm. wide. TUBERCLES closely set in 8 and 13 spirals, firm in texture, olive-gray-green, 4-sided at base, conic-cylindric, keeled ventrally, with watery sap, 7 mm. long, 5-7 mm. wide at base. AREOLES ovate, naked or with very scant tannish wool only in newly formed ones. AXILS naked (rarely an occasional bristle). CENTRAL SPINES 1-3, three types: (a) upper two subcentral, straight, 6 mm. long, nearly in the same plane as radials; lower one, either: (b) straight, 2-3 mm. long, subulate, porrect, or (c) hooked, 8-10 mm. long, acicular, porrect, all reddish brown, brown at tip. RADIAL SPINES 18-20, 5-8 mm. long, acicular, straight, smooth, stiff, yellowish tan to chalky gray, nearly horizontal. FLOWERS funnelform, near apex but in old axils, 20 mm. long and wide. *Outer perianth-segments* light greenish reddish brown wide mid-stripe, bordered with pale

M. guirocobensis

FIG. 201. *Mammillaria guirocobensis* x 1

green, greenish white margins, wide linear, tip obtuse, margins ciliate. *Inner perianth-segments* whitish, reddish brown fine mid-line to apex, bordered with pale green, lanceolate, tip acute, margins entire. *Filaments* bright purplish pink. *Anthers* bright orange. *Style* pink. *Stigma-lobes* 8, olive-green, 4 mm. long, overtop anthers 4 mm. FRUIT scarlet, clavate, 12x6 mm. SEEDS black, glossy, obovate with ventral hilum, pitted, 1x0.7 mm. ROOTS fibrous.

Distribution: SE. Sonora, NE. Sinaloa, SW. Chihuahua; Mexico.

Type locality: Rancho Guirocoba (Alamos District), Sonora.

Illustration: Fig. 201 is from a photograph of a plant we collected near Rancho Guirocoba.

This species represents a transition form between the straight and the hooked central spined types as both types of spines are found more or less regularly on the same plant.

Collections of this species were made by Mr. Howard Gentry at Guirocoba in 1934-1935, Quiriego in 1936, and Alamos in 1936; by Mr. W. T. Marshall near Guirocoba in 1935; by the author at Guirocoba in 1936, 1937 and 1939, Arroya Cuchijaqui and Navajoa in 1937; by Mr. George Lindsay and the author at Alamos in 1939.

The name of this species was suggested to us by Mr. Gentry who was the first botanist to do any extensive work in this vast and practically unknown territory.

The illustration in our article (Cact. Succ. Journ., 8:215, 1937), as *Mammillaria* sp. No. 625 from Batopilas, Chihuahua, is to be referred here. It showed the tubercles so arranged in vertical rows that it appeared to resemble an *Echinocereus*. This illusion was due to its dormant and shrunken condition but when the plant filled out and resumed growth, the normal characteristic spiral arrangement of the tubercles appeared.

Type material has been deposited in the Dudley Herbarium at Stanford University, California.

FIG. 202. *Mammillaria phitauiana* x 1

177. Mammillaria phitauiana (Baxter) Werdermann

Neomammillaria phitauiana Baxter, Journ. Cact. Succ. Soc., 2:471, 1931.
Nammillaria phitauiana Werdermann in Backeberg, Neue Kakteen, 96, 1931.

BODY cespitose from base, cylindric, rounded at apex, 17 cm. high, 6 cm. wide. TUBERCLES arranged in 8 and 13 spirals, grayish green, conic-cylindric, 4-sided base, flattened dorsally, keeled ventrally, with watery sap, 5-8 mm. long, 6-7 mm. wide at base. AREOLES pyriform to diamond shaped, slightly depressed, with very short white wool which persists. AXILS with as many as 20 white bristles longer than tubercles, persisting. CENTRAL SPINES 4, 8-10 mm. long, upper 3 straight, lower one usually hooked, sometimes straight, all acicular, stiff, smooth, slightly enlarged at base, some, reddish brown to base which form dark bands around the plant, others reddish brown only at tip with white below to base which form light bands, spreading porrect. RADIAL SPINES 18-24, 4-12 mm. long, upper shorter, all slender acicular, straight, smooth, white in lighter bands, reddish brown tip to outer half in darker bands, slightly ascending to horizontal. FLOWERS tubular, to 15 mm. long. *Outer perianth-segments* red mid-stripe which fades into white margins thus giving a pink shading, lanceolate, tip acute, margins serrate. *Inner perianth-segments* white, red mid-line, tip acute, margins nearly entire. FRUIT red, globose to clavate, 10 mm. long. SEEDS black, pyriform, 0.5 mm. long. ROOTS fibrous.

Distribution: Baja California, Mexico.

Type locality: Rancho del Chino in Sierra de la Laguna, 30 miles east of Todos Santos.

Illustration: Fig. 202 is a reproduction of an illustration in Blatter Kakteenf., 4, 1934, as *Mamillaria phitauiana*.

Fig. 203. *Mammillaria haehneliana*

178. Mammillaria haehneliana Boedeker

Mammillaria haehneliana Boedeker, Kakteenkunde, 27, 1934.

BODY simple and cespitose, globose, slightly sunken at apex, 4-5 cm. wide. TUBERCLES arranged in 8 and 13 spirals, light leaf-green, 4-sided at base, cylindric above, rounded at apex, with watery sap, 5-6 mm. long, 3-4 mm. wide at base. AREOLES with scant white wool, soon becoming naked. AXILS with several tortuous moderately long white hair-like bristles, but no wool. CENTRAL SPINES 5-7, to 8 mm. long, stout acicular, only those in top of plant are hooked, others straight, all pubescent, bright transparent yellow, dorsal ones more whitish yellow, all divergent from porrect. RADIAL SPINES 25, 7 mm. long, very slender, straight, smooth, white, horizontal to slightly ascending. FLOWERS funnelform, near top, 15 mm. long. *Onter perianth-segments* straw-yellow, brownish red ventral mid-stripe, lanceolate, tip acute, margins serrate, 7-12x2 mm. *Inner perianth-segments* whitish cream, very faint rose mid-line, oblong, tip acuminate, margins mostly entire. *Filaments* white. *Anthers* light yellow. *Style* green. *Stigma-lobes* 4, whitish cream, overtop anthers. FRUIT red, clavate, small. SEEDS dark brown, faintly glossy, curved pyriform with small white lateral hilum, 1 mm.

Distribution: San Luis Potosi, Mexico.

Type locality: 15-20 Km. south of San Luis Potosi city.

Illustration: Fig. 203 is a reproduction of an illustration in Kakteenk., 28, 1934, as *Mammillaria haehneliana*.

179. Mammillaria oliviae Orcutt

Mammillaria oliviae Orcutt, West Amer. Sci., **12**:50, 1902.
Neomammillaria oliviae Britton & Rose, The Cactaceae, 4:135, 1923.
Chilita oliviae Orcutt, Cactography, 2, 1926.

BODY simple and cespitose, globose to short cylindric, to 10 cm. high, 6-7 cm. wide. TUBERCLES arranged in 13 and 21 spirals, firm in texture, dull gray-green, 4-sided at base, ovate above, with watery sap, 6-10 mm. long, 5 mm. wide at base. AREOLES inverted pyriform, with very short wool, soon becoming naked. AXILS with very slight tuft of wool, but no bristles. CENTRAL SPINES 1-3 (4), lower one 3 mm. long, subulate, (occasionally elongate to 12 mm., slender, flexuous, and hooked, when present, it is more often on the offshoots than on the main stem), upper to 9 mm. long, acicular, straight, stiff, all smooth, white, usually only the lower one has a brown tip, lower porrect, uppers horizontal with radials. RADIAL SPINES 25-35, to 6 mm. long, upper shorter,

M. oliviae

Fig. 204. *Mammillaria oliviae* x 1

all acicular, straight, stiff, smooth, snow white, horizontal. FLOWERS funnelform, 2-3 cm. long and wide, July. *Outer perianth-segments* greenish below, reddish brown midstripe, lighter margins, linear-lanceolate, tip acuminate, margins ciliate. *Inner perianth-segments* rose, rose-pink margins, lanceolate, tip acute, margins entire. *Filaments* rose to magenta. *Anthers* yellow. Style white to pink-rose. *Stigma-lobes* 7, olive, green. FRUIT scarlet, clavate, to 25 mm. long. SEEDS black, glossy, globular obovate with slightly extending ventral hilum, pitted, 0.9x0.6 mm.

Distribution: Arizona, U. S.; Sonora, Mexico.

Type locality: None given, but collected west of Vail, near Tucson, (B. & R.), Santa Rita Mts., Canada del Oro in Santa Catalina Mts. (Gibson) in Arizona, Cedros (Gentry), Bacoachi to Arispe (Craig) in Sonora.

Illustration: Fig. 204 is from a photograph of a plant collected by Mr. Howard Gentry at Cedros, Sonora.

The occasional presence of a few hooked central spines, especially on the offshoots, may be due to the possible hybridization with *M. microcarpa* with which it is often associated or the possible presence of a recessive characteristic in the evolution of the species.

We are indebted to Mr. F. Gibson of the Boyce Thompson Arboretum of Superior, Arizona, for the data on the distribution and flowers of this species.

FIG. 205. *Mammillaria magallanii* x 1

180. X **Mammillaria magallanii** Schmoll sp. nov.

Corpus simplex, mamillis ad 13 et 21 seriebus ordinatis, cylindratis, textis firmis, suco aquario, axillis solum lanatis eziquis; spinis centralibus 0-1, 1-3 mm. longis, rectis ad flexis ad hamatis, subfulvis; spinis radialibus 70-75, 2-5 mm. longis, setaformis, rectis, subflavis; flores tubulatae infundibuliformes, sepalis luridis lutulentis, entegris, petalis albo flavis cum linea rufalba, entegris, stigmatibus 4, albis lutulentis; semina nigra, globosa ad obovata, punctata.

BODY simple, clavate, 4-5 cm. wide, 6 cm. high. TUBERCLES arranged in 13 and 21 spirals, firm in texture, light gray-green, cylindric, nearly terete, with watery sap, 6 mm. long, 4 mm. wide at base. AREOLES nearly round, small, with very scant dirty white wool only in youngest. AXILS with scant wool, but no bristles. CENTRAL SPINES 0-1, 1-3 mm. long, acicular, straight to curved to hooked, stiff, smooth, orange tan at base, brown at tip, porrect to nearly erect. RADIAL SPINES 70-75, 2-5 mm. long, slender acicular, straight to slight recurve, smooth, orange tan at base, becoming chalky white, brown at tip, slightly ascending in youth, soon horizontal and interlacing. FLOWERS lateral, tubular funnelform, 10 mm. long, 6 mm. wide. *Outer perianth-segments* pink to tannish mid-stripe, cream margins, clavate, tip obtuse to emarginate, margins entire. *Inner perianth-segments* tannish pink mid-line, wide cream margins, oblong, tip obtuse, margins entire. *Filaments* pink. *Anthers* sulphur yellow. *Style* very pale greenish cream. *Stigma-lobes* 4, very pale tan, overtop anthers, 0.5 mm. long. FRUIT unknown. SEEDS black, glossy, globular to obovate with ventral hilum, pitted 1.2x1 mm. ROOTS large tap.

Distribution: Coahuila, Mexico.

Type locality: Unknown.

Illustration: Fig. 205 is from a photograph of a plant we obtained from Sr. F. Schmoll of Cadereyta, Qro., in 1942.

X—Possible hybrid of *M. lasiacantha*.

FIG. 206. *Mammillaria hidalgensis* x 0.8

181. **Mammillaria hidalgensis** Purpus

Mamillaria hidalgensis Purpus, Monatsschr. Kakteenk., 17:118, 1907.

BODY simple or dividing by dichotomous branching, cylindric to clavate, rounded at apex, to 30 cm. or more high, to 13 cm. wide. TUBERCLES arranged in 8 and 13 spirals, dark green, separated, conic, nearly terete, with watery sap, 10 mm. or more long, 6 mm. wide at base. AREOLES eliptical, with flocculent white wool in youth, later becoming naked. AXILS with flocculent dirty white wool in youth, persisting somewhat. CENTRAL SPINES 2-4, 10 mm. long, acicular, straight to slightly recurved, stiff, smooth, grayish white to yellowish brown, reddish brown at tip, becoming all gray, when 2: divergent dorsally and ventrally, when 4: in cross formation, strongly spreading ascending. RADIAL SPINES usually none but sometimes 6-8, less than 1 mm. long, fine bristle-like, usually deciduous. FLOWERS wide funnelform, near top but not in apex, 18 mm. long, 22 mm. wide. *Outer perianth-segments* whitish to pale green at base, carmine to dark reddish purple tapering mid-stripe above, paler margins, linear-lanceolate, tip acuminate, margins serrate. *Inner perianth-segments* brighter carmine, brighter red at base, lanceolate, tip acuminate, margins entire or some short serrations. *Filaments* purplish red. *Anthers* yellowish white. *Style* pale yellow below, reddish above. *Stigma-*

lobes 4-6, reddish, very short, just overtop anthers. FRUIT bluish carmine, clavate, 15-20 mm. long, with dried perianth persisting. SEEDS light brownish, curved pyriform with lateral hilum at base, faintly reticulate, 1.5x0.7 mm.

Distribution: Hidalgo, Mexico.

Type locality: Ixmiquilpan.

Illustration: Fig. 206 is from a photograph of a plant we collected near Ixmiquilpan in 1942.

Britton & Rose referred this species to *M. polythele* Martius with a ? but it is definitely to be excluded from that species which has milky sap.

We collected specimens of this species on the rocky hills along the highway near Ixmiquilpan. It is not very common but the specimens that we found were quite large. It does well in cultivation and flowers quite freely for several months in the summer with a ring of reddish flowers near the top.

FIG. 207. *Mammillaria tetracantha* x 0.8

182. **Mammillaria tetracantha** Salm-Dyck

Mammillaria tetracantha Salm-Dyck in Pfeiffer, Enum. Cact., 18, 1837.
Mammillaria obconella Scheidweiler, Hort. Belge, 4:93, 1837.
Mammillaria obconella galeottii Scheidweiler, Hort. Belge, 4.93, 1837.
Mammillaria dolichocentra Lemaire, Cact. Aliq. Nov., 3, 1838.
Mammillaria tetracentra Otto in Förster, Handb. Cact., 214, 1846.
Mammillaria dolichocentra staminae Labouret, Monogr. Cact., 50, 1853.
Mammillaria dolichocentra phaeacantha Labouret, Monogr. Cact., 50, 1853.
Cactus dolichocentrus Kuntze, Rev. Gen. Pl., 1:260, 1891.
Cactus obconellus Kuntze, Rev. Gen. Pl., 1:261, 1891.
Cactus tetracanthus Kuntze, Rev. Gen. Pl., 1:261, 1891.
Cactus tetracentrus Kuntze, Rev. Gen. Pl., 1:261, 1891.
Mamillaria rigidispina Hildmann, Monatsschr. Kakteenk., 3:112, 1893.
Mamillaria dolichocentra brevispina Rünge, Monatsschr. Kakteenk., 3:112, 1893.
Neomammillaria tetracantha Britton & Rose, The Cactaceae, 4:106, 1923.

BODY simple and cespitose from base, globose to cylindric, to 30 cm. high, to 12 cm. wide. TUBERCLES arranged in 13 and 21 spirals, firm in texture, dark green, later brownish green, 4-sided, not sharply angled, rounded dorsally, keeled ventrally, with watery sap, 8-10 mm. long, 4 mm. wide at base. AREOLES somewhat 4-sided, with scant short white woolly felt in youth. AXILS with white wool. CENTRAL SPINES 4, seldom to 6, 10-30 mm. long (quite variable), lower longest, all acicular, straight to recurved, rigid, smooth, reddish in youth, chalky whitish horn-colored later, in cross formation, ascending, upper recurved over apex. RADIAL SPINES mostly absent or few bristles, usually deciduous. FLOWERS funnelform, short tube, 20 mm. long, 15 mm. wide. *Outer perianth-segments* reddish green mid-stripe, white base, carmine tip, linear-lanceolate, tip acute, margins very short ciliate. *Inner perianth-segments* carmine red mid-stripe, pale pink margins, lanceolate, tip acute, margins entire. *Filaments* carmine. *Anthers* yellow. *Style* white below, pink above. *Stigma-lobes* 4-5, rose-red. FRUIT dark red, clavate, 20-25 mm. long, with dried perianth persisting. SEEDS light brown, curved pyriform, faintly pitted, 1 mm.

Distribution: Hidalgo, and Queretaro, Mexico.

Type locality: None given but reported from Tembladora, Qro. (Schmoll).

Illustration: Fig. 207 is from a photograph of a plant sent to us by Sr. F. Schmoll of Cadereyta, Qro.

Britton & Rose described the spines as all radial but most of the other authors give them as centrals which is verified by the presence of very small deciduous bristles in the position usually represented by the radial spines.

M. tetracentra and *Cactus tetracentrus* which were referred to *M. magnimamma* by Britton & Rose are very probably referrable here. The specific name *tetracentra* was first listed but not described by Walper (Rep. Bot. Syst., 2:271, 1842).

M. longispina Reichenbach (Suppl. Terscheck Cact. Verz., 1) was referred here by Schumann. Britton & Rose were uncertain of its relationship, likewise we have not been able to obtain the original description and are uncertain of its relationship.

M. dolichacantha Förster (Handb. Cact., 213, 1846) is given as a synonym of *M. dolichocentra* without description.

Salm-Dyck (Hort. Dyck., 8-9, 1845) listed under *M. dolichochentra* the varieties *galeottii, phaeacantha, straminea* and *picta* without descriptions.

M. dolichocentra nigrispina is only a name used in Monatsschr. Kakteenk., 60, 1918.

a. var. galeottii (Scheidweiler) Borg

Mammillaria galeottii* Scheidweiler, Hort. Belge., 4:93, 1837.
Mammillaria dolichocentra galeotti Salm-Dyck in Förster, Handb. Cact., 213, 1846.
Neomammillaria galeottii Britton & Rose, The Cactaceae, 105, 1923.
Mammillaria tetracantha galeotti Borg, Cacti, 335, 1937.

TUBERCLES conic, lighter green. AXILS naked. CENTRAL SPINES 4-5, strongly recurved, flexuous, yellowish. RADIAL SPINES usually none, but sometimes 1-3 short deciduous bristles in lower part of areole.

Type locality: Jalapa.

---

*This was listed as *galeotii* which is very probably a typographical error, because the correct spelling was used later in the description.

Fig. 208. *Mammillaria decipiens* x 1

183. Mammillaria decipiens Scheidweiler

Mammillaria decipiens Scheidweiler, Bull. Acad. Sci. Brux., 5:496, 1838.
Mammillaria guilleminiana Lemaire, Cact. Gen. Nov. Sp., 48, 1839.
Mammillaria glochidiata inuncinata Lemaire, Cact. Gen. Nov. Sp., 102, 1839.
Mammillaria deficum Hort. ex Foerster, Handb. Cact., 185, 1846.
Mamillaria deficiens Hort. ex Salm-Dyck, Cact. Hort. Dyck. 1849, 7, 1850.
Cactus decipiens Kuntze, Rev. Gen. Pl., 1:260, 1891.
Cactus guilleminianus Kuntze, Rev. Gen. Pl., 1:260, 1891.
Neomammillaria decipiens Britton & Rose, The Cactaceae, 4:131, 1923.
Chilita decipiens Orcutt, Cactography, 2, 1926.

BODY cespitose from base and sides, globose, sunken at apex, to 10 cm. high. TUBERCLES arranged in 5 and 8 spirals, somewhat soft in texture, deep green, cylindric to somewhat ovate, terete, with watery sap, 11 mm. long, 3-5 mm. wide at base. AREOLES round, less than 1 mm. wide, with only a very faint trace of wool in the youngest. AXILS with very scant wool, and to 4 white to rose bristles. CENTRAL SPINES 1 (2), 12-18 mm. long, slender acicular, straight, stiff, smooth, not enlarged at base, yellowish at base, reddish brown above, porrect to a little descending. RADIAL SPINES 7-8, 7-10 mm. long, laterals shorter, fine acicular, straight, smooth, semi-flexuous, yellowish white, horizontal to appressed. FLOWERS broad funnelform, 15-20 mm. long. *Outer perianth-segments* green, reddish mid-stripe, lanceolate, tip acuminate, margins entire. *Inner perianth-segments* white, rose mid-line, lanceolate, tip acute, margins entire. *Filaments* white to rose above. *Anthers* yellowish. *Style* white to rose. *Stigma-lobes* 4-6, white to yellowish to pale pink. FRUIT green with reddish tint, cylindric clavate, 20x4 mm., with dried perianth persisting. SEEDS bright brown, very small.

Distribution: San Luis Potosi, Mexico.

Type locality: None given but reported from Hacienda de las Bocas.

Illustration: Fig. 208 is from a photograph of a plant sent to us by Sr. F. Schmoll of Cadereyta, Qro.

Fig. 209. *Mammillaria viereckii* x 1

184. Mammillaria viereckii Boedeker

Mamillaria viereckii Boedeker, Zeitschr. Sukkulentenk., 3:73, 1927.
Mammillaria viereckii brunispina Neal, Cact. Other Succ., 94, 1935.

BODY simple, occasionally cespitose from base only, globose, rounded at apex, 3-4 cm. wide. TUBERCLES arranged in 8 and 13 spirals, bright dark green, lighter at base, glossy, slender cylindric, conic at apex, with watery sap, 8-10 mm. long, 2-3 mm. wide at base. AREOLES round, 1-1.5 mm. wide, with scant white wool, later becoming naked. AXILS with white wool, and 8-10 white hair-like tortuous bristles, moderately long, often extending beyond the spines. CENTRAL SPINES more or less subcentral, not true centrals as some are in the same plane as the radials, 9-11, 12 mm. long, laterals a little shorter, slender acicular, straight, smooth, stiff, faintly enlarged at base, bright ochre yellow at base, transparent amber yellow above, peculiarly horizontal to slightly ascending. RADIAL SPINES 6-7 (8-10), 4-5 mm. long, very fine acicular, smooth, straight to somewhat tortuous, white, horizontal under the centrals. FLOWERS wide funnelform, 12 mm. long. *Outer perianth-segments* whitish cream, pale olive-green mid-line, lanceolate, tip acuminate, margins entire, 8x1.5 mm. *Inner perianth-segments* whitish cream, pale olive-green mid-line, lanceolate, tip acuminate, margins entire, 10x2-3 mm. *Filaments* white. *Anthers* bright chrome-yellow. *Style* pure white below, greenish white above. *Stigma-lobes* 6-7, white. FRUIT reddish brown, clavate, small. SEEDS black, glossy, curved pyriform, finely pitted, 1 mm. ROOTS tuberous.

Distribution: Tamaulipas, Mexico.

Type locality: Nogales.

Illustration: Fig. 209 is a reproduction of an illustration in Zeitschr. Sukkulentenk., 3:74, as *Mamillaria viereckii*.

M. viereckii brunea Hort. Cact. Succ. Journ., *10*:77, 1938, is only a name.

FIG. 210. *Mammillaria kelleriana* x 0.8

185. Mammillaria kelleriana Schmoll sp. nov.

Corpus globosus ad cylindratus, mamillis ad 13 et 21 seriebus ordinatis, textis firmis, viridis nigiribus, suco aquario, axillis lanatis exiquis ad nudis, spinis centralibus 0-1 (praesentia est sola semitempa), 15 mm., rectis, spadicibus, porrectis; spinis radialibus 6, 9-20 mm.; flores infundibuliformes, sepalis magentis, erratula levis, petalis magentis entegris, stigmatibus 4-5, magentis; fructus coccineus, clavatus; semina fulva.

BODY globular cylindric, rounded at apex and hardly sunken, 12 cm. high, 8 cm. wide. TUBERCLES arranged in 13 and 21 spirals, firm in texture, dull dark green, lighter in axils, cylindric conic, terete, with watery sap, 10 mm. long, 8 mm. wide at base. AREOLES oval, 3 mm. long, slightly sunken with white flocculent wool only in youngest. AXILS with scant white wool only in apex, naked otherwise. CENTRAL SPINES 0-1 (only about half of areoles have centrals), 15 mm. long, strong acicular, straight, smooth, stiff, purplish tan, dark brown end, porrect. RADIAL SPINES 6, 9-20 mm. long, lower ones longest, lateral shortest, all strong acicular to slender subulate, slightly recurved, smooth, purplish tan, dark brown at tip, stiff, somewhat flattened at base, strongly ascending. FLOWERS funnelform, 10 mm. long, 8 mm. wide, August. *Outer perianth-segments* brownish magenta mid-stripe, magenta margins, linear-lanceolate, tip acuminate, margins lightly serrate. *Inner perianth-segments* magenta, darker mid-line, lanceolate, tip acuminate, margins entire. *Filaments* magenta. *Anthers* tan. *Style* cream below, magenta top. *Stigma-lobes* 4-5, magenta, less than 1 mm. long. FRUIT scarlet, clavate, 20x5 mm., dried perianth persisting. SEEDS light tan, slightly curved pyriform with lateral hilum near base, faintly reticulate, 0.8x0.9 mm.

Distribution: Queretaro, Mexico.

Type locality: La Sierra San Moran.

Illustration: Fig. 210 is from a photograph of a plant sent to us by Sr. F. Schmoll of Cadereyta, Qro.

This species is named in honor of Dr. Kellar of Lugano, Switzerland.

Referrable here are the plants which Sr. Schmoll has been exporting under the names of *M. schmuckeri* and *M. nerispina* which have slightly heavier spines.

This species is close to and intergrades with *M. kewensis craigiana*.

186. Mammillaria fertilis Hildmann

Mamillaria fertilis Hildmann in Schumann, Gesamtb. Kakteen, 530, 1898.
Neomammillaria fertilis Britton & Rose, The Cactaceae, 4:131, 1923.
Chilita fertilis Orcutt, Cactography, 2, 1926.

BODY simple and cespitose from sides, globose to short cylindric, rounded at top and sunken in apex, 5-6 cm. wide. TUBERCLES closely set in 8 and 13 spirals, dark green, faintly 4-sided at base, slender cylindric, with watery sap, 7-8 mm. long. AREOLES round, 2 mm. wide, with scant white and somewhat curly wool, later becoming naked. AXILS with white wool in apex, later becoming very sparse. CENTRAL SPINES 1-2, to 10 mm. long, acicular, straight to somewhat bent, yellowish brown, darker at tip. RADIAL SPINES 8-10, 6 mm. long, thin acicular, very brittle, glossy, later gray. FLOWERS funnelform, near apex, 17-18 mm. long, 10-15 mm. wide. *Outer perianth-segments* brown mid-stripe, lighter red margins. *Inner perianth-segments* fiery carmine, linear-lanceolate, tip acute, margins scantly serrate. *Filaments* fiery carmine. *Anthers* yellow. *Style* yellow below, carmine above. *Stigma-lobes* 4, red, very short globular. FRUIT and SEEDS unknown.

Distribution: Mexico.

Type locality: None given.

We know this species only from the original description.

M. liebneriana is an unpublished name but referred here by Schumann (Monatsschr. Kakteenk., 130, 1904).

Mammillaria pyrrhochrantha Lemaire

Mammillaria pyrrhochrantha Lemaire, Cact. Gen. Nov. Sp., 51, 1839.
Cactus pyrrhochroacanthus Kuntze, Rev. Gen. Pl., 1:261, 1891.

BODY globose. TUBERCLES dark green, conic. AXILS woolly. CENTRAL SPINES 4, 6-8 mm. long, lower longer, golden red. RADIAL SPINES 8-9, 4-10 mm. FLOWERS, FRUIT and SEEDS unknown.

This species has been referred to *M. rhodantha* by many authors and was compared with it by the original author but the fewer radial spines would place it elsewhere. It may belong here but the incomplete description makes it uncertain, otherwise the older name would have priority.

Fig. 211. *Mammillaria picta*

187. **Mammillaria picta** Meinshausen

Mamillaria picta Meinshausen, Wöchenschr. Gärteneri Pflanz., 1:27, 1858.

BODY simple, globose to clavate, somewhat sunken at apex but not woolly, to 4 cm. wide. TUBERCLES closely set in 8 and 13 spirals, bright dark leaf-green, slender conic to cylindric, with watery sap, 7 mm. long, 3 mm. wide at base. AREOLES round, with scant wool only in the youngest, soon becoming naked. AXILS with no wool but scattered slender hair-like white tortuous bristles. CENTRAL SPINES 1 to occasionally 2, to 10 mm. long, acicular, straight, pubescent, stiff, yellow at base to white above, dark reddish brown at tip, porrect. RADIAL SPINES 12-14, 6-8 mm. long, lower longer, hair-like, tortuous, transparent white, upper stiff acicular, yellow at base, to white, dark reddish brown at tip, all pubescent, enlarged at base, slightly ascending. FLOWERS wide funnel-form, 9 mm. long, 11 mm. wide. *Outer perianth-segments* olive-green at base, cream near tip, brownish rose ventral stripe, linear-lanceolate, margins serrate. *Inner perianth-segments* whitish cream, bright olive-green mid-stripe, linear-lanceolate, margins serrate. *Filaments* whitish cream. *Anthers* bright chrome-yellow. *Style* whitish cream. *Stigma-lobes* 3, pale greenish cream, wide spreading, small. FRUIT red, clavate, small. SEEDS unknown. ROOTS tuberous.

Distribution: Tamaulipas, Mexico.

Type locality: Rio Blanco.

Illustration: Fig. 211 is a reproduction of an illustration in Kakteenk., 113, 1934, as *Mammillaria picta*.

This species has been unknown since the original plant was described and it has been listed by various authors as an unknown and doubtful species. Recently more material has been recollected and the plants have again become available. The original description gave the type locality as Rio Blanco without giving the state through which it flows. Inasmuch as this is a very common name for waterways it is uncertain as to the exact locality.

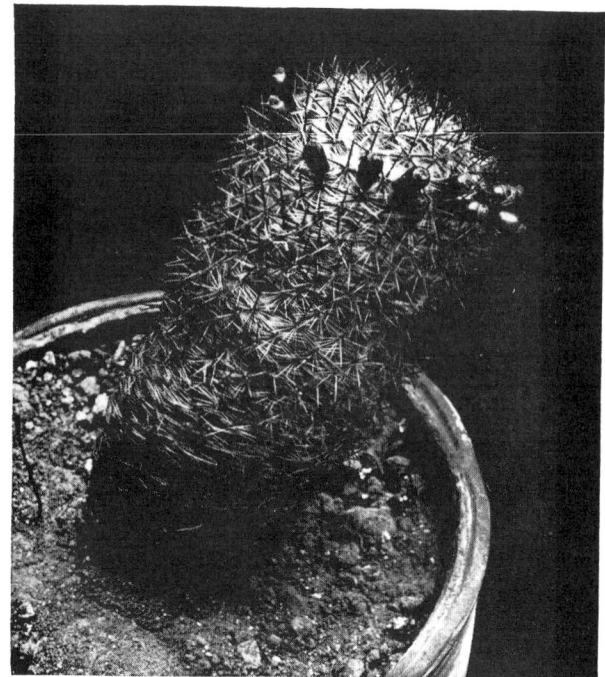

FIG. 212. *Mammillaria ruestii*

188. Mammillaria ruestii Quehl

Mamillaria rüstii Quehl, Monatsschr. Kakteenk., 15:173, 1905.
Mamillaria celsiana guatemalensis Eichlam, Monatsschr. Kakteenk., 19:59, 1909.
Neomammillaria ruestii Britton & Rose, The Cactaceae, 4:115, 1923.

BODY simple and later cespitose from base, elongate globular, sunken at apex and woolly, 6-7 cm. high, 4-5 cm. wide. TUBERCLES arranged in 13 and 21 spirals, firm in texture, bright leaf-green, conic, blunt truncate, with watery sap, 6-7 mm. long, 5 mm. wide at base. AREOLES oval, 4x2 mm., with white woolly felt in youth, soon becoming naked. AXILS with white wool and bristles. CENTRAL SPINES 4-(5), 7-8 mm. long, stout acicular to subulate, straight, stiff, enlarged at base, bright to dark chestnut-red, ascending in divergent cross formation, overtopping apex. RADIAL SPINES 16-20, unequal, to 6 mm. long, upper ones shortest, all slender acicular, straight, glossy white, nearly horizontal. FLOWERS campanulate, 20 mm. long, 15-20 mm. wide. *Outer perianth-segments* brownish red, whitish to reddish margins, lanceolate, tip acute, margins ciliate. *Inner perianth-segments* to 25, deep carmine red, white in throat, nearly colorless margins, lanceolate, tip acute, margins serrate. *Filaments* white below, reddish above. *Anthers* white with yellow luster. *Style* white. *Stigma-lobes* 7, green, overtop flowers. FRUIT red, clavate, dried perianth persisting. SEEDS brown, dull, curved pyriform with lateral hilum near base, faintly rugose, 1.7x0.6 mm.

Distribution: Honduras and Guatemala.

Type locality: Honduras.

Illustration: Fig. 212 is a reproduction of an illustration, Fig. 120, in Britton & Rose (Cact. 4:114) as *Neomammillaria ruestii*.

FIG. 213. *Mammillaria rhodantha* x 0.8

189. **Mammillaria rhodantha** Link & Otto

Mammillaria rhodantha Link & Otto, Icon. Pl. Rar., 51, 1829.
Mammillaria pulchra Haworth in Edwards, Bot. Reg., 16:1329, 1830.
Mammillaria fulvispina Haworth, Phil. Mag., 7:108, 1830.
Mammillaria inuncta Hoffmannsegg, Priess Verz., ed. 7, 23, 1833.
Mammillaria chrysacantha Pfeiffer, Enum. Cact., 28, 1837.
Mammillaria rhodantha prolifera Pfeiffer, Enum. Cact., 31, 1837.
Mammillaria rhodantha andreae* Otto in Pfeiffer, Enum. Cact., 31, 1837.
Mammillaria rhodantha wendlandii Pfeiffer, Enum. Cact., 31, 1837.
Mammillaria rhodantha neglecta Pfeiffer, Enum. Cact., 31, 1837.
Mammillaria rhodantha rubens Pfeiffer, Enum. Cact., 31, 1837.
Mammillaria ruficeps Lemaire, Cact. Gen. Nov. Sp., 37, 1839.
Mammillaria imbricata Wegner, Allg. Gartenz., 12:66, 1844.
?*Mammillaria rhodantha centrispina* Link in Foerster, Handb. Cact., 198, 1846.
Mammillaria robusta Otto in Foerster, Handb. Cact., 207, 1846.
Mamillaria fulvispina rubescens Salm-Dyck, Cact. Hort. Duck. 1849, 10, 1850.
Mamillaria rhodantha ruficeps Salm-Dyck, Cact. Hort. Dyck. 1849, 11, 1850.
Mamillaria chrysacantha fuscata Salm-Dyck, Cact. Hort. Dyck. 1849, 12, 1850.
Mamillaria rhodantha rubescens Salm-Dyck, Cact. Hort. Dyck. 1849, 97, 1850.
Mamillaria russea Dietrich, Allg. Gartenz., 19:347, 1851.
Cactus chrysacanthus Kuntze, Rev. Gen. Pl., 1:260, 1891.
Cactus rhodanthus Kuntze, Rev. Gen. Pl., 1:261, 1891.
Cactus ruficeps Kuntze, Rev. Gen. Pl., 1:261, 1891.
Mamillaria rhodantha rubra Schumann, Gesamtb. Kakteen, 550, 1898.
Mamillaria rhodantha ruberrima Schumann, Gesamtb. Kakteen, 550, 1898.
Mamillaria rhodantha pyramidalis Schumann, Gesamtb. Kakteen, 550, 1898.
Mamillaria rhodantha callaena Schumann, Gesamtb. Kakteen, 550, 1898.
Mmillaria rhodantha crassispina Schumann, Gesamtb. Kakteen, 550, 1898.
Mamillaria rhodantha chrysacantha Schumann, Gesamtb. Kakteen, 550, 1898.
Mamillaria rhodantha fulvispina Schelle, Handb. Kakteenk., 257, 1907.
Neomammillaria rhodantha Britton & Rose, The Cactaceae, 4:121, 1923.

**Mamillaria andreae* was listed by Schumann, Gesamtb. Kakteen, 598, 1898.

BODY simple and cespitose from the base and by dichotomous branching, globose to cylindric, 10-30 cm. high, 7-10 cm. wide. TUBERCLES closely set in 13 and 21 spirals, dark dull green, firm in texture, cylindric conic, terete, with watery sap, 8-12 mm. long, 6-8 mm. wide at base. AREOLES round to oval, with short white wool in youth. AXILS with white wool and bristles. CENTRAL SPINES 4-7, 10-25 mm. long, straight to slightly curved, stout acicular, smooth, stiff, color variable: white, yellow, red, or brown, ascending to spreading porrect. RADIAL SPINES 16-20, 6-10 mm. long, lower longer, all thin acicular, smooth, white or yellow, horizontal to little ascending. FLOWERS wide funnelform, 20 mm. long, 16 mm. wide. *Outer perianth-segments* tannish green at base, greenish light brown mid-stripe above, whitish margins, linear-lanceolate, tip acute, margins serrate. *Inner perianth-segments* deep purplish pink, lanceolate, tip acute, margins entire. *Filaments* whitish below, deep purplish pink above. *Anthers* yellowish white. *Style* light green below, pink at top. *Stigma-lobes* 4-5, light pink to tan, darker ventral groove, 1 mm. long, same level with anthers. FRUIT reddish, clavate, 25 mm. long, with dried perianth persisting. SEEDS light brown, rounded pyriform with lateral hilum from base, pitted, 1x0.8 mm.

Distribution: Valley of Mexico, Hidalgo, Queretaro; Mexico.

Type locality: None given but reported from Sierra de Guadalupe, Mex., Pachuca, Hgo., San Juan del Rio, Qro.

Illustrations: Fig. 213 is from a photograph of a plant sent to us by Sr. F. Schmoll of Cadereyta, Qro., as *M. rhodantha rubra*. Fig. 214 is also from the same source as *M. rhodantha gigantea.*

This species from the central plateau region is extremely variable in spine color and length. It merges into several closely related species. Many species and varieties have been described largely on the basis of the spine coloration and this resulted in much confusion since many of the names were used for both the species and the varieties. We have had several of these variations and some have bloomed very freely for us.

Some of the trinomials of Schumann and Schelle may be referrable elsewhere but it is uncertain where to place them because of the incomplete descriptions and the lack of information concerning the basis for these varieties.

M. aleodantha is given by Dietrich (Symp. Plant., 93, 1843) as a variety of *M. ruficeps* without description.

FIG. 214. *Mammillaria rhodantha* x 0.5

M. aurata DeCandolle (Prodrom. Syst. Nat., 460, 1828) is only a name in Berlin Bot. Garden, later referred to *M. rhodantha* but not described.

"*M. aurea* Pfeiffer (Foerster, Handb. Cact., 200, 1846). *M. rhodantha aurea* Salm-Dyck (Cact. Hort. Dyck. 1849, 11, 1850) is referrable here. We have found no description of it. *M. odieriana aurea* Salm-Dyck (Hort. Dyck. 1844, 7, 1845), also undescribed, may be the same." Britton & Rose, Cact., 4:123, 1923.

M. erinacea Steudel (Nom. Bot., 97, 1841) credits the author as Wendland with name only. Dietrich in 1843 refers it to *M. rhodantha wendlandii*. The specific name was used later by Poselger for another species from Saltillo, Coahuila, which has been referred to *M. melanocentra*.

M. floccigera Salm-Dyck (Hort. Dyck. 1844, 8, 1845) is listed as name only. Later in 1850 the same author referred it to *M. crassispina* but it was not described. Schumann (Gesamtb. Kakteen, 550, 1898) referred it to *M. rhodantha*.

M. floccigera longispina Förster (Handb. Cact., 254, 1846) is only a name.

M. fulvispina Haworth (Phil. Mag., 108, 1830) was described as from Brazil with flowers like *M. rhodantha* but larger. Pfeiffer (Enum. Cact., 30, 1837) gives the habitat as Brazil and Mexico but in another publication (Beschr. Syn., 28, 1837) he lists only Mexico. Dietrich (Allg. Gartenz., 242, 1838) describes a typical *M. rhodantha* flower for this species. Salm-Dyck (Hort. Dyck. 1845) lists the varieties *media* and *minor* and in 1850 he refers to the variety *rubescentem*. Labouret referred to the latter variety as *rubescens* and also credits Salm-Dyck with the variety *pyrrhocentra* but this was referred to as only near var. *rubescentem*. Blanc in 1893 gave it as a synonym of *M. beguinii*. Schumann in 1898 listed it as a synonym of *M. rhodantha*. Schelle in 1907 referred it to *M. rhodantha* var. *fulvispina*.

M. hybrida (Index Spec. Hort. Bot. Berl., 430, 1829) is only a name listed as a synonym of *M. rhodantha* but was never described.

M. neglecta was given as a synonym of *M. rhodantha neglecta* by Salm-Dyck (Cact. Hort. Dyck. 1849, 11, 1850).

M. pyramidalis Link & Otto (Verh. Ver. Bedford., 6:429, 1830) is a name only. Steudel in 1841 referred it to *M. rhodantha* but did not describe it. Schumann in 1898 listed it as *M. rhodantha* var. *pyramidalis*.

"*M. pyrrhocentra* Otto, its var. *gracilior* Salm-Dyck (Cact. Hort. Dyck., 8, 1850) and *M. fulvispina pyrrhocentra* Salm-Dyck (Cact. Hort. Dyck., 10, 1850) were referred as synonyms of *M. rhodantha* by Schumann, but were not described at the places cited." Britton & Rose (Cact., 4:123).

M. radula is only a name in Allg. Gartenz., 286, 1841. Dietrich in 1843 referred it as a synonym of *M. phaeacantha*. Schumann in 1898 referred it to *M. rhodantha*.

"*M. recurvispina* Hildmann in Schelle (Handb. Kakteenk., 257, 1907) is given without synonymy or description." B. & R. (Cact., 4:123).

"*M. rhodantha* var. *inuncta* Hoffmannsegg was listed by Labouret (Monogr. Cact., 45, 1853) as one of the synonyms of *M. rhodantha*. *M. rhodantha rubra* was given by Rumple in Förster (Handb. Cact. ed. 2, 1885) as a synonym of *M. rhodantha ruficeps*, but afterwards it was formally published by Schumann." Britton & Rose (Cact., 4:123).

M. rhodantha varieties *quadrispina, esperanza* and *isabelliana* are listed by Schelle (Kakteen, 314, 1926) as Hort. var. without descriptions.

M. schochiana Walton (Cact. Price List, 24, 1898) is a name only. Schelle in 1907 listed but did not describe it as *M. rhodantha schochiana* with *M. schochiana* Hort. as a synonym.

"*M. tentaculata conothele* Monville is given by Labouret (Monogr. Cact., 55, 1853) as a synonym of *M. stueberi*, while he refers *M. tentaculata fulvispina* (Monogr. Cact., 44, 1853) to *M. fulvispina*. *M. tentaculata rubra* Förster (Handb. Cact., 207, 1846) was given as a synonym of *M. tentaculata ruficeps*." B. & R. (Cact., 4:123). *M. tentaculata longispina* Rebut, (Catal., 7, 1896) is only a name.

Fig. 215. *Mammillaria phaeacantha* x 1

190. **Mammillaria phaeacantha** Lemaire

Mammillaria phaeacantha Lemaire, Cact. Gen. Nov. Sp., 47, 1839.
Mamillaria nigricans Fennel, Allg. Gartenz., 15:66, 1847.
Cactus nigricans Kuntze, Rev. Gen. Pl., 1:261, 1891. Not Haworth, 1803.
Cactus phaeacanthus Kuntze, Rev. Gen. Pl., 1:261, 1891.
Neomammillaria phaeacantha Britton & Rose, The Cactaceae, 4:116, 1923.

BODY simple, globose, slightly sunken at apex. TUBERCLES arranged in 8 and 13 spirals, firm in texture, green, nearly cylindric, obtuse, compressed a little laterally, with watery sap, 6-8 mm. long, 6-8 mm. wide at base. AREOLES round, with short yellowish wool in youth, soon becoming naked. AXILS with white wool and long tortuous setaceous bristles. CENTRAL SPINES 4, two lateral ones 8-9 mm. long, upper one 14 mm. long, lower one 10-12 mm. long, all subulate, straight, smooth, stiff, enlarged at base, black, in new growth reddish, in cross formation, ascending. RADIAL SPINES 16-20, 4-5 mm. long, setaceous, straight, stiff, slightly enlarged at base, white, darker at base, horizontal. FLOWERS funnelform. *Outer perianth-segments* pale green at base, reddish brown wide mid-stripe above, margins tan, lanceolate, tip acuminate, margins entire. *Inner perianth-segments* deep pink to reddish, darker mid-line, lanceolate, tip acute, margins entire. *Filaments* deep pink. *Anthers* light tan. *Style* nearly white, pale pink at very top. *Stigma-lobes* 4, deep pink, 1 mm. long, overtop anthers. FRUIT pink, clavate, 15x4 mm., with dried perianth persisting. SEEDS tan, curved pyriform with lateral hilum near base, faintly rugose, 0.9x0.5 mm.

Distribution: Mexico.

Type locality: None given but reported from San Toro and Regla.

Illustration: Fig. 215 is from a photograph of a seedling plant in our garden.

Fig. 216. *Mammillaria inaiae* x 1

191. **Mammillaria inaiae** Craig

Mammillaria inaiae Craig, Cact. Succ. Journ., 10:111, 1939.

BODY simple or occasionally branching from base, cylindric, rounded at apex, 20 cm. high, to 6 cm. wide. TUBERCLES arranged in 13 and 21 spirals, firm in texture, light green, rounded pyramidal, terete at apex, grooved ventrally from areole for 1-2 mm., with clear watery sap, 8 mm. long, 5 mm. wide at base. AREOLES oval, with wool in very youngest. AXILS with occasional white bristles and little white wool in youth. CENTRAL SPINES 2 or occasionally 3, 9 mm. long, slender subulate, smooth, straight or occasionally slightly bent at apex but not hooked, not enlarged at base, reddish tan to purplish to nearly black, darker at tip, lower porrect, upper one ascending. RADIAL SPINES 17-24 (mostly 20), 4-6 mm. long, acicular, straight, smooth, stiff, white to tip, horizontal. FLOWERS funnelform, 18 mm. long, 20 mm. wide, July, open for several days. *Scales* light grass-green, whitish green narrow margins, lanceolate, tip acute, margins ciliate. *Outer perianth-segments* reddish brown mid-stripe, 1.5 mm. wide, not tapering, more prominent ventrally, pinkish white margins, clavate, tip acute, margins serrate (more so on lower part). *Inner perianth-segments* white to cream, pink at base, pinkish brown mid-line, wider and pinker at base, eliptical rhomboid, tip acute, margins entire. *Filaments* deep pink in upper half, light green in lower half. *Anthers* orange. *Style* white. *Stigma-lobes* 8, light green, 3 mm. long, overtop anthers 3 mm. FRUIT scarlet, ovoid, 11x5 mm., with dried perianth persisting. SEEDS black, glossy, pyriform, faintly pitted, 0.8x0.5 mm.

Distribution: Sonora, Mexico.

Type locality: San Carlos Bay near Guaymas on Gulf of California.

Illustration: Fig. 216 is a reproduction of an illustration in Cact. Succ. Journ., *10*:105, as *Mammillaria inaiae*.

Habitat: Shade or partial shade in decomposed sea shell and leaf mold, salty soil, very near the edge of the sea water.

Fig. 217. *Mammillaria sphacelata* x 1

192. Mammillaria sphacelata Martius

Mammillaria sphacelata Martius, Hort. Reg. Monac., 127, 1829. Nomen.
Mammillaria sphacelata Martius, Nov. Act. Nat. Cur., 16:339, 1832.
Echinocactus sphacelatus Poselger, Allg. Gartenz., 2:107, 1853.
Cactus sphacelatus Kuntze, Rev. Gen. Pl., 1:261, 1891.
Neomammillaria sphacelata Britton & Rose, The Cactaceae, 4:138, 1923.
Chilita sphacelata Orcutt, Cactography, 2, 1926.

BODY cespitose from base and body, cylindric, heads 10-20 cm. high, 2-3 cm. wide. TUBERCLES arranged in 5 and 8 spirals, hidden by spines, bright green, conic, obtuse to rounded at apex, terete, with watery sap, 7 mm. long, 5-6 mm. wide at base. AREOLES round, with very scant woolly mat, very soon becoming naked. AXILS with slight tuft of wool (rarely occasional hair-like bristle, not typical). CENTRAL SPINES 1-4, 4-6 mm. long, slender subulate, straight, smooth, cream, blood-red at apex, later dark brown to gray, 1 porrect and stouter, others dorsally and subcentral. RADIAL SPINES 10-15, 5-8 mm. long, stout acicular, straight, stiff, smooth, creamy white, blood-red at apex, later dark brown, in age chalky white, nearly horizontal. FLOWERS funnelform, lateral in upper part of body, 15 mm. long, 7-8 mm. wide. *Outer perianth-segments* reddish brown wide mid-stripe, narrow greenish cream margins, obtuse to broad lanceolate, tip acute, margins short ciliate. *Inner perianth-segments* dark red, darker mid-line, linear-lanceolate, tip acute, margins entire. *Filaments* purplish pink. *Anthers* orange to yellow. *Style* pink. *Stigma-lobes* 4-5, bright yellowish green, overtop anthers. FRUIT red, curved clavate, with dried perianth persisting. SEEDS black, glossy, nearly smooth, globular with extending ventral hilum, 1.5x0.8 mm.

Distribution: Puebla and Oaxaca, Mexico.

Type locality: None given but reported from near Tehuacan, Puebla.

Illustration: Fig. 217 is from a photograph of a plant collected for us by Mr. Theodore Hutchison.

The original description is supplemented with data from plants collected by Mr. Hutchison of Azusa, California, in 1937, at Tehuacan and from plants collected by us in the low rolling hills near Tehuacan and particularly at K. 227 on the Puebla-Tehuacan Highway in 1941 and 1942.

Britton & Rose report the seeds as "deeply pitted" but the specimens that we obtained had only nearly smooth seed.

FIG. 218. *Mammillaria mazatlanensis* x 0.8

193. **Mammillaria mazatlanensis** Schumann

Mamillaria mazatlanensis Rebut, Catal. 7, 1896. Nomen.
Mamillaria mazatlanensis Schumann, Monatsschr. Kakteenk., 11:154, 1901.
Mammillaria littoralis K. Brandegee, Monatsschr. Kakteenk, 17:80, 1907.
Neomammillaria mazatlanensis Britton & Rose, The Cactaceae, 4:138, 1923.
Chilita mazatlanensis Orcutt, Cactography 2, 1926.

BODY cespitose from base and body, globular to cylindric, rounded at apex, to 12 cm. high, 3-4 cm. wide. TUBERCLES widely separated in 5 and 8 spirals, more or less firm in texture, gray-green, lighter at base, short broad conic, little flattened dorsally, with watery sap, 4-8 mm. long, 10 mm. wide at base. AREOLES round to oval, to 2 mm. wide, with very scant pale tan woolly mat only in youngest. AXILS naked or with 1-2 short bristles. CENTRAL SPINES 3-4, 8-10 (-15) mm. long, acicular, upper ones more slender and more subcentral, all straight, stiff, smooth, lower and lateral with slightly enlarged base, all with cream base, reddish brown above, spreading from porrect, upper ones nearly in radial plane. RADIAL SPINES 13-15, 5-10 mm. long, slender acicular, straight, smooth, stiff, chalky creamy white, horizontal. FLOWERS funnelform, near top, 3-4 mm. long. *Outer perianth-segments* (10) transparent white margins, wide brownish red tapering mid-stripe, lanceolate to spatulate, tip obtuse, margins somewhat serrate. *Inner perianth-segments* (9) carmine red, lanceolate, tip acuminate, margins serrate at apex, 30x5 mm. *Filaments* carmine-red, 10 mm. long. *Anthers* chrome-yellow, slender. *Style* rose. *Stigma-lobes* 7 (5-8), bright green, slender, spreading, 6 mm. long. FRUIT brown becoming reddish yellow, clavate, 7x20 mm., with dried perianth persisting, with approx. 30 seeds per pod. SEEDS black, glossy, flattened globular with protruding basal hilum, pitted, 1 mm. ROOTS fibrous.

Distribution: Sinaloa and Sonora, Mexico.

Type locality: Mazatlan, Sinaloa.

Illustration: Fig. 218 is from a photograph of a plant collected by Mr. E. M. Baxter in the spring of 1940 on a rocky hill back of the city of Maxatlan.

The place of the description of *M. littoralis* is somewhat uncertain. The place usually given is in the Kew Bull. Misc. Imf. *1908* App. 91, 1908, but the same information contained therein is also to be found in the Monatsschr. Kakteenk., *17*:80, 1907, in an unsigned article comparing *M. mazatlanensis* K. Schum. and *M. littoralis* K. Brand., but it gives no previous reference for the latter species.

Mammillaria mazatlensis is a name used by Haage without description, is probably referrable here.

Associated with this species and often confused with it is the type which has 1-4 hooked central spines but we have referred it to *M. occidentalis*.

a. var. **monocentra** var. nov.

Spinis centralibus 1, 7 mm. longis, rectis ad curvis; spinis radialibus 15-16.

BODY cespitose from base. TUBERCLES short broad conic, with watery sap, 4 mm.

Fig. 219. *Mammillaria mazatlanensis var. monocentra* x 1

long, 6-7 mm. wide at base. AXILS naked. CENTRAL SPINES 1, 7 mm. long, acicular, straight to bent to occasionally an incomplete hook, reddish brown porrect. RADIAL SPINES 15-16, 6 mm. long, equal, fine acicular, white, horizontal. FLOWERS unknown. FRUIT scarlet, clavate, 18x6 mm. SEEDS black, glossy, globular obovate with basal hilum, lightly pitted.

Distribution: Sonora, Mexico.

Type locality: Yaqui Valley.

Illustration: Fig. 219 is from a photograph of a plant collected in 1936 by John Hilton and the author in the lower delta land of the Rio Yaqui.

This variety may be a distinct species when the flower is known but with the present information it appears to be close enough to the above type species to be listed as a possible variety until complete data is available.

FIG. 220. *Mammillaria wiesingeri*

194. Mammillaria wiesingeri Boedeker

Mammillaria wiesingeri Boedeker, Kakteenk., 1:204, 1933.

BODY simple, very flattened globular, slightly sunken at apex, 4 cm. high, 8 cm. wide. TUBERCLES arranged in 16 and 26 spirals, flabby in texture, dull leaf green, slender pyramidal, weakly angled, with watery sap, to 10 mm. long, 3-4 mm. wide at base. AREOLES round, 2 mm. wide, with white wool only in top of plant, soon becoming naked. AXILS with no wool, but only occasional 1-2 short thin white bristles. CENTRAL SPINES 4 (rarely 5-6), 5-6 mm. long, somewhat strong acicular, straight, smooth, slightly enlarged base, reddish brown, spreading from porrect. RADIAL SPINES 18-20, 5-6 mm. long, fine acicular, straight, smooth, glossy white, horizontal. FLOWERS campanulate, in crown near top, 12 mm. long and wide. *Outer perianth-segments* carmine-red, rose margins, linear-lanceolate, tip acuminate, margins serrate. *Inner perianth-segments* rose, darker carmine-rose mid-stripe, more linear-lanceolate, tip acuminate, margins serrate.

Filaments carmine-red. *Anthers* bright yellow. *Style* rose. *Stigma-lobes* 5, white, overtop anthers. FRUIT carmine-red, slender clavate, 10 mm. long, with dried perianth persisting. SEEDS brown, curved pyriform with lateral hilum near base, smooth to wrinkled, 1 mm. ROOTS tuberous, 15 mm. thick.

Distribution: Hidalgo, Mexico.

Type locality: Near Metzquititlan, at height of 2000 meters in the obsidian rocks.

Illustration: Fig. 220 is a reproduction of an illustration in Kakteenk., 1:204, as *Mammillaria wiesingeri*.

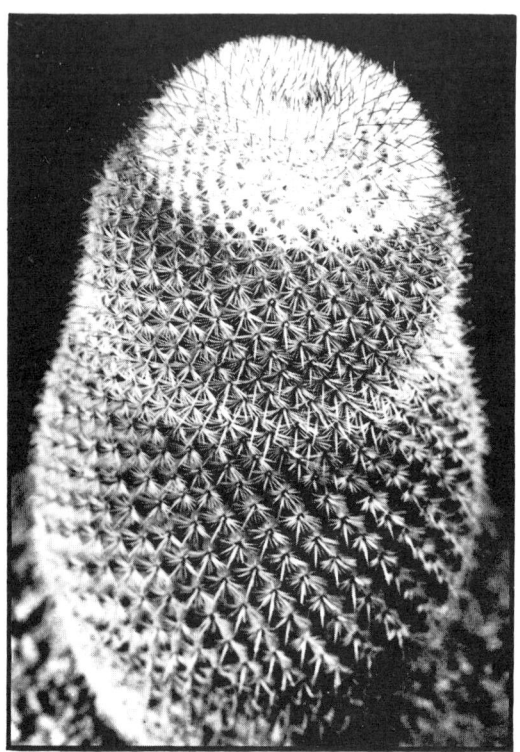

Fig. 221. *Mammillaria perbella* x 0.8

195. Mammillaria perbella Hildmann

Mamillaria perbella Hildmann in Schumann, Gesamtb. Kakteen, 567, 1898.
Neomammillaria perbella Britton & Rose, The Cactaceae, 4:111, 1923.

BODY simple or dichotomous branching, depressed globose to cylindric, sunken at apex, with very little wool, to 6 cm. wide. TUBERCLES arranged in 13 and 21 spirals, gray-green, later bluish, slender conic, with watery sap, 5 mm. long, 4 mm. wide at base. AREOLES round, 2 mm. wide, with short white slightly curly wool. AXILS with some white wool. CENTRAL SPINES 2, 4-6 mm. long, upper one longer, all strong to moderate subulate, straight, stiff, smooth, ivory white, in youth reddish chalky white, divergent porrect. RADIAL SPINES 14-18, 1.5-3 mm. long, bristle-like to thin acicular, straight, stiff, smooth, white, lower ones with black tip. FLOWERS broad funnelform in crown, 9-10 mm. long, 8 mm. wide. *Outer perianth-segments* greenish brown mid-stripe, cream margins, lanceolate, tip acute, margins fine ciliate. *Inner perianth-segments* carmine-red mid-stripe, pale pink margins, linear-lanceolate, tip acute to acuminate, margins mostly

Fig. 222. *Mammillaria perbella*

entire. *Filaments* bright carmine. *Anthers* yellow to tan. Style greenish below, to white, rose-red above. *Stigma-lobes* 3, rose-red, just overtop anthers. FRUIT carmine-red, cylindric, 15 mm. long, very few seeds. SEEDS reddish brown, curved pyriform with lateral hilum, smooth, 1.3x0.8 mm.

Distribution: Queretaro, Hidalgo, state of Mexico; Mexico.

Type locality: None, but reported from San Pablo, Qro., Toleman, Hgo.

Illustrations: Fig. 221 is from a photograph of a plant sent to us by Sr. F. Schmoll of Cadereyta, Qro., as *Mammillaria perbella lanata*. Fig. 222 is from a photograph of a plant from the same source as *Mammillaria perbella fina,* but these two varieties have never been described as such.

196. **Mammillaria lesaunieri** Rebut

Mamillaria lassonneri Rebut, Catalogue, 7, 1896.
Mamillaria lesaunieri Schumann, Gesamtb. Kakteen, 553, 1898.

BODY simple, semi-globose to short cylindric, rounded above, sunken at apex, with short white wool. TUBERCLES arranged in 13 and 21 spirals, firm in texture, dull dark green, slender conic, with watery sap, 6-10 mm. long, 7 mm. wide at base. AREOLES inverted pyriform, with very scant short white wool only in youngest. AXILS with very scant wool, but no bristles. CENTRAL SPINES 1-4, 5-8 mm. long, acicular, straight, smooth, stiff, dull purplish brown, porrect to spreading. RADIAL SPINES 11-13, 4-8 mm. long, lower ones longer, all slender acicular, straight to slightly bent, smooth, white becoming yellowish, horizontal to ascending. FLOWERS recurved funnelform in crown, 25 mm. long, 15 mm. wide, June and July. *Outer perianth-segments* light green base, purplish dark brown tapering mid-stripe above, narrow pink margins, lanceolate, tip acute, margins entire. *Inner perianth-segments* fiery carmine-red, linear-lanceolate, tip acuminate, margins entire. *Filaments* carmine-red. *Anthers* yellow-tan to rose. *Style* creamy white below, rose-red above. *Stigma-lobes* 4-5, tan to rose, 0.5 mm. long. FRUIT red, clavate, 15x5 mm., with dried perianth persisting. SEEDS dull light brown, curved pyriform with lateral hilum near tip, slightly rugose, 1.5x0.6 mm.

Distribution: Vera Cruz (?), Mexico.

Type locality: None given but reported from Cerro Gordo near Gozatlan (Schmoll).

This species has been in the trade for some time but the origin is unknown. It has been referred to as near *M. heyderi* and *M. applanata,* but these have a milky sap. It has also been tentatively reported from Baja California, but reports from explorers of this territory indicate nothing of this nature from there. It is possible that it is a garden hybrid as it was first reported in a seed catalogue. The name has had various spellings; the original of Rebut was *lassonneri* but this was corrected by Schumann to *lesaunieri* since it was named after Herr Lesaunier. Various other spellings have been reported, such as *lassomeri, lesaunierii, and lasonnieri* but they are all referable here.

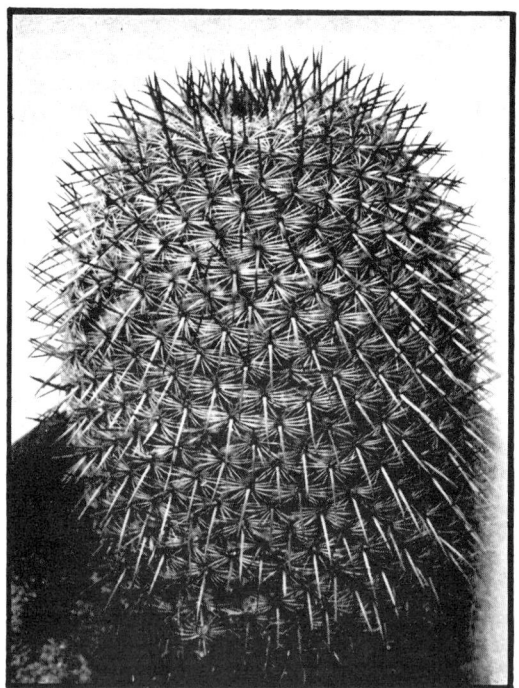

Fig. 223. *Mammillaria haageana* x 1

197. **Mammillaria haageana** Pfeiffer

Mammillaria haageana Pfeiffer, Allg. Gartenz., 4:257, 1836.
Mammillaria diacantha nigra Haage, Catal. 1836. Nomen.
Mammillaria haageana validior Monville in Labouret, Monogr. Cact., 54, 1853.
Cactus haageanus Kuntze, Rev. Gen. Pl., 1:260, 1891.
Neomammillaria haageana Britton & Rose, The Cactaceae, 4:110, 1923.

BODY simple and occasionally cespitose from base, globose to cylindric, rounded above, sunken at apex, 6 cm. high, 4 cm. wide. TUBERCLES closely set in 13 and 21 spirals, firm in texture, bluish green, conic, 4-sided, with watery sap, 5 mm. long, 3 mm.

wide at base. AREOLES oval, 2.5 mm. wide, with white wool, later becoming naked. AXILS with flocculent white wool, later becoming naked. CENTRAL SPINES 2, 6-8 mm. long, lower one longer, all stout acicular, slightly recurved, smooth, reddish brown to black, divergent dorsally and ventrally. RADIAL SPINES 18-20, 3 mm. long, slender acicular, straight to slightly recurved, smooth, white, slightly arched ascending. FLOWERS funnelform, lateral, 12 mm. long. *Outer perianth-segments* pale tannish pink mid-stripe, pale tan margins, very pale greenish cream base, linear, tip acute to obtuse, margins finely serrate, 1-1.5 mm. wide. *Inner perianth-segments* purplish pink, darker mid-line at tip, greenish white base, linear, tip obtuse, margins nearly all entire, 2 mm. wide. *Filaments* nearly white below, pale pink above. *Anthers* yellow. *Style* pale cream below, very pale pink above. *Stigma-lobes* 3, pale olive-green, less than 1 mm. long, overtop anthers 1-2 mm. FRUIT bright red above, pink below, cylindric clavate, 11x4 mm., with dried perianth persisting. SEEDS olive-green light brown, curved pyriform with lateral hilum at base, 1 mm.

Distribution: Very Cruz, Mexico.

Type locality: None given but reported from Perote near Orizaba.

Illustration: Fig. 223 is from a photograph of a plant sent to us by Sr. F. Schmoll.

Pfeiffer (Enum. Cact., 26, 1837) listed as synonyms of this species *M. diacentra nigra* which Haage only listed in his catalogue of 1836 and *M. perote* (Allg. Gartenz., 4:257, 1836) which is only a garden name without author or description. The specific name *diacantha* has been much confused. Haage's plant (1836) was referred to *M. haageana*. Lemaire, 1838, used the name for a plant similar to *M. sempervivi*. Haage (1900) reported a third plant under this name as a variety of *M. centricirrha* (*M. magnimamma*).

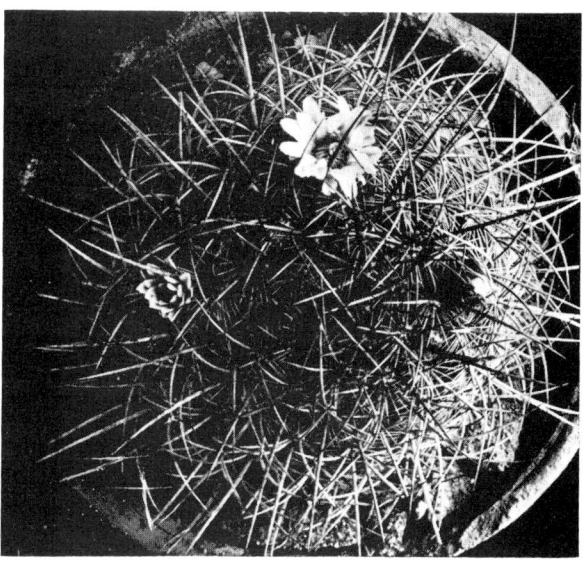

FIG. 224. *Mammillaria amoena*

198. Mammillaria amoena Hopffer

Mamillaria amoena Salm-Dyck, Hopfer in Hort. Dyck, 6, 1845. Nomen.
Mamillaria amoena Hopfer in Salm-Dyck, Cact. Hort. Dyck. 1849, 99, 1850.
Neomammillaria amoena Britton & Rose, The Cactaceae, 4:120, 1923.

BODY simple and cespitose from base, globose to cylindric, sunken at apex, to 10 cm. high, to 12 cm. wide. TUBERCLES somewhat separated in 8 and 13 spirals, bluish milky green, conic to ovate, blunt at apex, laterally compressed at base, with watery sap,

5-7 mm. long. AREOLES round to oval, 3 mm. wide, with little white wool in youth. AXILS with little white wool, but no bristles. CENTRAL SPINES 2, 8-15 mm. long, upper one longest, all acicular, straight to slightly recurved, rigid, yellow-brown to red, divergent dorsally and ventrally from porrect. RADIAL SPINES 16-20, 2-5 mm. long, upper ones shortest, all setaceous, straight, white, horizontal to little ascending. FLOWERS wide funnelform, in crown, 20 mm. long, 15 mm. wide. *Outer perianth-segments* brownish green, red margins, lanceolate, tip acuminate, margins ciliate. *Inner perianth-segments* reddish brown mid-stripe, nearly white margins, spatulate, tip obtuse, margins entire. *Filaments* white below, to carmine-red above. *Anthers* yellow to red. *Style* pale green. *Stigma-lobes* 4, green. FRUIT red, clavate. SEEDS brown.

Distribution: Morelos, Hidalgo; Mexico.

Type locality: None given but reported from Cuernavaca, Mor., Pachuca, Hgo.

Illustration: Fig. 224 is a reproduction of an illustration, Fig. 130, in Britton & Rose (Cact, 4:121), as *Neomammillaria amoena*.

Schumann confused this species with *M. rhodantha*.

The spelling of the author's name has been listed in different ways. Salm-Dyck listed the abbreviation "Hopfr.", Schumann as "Hopffer," Britton & Rose as "Hoppfer."

Mammillaria stueberi Foerster

Mammillaria stüberi Förster, Handb. Cact., 517, 1846.
Mamillaria tentaculata conothele Monville in Labouret, Monogr. Cact., 55, 1853.
Cactus stüberi Kuntze, Rev. Gen. Pl., 1:261, 1891.

BODY globose. TUBERCLES conic. AXILS with white wool, later naked. CENTRAL SPINES 2, 12 mm. long, red-brown. RADIAL SPINES 12-14, 6-8 mm. long, red-brown. FLOWERS, FRUIT and SEEDS unknown.

This species has been referred to *M. rhodantha* but the fewer central spines exclude it from there. It may belong with the above species but inasmuch as the incomplete description makes it uncertain, otherwise the older name would have priority.

Mammillaria conothele Salm-Dyck (?) is only a name listed by Labouret (Des Cactees, 55, 1853) as a synonym of this species.

FIG. 225. *Mammillaria mundtii*

199. Mammillaria mundtii Schumann

Mamillaria mundtii Schumann, Monatsschr. Kakteenk., 13:141, 1903.
Neomammillaria mundtii Britton & Rose, The Cactaceae, 4:112, 1923.

BODY globose to cylindric, rounded above, sunken at apex, 6-7 cm. wide. TUBERCLES arranged in 8 and 13 spirals, semi-firm in texture, dark green, paler in axils, conical, with watery sap, 6-7 mm. long, 5 mm. wide at base. AREOLES round, with scant white then yellowish woolly felt, later becoming naked. AXILS naked. CENTRAL SPINES 2-(4), to 10 mm. long, stout acicular, straight, stiff, brown, darker at tip, becoming gray, divergent dorsally and ventrally. RADIAL SPINES 10-12, 5 mm. long, lower longer, all thin acicular, straight, smooth, white with darker tips, ascending. FLOWERS campanulate, 12 mm. long, 14 mm. wide. *Outer perianth-segments* brownish red mid-stripe, rose margins, lanceolate, tip acute, margins entire, tips sometimes serrate. *Inner perianth-segments* deep carmine, glossy, lanceolate, tip acute, margins entire, recurved. *Filaments* carmine. *Anthers* sulphur-yellow. *Style* white to carmine. *Stigma-lobes* 4, carmine* (white), 1 mm. long overtop anthers 1 mm. FRUIT unknown. SEEDS brown, curved pyriform with lateral hilum, slightly roughened but not pitted, 1.3x0.7 mm. ROOTS large bulky tuberous.

Distribution: Queretaro, Mexico (Schmoll).

Type locality: None given but reported from La Fosiquin (Schmoll).

Illustration: Fig. 225 is a reproduction of an illustration in Monatsschr. Kakteenk., 13:142, as *Mammillaria mundtii*.

FIG. 226. *Mammillaria collina*

200. Mammillaria collina Purpus

Mamillaria collina Purpus, Monatsschr. Kakteenk., 22:162, 1912.
Neomammillaria collina Britton & Rose, The Cactaceae, 4:111, 1932.

BODY simple, seldom branching, globose, a little sunken at apex, with little wool, 13 cm. wide. TUBERCLES arranged in 13 and 21 spirals, gray-green, cylindric, blunt

---

*Schumann reports the stigma-lobes as white but our plant which was imported from Mexico has deep carmine lobes.

at apex, obscurely 4-angled, keeled ventrally, with watery sap, 10 mm. long. AREOLES oval, with white wool in youth, later becoming naked. AXILS with wool in youth. CENTRAL SPINES 2, often only 1, 5-8 mm. long, stout acicular to subulate, straight or little bent, smooth, base enlarged, gray-brown, darker at tip, grayish white at base, divergent dorsally and ventrally. RADIAL SPINES 16-18, 1-5 mm. long, lateral longest, all slender acicular, somewhat curved, stiff, white, horizontal to somewhat ascending. FLOWERS funnelform, in crown near apex, 15-20 mm. long. *Outer perianth-segments* white at base, deep pink wide tapering mid-stripe above, very pale margins, lanceolate, tip acute, margins entire. *Inner perianth-segments* bright rose-red, darker mid-stripe, lanceolate, tip acute, margins entire. *Filaments* white below to pale pink above. *Anthers* pale yellow. *Style* white below to very pale pink above. *Stigma-lobes* 3-5, very pale greenish white, less than 1 mm. long, overtop anthers. FRUIT carmine-red, clavate, to 25 mm. long, with dried perianth persisting. SEEDS brown, finely pitted.

Distribution: Puebla, Mexico.

Type locality: Esperanza.

Illustration: Fig. 226 is a reproduction of an illustration in Monatsschr. Kakteenk., 23:99, as *Mammillaria collina.*

We are referring here the several variations of this species that we collected along the highway between Tehuacan and Los Combres and also near Puebla. They are usually found on the hillsides amongst the rocks.

FIG. 227. *Mammillaria donatii*

201. Mammillaria donatii Berge

Mamillaria donatii Berge in Schumann, Gesamtb. Kakteen Nachtr., 135, 1903.
Neomammillaria donatii Britton & Rose, The Cactaceae, 4:111, 1923.

BODY simple and cespitose, globose, rounded above, faintly sunken and woolly in apex, 8-9 cm. wide. TUBERCLES loosely arranged in 13 and 21 spirals, light bluish green, conic, obtuse at tip, with watery sap, 8 mm. long. AREOLES round, 2 mm. wide, with white wool but very soon becoming naked. AXILS with white wool in youth, soon becoming naked. CENTRAL SPINES 2, to 10 mm. long, lower longer, all stout acicular, straight, smooth, stiff, dark brown, later gray, divergent dorsally and ventrally. RADIAL SPINES 16-18, 6-8 mm. long, later longer, all slender acicular, straight, smooth, semi-flexuous, transparent white, ascending. FLOWERS funnelform, near top, 15 mm. long, 13-14 mm. wide. *Outer perianth-segments* fiery carmine-red, lanceolate, tip acute, margins

(?). *Inner perianth-segments* fiery carmine red, lanceolate, tip acute, margins (?). *Filaments* white. *Anthers* yellow. *Style* white. *Stigma-lobes* 4, golden, overtop anthers. FRUIT red, clavate, 12-15x4 mm. SEEDS brown, curved pyriform.

Distribution: Puebla, Mexico.

Type locality: None given but reported from near Boca del Monte near Esperanza.

Illustration: Fig. 227 is a reproduction of an illustration, Fig. 114, in Britton & Rose (Cact., 4:111) as *Neomammillaria donatii*. In the B. & R. reprint the caption is given in error as Fig. 115 *Neomammillaria collina*.

This species is very close to and may be synonymous with one of the variations of *M. collina* which is found in the same general locality. The tubercles are less angled and the radial spines are more ascending in the above species. Further investigations at the type locality will be necessary to clear up the question.

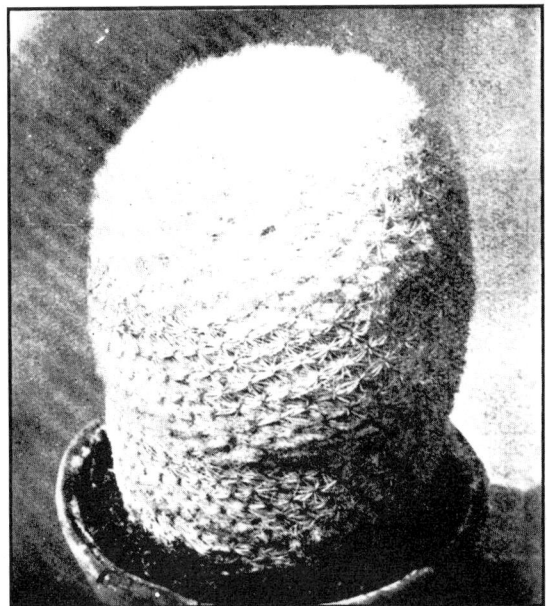

FIG. 228. *Mammillaria albilanata*

202. **Mammillaria albilanata** Backeberg

Mamillaria albilanata Backeberg, Kakteenk., 47, 1939.

BODY simple to occasionally branching, globose to short cylindric, sunken at apex, to 15 cm. high, to 8 cm. wide. TUBERCLES arranged in 13 and 21 spirals, firm in texture, gray-green, obscurely 4-sided at base, cylindric-conic above, with watery sap, 5-6 mm. long, 3-4 mm. wide at base. AREOLES round to oval, with short white wool only in the youngest. AXILS with white very coarse curly woolly hair, persisting. CENTRAL SPINES 2-(4), 2-3 mm. long, subulate, straight, stiff, smooth, enlarged at base, white to cream, brown at very tip, when 2: divergent dorsally and ventrally, when 4: in cross formation. RADIAL SPINES 15-20, 2-4 mm. long, longer laterally, all acicular, straight to slightly curved, stiff, smooth, chalky white, horizontal to slightly ascending. FLOWERS small, 7 mm. long. *Outer perianth-segments* unknown. *Inner perianth-segments* deep carmine red. FRUIT pink to red, with dried perianth persisting. SEEDS pale brown, curved pyriform with lateral hilum at base, faintly roughened but not pitted. 1.2x0.6 mm.

Fig. 229. *Mammillaria albilanata* (*martinezii*)

Distribution: Guerrero, Oaxaca (?); Mexico.

Type locality: Iguala to Chilpancingo.

Illustrations: Fig. 228 is a reproduction of an illustration in Kakteenk., 47, 1939, as *Mamillaria albilanata*. Fig. 229 is from a photograph of a plant received from Sr. F. Schmoll of Cadereyta, Qro., as *M. martinezii*.

Referrable here may be the plants from Mixteca, Oaxaca, which were named but not described by Ernst Tiegel of Germany and distributed by Sr. F. Schmoll under the name of *M. martinezii*. Neal (Cact. Other Succ., 90, 1935) very briefly described a plant under this name as having no central spines and pink flowers but this does not correspond with the plants we have received from Sr. F. Schmoll. This latter form differs from the type in the lack of the abundant axillary wool as shown in the illustration accompanying the original description. It has not flowered for us as yet but in general the description would indicate that it belongs here but the final determination will have to depend upon the flower characteristics. Backeberg made no mention of the type of sap in the tubercles but on the strength of the similarity in most respects of the type and the other form, the latter having watery sap, so we are placing it in this group. Some authors have placed it in the group with the milky sap but with no explanation for doing so. The fruit and seed data is from the related form as no data were given by Backeberg.

Although we have collected in the type locality on two occasions, 1941 and 1942, we were not fortunate enough to be able to locate this species.

From the data available, it appears as if this species might only be a geographical variation of *M. haageana* but the final determination will depend upon the full flower description of this species, which to date is unknown to us.

Fig. 230. *Mammillaria echinaria* x 1

203. **Mammillaria echinaria** DeCandolle

Mammillaria echinaria DeCandolle, Mem. Mus. Hist. Nat. Paris, 17:110, 1828.
Mammillaria densa Link & Otto, Icon. Pl. Rar., 69, 1830.
Mammillaria echinata DeCandolle, Mem. Cact., 3, 1834.
Mammillaria echinata densa Pfeiffer, Enum. Cact., 6, 1837.
Mammillaria gracilis Pfeiffer, Allg. Gartenz., 6:275, 1838.
?Echinocactus densus Steudel, Nom. ed. 2, 1:536, 1840.
Mamillaria anguinea Otto in Salm-Dyck, Cact. Hort. Dyck. 1849, 101, 1850.
Mamillaria subechinata Salm-Dyck, Cact. Hort. Dyck. 1849, 101, 1850.
Cactus anguineus Kuntze, Rev. Gen. Pl., 1:260, 1891.
Cactus echinarius Kuntze, Rev. Gen. Pl., 1:260, 1891.
Cactus gracilis Kuntze, Rev. Gen. Pl., 1:260, 1891.
Mamillaria elongata anguinea Schumann, Gesamtb. Kakteen, 521, 1898.
Mamillaria elongata echinata Schumann, Gesamtb. Kakteen, 521, 1898.
Neomammillaria echinaria Britton & Rose, The Cactaceae, 4:136, 1923.
Mammillaria elongata subechinata Schelle, Kakteen, 300, 1926.
Chilita echinaria Orcutt, Cactography, 2, 1926.
Mamillaria gracilis fragilis Berger, Kakteen, 304, 1929.

BODY cespitose from base, forming large clumps, cylindric, 7-8 cm. high, 1-1.5 cm. wide. TUBERCLES widely separated in 5 and 8 spirals, firm in texture, yellowish gray-green, short rounded conic, broad base, obtuse apex, with watery sap, 4-5 mm. long, 6-7 mm. wide at base. AREOLES oval, with very scant tan wool in youth. AXILS naked, wide. CENTRAL SPINES 1-2, to 10 mm. long, acicular, straight, stiff, smooth, chalky yellow at base, brownish yellow at tip, divergent porrect. RADIAL SPINES 16-18, 6-8 mm. long, slender acicular, stiff, smooth, straight to slightly recurved, clear yellow to nearly white, little ascending to horizontal. FLOWERS funnelform, spiny base, 12 mm. long, 10 mm. wide. *Outer perianth-segments* pale salmon tapering mid-stripe, clear yellow margins, lanceolate, tip obtuse, margins entire. *Inner perianth-segments* pale salmon mid-stripe, pale yellowish tan margins, lanceolate, tip acute to obtuse, margins entire. *Filaments* nearly white. *Anthers* yellow. *Style* pale yellowish green. *Stigma-lobes* 4, greenish yellow, 1.5 mm. long, overtop anthers 2 mm. FRUIT scarlet to pink, clavate, 16x4 mm., with dried perianth persisting. SEEDS light brown, glossy, curved pyriform with lateral hilum, faintly pitted, 1.3x0.7 mm.

Distribution: Hidalgo, Mexico.

Type locality: None given but reported from near Ixmiquilpan.

Illustration: Fig. 230 is a reproduction of an illustration, Fig. 148, in Britton & Rose (Cact., 4:136) as *Neomammillaria echinaria*.

This species is closely related to and often intergrading with *M. elongata* DeCandolle. The distinction is made on the predominating presence of one or more central spines as compared to the usual absence or only occasionally presence of any central spine in *M. elongata*.

M. gracilis Pfeiffer which is very often confused with *M. fragilis* Salm-Dyck is without much question referrable to the above species as was done by Britton & Rose because the original description of *M. gracilis* calls for "plants cylindric, slender, branching" and "belonging to the group of these slender plants."

M. gracilis var. *monville* Cact. Succ. Journ., *10*:77, 1938, is only a name without description but might be referrable here.

Near Ixmiquilpan we found several specimens of this species and one small cluster of which was growing in the top of a large plant of *Echinocactus ingens* whose top had been injured and in which a small amount of soil had collected.

a. var. **rufocrocea** (Salm-Dyck) Schumann

Mamillaria rufo-crocea Salm-Dyck, Cact. Hort. Dyck. 1849, 102, 1850.
Cactus rufocroceus Kuntz. Rev. Gen. Pl., 1:261, 1891.
Mamillaria elongata rufocrocea Schumann, Gesamtb. Kakteen, 521, 1898.

BODY cylindric, 3 cm. wide. TUBERCLES arranged in 5 and 8 spirals. CENTRAL SPINES 1 to rarely 2. RADIAL SPINES 14-16, acicular, yellow-orange, red orange in outer half, especially in upper part of plant, base orange.

Distribution: The same.

204. **Mammillaria graessneriana** Boedeker

Mamillaria grässneriana Boedeker, Monatsschr. Kakteenk., 30:84, 1920.
Neomammillaria graessneriana Britton & Rose, The Cactaceae, 4:117, 1923.

BODY simple and sparingly cespitose from base, globular to ovate, sunken at apex 8 cm. high, 6 cm. wide. TUBERCLES arranged in 21 and 34 spirals, dark bluish green, rounded 4-sided pyramidal, blunt at apex, with watery sap, 6-8 mm. long, 3-4 mm. wide at base. AREOLES round, to 3 mm. wide, with rich snow white wool, very soon becoming naked. AXILS with short snow white wool in apex, often persisting, but no bristles. CENTRAL SPINES 2-4, at first only 2, later 4, to 8 mm. long, stout acicular, straight, stiff, not enlarged at base, dull reddish brown, lighter at base, when 2: divergent dorsally and ventrally, when 4: in cross formation, wide spreading. RADIAL SPINES 18-20, 6-8 mm. long, acicular, straight, stiff, glossy white, horizontal. FLOWERS lateral, small. *Perianth-segments* reported to be red (?). FRUIT and SEEDS unknown.

Distribution: Central states of Mexico.

Type locality: None given.

Illustration: Fig. 231 is a reproduction of an illustration in Monatsschr. Kakteenk., *30*:85, as *Mammillaria grässneriana* and a reproduction of the same in Britton & Rose (Cact., 4:116) as *Neomammillaria graessneriana*.

We have not been able to recollect this species, so we know it only from the original description. It is reported that the flowers are similar to those of *M. elegans*.

M. schulzeana is only a name mentioned by Boedeker in Berger (Kakteen, 308, 1929) under *M. graessneriana*.

Fig. 231. *Mammillaria graessneriana*

Mammillaria rutila Zuccarini

Mammillaria rutila Zuccarini in Pfeiffer, Enum. Cact., 29, 1837.
Mammillaria rutila octospina Scheidweiler, Bull. Acad. Sci. Brux., 6:91, 1839.
Mamillaria eugenia Lemaire, Cact. Mon., 99, 1839. (?Hort. Belg.) Nomen.
Mamillaria rutila pallidior Salm-Dyck, Cact. Hort. Dyck. 1849, 11, 1850.
Cactus rutilus Kuntze, Rev. Gen. Pl., 1:261, 1891.

BODY simple, globose, sunken at apex. TUBERCLES dark green, compressed conic, 10 mm. long, 6 mm. wide at base. AREOLES with wool in youth. AXILS nearly naked. CENTRAL SPINES 4-5 (6), 8-12 mm. long, lower longer, all rigid, curved, reddish brown, spreading. RADIAL SPINES 14-16 (20-24), 4-8 mm. long, upper 5-6 shorter, all setaceous, white, horizontal. FLOWERS (by Dietrich) funnelform, 12 mm. long. *Outer perianth-segments* smudgy pale purplish red, lanceolate. *Inner perianth-segments* purplish red, lanceolate, tips split, margins entire. *Filaments* bright purplish red, curved to hooked. *Anthers* yellow. *Style* bright purplish red. *Stigma-lobes* 4, purplish red. FRUIT and SEEDS unknown.

Distribution: Hidalgo (?), Mexico.

Type locality: Atotonilco el Chico, Sierra Rosa, 8000 feet altitude.

The greater spine-count is recorded in a subsequent publication by Zuccarini in which he probably confused it with *M. celsiana*. There is no mention made of the sap but we are assuming that it is watery on the strength that it was compared with *M. celsiana* by Salm-Dyck. Schumann refers it to *M. coronaria* but that species is supposed to have hooked central spines. Britton & Rose placed it under the uncertain species.

This uncertain species may be referrable to *M. graessneriana* whose flower is unknown. If, when the flower is known and it corresponds with this description, it will give priority to the older name.

Fig. 232. *Mammillaria hoffmanniana*

205. **Mammillaria hoffmanniana** (Tiegel) Bravo

Neomammillaria hoffmanniana Tiegel, An. Inst. Biol. Mex., 5:269, 1934.
Mammillaria hoffmanniana Bravo, Las Cact. Mex., 687, 1937.

BODY simple, globose to short cylindric, flattened and sunken at apex, 30 cm. high, 12 cm. wide. TUBERCLES arranged in 13 and 21 spirals, firm in texture, bright green, rounded 4-sided, cylindric, rounded at apex, flattened dorsally, with watery sap, 10 mm. long, 10 mm. wide at base. AREOLES oval to round, 3 mm. wide, with abundant white wool in youth and persisting for some time. AXILS with dense white wool, almost to top of tubercles. CENTRAL SPINES 4-6 or 7, to 25 mm. long, upper and lower longer, all thin subulate, recurved and tortuous, stiff, smooth, slightly enlarged at base, white to bright cream, brown at very tip, nearly horizontal to slightly ascending, overtopping apex, interlacing, occasionally an extra one that is mid-center and porrect. RADIAL SPINES 18-20, often deciduous, less than 1 to 3 mm. long, very fine setaceous, straight to somewhat tortuous and slightly recurved, semi-flexuous, white, horizontal to somewhat ascending, sometimes more so than the centrals. FLOWERS campanulate, near apex, 10 mm. long and wide, April. *Outer perianth-segments* light yellow-green base, brownish above, carmine-red mid-stripe, lanceolate, tip acute, margins finely serrate. *Inner perianth-segments* magenta, pale green base, linear lanceolate, ends split, margins entire, little longer and wider than outer perianth-segments. *Filaments* magenta, paler at base, short. *Anthers* pale rose to magenta with yellow pollen. *Style* white below, reddish above, heavy. *Stigma-lobes* 4-5, reddish-carmine, small, overtop perianth-segments. FRUIT carmine-red, elongate clavate, 20 mm. long, with dried perianth persisting. SEEDS yellowish light tan, curved pyriform with narrow lateral hilum, smooth, 1x0.5 mm.

Distribution: Queretaro, Guanajuato; Mexico.

Type locality: None given but reported from La Templadora, Qro.

Illustration: Fig. 232 is from a photograph of a plant obtained from Sr. F. Schmoll of Cadereyta, Qro.

Tiegel says that it has watery sap and differs from the *M. polythele* group to which it appears to be like in appearance and by the character of the sap. Bravo places it in the section of Galactochylus, or milky group, with no explanation for doing so. Our plants were obtained from the same source as were those used by Tiegel and they were found to have a watery sap in both the tubercles and the body.

FIG. 233. *Mammillaria fuliginosa* x 0.8

206. **Mammillaria fuliginosa** Salm-Dyck

Mamillaria fuliginosa Salm-Dyck, Cact. Hort. Dyck. 1849, 93, 1850.
Mammillaria fuliginosa longispina Haage, Cateen Cult., 133, 1900.

BODY globose, somewhat sunken at apex, to 8 cm. wide. TUBERCLES arranged in 21 and 34 spirals, firm in texture, dull dark green, conic, with watery sap, 6-7 mm. long, 5 mm. wide at base. AREOLES inverted pyriform, with white wool persisting for some time. AXILS with scant wool but no bristles. CENTRAL SPINES 4, 8-10 mm. long, upper and lower longer, all stout acicular, straight, smooth, stiff, chalky white, black tips, strongly spreading in cross formation. RADIAL SPINES 16, 1-2 mm. long, setaceous, straight, white, somewhat ascending. FLOWERS tubular, 15 mm. long, 10 mm. wide. *Outer perianth-segments* reddish brown wide tapering mid-stripe, deep pink margins, lanceolate, tip acute, margins lightly serrate. *Inner perianth-segments* deep pink, darker mid-line, lanceolate, tip acuminate, margins entire. *Filaments* deep pink. *Anthers* tan. *Style* cream. *Stigma-lobes* 5, deep pink, 1 mm. long, just overtop anthers. FRUIT carmine-red, cylindric clavate, 20x4 mm., with dried perianth persisting. SEEDS light tan, curved pyriform with lateral hilum near base, very faintly wrinkled, 1x0.5, very few seeds per pod.

Distribution: Uncertain.

Type locality: None given but reported from Caracas (Venezuela)?

Illustration: Fig. 233 is from a photograph of a seedling plant in our garden.

This species, as described by Salm-Dyck from plants received from Scheer of England, is of uncertain nature. He compared it with but distinguished it from *M. fulvispina*. The original description gave no type locality or distribution but Labouret in 1853 referred it to Caracas with a ?, which probably refers to a locality in Venezuela by that name. Britton & Rose, probably on the strength of this reported distribution, refer it to *M. mammillaris* but this latter species differs in that it has a milky sap. The plant in the trade coincides very closely with the described species.

Mammillaria obvallata Otto

Mammillaria obvallata Otto in Dietrich, Allg. Gartenz., 14:308, 1846.
Cactus obvallatus Kuntze, Rev. Gen. Pl., 1:261, 1891.

BODY ovate, sunken at apex. TUBERCLES bluish green, conic. AREOLES with white wool, later becoming naked. AXILS with wool. CENTRAL SPINES 4, 25 mm. long, lower one longest to 35 mm., heavier, all acicular, curved, deep yellow in youth, brown in age, in cross formation. RADIAL SPINES 16, fine bristle-like, white. FLOWERS 10 mm. long. *Outer perianth-segments* 8-10, brownish red, lanceolate, tip acute. *Inner perianth-segments* purplish red, lanceolate, tip acuminate, margins entire. *Filaments* purplish red. *Anthers* yellow. *Style* purplish red. *Stigma-lobes* 3-4, purplish red, short, thick. FRUIT and SEEDS unknown.

Distribution: Mexico.

Type locality: Unknown.

Inasmuch as this species is incompletely described, its association is not certain. It appears to be somewhat similar to the above species but if complete data were available, it might have priority as it is the older name.

207. Mammillaria ochoterenae (Bravo) Werdermann

Neomammillaria ochoterenae Bravo, An. Inst. Biol. Mex., 2:127, 1931.
Mammillaria ochoterrenae Werdermann in Backeberg, Neue Kakteen, 98, 1931.

BODY simple, depressed globose, 8 cm. wide. TUBERCLES arranged in 13 and 21 spirals, hidden by spines, firm in texture, bright green, conic, keeled ventrally, with watery sap, 6 mm. long, 5 mm. wide at base. AREOLES with scant white wool in youth. AXILS naked. CENTRAL SPINES 5-6, 10-20 mm. long, lower longer and heavier, all heavy acicular to subulate, straight to recurved, stiff, smooth, yellowish gray with reddish tip in youth, later brown with black tip, lower depressed porrect, uppers ascending. RADIAL SPINES 17-18, 4-9 mm. long, slender acicular, straight, smooth, white to light amber, horizontal to somewhat ascending, interlacing. FLOWERS lateral in crown, 10 mm. long. *Outer perianth-segments* brownish red mid-stripe, white margins, lanceolate, margins entire. *Inner perianth-segments* rose mid-stripe, white margins, linear, tip acuminate, margins entire. *Filaments* bright rose. *Anthers* yellow. *Stigma-lobes* 5-6, bright yellow. FRUIT red, globular clavate, 20 mm. long, with dried perianth persisting. SEEDS bright brown, pyriform with lateral hilum near base, faintly roughened, 1 mm. long.

Distribution: Oaxaca, Mexico.

Type locality: None given.

FIG. 234. *Mammillaria ochoterenae*

Illustration: Fig. 234 is a reproduction of an illustration in An. Inst. Biol. Mex., 2:127, as *Neomammillaria ochoterenae*.

Werdermann places this species in the Galactochylus (milky) Section but states that the tubercles are not milky on injury. Inasmuch as this is our only means of determining the milkiness of the tubercle, we are placing it in the group with the watery sap as was done by Bravo in the original description.

Schmoll says that it is from Esperanza, Puebla, and believes it to be similar to and probably the same as *M. esperanzaenis* Boedeker.

208. **Mammillaria pringlei** (Coulter) Brandegee

Cactus pringlei Coulter, Contr. U. S. Nat. Herb., 3:109, 1894.
Mamillaria pringlei K. Brandegee, Zoe, 5:7, 1900.
Neomammillaria pringlei Britton & Rose, The Cactaceae, 4:115, 1923.

BODY simple, globose to short cylindric, to 16 cm. high, to 7 cm. wide. TUBERCLES arranged in 13 and 21 spirals, firm in texture, yellowish gray-green, conic, blunt apex, with watery sap, 6-10 mm. long, 7 mm. wide at base. AREOLES oval to 2 mm., with yellowish wool in youngest. AXILS with white wool, (occasional short white bristle but not typical). CENTRAL SPINES 6 (5-7), 18-20 mm. long, stout acicular, recurved, smooth, stiff, yellow, strongly spreading from porrect, overtopping apex. RADIAL SPINES 15-20, 5-8 mm. long, fine acicular, stiff, smooth, straight or slightly bent, yellow, nearly horizontal. FLOWERS funnelform, lateral 8-10 mm. long. *Outer perianth-segments* brownish red, paler margins, linear, tip acute, margins serrate. *Inner perianth-segments* deep red, linear lanceolate, tip acute, margins nearly entire. *Filaments* deep pink. *Anthers* yellow. *Style* pink above, nearly white below. *Stigma-lobes* 3, tan, 1 mm. long. FRUIT reddish, 12-15 mm. long, elongate clavate, with dried perianth persisting. SEEDS brown, curved pyriform with lateral hilum near base, 1x0.7 mm.

Distribution: San Luis Potosi, state of Mexico; Mexico.

Fig. 235. *Mammillaria pringlei* x 1.

Type locality: None given but reported from Tultenango Canyon, state of Mexico.

Illustration: Fig. 235 is from a photograph of a plant sent to us by Sr. F. Schmoll of Cadereyta, Qro.

a. var. **columnaris** Schmoll var. nov.

Corpus cylindratior, omnes spinis tenuioris acicularibus et brevioris.

Body more cylindric, all spines more slender acicular and shorter.

209. Mammillaria discolor Haworth

Mammillaria discolor Haworth, Syn. Pl. Succ., 177, 1812.
Cactus depressus DeCandolle, Cact. Hort. Monsp., 84, 1813. Not Haworth 1812.
Cactus pseudomammillaris Salm-Dyck, Liste Pl. Gr., 1:1, 1815.
Cactus spinii Colla, l'Antol. Bot. Opera, 6:400, 1813.
Mammillaria pseudomammillaris Desfontain, Tabl. Ecol. Bot. Roi, 191, 1815. Nomen.
Mammillaria discolor prolifera Pfeiffer, Enum. Cact., 28, 1837.
Mammillaria albida Haage in Pfeiffer, Enum. Cact., 28, 1837.
Mammillaria aciculata Otto in Pfeiffer, Enum. Cact., 29, 1837.
Mammillaria discolor monstrosa Monville in Lemaire, Cact. Gen. Nov. Sp., 99, 1839.
Mamillaria discolor albida Salm-Dyck, Hort. Dyck. 1844, 7, 1845.
Mamillaria discolor rhodacantha Salm-Dyck, Hort. Dyck. 1844, 8, 1845.
?*Mamillaria polythele aciculata* Salm-Dyck, Hort. Dyck. 1844, 9, 1845.
Mamillaria curvispina Otto in Dietrich, Allg. Gartenz., 14:204, 1846.
Mamillaria curvispina parviflora Otto in Dietrich, Allg. Gartenz., 14:204, 1846.
?*Mamillaria discolor puchella* Otto in Foerster, Handb. Cact., 206, 1846.
Mamillaria nitens Otto in Linke, Allg. Gartenz., 16:331, 1848.
Mamillaria puchella Otto in Linke, Allg. Gartenz., 16:331, 1848.
Mamillaria discolor aciculata Salm-Dyck, Cact. Hort. Dyck. 1849, 11, 1850.

Mamillaria discolor coniflora Salm-Dyck, Cact. Hort. Dyck. 1849, 11, 1850.
Mamillaria discolor curvispina Salm-Dyck, Cact. Hort. Dyck. 1849, 11, 1850.
Mamillaria discolor nitens Salm-Dyck, Cact. Hort. Dyck. 1849, 11, 1850.
Mamillaria rhodacantha Salm-Dyck, Cact. Hort. Dyck. 1849, 96, 1850.
Mamillaria pulchella flore pallidiore Salm-Dyck ex Labouret, Monogr. Cact., 40, 1853.
Mamillaria puchella nigricans Monville in Labouret, Monogr. Cact., 40, 1853.
Mamillaria porphyracantha Jacobi, Allg. Gartenz., 24:81, 1856.
Cactus aciculatus Kuntze, Rev. Gen. Pl., 1:260, 1891.
Cactus discolor Kuntze, Rev. Gen. Pl., 1:260, 1891.
Cactus pulchellus Kuntze, Rev. Gen. Pl., 1:261, 1891.
Neomammillaria discolor Britton & Rose, The Cactaceae, 4:132, 1923.
Chilita discolor Orcutt, Cactography, 2, 1926.

FIG. 236. *Mammillaria discolor* x 0.8

BODY simple, globose to ovate, later cylindric, rounded at apex, to 8 cm. wide. TUBERCLES arranged in 8 and 13 spirals, firm in texture, bluish green, ovoid conic, with watery sap, 6-7 mm. long, 5-7 mm. wide at base. AREOLES oval, 2-3 mm. wide, with short white wool in youth, later becoming naked. AXILS with very scant wool to naked. CENTRAL SPINES 6 (8), 10 mm. long, lower longer, all stout acicular, straight, stiff, smooth, amber-yellow, black at tip, white at base, later ashen, ascending spreading. RADIAL SPINES 16-20 (sometimes more), to 10 mm. long, lower longer, all thin acicular, somewhat stiff, straight, snow white, horizontal. FLOWERS funnelform, lateral, 20 mm. long, 16 mm. wide. *Outer perianth-segments* old rose to dark reddish mid-stripe, whitish margins, linear, 2 mm. wide, tip obtuse, margins entire. *Inner perianth-segments* white, rose-red mid-line, linear, tip acute to obtuse, margins entire. *Filaments* white to pink. *Anthers* yellow. *Style* white to pale yellow. *Stigma-lobes* 6-7, pale green. FRUIT reddish tan, clavate, 25 mm. long, with dried perianth persisting. SEEDS brown, flattened curved pyriform with lateral hilum, faintly pitted, 1x0.6 mm.

Distribution: Puebla, Mexico.

Type locality: None given.

Illustration: Fig. 236 is from a photograph of a seedling plant obtained from Johnson's Water Gardens of Hynes, California.

M. aciculata Otto and *M. polythele aciculata* Salm-Dyck were referred by Schumann to *M. polythele* but the greater spine count probably places it here.

"*M. depressa* was credited by mistake to DeCandolle by Pfeiffer (Enum. Cact., 28, 1837) in listing the synonyms of *M. discolor*.

"*M. confinis* Haage, according to Pfeiffer (Enum. Cact., 28, 1837), appeared in 'Haage, Catal. Cact., 1836' and he lists it as a synonym of *M. albida*.

"*M. canescens* Hort. in Pfeiffer (Enum. Cact., 28, 1837), was given as a synonym of *M. discolor*. This is different from *M. canescens* Jacobi (Allg. Gartenz., 24:98, 1859) which Schumann lists among his unknown plants. (See also below.)

"*M. coniflora* Hort. and *M. discolor coniflora* Salm-Dyck (Cact. Hort. Dyck. 1849, 11, 1850) are only names which belong here.

"*M. discolor fulvescens* Salm-Dyck (Hort. Dyck. 1844, 7, 1845) was not published at the place cited.

"*M. discolor breviflora* Foerster (Handb. Cact., 206, 1846), although not described at the place here cited, is usually referred here.

"*Cactus pseudomammillaris* appeared simply as a name in 1815 (Desfontaines, Tab. Bot., ed. 2, 191) and again in Pfeiffer's (Enum. 28, 1837), as a synonym of *Mammillaria discolor prolifera*. Pfeiffer credits the name to Salm-Dyck and gives the reference to Allg. Gartenz. (3:57, 1835), but the name appeared there under *Mammillaria* along with *spinii* and *canescens*. *M. spinii* credited to Colla, is given by Salm-Dyck (Cact. Hort. Dyck. 1849, 11, 1850) as a synonym of *M. discolor*." Britton & Rose (Cact., 4:133, 1923).

M. rhodantha droegeana K. Schumann (Gesamtb. Kakteen, 550, 1898) was listed as a variety without seeing the flower. Quehl (Monatsschr. Kakteenk., 48, 1915) reports the bloom as yellowish and refers it as a variety of *M. discolor*. Hildmann in Schelle (Handb. Kakteenk., 257, 1907) raises it to specific rank.

M. canescens was very briefly described by DeCandolle (Prodr. Syst. Nat., 459, 1828) and referred to it as a synonym of *M. lanifera* but Pfeiffer later referred it to *M. discolor* and also to *M. grandiflora*.

Fig. 237. *Mammillaria albicans*

210. **Mammillaria albicans** (B. & R.) Berger

Neomammillaria albicans Britton & Rose, The Cactaceae, 4:138, 1923.
Neomammillaria slevinii Britton & Rose, The Cactaceae, 4:139, 1923.
Chilita albicans Orcutt, Cactography, 2, 1926.
Chilita slevinii Orcutt, Cactography, 2, 1926.
Mamillaria albicans Berger, Kakteen, 308, 1929.
Mammillaria slevinii Boedeker, Mammill. Vergl. Schluss., 44, 1933.

BODY simple and cespitose, cylindric, 10-20 cm. high, to 6 cm. wide. TUBERCLES firm in texture, pale green, broad conic, with watery sap, 4-5 mm. long, 6 mm. wide at base. AREOLES with sometimes white wool in youth. AXILS with more or less dense wool. CENTRAL SPINES 3-4 (6), 8-10 mm. long, straight, stout acicular, stiff, brownish with black tip, widely spreading. RADIAL SPINES 14-17, to 8 mm. long, acicular, straight, smooth, stiff, white to pinkish brown to black at tips, nearly horizontal. FLOWERS 20 mm. wide. *Outer perianth-segments* pinkish. *Anthers* yellowish. *Style* nearly white. *Stigma-lobes* nearly white. FRUIT red, 10-18 mm. long, clavate, with dried perianth persisting. SEEDS black, nearly globular with large basal hilum projecting beyond the body.

Distribution: Islands in Gulf of California off coast of Baja California, Mexico.

Type locality: Santa Cruz and San Jose Islands.

Illustration: Fig. 237 is a reproduction of an illustration, Fig. 153, in Britton & Rose (Cact., 4:139) as *Neomammillaria slevinii*.

Mammillaria slevinii is referred here because the presence of more or less wool in the axils is hardly a substantial distinguishing factor in the separation of species as its presence or absence in many species varies according to the soil and moisture conditions and the time of flowering.

211. Mammillaria esperanzaensis Boedeker

Mammillaria fuscata-esperanza Boedeker, Mammill. Vergl. Schluss., 40, 1933.
Mammillaria esperanzaensis Boedeker, in Backeberg and Knuth, Kaktus A.B.C. 392, 1935.

BODY globular to clavate, 80 mm. wide. TUBERCLES cylindric to ovoid, with watery sap. AXILS naked. CENTRAL SPINES 4-7, lower longer and bent downward, thickened at base, stiff, amber-brown with metalic luster. RADIAL SPINES to 20, thin, straight, stiff, bright yellow, metalic luster. FLOWERS 25 mm. wide. *Inner perianth-segments* white with red mid-stripe. *Filaments* white. *Style* yellowish white. *Stigma-lobes* 5, bright yellow. FRUIT and SEEDS unknown.

Distribution: Puebla, Mexico (Bravo).

Type locality: None given but reported from near Esperanza (Bravo).

Boedeker cited Purpus as the original author but gave no reference and we have not been able to find any.

Schmoll suggested to us that this species is really *M. ochoterenae* of the type from Esperanza, Puebla. Inasmuch as we have not been able to visit this locality to study the plants it will have to remain uncertain until further study can be made.

212. Mammillaria halbingeri Boedeker

Mammillaria halbingeri Boedeker, Kakteenk., 1:9, 1933.

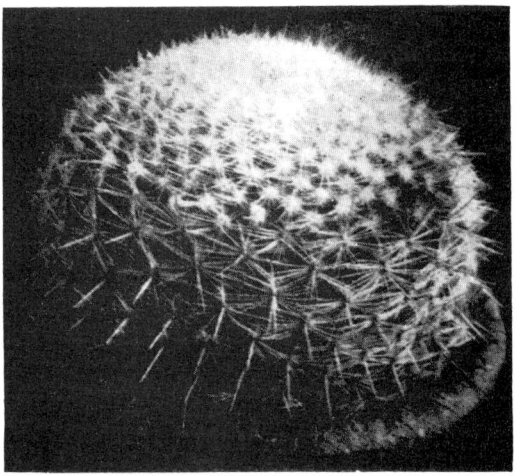

FIG. 238. *Mammillaria halbingeri*

BODY simple, flattened globular, rounded not sunken, but woolly at apex. TUBERCLES loosely arranged in 13 and 21 spirals, glossy, light green, conic to ovate, with watery sap, 4-5 mm. long, 3 mm. wide at base. AREOLES round, to 1.5 mm. wide, with short white wool only in youngest. AXILS with several very thin hair-like bristles. CENTRAL SPINES 2, 5-6 mm. long, thin subulate, straight, smooth, slightly enlarged at base, yellowish white, pale brown at tip, divergent dorsally and ventrally. RADIAL SPINES to 25, 5-7 mm. long, thin acicular, straight, smooth, white, nearly horizontal. FLOWERS funnelform, in upper part of plant, to 12 mm. long. *Outer perianth-segments* greenish yellow, bright green ventral-stripe, margins pale yellow, lanceolate, tip acute, margins serrate. *Inner perianth-segments* sulphur yellow, paler margins, wider lanceolate, tip obtuse, margins serrate. *Filaments* bright yellow. *Anthers* orange yellow. *Style* white. *Stigma-lobes* 6, white, small, moderately extending. FRUIT white (?), clavate, small. SEEDS bright reddish brown, faintly glossy, smooth, with lateral hilum, 1 mm.

Distribution: Oaxaca, Mexico.

Type locality: SE. part of state.

Illustration: Fig. 238 is a reproduction of an illustration in Kakteenk. 1:9 as *Mammillaria halbingeri*.

213. **Mammillaria conspicua** Purpus

Mammillaria conspicua Lemaire, Les Cactees 34, 1868. Nomen.
Mamillaria conspicua Purpus, Monatsschr. Kakteenk., 22:163, 1912.
Neomammillaria conspicua Bravo, An. Inst. Biol. Mex., 122, 1930.

BODY simple, globose to cylindric, rounded above, with apex little sunken, 14 cm. high, 10 cm. wide. TUBERCLES closely set, gray-green, conic to bluntly 4-angled, with watery sap, 6-7 mm. long. AREOLES round to oval, small with wool only in apex. AXILS with wool and bristles. CENTRAL SPINES 2, 10 mm. long, heavy acicular, straight, to somewhat bent, white with brown tip, reddish brown in youth, divergent dorsally and ventrally. RADIAL SPINES 16-25, to 6 mm. long, lateral longer, setaceous

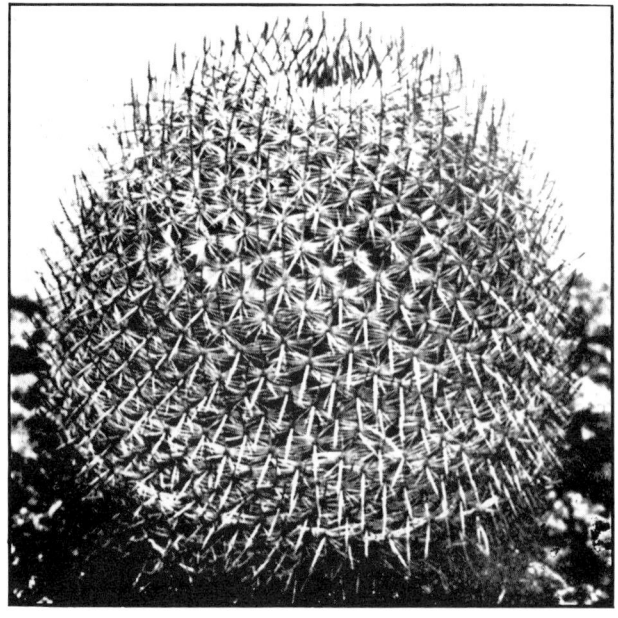

FIG. 239. *Mammillaria conspicua*

to acicular, straight, white, horizontal. *Inner perianth-segments* rose with darker mid-stripe. *Filaments* white. *Anthers* yellow. *Style* white. *Stigma-lobes* yellow. FRUIT carmine-red, cylindric clavate. SEEDS light brown, glossy curved pyriform, hardly 1 mm.

Distribution: Puebla, Mexico.

Type locality: Zapotitlan.

Illustration: Fig. 239 is a reproduction of an illustration in Monatsschr. Kakteenk. 24:37 as *Mammillaria conspicua.*

Lemaire (Les Cactees, 34, 1868) only mentioned the name. Britton & Rose (Cact. 4:108, 1923) refers this species as related to *Neomammillaria elegans* but it differs from it in that it has bristles in the axils. Bravo redescribes the species under *Neomammillaria* (An. Inst. Biol. Mex. 112, 1930) but does not even mention it in her later book (Las Cactaceas de Mexico, 1937).

FIG. 240. *Mammillaria nunezii* x 1

214. Mammillaria nunezii (B. & R.) Orcutt

Neomammillaria nunezii Britton & Rose, The Cactaceae, 4:120, 1923.
Mammillaria nunezii Orcutt, Cactography 8, 1926.

BODY simple to occasionally branching from body, globose to cylindric, to 15 cm. high, 6 8 cm. wide. TUBERCLES arranged in 13 and 21 spirals, firm in texture, dull green, conic cylindric, blunt rounded apex, with watery sap, 7 mm. long, 6-7 mm. wide at base. AREOLES oval, with creamy wool in youth. AXILS with 8-10 white bristles. CENTRAL SPINES 2 to mostly 4, occasionally 5-6, 8-15 mm. long, subulate, smooth, stiff, only slightly enlarged at base, yellowish cream, reddish brown to black at tip, spreading porrect. RADIAL SPINES 25-30, 5-7 mm. long, slender acicular, straight, smooth, stiff, white, nearly horizontal. FLOWERS funnelform, 15 mm. wide. *Outer perianth-segments* brownish red mid-stripe, lighter margins, lanceolate, tip acute, margins lightly serrate. *Inner perianth-segments* deep rose to reddish magenta, lanceolate, tip acute, margins entire. *Filaments* whitish below, rose above. *Anthers* yellow. *Style* very pale pinkish cream. *Stigma-lobes* 5, greenish yellow. FRUIT pale greenish white, tinged with pink near tip, clavate, 23x7 mm., with dried perianth easily detached. SEEDS brown, pyriform with lateral hilum at tip, faintly rugose, 1.2x0.8 mm. ROOTS fibrous.

Distribution: Guerrero, Mexico.

Type locality: Buenavista de Cuellar near Iguala, also reported from Taxco.

Illustration: Fig. 240 is from a photograph of a plant that we collected near Taxco in 1942.

This species is very closely related to the hooked spine type of *M. solisii* (cf.) with which it is found associated and with which it readily hybridizes as we found both species growing in the same canyons near Taxco.

Orcutt (Cactography, 2, 1926) confused this species with *M. guerreronis*.

FIG. 241. *Mammillaria ortiz rubiona* x 1

215. Mammillaria ortiz rubiona (Bravo) Werdermann

Neomammillaria ortiz rubiona Bravo, An. Inst. Biol. Mex., 2:193, 1931.
Mammillaria ortiz-rubiana Werdermann in Backeberg Neue Kakteen, 95, 1931.

BODY very cespitose to large clumps, individual heads depressed globose, slightly sunken and white woolly at apex, 8-10 cm. wide. TUBERCLES arranged in 8 and 13 spirals, dull gray green, cylindric to clavate, rounded at apex, with watery sap, 15-18 mm. long, 10 mm. wide at base. AREOLES oval, with short white wool in youth. AXILS with numerous white setaceous hairs extending beyond tubercles. CENTRAL SPINES 4-6, 12-15 mm. long, acicular, straight, smooth, stiff, brittle, white sometimes with rose tip, one true is porrect, others nearly in same plane with radials and differing only in size. RADIAL SPINES 25-30, 4-15 mm. long, very fine acicular to hair-like, straight, smooth, semi-flexuous, white, horizontal and interlacing. FLOWERS funnelform, 35 mm. long, 30 mm. wide. *Outer perianth-segments* light gray-green wide mid-stripe, pale pink mid-line, very pale pink margins, wide spatulate, tip acute, margins ciliate. *Inner perianth-segments* very pale greenish white to very pale pink, pink mid-line, spatulate, 4 mm. wide, tip acute, margins entire. *Filaments* very pale pink, hooked. *Anthers* deep orange. *Style* pale pink. *Stigma-lobes* 6-8, tannish pink, later to deep pink, 3 mm. long, overtop anthers 3-4 mm. FRUIT carmine, clavate, with dried perianth persisting. SEEDS black, glossy, pitted regularly in rows, 1.5 mm.

Distribution: Guanajuato and Queretaro, Mexico.

Type locality: None given but reported from along the border between the two states.

Illustration: Fig. 241 is from a photograph of a plant obtained from Sr. Schmoll of Cadereyta, Qro.

FIG. 242. *Mammillaria neopalmeri* x 1

216. **Mammillaria neopalmeri** (Coulter) nom. nov.

Cactus palmeri Coulter, Contr. U. S. Nat. Herb., 3:108, 1894.
Mamillaria dioica insularis K. Brandegee, Erythea, 5:115, 1897.
Neomammillaria palmeri Britton & Rose, The Cactaceae, 4:140, 1923.
Chilita palmeri Orcutt, Cactography, 2, 1926.
Mammillaria palmeri Boedeker, Mammill. Vergl. Schluss., 43, 1933. Not Jacobi, 1856.

BODY simple, cylindric, rounded at top, somewhat sunken at apex, 9 cm. high, 5 cm. wide. TUBERCLES arranged in 8 and 13 spirals, firm in texture, dull whitish grayish green, to bluish green, blunt conic, 4-sided at base, rounded and broader dorsally, obtuse at tip, with watery sap, 6 mm. wide at base, 4 mm. high. AREOLES oval, 1 mm. wide, with very dense white wool, persisting for some time, later becoming naked. AXILS filled with white wool, especially in apex, also white tortuous bristles, not as long as tubercles, hidden in wool. CENTRAL SPINES 3-5, usually 4, 7-8 mm. long, slender acicular, straight, smooth, stiff, not enlarged at base, brownish with darker tips, lower porrect, upper erect just above the plane of radials, laterals divergent dorsally and more ascending. RADIAL SPINES 25-30, 5-6 mm. long, lower longer, all very slender acicular, straight, smooth, stiff, white to tip, somewhat ascending, interlacing. FLOWERS funnelform, wide throat, 12 mm. long, 10 mm. wide. *Outer perianth-segments* reddish tan mid-stripe, tan margins, oblong, tip obtuse, margins entire. *Inner perianth-segments* very pale greenish white to light cream, olive-green mid-line, sometimes with rose tint, wide spatulate, 3-4 mm. wide, tip obtuse, margins entire. *Filaments* whitish. *Anthers* orange yellow. *Style* whitish. *Stigma-lobes* 5-6 tannish olive-green to yellowish, 3 mm. long, overtop anthers. FRUIT scarlet, clavate, 13x5 mm., with dried perianth persisting. SEEDS black, glossy, pyriform with lateral hilum, deeply pitted.

Distribution: Islands off the west coast of Baja California, Mexico.

Type locality: San Benito Island, also reported from Guadalupe Island.

Illustration: Fig. 242 is from a photograph of a plant from Guadalupe Island.

The name *Mammillaria palmeri* as used by Boedeker is not allowable, because it was previously used by Jacobi (Allg. Gartenz., 24:82, 1856) for a different plant, so in placing Coulter's species in the genus *Mammillaria* a new name becomes necessary. In order to preserve as near an approximation as possible, the new name of *Mammillaria neopalmeri* is offered.

The original description is supplemented with data from plants from Guadalupe Island. Our specimen died before it flowered but we were able to obtain the flower description from plants in the garden of Mr. A. F. Wilhite of Corona, California.

217. **Mammillaria spinosissima** Lemaire

Mammillaria spinosissima Lemaire, Cact. Aliq. Nov., 4, 1838.
Mammillaria polycentra Berg, Allg. Gartenz., 8:130, 1840.
Mamillaria auricoma Dietrich, Allg. Gartenz., 14:308, 1846.
Mamillaria polyacantha Ehrenberg, Allg. Gartenz., 16:265, 1848.
Mamillaria polyactina Ehrenberg, Allg. Gartenz., 16:266, 1848.
Mamillaria hepatica Ehrenberg, Allg. Gartenz., 16:267, 1848.
Mamillaria pomacea Ehrenberg, Allg. Gartenz., 16:267, 1848.
Mamillaria pulcherrima Ehrenberg, Allg. Gartenz., 17:249, 1849.
Mamillaria pretiosa Ehrenberg, Allg. Gartenz., 17:250, 1849.
Mamillaria caesia Ehrenberg, Allg. Gartenz., 17:251, 1849.
Mamillaria mirabilis Ehrenberg, Allg. Gartenz., 17:251, 1849.
Mamillaria pruinosa Ehrenberg, Allg. Gartenz., 17:261, 1849.
Mamillaria seegeri Ehrenberg, Allg. Gartenz., 17:261, 1849.
Mamillaria herrmannii Ehrenberg, Allg. Gartenz., 17:303, 1849.
Mamillaria aurorea Ehrenberg, Allg. Gartenz., 17:303, 1849.
Mamillaria haseloffii Ehrenberg, Allg. Gartenz., 17:303, 1849.
Mamillaria linkeana Ehrenberg, Allg. Gartenz., 17:308, 1849.
Mamillaria vulpina Ehrenberg, Allg. Gartenz., 17:308, 1849.
Mamillaria eximia Ehrenberg, Allg. Gartenz., 17:309, 1849.
Mamillaria isabellina Ehrenberg, Allg. Gartenz., 17:309, 1849.
Mamillaria spinosissima flavida Salm-Dyck, Cact. Hort. Dyck. 1849, 8, 1850.
Mamillaria spinosissima rubens Salm-Dyck, Cact. Hort. Dyck. 1849, 8, 1850.
Mamillaria spinosissima brunnea Salm-Dyck, Cact. Hort. Dyck. 1849, 8, 1850.
Mamillaria herrmanni flavicans Salm-Dyck, Cact. Hort. Dyck. 1849, 8, 1850.
Mamillaria seegeri gracilispina Salm-Dyck, Cact. Hort. Dyck. 1849, 8, 1850.
Mamillaria seegeri pruinosa Salm-Dyck, Cact. Hort. Dyck. 1849, 8, 1850.

FIG. 243. *Mammillaria spinosissima* x 1

Mamillaria uhdeana Salm-Dyck, Cact. Hort. Dyck. 1849, 8, 1850.
Mammillaria spinosissima hepatica Labouret, Monogr. Cact., 35, 1853.
Mammillaria castaneoides Lemaire in Labouret, Monogr. Cact., 37, 1853.
Mammillaria castanea Labouret, Monogr. Cact., 37, 1853.
Mammillaria castanaeformis Labouret, Monogr. Cact., 37, 1853.
Mammillaria seegeri mirabilis Labouret, Monogr. Cact., 37, 1853.
Mammillaria sanguinea Haage Jr. in Regel, Act. Hort Petrop., **8**:276, 1883.
Mammillaria poselgeriana Haage in Förster, Handb. Cact. ed. 2, 269, 1885.
Mammillaria pretiosa cristata Hildmann in Förster, Handb. Cact. ed. 2, 273, 1885.
Cactus auricomus Kuntze, Rev. Gen. Pl., **1**:260, 1891.
Cactus auroreus Kuntze, Rev. Gen. Pl., **1**:260, 1891.
Cactus eximus Kuntze, Rev. Gen. Pl., **1**:260, 1891.
Cactus isabellinus Kuntze, Rev. Gen. Pl., **1**:260, 1891.
Cactus linkeanus Kuntze, Rev. Gen. Pl., **1**:260, 1891.
Cactus mirabilis Kuntze, Rev. Gen. Pl., **1**:260, 1891.
Cactus polycentrus Kuntze, Rev. Gen. Pl., **1**:261, 1891.
Cactus pomaceus Kuntze, Rev. Gen. Pl., **1**:261, 1891.
Cactus pretiosus Kuntze, Rev. Gen. Pl., **1**:261, 1891.
Cactus pulcherrimus Kuntze, Rev. Gen. Pl., **1**:261, 1891.
Cactus spinosissimus Kuntze, Rev. Gen. Pl., **1**:261, 1891.
Cactus vulpinus Kuntze, Rev. Gen. Pl., **1**:261, 1891.
Mammillaria spinosissima sanguinea Haage in Brandegee, Cycl. Amer. Hort. Bailey, **2**:976, 1900.
Mammillaria spinosissima aurorea Gürke, Bluhende Kak., **2**: under pl. 71, 1905.
Mammillaria spinosissima auricoma Gürke, Bluhende Kak., **2**: under pl. 71, 1905.
Mammillaria spinosissima eximia Gürke, Bluhende Kak. **2**: under pl. 71, 1905.
Mammillaria spinosissima haseloffii Gürke, Bluhende Kak., **2**: under pl. 71, 1905.
Mammillaria spinosissima herrmannii Gürke, Bluhende Kak., **2**: under pl. 71, 1905.
Mammillaria spinosissima isabellina Gürke, Bluhende Kak., **2**: under pl. 71, 1905.
Mammillaria spinosissima linkeana Gürke, Bluhende Kak., **2**: under pl. 71, 1905.
Mammillaria spinosissima mirabilis Gürke, Bluhende Kak., **2**: under pl. 71, 1905.
Mammillaria spinosissima pruinosa Gürke, Bluhende Kak., **2**: under pl. 71, 1905.

Mammillaria spinosissima pulcherrima Gürke, Bluhende Kak., 2: under pl. 71, 1905.
Mammillaria spinosissima seegeri Gürke, Bluhende Kak., 2: under pl. 71, 1905.
Mammillaria spinosissima vulpina Gürke, Bluhende Kak., 2: under pl. 71, 1905.
Mammillaria spinosissima pretiosa Schelle, Handb. Kakteenk., 253, 1907.
Neomammillaria spinosissima Britton & Rose, The Cactaceae, 4:117, 1923.
Mammillaria spinosissima castenoides Borg, Cacti, 328, 1937.

BODY simple, columnar, rounded at apex but not sunken, to 30 cm. high, to 10 cm. wide. TUBERCLES arranged in 13 and 21 spirals, firm in texture, dark bluish green, oval conic, obscurely 4-sided at base, compressed laterally above, with watery sap, 4-6 mm. long, 4 mm. wide at base. AREOLES round, 2-3 mm. wide, with short white wool, soon becoming naked. AXILS with wool and bristles. CENTRAL SPINES variable in number, 12-15 (7-10), 10-20 mm. long, acicular, semi-flexuous, straight (occasionally 1 hooked, not typical), smooth, little enlarged at base, white to ruby red to rose to reddish brown, spreading from porrect. RADIAL SPINES 20-30, 4-10 mm. long, setaceous, straight, smooth, semi-flexuous, white, horizontal interlocking. FLOWERS wide funnelform, near top, 15-20 mm. long, 15 mm. wide. *Outer perianth-segments* light green midstripe below, reddish to tan above, pinkish margins, lanceolate, tip acute, margins finely serrate (under magnification), 2-3 mm. wide. *Inner perianth-segments* reddish pink to purplish, slightly darker mid-line, more brownish ventrally, linear, tip acute to acuminate, margins mostly entire but with some fine serrations, 2 mm. wide. *Filaments* white to pale greenish below. *Anthers* yellow. *Style* white below, very pale greenish pink above. *Stigma-lobes* 5-(8), light greenish yellow to reddish yellow, less than 1 mm. long, overtop anthers 3 mm. FRUIT red, clavate, 20 mm. long. SEEDS reddish brown, glossy, curved pyriform with lateral hilum near base, finely pitted, 1.3x0.8 mm.

Distribution: Morelos, State of Mexico, Hidalgo, Oaxaca, Guerrero; Mexico.

Type locality: None given but reported from Carantla, Totlapan and Tleyacapa, Mor.; Real del Monte, Hgo.; between Tehuacan and Oaxaca on the railroad; Taxco, Gro.

Illustration: Fig. 243 is from a photograph of a plant sent to us by Sr. F. Schmoll of Cadereyta, Qro.

The many varieties of this species have been erected largely on the strength of the variation in the spine coloration. Our investigations in the field have revealed that there are all types of spine coloration growing together and if a separate variety is to be erected for each slight variation, the number would be unlimited, so we are referring them all to the type as coming within its limits.

The original description is supplemented with data from plants sent to us by Sr. F. Schmoll, plants collected for us by Mr. George Lindsay in 1937 at Taxco, Gro., and also between Tehuacan and Oaxaca, and from our collections near Taxco in 1941 and 1942.

The plant with the occasional hooked central spine which Mr. George Lindsay collected near Taxco is without doubt a natural hybrid with *M. solisii* as we found both species and their natural hybrids growing together in the same canyon.

218. Mammillaria albicoma Boedeker

Mamillaria albicoma Boedeker, Monatsschr. Deutsch. Kakt. Ges. 1:241, 1929.

BODY cespitose from base, globose to elongate, not sunken at apex, 5 cm. high, 3 cm. wide, densely covered with spines. TUBERCLES arranged in 8 and 13 spirals, bright glossy leaf-green, conic to cylindric, truncate at apex, with watery sap, to 7 mm. long, 2-3 mm. wide at base. AREOLES round, 1.5 mm. wide, with white wool persisting for some time. AXILS with white woolly mat, and longer white tortuous fine hair-like bristles. CENTRAL SPINES 1-4, often missing, hardly noticeable, 4-10 mm. long, slender acicular, straight, stiff, smooth, white, reddish brown tip, nearly porrect. RADIAL SPINES 30-40, 8-10 mm. long, very thin hair-like, soft, more or less tortuous, pure white, strongly

M. albicoma

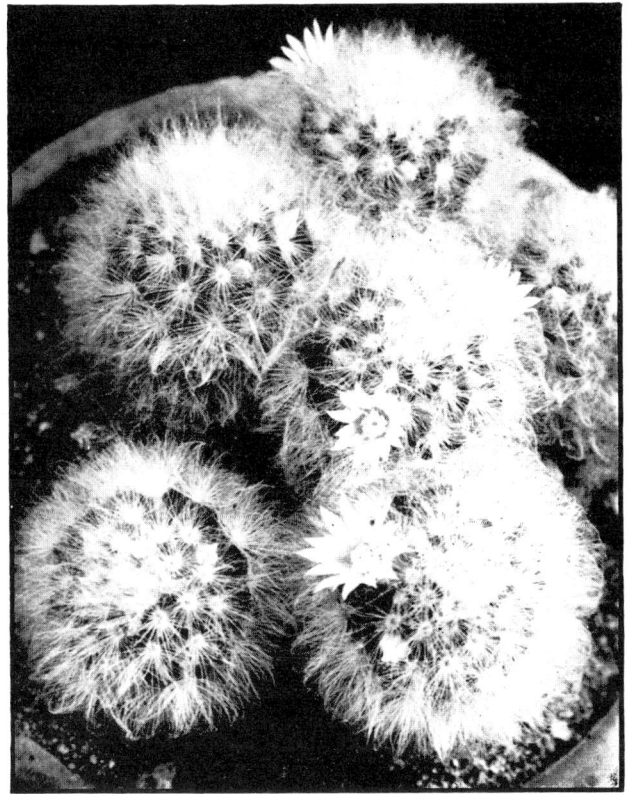

FIG. 244. *Mammillaria albicoma* x 1

ascending. FLOWERS spreading funnelform, somewhat lateral in upper axils, 10-15 mm. long and wide. *Outer perianth-segments* light green at base, pale greenish yellowish white above, light reddish brown to olive brown tapering mid-line, lanceolate, tip acute, margins serrate, 5x2 mm. *Inner perianth-segments* very pale greenish yellow to whitish cream, sometimes light greenish brown mid-line, lanceolate, tip acuminate, margins serrate. *Filaments* very pale green to whitish. *Anthers* bright yellow. *Style* pale greenish yellow. *Stigma-lobes* 3-4, yellowish green, 1 mm. long, overtop anthers. FRUIT small. SEEDS medium to dark gray (?), curved pyriform with lateral hilum, smooth, hardly 1 mm. ROOTS robust tuberous.

Distribution: Tamaulipas, Mexico.

Type locality: Jaumave.

Illustration: Fig. 244 is from a photograph of a plant sent to us by Sr. F. Schmoll of Cadereyta, Qro.

The color of the seeds as given by Boedeker is open to question as it is not in common with the other related species. We have not seen any seed of this species, so can not be sure.

Fig. 245. *Mammillaria candida*

219. **Mammillaria candida** Scheidweiler

Mammillaria candida Scheidweiler, Bull. Acad. Sci. Brux., 5:496, 1838.
Mammillaria sphaerotricha Lemaire, Cact. Gen. Nov. Sp., 33, 1839.
Mamillaria sphaerotricha rosea Salm-Dyck, Cact. Hort. Dyck. 1849, 85, 1850.
Cactus sphaerotrichus Kuntze, Rev. Gen. Pl., 1:261, 1891.
Mamillaria candida roseo Salm-Dyck in Schumann, Gesamtb. Kakteen, 525, 1898.
Mammillaria candida sphaerotricha Schelle, Handb. Kakteenk., 248, 1907.
Neomammillaria candida Britton & Rose, The Cactaceae, 4:130, 1923.
Chilita candida Orcutt, Cactography, 2, 1926.

BODY simple and cespitose from base and body, globose, later cylindric, sunken at apex, 5-7 cm. wide. TUBERCLES arranged in 13 and 21 spirals, pale bluish green, cylindric, nearly clavate, obtuse at tip, with watery sap, 10 mm. long, 5 mm. wide at base. AREOLES round, with scant white wool, later becoming naked. AXILS with 4-7 white bristles to length of tubercles. CENTRAL SPINES 8-12, 4-7 mm. long, acicular, little heavier than radials, straight, smooth, stiff, white, brown at tip, middle one porrect, others spreading. RADIAL SPINES over 50, 5-9 mm. long, slender acicular, straight, smooth, white, horizontal to somewhat ascending in confusion. FLOWERS funnelform, near top, 20 mm. long, 15 mm. wide. *Outer perianth-segments* greenish light brown to nearly white, rose-red to brownish mid-stripe, darker ventrally, lanceolate, tip acuminate, margins ciliate, 10 mm. long. *Inner perianth-segments* muddy rose mid-stripe, white margins, lanceolate, tip acute, margins serrate, 20 mm. long. *Filaments* and *Style* rose-red. *Anthers* golden yellow. *Stigma-lobes* 6, purple-red. FRUIT red, elongate, SEEDS black, glossy.

Distribution: San Luis Potosi, Mexico.

Type locality: Near San Luis Potosi, S.L.P.

Illustration: Fig. 245 is a reproduction of an illustration, Fig. 141, in Britton & Rose (Cact., 4:130) as *Neomammillaria candida*.

Fig. 246. *Mammillaria multiceps* x 0.8

220. **Mammillaria multiceps** Salm-Dyck

Mamillaria multiceps Salm-Dyck, Cact. Hort. Dyck. 1849, 81, 1850.
Mamillaria multiceps elongata Meinshausen, Wöchensch. Gärtn. Pflanz., 1:27, 1858.
Mamillaria multiceps humilis Meinshausen, Wöchensch. Gärtn. Pflanz., 1:27, 1858.
Mamillaria multiceps perpusilla Meinshausen, Wöchensch. Gärtn. Pflanz., 1:27, 1858.
Mamillaria multiceps grisea Meinshausen, Wöchensch. Gärtn. Pflanz., 1:27, 1858.
Mamillaria pusilla texana Engelmann, Cact. Mex. Bound., 5, 1859.
Mamillaria texana Poselger in Young, Fl. Texas, 279, 1873.
Cactus multiceps Kuntze, Rev. Gen. Pl., 1:260, 1891.
Cactus stellatus texanus Coulter, Contr. U. S. Nat. Herb., 3:108, 1894.
Cactus texanus Small, Fl. Southeast. U. S., 812, 1903.
Neomammillaria multiceps Britton & Rose, The Cactaceae, 4:125, 1923.
Chilita multiceps Orcutt, Cactography, 2, 1926.
Mammillaria prolifera texana Borg, Cacti, 316, 1937.
Mammillaria prolifera multiceps Borg, Cacti, 316, 1937.

BODY very cespitose from base and body forming large clumps, globose to short cylindric, 10-20 mm. wide. TUBERCLES arranged in 8 and 13 spirals, soft in texture, dark gray-green, compressed cylindric, conic at apex, with watery sap, 4 mm. long and wide at base. AREOLES round, with white wool in youth, later becoming naked. AXILS with long tortuous hair-like bristles. CENTRAL SPINES 6-8 (10-12), 6-8 mm. long, slender acicular, straight, pubescent, rigid, enlarged at base, whitish below, reddish yellow to reddish brown above, spreading. RADIAL SPINES 30-50, 2-5 mm. long, fine hair-like, pubescent (but not as much so as centrals), straight to tortuous, flexuous, white horizontal. FLOWERS funnelform, lateral, 14-20 mm. long, April to June. *Outer*

Fig. 247. *Mammillaria multiceps*

perianth-segments pale greenish tapering mid-stripe below, pale dirty pinkish green above, pale greenish tan margins, oblong, linear, tip acute, margins serrate. *Inner perianth-segments* brownish pink mid-stripe, pale greenish yellow margins, wide lanceolate, tip obtuse to emarginate, margins somewhat serrate, straight to recurved. *Filaments* white. *Anthers* golden yellow. *Style* very pale green below, more olive-green above. *Stigma-lobes* 4-8, greenish yellow, overtop anthers. FRUIT scarlet, elongate clavate to nearly cylindric, 8-12 mm. long, with dried perianth persisting. SEEDS black, glossy, obovate with linear basal hilum, pitted, 1 mm.

Distribution: Texas, U. S.; Coahuila, Nuevo Leon, and Tamaulipas; Mexico.

Type locality: Near Eagles Pass, Texas, along the Rio Grande (Engelmann).

Illustrations: Fig. 246 is from a photograph of a plant that we colected near Linares, N.L., Mexico, in 1942. Fig. 247 is a reproduction of an illustration, Fig. 133, in Britton & Rose (Cact., 4:125) as *Neomammillaria muticeps.*

M. pusilla multiceps Regel (Ind. Sem. Petrop., 526, 1898) which is listed, but without description, may belong here.

M. pusilla mexicana offered for sale by Grässner (Monatsschr. Kakteenk., Feb., 1920) may probably be the type that we collected in Nuevo Leon.

"*M. caespititia* Hort. was referred by Salm-Dyck as a synonym of *M. multiceps. M. pusilla caespititia* (Schelle, Handb. Kakteenk. 249, 1907) is the same.

"*M. parvissima* Karwinsky (Wöchenschr. Gärtn. Pflanz., 1:27, 1858) is sometimes credited to Meinshausen, but seems never to have been described. *M. perpusilla* Meinshausen, given only as a synonym, belongs here." Britton & Rose (Cact., 4:126, 1923).

M. multiceps albida, Monat. Kakteenk. 70, 1898, is only a name.

M. pusilla neomexicana mentioned in Nickel's Catalogue 8 may be referrable here.

221. **Mammillaria prolifera** (Miller) Haworth

Cactus proliferus Miller, Gard. Dict., ed. 8, No. 6, 1768.
Cactus glomeratus Lamark, Encycl., 1:537, 1783.
Cactus mammillaris prolifer Aiton, Hort. Kew., 2:150, 1789.
Mammillaria prolifera Haworth, Syn. Pl. Succ., 177, 1812.
Cactus pusillus DeCandolle, Cact. Hort. Monsp., 184, 1813. Not Haworth, 1803.
Cactus stellatus Willdenow, Enum. Pl. Suppl., 30, 1813.
Mammillaria stellaris Haworth, Suppl. Pl. Succ., 72, 1819.
Mammillaria pusilla Sweet, Hort. Brit., 171, 1826.
Mammillaria stellata Sweet, Hort. Brit., 171, 1826.
Mammillaria glomerata DeCandolle, Prodr., 3:459, 1828.
Mammillaria pusilla major Pfeiffer, Enum. Cact., 36, 1837.

FIG. 248. *Mammillaria prolifera* x 1

Cactus haworthianus Kuntze, Rev. Gen. Pl., 1:259, 1891.
Cactus prolifer Kuntze, Rev. Gen. Pl., 1:259, 1891.
Mamillaria pusilla haitiensis Schumann, Blühende Kakteen, 1, under pl. 46, 1904.
Neomammillaria prolifera Britton & Rose, The Cactaceae, 4:124, 1923.
Chilita prolifera Orcutt, Cactography, 2, 1926.
Mammillaria prolifera haitiensis Borg, Cacti, 316, 1937.

BODY very cespitose, globular to short cylindric, rounded at apex, 4-6 cm. long, 3-4 cm. wide. TUBERCLES closely set in 5 and 8 spirals, soft in texture, dark green, conic to oval to nearly rounded, with watery sap, 5-8 mm. long, 4-5 mm. wide at base. AREOLES nearly round, with scant whitish felt, soon becoming naked. AXILS with white bristles longer than the tubercles, and very short white woolly mat. CENTRAL SPINES 5-9 (12), 6-8 mm. long, slender acicular, straight, pubescent, enlarged at base, pale yellow, one more or less porrect, others ascending to horizontal. RADIAL SPINES to 40, 6-10 mm. long, bristle-like to fine hair-like, straight to very tortuous, smooth, flexuous, white, mostly horizontal. FLOWERS funnelform, near apex, to 14 mm. long, May. *Outer perianth-segments* tannish lavender mid-stripe, tannish cream to greenish white margins, lanceolate to eliptical, tip acute, margins finely serrate. *Inner perianth-segments* yellow cream, reddish tan mid-line, lanceolate, tip acute, margins entire. *Filaments* cream to pale rose. *Anthers* yellowish orange. *Style* very pale yellowish green. *Stigma-lobes* 3-4, light yellowish green, 2 mm. long, just overtop anthers. FRUIT orange-red, somewhat curved clavate, 10 mm. long, with dried perianth persisting, with very few seeds. SEEDS black, glossy, globular, elongated ventrally to hilum which extends a little laterally.

Distribution: West Indies.

Type locality: None given but found in Cuba, Hispanola.

Illustration: Fig. 248 is from a photograph of a plant obtained from the West Indies by Mr. Scott Haselton.

Plumier (Nov. Plant. Amer., 19, 1703) and also in Burmann (Plant. Amer., 1758) described a *M. glomerata* but it does not appear to belong here.

Regel & Klein (Ind. Sem. Petrop., 19, 1860) listed, but without description, the following varieties of *M. pusilla: elongata, gemina* and *humilis.*

M. pusilla albida Schelle (Handb. Kakteenk., 249, 1907) is only a name but may be referrable here.

Borg attributes *M. prolifera haitiensis* to Schumann but the latter described it as *M. pusilla haitiensis.*

Fig. 249. *Mammillaria calacantha* x 1

222. Mammillaria calacantha Tiegel

Mammillaria calacantha Tiegel, Kakteenk., 232, 1933.
Mammillaria calacantha rubra (?) Neal, Cacti Other Succ., 84, 1935.

BODY simple, globular to cylindric, sunken at apex, 7-12 cm. high, 6-9 cm. wide TUBERCLES arranged in 13 and 21 spirals, firm in texture, light green, glossy, somewhat 4-sided, not sharply angled, with watery sap, 8-12 mm. long, 6-8 mm. wide at base. AREOLES elongate, 1.5x2.5 mm., with white wool in youth, later becoming naked. AXILS with short white wool, soon becoming naked. CENTRAL SPINES 2-4, 10-15 mm. long, stout acicular, straight to slightly recurved, smooth, stiff, light reddish tan, brown tip, purplish in youth, when 2: divergent dorsally and ventrally, when 4: in cross formation. RADIAL SPINES 25-35, 5-7 mm. long, slender acicular, straight, smooth, pale yellow becoming grayish, horizontal to slightly ascending. FLOWERS campanulate, very short tube, often beneath the spines, 8 mm. long, 12-14 mm. wide. *Outer perianth-segments* pale greenish white to cream below, orange-pink tip, margins becoming paler, linear-lanceolate, tip acuminate, margins entire to few fine serrations. *Inner perianth-segments* pale greenish white base, pale carmine-pink margins, darker carmine tip and mid-line, linear-lanceolate, tip acute, margins mostly entire. *Filaments* very pale pink. *Anthers* light yellow. *Style* pale pink above, whitish base. *Stigma-lobes* 4-5, carmine, overtop anthers a little. FRUIT pink, light green at apex, clavate, 21x5 mm. SEEDS light yellowish tan, curved pyriform with darker lateral hilum, slightly rugose, not pitted, 1.6x0.8 mm.

Distribution: Queretaro, Guanajuato; Mexico.

Type locality: None given but reported from Augosbura de Charcos (Schmoll).

Illustration: Fig. 249 is from a photograph of a plant sent to us by Sr. F. Schmoll of Cadereyta, Qro.

Fig. 250. *Mammillaria pseudoperbella*

223. Mammillaria pseudoperbella Quehl

Mamillaria pseudoperbella Quehl, Monatsschr. Kakteenk., 19:188, 1909.
Mamillaria pseudoperbella rufispina Quehl, Monatsschr. Kakteenk., 26:94, 1916.
Neomammillaria pseudoperbella Britton & Rose, The Cactaceae, 4:109, 1923.

BODY simple and cespitose by dichotomous branching, flattened globose to short cylindric, rounded above, sunken and somewhat woolly at apex, 10-15 cm. wide. TUBERCLES arranged in 13 and 21 spirals, leaf-green, cylindric, truncate, with watery sap, 6-7 mm. long, 2-3 mm. wide at base. AREOLES round, with white woolly felt in youth, later becoming naked. AXILS with scant wool. CENTRAL SPINES 2, upper 5 mm. long, lower shorter, all subulate, straight, stiff, smooth, brown with reddish brown tip, somewhat transparent, upper erect, lower porrect. RADIAL SPINES 20-30, 2-3 mm. long, thin bristle-like, straight, smooth, pure white, horizontal, mostly lateral. FLOWERS funnelform, near top, 15 mm. long. *Scales* greenish white, 2 mm. long. *Outer perianth-segments* dark carmine-red with darker ventral stripe, wide lanceolate, tip acute, margins serrate at tip. *Inner perianth-segments* carmine-red, darker mid-stripe, lanceolate, tip often split, margins (?). *Filaments* pale rose above, white below. *Anthers* chrome-yellow. *Style* pale rose above, white below. *Stigma-lobes* 3-5, red. FRUIT bright red, clavate. SEEDS dull brown, curved pyriform with lateral hilum at base, faintly roughened, 1.2x0.7 mm.

Distribution: Queretaro, Central Mexico, and Oaxaca; Mexico.

Type locality: None given but reported from Higuerillas, Qro.

Illustration: Fig. 250 is a reproduction of an illustration. Fig. 110, in Britton & Rose (Cact., 4:109) as *Neomammillaria pseudoperbella*.

FIG. 251. *Mammillaria celsiana*

224. Mammillaria celsiana Lemaire

Mammillaria celsiana Lemaire, Cact. Gen. Nov. Sp., 41, 1839.
Mamillaria muehlenpfordtii Förster, Allg. Gartenz., 15:49, 1847.
Mamillaria schaeferi Fennel, Allg. Gartenz., 15:66, 1847.
Mamillaria tomentosa Ehrenberg, Allg. Gartenz., 17:262, 1849.
Mamillaria schaeferi longispina Haage, Hamb. Gartenz., 17:160, 1861.
Cactus celsianus Kuntze, Rev. Gen. Pl., 1:260, 1891.
Cactus muehlenpfordtii Kuntze, Rev. Gen. Pl., 1:260, 1891.
Cactus schaeferi Kuntze, Rev. Gen. Pl., 1:261, 1891.
Mammillaria celsiana longispina Haage, Cact. Cult., 130, 1900.
Mamillaria perringii* Hildmann, Gartenwelt, 10:250, 1906.
Neomammillaria celsiana Britton & Rose, The Cactaceae, 4:112, 1923.

BODY simple and cespitose from base, globular to cylindric, rounded above, with apex sunken, to 12 cm. high, 8 cm. wide. TUBERCLES arranged in 13 and 21 spirals, firm in texture, bluish green, conic, somewhat compressed base, with watery sap, 6-10 mm. long, 8 mm. wide at base. AREOLES oval to round, 2-3 mm. wide, with abundant white wool, soon becoming naked. AXILS with white wool, later becoming naked. CENTRAL SPINES 4-6, upper and lateral 8-14 mm. long, lower to 30 mm. long, all stout acicular, straight, or upper a little recurved toward apex, smooth, pale to dark yellow, tip brown, spreading from porrect. RADIAL SPINES 24-30, 6-8 mm. long, lateral longer, all slender acicular, straight, stiff, smooth, white transparent, horizontal.

*Name listed under *M. acanthoplegma* and *M. shaeferi* (Monatsschr. Kakteenk. 147, 1897). The description also appears in Monatsschr. Kakteenk. 49, 1906.

FLOWERS funnelform, 11 mm. long. *Outer perianth-segments* reddish brown, lanceolate, tip acute, margins serrate. *Inner perianth-segments* rose to fiery carmine, lanceolate, tip acute, margins entire. *Filaments* white below, reddish above. *Anthers* yellow. *Style* white below, rose above. *Stigma-lobes* 4, rose to red. FRUIT red, clavate, with dried perianth persisting. SEEDS light tan, dull, curved pyriform, with lateral hilum at tip, slightly rugose, 1.2x0.7 mm.

Distribution: San Luis Potosi, Queretaro, Guanajuato, State of Mexico, Oaxaca?; Mexico.

Type locality: None given but reported from San Felipe, S.L.P., north of Mexico City, State of Mexico; District of Cuicalan, Oaxaca (?)

Illustration: Fig. 251 is from a photograph of a plant sent to us by Sr. F. Schmoll of Cadereyta, Qro.

Mammillaria lanifera Haworth (Phil. Mag., 63:41, 1824) was referred here by Schumann but this was questioned by Britton & Rose. It was described very briefly and indefinitely. DeCandolle (Prodr., 3:459, 1828) referred *Cactus canescens* Mocin & Sesse to *M. lanifera*. A subsequent article in Phil. Mag., 1830, says that the DeCandolle plant is different. Salm-Dyck in 1834 referred it to *M. geminispina* and in 1850 he referred it to *M. rhodantha*. *M. monacantha* was given by Loudon (Gard. Mag., 313, 1841) as a synonym of *M. lanifera*.

225. Mammillaria fuscata Pfeiffer

Mammillaria fuscata Pfeiffer, Enum., Cact., 28, 1837.
Mammillaria tentaculata Hort. Berl. Pfeiffer, Enum. Cact., 29, 1837.
Mammillaria odieriana Lemaire, Cact. Gen. Nov. Sp., 46, 1839.
Mammillaria pfeifferi Booth in Scheidweiler, Bull. Acad. Sci. Brux., 6:93, 1839.
Mammillaria pfeifferi fulvispina Scheidweiler, Bull. Acad. Sci. Brux., 6:93, 1839.
Mammillaria pfeifferi dichotoma Scheidweiler, Bull. Acad. Sci. Brux., 6:93, 1839.
Mammillaria pfeifferi altissima Scheidweiler, Bull. Acad. Sci. Brux., 6:93, 1839.
Mammillaria pfeifferi flaviceps Scheidweiler, Bull. Acad. Sci. Brux., 6:93, 1839.
Mammillaria pfeifferi variabilis Scheidweiler, Bull. Acad. Sci. Brux., 6:93, 1839.
Mammillaria crassispina Pfeiffer, Allg. Gartenz., 8:406, 1840.
Mamillaria odieriana aurea Salm-Dyck, Hort. Dyck. 1844, 7, 1845.
Mamillaria crassispina gracilior Salm-Dyck, Hort. Dyck. 1844, 8, 1845.
Mamillaria sulphurea Senke in Förster, Handb. Cact., 200, 1846.
Mamillaria odieriana rigidior Salm-Dyck, Cact. Hort. Dyck. 1849, 98, 1850.
Mamillaria odieriana rubra Sencke in Förster, Handb. Cact. ed. 2, 295, 1885.
Mamillaria tentaculata picta Förster, Handb. Cact. ed. 2, 309, 1885.
Mamillaria crassispina rufa Rümpler in Förster, Handb. Cact. ed. 2, 311, 1885.
Cactus crassispinus Kuntze, Rev. Gen. Pl., 1:260, 1891.
Cactus fuscatus Kuntze, Rev. Gen. Pl., 1:260, 1891.
Cactus odierianus Kuntze, Rev. Gen. Pl., 1:261, 1891.
Cactus tentaculatus Kuntze, Rev. Gen. Pl., 1:261, 1891.
Cactus rhodanthus sulphureospinus Coulter, Contr. U. S. Nat. Herb., 3:107, 1894.
Mamillaria rhodantha pfeifferi Schumann, Gesamtb. Kakteen, 550, 1898.
Mamillaria rhodantha sulphurea Schumann, Gesamtb. Kakteen, 550, 1898.
Mamillaria rhodantha fuscata Schumann, Gesamtb. Kakteen, 551, 1898.
Mamillaria sulphurea longispina Haage, Cact. Cult., 139, 1900. Nomen.
Mamillaria rhodantha odieriana Schelle, Handb. Kakteenk., 257, 1907.
Mamillaria rhodantha tentaculata Schelle, Handb. Kakteenk., 257, 1907.

BODY simple, globose. TUBERCLES firm in texture, pale bluish green, conic 4-sided at base, with watery sap, 8-10 mm. long, 6 mm. wide at base. AREOLES oval, with white wool in youth. AXILS with some white wool to naked. CENTRAL SPINES 4-6, 10-30 mm. long, stout acicular, straight to recurved, smooth, stiff, golden yellow to brown, spreading porrect. RADIAL SPINES 25-28, 6-8 mm. long, slender acicular, whitish to golden, horizontal. FLOWERS 16 mm. wide, May and June. *Perianth-segments* purplish, linear, tip acute, margins serrate. *Filaments* purplish red. *Anthers* yellow. *Stigma-lobes* 5, short, pale pinkish. FRUIT purplish red, slender clavate, 20 mm. long. SEEDS light tan, curved pyriform with lateral hilum near base.

Distribution: Valley of Mexico and central plateau.

Type locality: None given but reported from Mesa de Magdalena (Ehrenb.).

This species is very close to and intergrades with the variations of *M. rhodantha* but it differs from it chiefly in the greater number of radial spines.

Link & Otto (Uber Gatt. Melocact. Echinocact., 21, 1827) mentioned the name *M. fuscata* but did not describe it.

M. flavisceps was referred by Labouret to *M. crassispina*. Britton & Rose referred it to *M. rhodantha* but it may belong here.

"*M. olivacea* was cited by Pfeiffer (Enum. Cact., 180, 1837) as a synonym of *M. tentaculata.*" Britton & Rose (Cact., 4:123).

M. oliveriana Förster (Handb. Kakteenk., 301, 1892) is probably a mispelling of *M. odieriana*.

FIG. 252. *Mammillaria elegans* x 1

226. Mammillaria elegans DeCandolle

Mammillaria geminispina DeCandolle, Mem. Mus. Hist. Nat. Paris, 17:30, 1828. Not Haworth 1824.
Mammillaria elegans DeCandolle, Mem. Mus. Hist. Nat. Paris, 17:111, 1828.
Mammillaria elegans minor DeCandolle, Mem. Mus. Hist. Nat. Paris, 17:111, 1828.
?*Mammillaria elegans globosa* DeCandolle, Mem. Mus. Hist. Nat. Paris, 17:111, 1828.
Mammillaria acanthophlegma Lehmann, Del. Sem. Hamb., 1832.
Mammillaria dyckiana Zuccarini in Pfeiffer, Enum. Cact., 26, 1837.
Mammillaria elegans micracantha Lemaire, Cact. Gen. Nov. Sp., 100, 1839.
Mammillaria geminispina tetracantha Lemaire, Cact. Gen. Nov. Sp., 100, 1839.
Mammillaria klugii Ehrenberg, Bot. Zeit., 2:834, 1844.
Mammillaria meissnerii Ehrenberg, Bot. Zeit., 2:834, 1844.
Mammillaria kunthii Ehrenberg, Bot. Zeit., 2:835, 1844.
?*Mammillaria micans* Dietrich, Allg. Gartenz., 16:330, 1848.
Mammillaria splendens Ehrenberg, Allg. Gartenz., 17:242, 1849.
Mamillaria elegans klugii Salm-Dyck, Cat. Hort. Dyck. 1849, 9, 1850.
Mamillaria acanthophlegma decandollei Salm-Dyck, Cact. Hort. Dyck. 1849, 9, 1850.
Mamillaria acanthophlegma meissneri Salm-Dyck, Cact. Hort. Dyck. 1849, 9, 1850.
Mamillaria acanthophlegma elegans Monoville in Labouret, Monogr. Cact., 63, 1853.

Mamillaria acanthophlegma monacantha Monoville in Labouret, Monogr. Cact., 63, 1853.
Mamillaria acanthophlegma leucocephala Monoville in Labouret, Monogr. Cact., 63, 1853.
Mamillaria acanthophlegma abducta Monoville in Labouret, Monogr. Cact., 64, 1853.
Cactus acanthophlegmus Kuntze, Rev. Gen. Pl., 1:260, 1891.
Cactus dyckianus Kuntze, Rev. Gen. Pl., 1:260, 1891.
Cactus elegans Kuntze, Rev. Gen. Pl., 1:260, 1891.
Cactus kunthii Kuntze, Rev. Gen. Pl., 1:260, 1891.
Cactus klugii Kuntze, Rev. Gen. Pl., 1:260, 1891.
Cactus meissneri Kuntze, Rev. Gen. Pl., 1:261, 1891.
Neomammillaria elegans Britton & Rose, The Cactaceae, 4:107, 1923.
Mammillaria elegans nigrispina Borg, Cacti, 336, 1937.
Mammillaria elegans aureispina Borg, Cacti, 336, 1937.

BODY simple and cespitose from base, globose to cylindric, sunken at apex, to 15 cm. high, to 6 cm. wide. TUBERCLES arranged in 13 and 21 spirals, glossy, medium dark green, conic, terete, with watery sap, 5 mm. long, 4 mm. wide at base. AREOLES oval, 1.5-2 mm. wide, with considerable white wool in youth, soon becoming naked. AXILS with white wool to the length of tubercles, persisting for some time. CENTRAL SPINES 1-4, 4-7 mm. long, upper one longest, all stout acicular to nearly subulate, straight or slightly recurved, smooth, stiff, chalky white, dark brown to black tip, when 2: divergent dorsally and ventrally, lower porrect, upper erect and overtopping apex, when 4: in cross formation. RADIAL SPINES 20-30, 3-6 mm. long, longer laterally, all slender acicular, mostly straight, smooth, stiff, chalky white, horizontal interlacing. FLOWERS campanulate, broad throat, 13-15 mm. long, April. *Outer perianth-segments* whitish at base, bright rose-red to brownish purplish pink, darker mid-line, linear-lanceolate, tip acute to obtuse, margins entire, 1 mm. wide. *Inner perianth-segments* carmine-red to deep purplish pink, darker mid-stripe, linear-lanceolate, tip acuminate, margins entire, ends may be split, 1.5 mm. wide. *Filaments* white below, pale yellow to purplish pink above. *Anthers* chrome-yellow. *Style* white to tannish yellow to faint pink. *Stigma-lobes* 3-4, pale greenish yellow, overtop anthers slightly. FRUIT dark carmine-red, clavate, 20x4 mm., with dried perianth persisting. SEEDS tan, dull, curved pyriform with lateral hilum near base, faintly and finely pitted, 1x0.7 mm.

Distribution: Central plateau, Valley of Mexico, Puebla, Oaxaca, Hidalgo; Mexico.

Type locality: None given but reported from Tehuacan, Puebla; San Antonio, O. (Hutchison); Ixmiquilpan and Zimapan, Hgo. (Ochoterena).

Illustration: Fig. 252 is from a photograph of a plant collected for us near Tehuacan, P. by Mr. Theodore Hutchison.

M. leucocephala Hort. was referred by Pfeiffer (Beschr. Synon. 24, 1837) as a synonym of *M. acanthophlegma*.

M. recta Miquel in Labouret (Monogr. Cact., 63, 1853) was also reported as a synonym of the latter species.

M. obdusta Walpers (Repert. Bot. Syst., 2:273, 1843) is probably a misprint of *M. abdusta* under *M. acanthophlegma*.

This species is quite variable and it merges into several closely related species. Many authors consider these variations as distinct species but the divisions would have to be made on very minor characteristics which would hardly justify the separations.

a. var. **dealbata** Schumann. Not Dietrich 1846.

Mamillaria elegans dealbata Schumann, Monatsschr. Kakteenk., 149, 1897.
Neomammillaria dealbata Britton & Rose, The Cactaceae, 4:110, 1923.

The plants differs from the type in being more cylindric, central spines usually only 1 and erect, occasionally 2.

This name was confused by Schumann in applying it to this variety when it was originally described as belonging to a very different plant which was referred to as being near *M. parkinsonii*. This species was later raised to specific rank and as a result the plants that have been described by later authors under the name *M. dealbata* do not conform with the original description.

Type locality: Valley of Mexico.

Illustration: Britton & Rose (Cact. 4:109) fig. 111, as *Neomammillaria dealbata*.

M. peacockii Rumpler in Förster (Handb. Cact. ed. 2, 286, 1885) is a garden name for a plant that is reported to have come from the Valley of Mexico and it is probably to be referred here.

FIG. 253. *Mammillaria elegans* var. *supertexta* x 1

b. var. **supertexta** (Martius) Schelle

Mammillaria supertexta Martius, Hort. Reg. Monac., 128, 1829. Nomen.
Mammillaria supertexta Martius in Zuccarini, Plant. Nov. Monac., 706, 1837.
Mamillaria supertexta dichotoma Salm-Dyck, Cact. Hort. Dyck. 9 and 88, 1850.
Cactus supertextus Kuntze, Rev. Gen. Pl., 1:261, 1891.
Mammillaria elegans supertexta Schelle, Handb. Kakteenk., 261, 1907.

This variety has been considered by many to be a distinct species but the original description would indicate a synonomy or at the most only a variety. The plants in the trade appear to show some differences and do not exactly coincide with the original description and probably do represent a distinct variation. They appear to be an intermediate form between *M. elegans* and *M. lanata* in that they have the smaller tubercles and areole arrangement of *M. lanata* but they have 0-1-2 central spines of 1-2 mm.

Type locality: San Jose del Oro, also reported from Ixmiquilpan, Hgo., also from Oaxaca.

Illustration: Fig. 253 is from a photograph of a plant sent to us by Sr. F. Schmoll of Cadereyta, Qro.

M. supertexta caespitosa Monville in Salm-Dyck (Hort. Dyck., 6, 1845) was used as a name only. *M. supertexta tetracantha* Lemaire in Salm-Dyck (Cact. Hort. Dyck., 6, 1850) was used as a name only. *M. supertexta dichotoma* Salm-Dyck is based on *M.*

polycephala. *M. supertexta compacta* Scheidweiler in Labouret (Monogr. Cact., 61, 1853) was given as a synonym of *M. supertexta teracantha* but later Förster (Handb. Cact., 279, 1892) referred the var. *compacta* to *M. supertexta caespitosa*. *M. supertexta longioribus* Scheidweiler (Bull. Acad. Sci. Brux., 496, 1838) is to be referred elsewhere as the sap is reported to be milky.

FIG. 254. *Mammillaria elegans* var. *Schmollii* x 1

c. var. **schmollii** var. nov.

Corpus clavatus ad cylindratus, axillis lanissimis albis; spinis centralibus 2, ad 7 mm., albis; spinis radialibus 20, 4 mm., albis; flores color pallidior.

BODY slender clavate to cylindric. AXILS with dense white wool longer than the tubercles and persisting. CENTRAL SPINES 2, to 7 mm. long, stout acicular, white, divergent dorsally and ventrally. RADIAL SPINES 20, 4 mm. long, chalky white. FLOWERS same as type except of paler color.

Distribution: Puebla, Mexico.

Type locality: San Andreas.

Illustration: Fig. 254 is from a photograph of a plant obtained from Sr. F. Schmoll of Cadereyta, Qro.

This is the plant that has been offered erroneously in the trade as *Mammillaria schmollii* (Bravo) Werd. cf. One collector reported to us that Miss H. Bravo H. had varified his collected material of this variety as the true *M. schmollii* but we are prone to question this unless there is some error in the original description of that species which is described as having 10-15 central spines.

Fig. 255. *Mammillaria columbiana*

227. **Mammillaria columbiana** Salm-Dyck

Mamillaria columbiana Salm-Dyck, Cact. Hort. Dyck. 1849, 99, 1850.
Mammillaria bogotensis Werdermann in Backeberg, Neue Kakteen, 98, 1931.
Mammillaria henissii Boedeker, Monatsschr. Deutsch. Kakt. Ges., 7, 1932.

BODY simple, globose to short clavate, slightly sunken at apex with white wool and with spines overtopping, 10-25 cm. high, 5-9 cm. wide. TUBERCLES loosely arranged in 13 and 21 spirals, bright green to grayish, short conic, not angled, with watery sap, 4-6 mm. long, 6-7 mm. wide at base. AREOLES oval, to 2 mm. wide, with white wool in apex, then yellowish white mat, later becoming nearly naked. AXILS with thick white wool, but no bristles. CENTRAL SPINES 3-6, 6-8 mm. long, acicular, straight, usually smooth, slightly enlarged at base, mostly yellow to brown to reddish brown, darker in youth, darker at tip, one is more centrally placed and porrect, others spreading from porrect. RADIAL SPINES 20-30, 4-6 mm. long, fine acicular to bristle-like, straight, smooth, glossy white, yellowish at base, somewhat ascending. FLOWERS tubular to funnelform, 8-10 mm. long, 5-10 mm. wide. *Outer perianth-segments* olive green below, to carmine-red at tip, linear-lanceolate, tip acute, margins coarse serrations. *Inner perianth-segments* deep pink, darker mid-line, greenish yellow in throat, lighter margins, lanceolate, tip acute, margins serrate. *Filaments* light yellowish green below, carmine-red above. *Anthers* orange. *Style* yellow. *Stigma-lobes* 5, greenish yellow, slender, below anthers. FRUIT orange-red, clavate. SEEDS light brown, glossy, curved pyriform with lateral hilum near base, pitted, 1.6x0.8 mm.

Distribution: Columbia and Venezuela.

Type locality: Sogamosa north of Bogota, Columbia.

Illustration: Fig. 255 is a reproduction of an illustration in Monatsschr. Deutsch. Kakt. Ges., 7, 1932, as *Mammillaria henissii*.

This species was originally described by Salm-Dyck but without any data on flower, fruit or seed. He compared it as similar to *M. eriacantha* Link & Otto but stated definite differences: tubercles larger, radial spines stronger, central spines 4-5, smooth and not pubescent, which separate it from *M. eriacantha* which has 1-2 central spines which are strongly pubescent. Only Werdermann reports them as "almost smooth."

Doubt was cast upon the distribution of this species by some of the earlier authors because no members of this genus except *M. mammillaris* were supposed to exist in South America until it was rediscovered near Bogota, Columbia, and described as *M. bogotensis* by Werdermann and as *M. henissii* by Boedeker.

FIG. 256. *Mammillaria yucatanensis*

228. Mammillaria yucatanensis (B. & R.) Orcutt

Neomammillaria yucatanensis Britton & Rose, The Cactaceae, 4:114, 1923.
Mamillaria yucatanensis Orcutt, Cactography, 8, 1926.
Mammillaria jobeana Boedeker, Kakteenk., 154, 1933.

BODY simple and cespitose, to cylindric, 10-15 cm. high, 3-7 cm. wide, slightly sunken at apex. TUBERCLES loosely arranged in 13 and 21 spirals, light yellowish green, slightly glossy, short conic, not angled, with watery sap, 4-7 mm. long, 3-5 mm. wide at base. AREOLES nearly round, with scant wool in youth. AXILS with little white wool only in youth, and no bristles. CENTRAL SPINES 3-6, 4-8 mm. long, stout acicular, straight, smooth, tannish yellow, brown at tip, spreading from porrect. RADIAL SPINES 20-30 (25), 3-5 mm. long, slender acicular, straight, smooth, creamy white, horizontal to slightly ascending. FLOWERS short campanulate, to 12 mm. wide. *Outer perianth-segments* pale green at base, pink to dark rose tapering mid-stripe above, pale tan margins, linear-lanceolate, tip acuminate, margins serrate. *Inner perianth-segments* nearly white at base, wide pink to rose mid-line above, very pale pink margins, lanceolate, tip acute to obtuse, margins entire to serrate at top. *Filaments* deep purplish pink to rose. *Anthers* orange-brown. *Style* white below, very pale pink above. *Stigma-lobes* 4-6, pale yellowish olive green, 2 mm. long, recurved, slender. FRUIT orange-red, clavate, 16x4 mm., with dried perianth missing. SEEDS brown, curved pyriform with lateral hilum near base, faintly rugose.

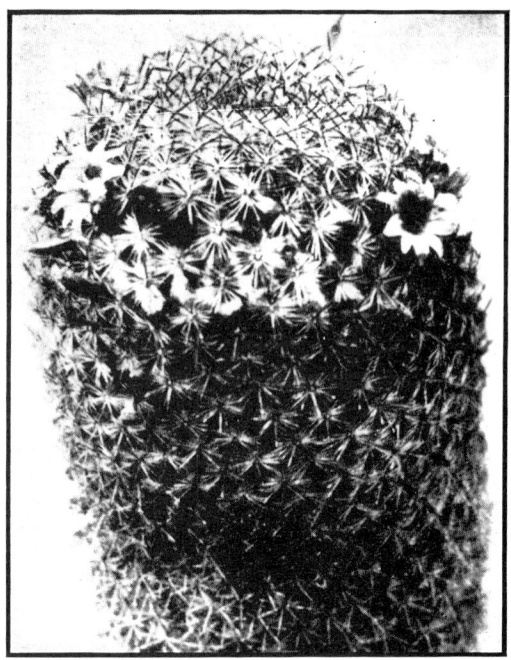

FIG. 257. *Mammillaria yucatanensis*

Distribution: Yucatan, Mexico.

Type locality: Progresso.

Illustration: Fig. 256 is a reproduction of an illustration, Fig. 119, in Britton & Rose (Cact., 4:114) as *Neomammillaria yucatanensis*. Fig. 257 is a reproduction of an illustration in Kakteenk., 155, 1933, as *Mammillaria fobeana*.

We are referring *M. fobeana* here because the description is practically identical with this species. As the distribution was given only as Mexico, it makes recollection of any additional material for study very uncertain.

229. Mammillaria eriacantha Link & Otto

Mammillaria eriacantha Hort. ex Sweet, Hort. Brit. ed. 3, 281, 1826. Nomen.
Mammillaria eriacantha Link & Otto in Zuccarini, Plant. Nov. Monac., 704, 1837.
Cactus eriacanthus Kuntze, Rev. Gen. Pl., 1:260, 1891.
Neomammillaria eriacantha Britton & Rose, The Cactaceae, 4:127, 1923.
Chilita eriacantha Orcutt, Cactography, 2, 1926.

BODY simple and cespitose, elongate cylindric, rounded at apex, 15 cm. high, 4-5 cm. wide. TUBERCLES closely set in 8 and 13 spirals, bright emerald green, conic, compressed to 4-sided at base, with watery sap, 6-8 mm. long, 4-6 mm. wide at base. AREOLES oval, with very slight white wool in only the youngest. AXILS with band of whitish wool in flowering area, longer than tubercles. CENTRAL SPINES 2, 8-10 mm. long upper shorter, all strong acicular, straight, stiff, pubescent, golden brown in youth, becoming lighter, spreading dorsally and ventrally, overtopping apex. RADIAL SPINES 20-25, in several series, 3-6 mm. long, upper shorter, all very fine acicular, straight, pubescent, golden yellow, horizontal, interlacing. FLOWERS short funnelform, lateral in middle of plant, 15 mm. long and wide. *Outer perianth-segments* greenish yellow, wide lanceolate, tip acute, margins (?). *Inner perianth-segments* straw-yellow to muddy canary-

FIG. 258. *Mammillaria eriacantha* x 0.8

yellow, linear-lanceolate, tip acute to acuminate, margins (?). *Anthers* sulphur-yellow. *Stigma-lobes* 4. FRUIT greenish white to orange to reddish, clavate, 10 mm. long. SEEDS pale yellow to golden, oboclavate, pitted, 1 mm.

Distribution: Vera Cruz, Mexico.

Type locality: None given but reported from Cerra Gordo and Jalapa.

Illustration: Fig. 258 is from a photograph of a plant collected for us by Mr. Theodore Hutchison at Cerro Gordo, V. C.

The credit for the original description is confusing because the description appeared in three publications in the same year. Which one actually appeared first, we have not been able to determine for certain. Link & Otto described it in Zuccarini (Plant. Nov. Monac., 704, 1837). Pfeiffer described it in two of his publications (Enum. Cact., 32, 1837 [in Latin]) and in (Besch. Symon., 30, 1837 [in German]), both as "H. Berol." Pfeiffer & Otto (Abild. Bluh. Cact., 25, 1843) in listing the references give preference to the Link & Otto publication. Labouret (Des Cact., 88, 1853) first confused the synonymy by referring to the Link & Otto description as occurring in Pfeiffer's Enum. Cact. and was followed likewise by Schumann and Britton & Rose.

M. columbiana (cf.) was referred here by Britton & Rose and others because it was originally compared to the above species by Salm-Dyck and also on the strength that they knew of no such plant from Columbia, but recent discoveries have substantiated the original distribution citation, so we have restored it to its specific rank.

M. eriantha Hort. in Pfeiffer (Enum. Cact., 32, 1837) was referred by him as a synonym of this species but without description.

M. cylindracea DeCandolle (Mem. Mus. Hist. Nat. Paris, *17*:111, 1828) and *M. cylindrica* Don (Gen. Syst. Gard. Bot., 158, 1834) were referred here by Pfeiffer, Otto, Schumann and others but the original description calls for bristles in the axils which would exclude it from this species.

Fig. 259. *Mammillaria densispina* x 1

230. **Mammillaria densispina** (Coulter) Orcutt

Cactus densispinus Coulter, Contr. U. S. Nat. Herb., 3:96, 1894.
Mamillaria pseudofuscata Quehl, Monatsschr. Kakteenk., 24:114, 1914.
Neomammillaria densispina Britton & Rose, The Cactaceae, 4:119, 1923.
Mamillaria densispina Orcutt, Cactography, 7, 1926.
Mammillaria (Neomammillaria) mieheana Tiegel, Möller Deutsch. Gartner. Zeit., 48:397, 1933.
Mammillaria (Neomammillaria) mieheana globosa Tiegel, Möller Deutsch. Gartner. Zeit., 48:398, 1933.

BODY simple and cespitose from base, globose to short cylindric, rounded at apex, to 10 cm. wide. TUBERCLES arranged in 8 and 13 spirals, firm to somewhat flabby in texture, dark green, conic, 4-sided at base, flattened dorsally, with watery sap, 5 mm. long, 5 mm. wide at base. AREOLES oval, with white wool in youth, later becoming naked. AXILS with white wool in youth and in flowering area, later becoming naked. CENTRAL SPINES 5-6, 10-20 mm. long, upper longer, all acicular, straight to a little bent, stiff, smooth, enlarged at base, white at base, becoming mostly yellow with red to black at very tip, spreading from porrect. RADIAL SPINES 20-25, unequal 8-13 mm. long, lower 5 shorter, all slender acicular, straight, stiff, smooth, enlarged at base, white to yellow, darker in age, ascending and interlocking so as to cover the plant. FLOWERS funnelform, near top, 20 mm. long, 10 mm. wide. *Outer perianth-segments* greenish yellow at base, pale yellow above, tinged with pink at apex, linear-lanceolate, tip acute to obtuse, margins entire, 2-8x1-2 mm. *Inner perianth-segments* sulphur-yellow, lanceolate, longer and wider than the outer, tip acute, margins entire. *Filaments* white to pale greenish yellow. *Anthers* yellow. *Style* very pale green. *Stigma-lobes* 4-5, green to pale greenish yellow. FRUIT red, short clavate, with dried perianth persisting. SEEDS reddish brown, glossy, obovate, pitted, 1 mm. long.

Distribution: San Luis Potosi, Queretaro, Guanajuato; Mexico.

Type locality: San Luis Potosi.

Illustration: Fig. 259 is from a photograph of a plant obtained from Sr. F. Schmoll of Cadereyta, Qro.

M. mieheana is tentatively referred here until more data can be obtained on the flower description. It is more slender in its body growth habit and slightly more flabby in texture.

M. esausseri is reported to have been named by Fric, but as far as we have been able to determine, it has not been described. F. Schmoll used the name in his catalogue for the plants from Queretaro.

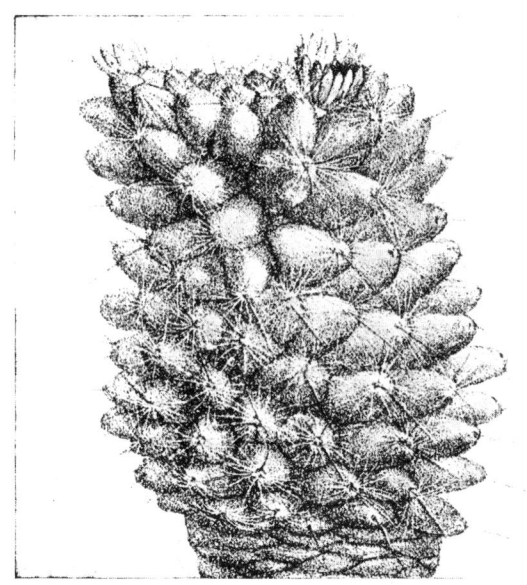

Fig. 260. *Mammillaria vetula*

231. **Mammillaria vetula** Martius

Mammillaria vetula Martius, Hort. Reg. Monac., 128, 1829. Nomen.
Mammillaria vetula Martius, Nov. Act. Nat. Cur., 16:338, 1832.
Cactus vetulus Kuntze, Rev. Gen. Pl., 1:261, 1891.
Neomammillaria vetula Britton & Rose, The Cactaceae, 4:130, 1923.
Chilita vetula Orcutt, Cactography, 2, 1926.

BODY simple and cespitose from body, cylindric to clavate, rounded to flattened at apex, 20 cm. high, 3-5 cm. wide. TUBERCLES arranged in 7 and 18 also 11 and 29 (?) spirals, glossy bluish green, conic, with watery sap, 8-9 mm. long. AREOLES round, 2 mm. wide, short white woolly mat. AXILS with small tuft of wool or naked. CENTRAL SPINES 1-2, 10 mm. long, upper shorter, all stout acicular, straight, dark reddish, later becoming gray, divergent. RADIAL SPINES 25-30 (later to 50?), 3-10 mm. long, thin acicular, straight, smooth, white, horizontal, interlocking. FLOWERS funnelform, 12-15 mm. long and wide. *Outer perianth-segments* citron-yellow with red ventral stripe, lanceolate, tip acute, margins (?). *Inner perianth-segments* citron-yellow, nearly reddish mid-stripe, lanceolate, tip acute, margings (?). *Filaments* greenish yellow. *Anthers* yellow. *Style* greenish yellow. *Stigma-lobes* 5, white to yellow. FRUIT and SEEDS unknown.

Distribution: Hidalgo, Mexico.

Type locality: None given but reported from San Jose del Oro.

Illustration: Fig. 260 is a reproduction of an illustration in Nov. Act. Cur., *16*:pl. 24 as *Mammillaria vetula*, the same was reproduced in Britton & Rose (Cact., *4*:131) Fig. 143 as *Neomammillaria vetula*.

"*M. vetula major* Salm-Dyck in Walpers (Repert. Bot., *2*:270, 1843) is said to be the same as *M. grandiflora* Hort. If so, this must be different from *M. grandiflora* Otto, which we have referred to *Neolloydia conoidea*." Britton & Rose (Cact., *4*:131).

We have not been able to recollect or obtain this species to complete the description.

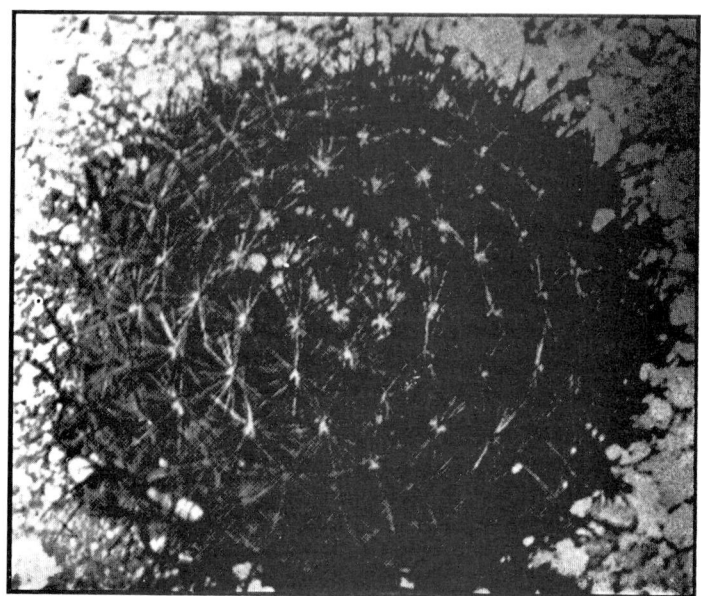

Fig. 261. *Mammillaria schmollii*

232. Mammillaria schmollii (Bravo) Werdermann

Neomammillaria schmollii Bravo, An. Inst. Biol. Mex., 2:128, 1931.
Mammillaria schmollii Werdermann in Backeberg, Neue Kakteen, 98, 1931.

BODY simple, depressed globose, sunken at apex, 7 cm. wide. TUBERCLES arranged in 8 and 13 spirals, olive-green, conic, with watery sap. AREOLES with scant white wool in youth. AXILS naked. CENTRAL SPINES 11-15, 7-10 mm. long, one longer and little more central, all stout acicular, straight, honey yellow, spreading from porrect. RADIAL SPINES to 25, 4-5 mm. long, slender acicular, straight, glossy white, nearly horizontal. FLOWERS near apex, yellow. FRUIT and SEEDS unknown.

Distribution: Oaxaca, Mexico.

Type locality: Near Mitla.

Illustration: Fig. 261 is a reproduction of an illustration in An. Inst. Biol. Mex., 2:128, as *Neomammillaria schmollii*.

Boedeker (Mammill. Vergl. Schluss., 57, 1933) referred this species to the Galactochylus section (milky sap) but the original description calls for a watery sap.

This species was very incompletely described and as such it is quite uncertain. We visited the type locality of Mitla in June of 1942 and scouted the hills with no success in finding any specimens. The guard at the archeological ruins had a small collection of cacti but he told us that he had never seen any plants corresponding to this description in that vicinity. He did report that a party had been there a few years previous and had made a trip into the sierras on horseback and brought some plants but he did not see them.

The plants in the trade under this name are not to be referred here but to one of the varieties of M. *elegans.*

Fig. 262. *Mammillaria baumii*

233. **Mammillaria baumii** Boedeker

Mamillaria baumii Boedeker, Zeitschr. Sukkulentenk., 2:238, 1926.

BODY cespitose from body and base, long ovoid, sunken at apex, 8 cm. high. to 6 cm. wide. TUBERCLES arranged in 8 and 13 spirals, glossy medium green, lighter in axils, thick cylindric to short conic, nearly terete, rounded at apex, with watery sap, 8-10 mm. long, 5 mm. wide at base. AREOLES round, 1 mm. wide, with white wool in youth, soon becoming naked. AXILS with scant wool only in youngest. CENTRAL SPINES 5-6, to 18 mm. long, thin acicular, straight, smooth, faint yellowish, opalescent, brownish at base, horizontal to slightly ascending. RADIAL SPINES 30-35, 6-15 mm. long, very thin hair-like, tortuous, smooth, transparent, white, horizontal, interlacing. FLOWERS wide funnelform, in crown, near apex, 25-30 mm. long and wide, April. *Outer perianth-segments* greenish yellow with rose mid-stripe only ventrally but as mid-line dorsally, outer scales more brownish rose, linear-lanceolate, tip acute, margins entire, 20x2-3 mm. *Inner perianth-segments* bright dark sulphur-yellow, glossy, lanceolate, tip acute, margins entire to slightly serrate at tip, 20x3 mm. *Filaments* whitish below, yellow above. *Anthers* bright orange-yellow. *Style* greenish. *Stigma-lobes* 4-5, bright green, flattened, 5 mm. long. FRUIT grayish green, slightly dirty rose tinge at base, large ovate (like a *Coryphantha*), 15x10 mm. SEEDS dark brown, curved pyriform with lateral hilum, glossy, 1 mm. long.

Distribution: Tamaulipas, Mexico.

Type locality: Near San Vicente.

Illustration: Fig. 262 is a reproduction of an illustration in Zeitschr. Sukkulentk., 2:239, as *Mammillaria baumii.*

Fig. 263. *Mammillaria radiaissima* x 1

234. **Mammillaria radiaissima** Lindsay sp. nov.

Corpus cespitosus, globosus, mamillis textis mollis, succo aquario, axillis nudis; spinis centralibus 1-8, 6 mm. longis, rectis; spinis radialibus 50, ad 18 mm. longis, rectis, setiformis, albis; flores infundibuliformes, sepalis 8, flavis, cum lineis ruberis, entegris, petalis 12, subflavis, cum lineis ruberis, entegris, 10-14x4 mm., stigmatibus 4-6, fluctuosis, fructus et semina ignoti.

BODY cespitose, flattened globose, to 50 mm. wide and high. TUBERCLES soft in texture, with watery sap, 10 mm. long, 6 mm. wide at base. AREOLES round, 1.5 mm. wide, with white wool only in the youngest. AXILS naked or with slight wool. CENTRAL SPINES 1-8, to 6 mm. long, straight, very little different from radials, pale brown, spreading porrect. RADIAL SPINES to 50, to 18 mm. long, bristle-like, straight, white, horizontal (little ascending in youth). FLOWERS 30 mm. long, 15 mm. wide, funnelform. *Outer perianth-segments* 8, yellow, tinged on ventral with ochre-red, broad, tip acute, margins entire, 10-14x4 mm. *Inner perianth-segments* 12, citron-yellow, tinged with ochre-red along central outer tip, broadly lanceolate, tip acute, margins entire, 10-14x4 mm. *Filaments* white, 3-7 mm. long. *Anthers* yellow, 5 mm. long, splitting along both sides. *Stigma-lobes* 4-6, not linear or distinct but revolute wavy or crest-like. *Tube* 6 mm. long. FRUIT and SEEDS unknown.

Distribution: Baja California, Mexico.

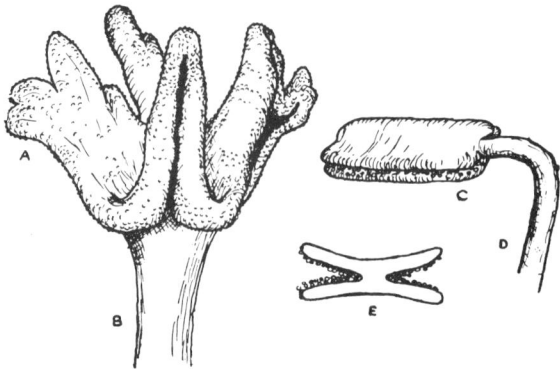

Fig. 264. *Mammillaria radiaissima*

Type locality: Puente Escondido, 110 miles north of La Paz.

Illustrations: Fig. 263 is from a photograph by Mr. George Lindsay. Fig. 264 is from a drawing by Mr. Lindsay: (a) stigma surface showing the peculiar convoluted lobes, a characteristic apparently unique with this species, (b) style, (c) anthers, showing the lateral dehiscence, (d) filament, (e) cross section of the anther.

This plant was discovered north of La Paz on a mountain slope facing the Gulf of California by Mr. Robert Leslie of Santa Barbara, California. The flower is very unusual in the structure of the stigma-lobes and the anthers as well as in the spine count. Several flowers were open the day that these observations were made and all of them showed the same unusual structure. The plants No. 792 in the collection of the Boyce Thompson Arboretum of Superior, Arizona, were used by Mr. Fred Gibson and Mr. George Lindsay in preparing the description.

FIG. 265. *Mammillaria microhelia* x 1

235. Mammillaria microhelia Werdermann

Mamillaria microhelia Werdermann, Monatsschr. Deutsch. Kakt. Ges., 2:236, 1930.
Mamillaria microhelia albiflora Backeberg & Knuth, Kaktus A.B.C., 381, 1935.

BODY simple and occasionally branching from base, cylindric to clavate, rounded at apex, 15 cm. high, 35-45 mm. wide. TUBERCLES arranged in 8 and 13 spirals, firm in texture, bright medium light green, short blunt conic, rounded at tip, flattened dorsally, somewhat keeled ventrally, with watery sap, 6 mm. long, 6-7 mm. wide at base. AREOLES inverted pyriform, 2x1 mm., with scant white wool in youth, later becoming naked, with groove on lower border into which the radial spines are depressed. AXILS with scant wool in youth, later becoming naked. CENTRAL SPINES 0-4, mostly 1-2, frequently missing in lower part of plant, increasing in number in older plants, to 11 mm.

long, when more than one: lower is longer, strong acicular, lower slightly recurved, upper straight, all smooth, stiff, enlarged at base, ruby-red to dark brown, lighter at base, uppers nearly erect, lower little dorsally from porrect. RADIAL SPINES nearly 50, 4-6 mm. long, upper shorter, all thin acicular to bristle-like, recurved, smooth, stiff, shining golden yellow to red-brown to red at base, which forms a definite band around the areole, pure white above, uniformly distributed in all directions, slightly ascending and arched backward, interlacing, so hiding the plant. FLOWERS campanulate, lateral, 16 mm. long and wide. *Outer perianth-segments* greenish white base, broad greenish yellow mid-stripe, often pinkish at tip, lanceolate, tip acute, margins mostly entire (here and there a few serrations), 5-9x1 mm. *Inner perianth-segments* whitish to cream to pale greenish yellow, silken glossy, oblong, tip acute, margins entire, 5x2 mm. *Filaments* pure white, curved. *Anthers* pale yellow. *Style* white below, pale green above. *Stigma-lobes* 5-6, pale green, 1 mm. long, overtop anthers. FRUIT pink, clavate, 11x4 mm., with dried perianth persisting. SEEDS golden brown, glossy, pyriform, with lateral oval hilum, pitted reticulate, hardly 1 mm. long.

Distribution: Queretaro, Mexico.

Type locality: Sierra San Moran.

Illustration: Fig. 265 is a reproduction of an illustration in Monatsschr. Deutsch. Kakt. Ges., 2:237, as *Mamillaria microhelia*. Fig. 265a is from a photograph of a plant received from Sr. F. Schmoll.

Neomammillaria microhelia rubispina (Cact. Succ. Journ., 4:381, 1933) is briefly described without author but is well illustrated and is referrable here.

FIG. 265a. *Mammillaria microhelia*

Fig. 266. *Mammillaria droegeana*

236. Mammillaria droegeana Hildmann

?*Echinocactus drageanus* Moeder, Rev. Hort., 67:186, 1895.
Mamillaria droegeana Rebut, Catal., 7, 1896. Nomen.
?*Mamillaria deluelii* Rebut, Catal., 7, 1896. Nomen.
?*Echinocactus droegeanus* Hildmann in Schumann, Gesamtb. Kakteen., 438, 1898.
Mamillaria rhodantha droegeana Schumann, Gesamtb. Kakteen., 550, 1898.
Mamillaria droegeana Hildmann in Schelle, Handb. Kakteenk., 257, 1907.
Mammillaria discolor droegeana Quehl, Monatsschr. Kakteenk., 48, 1915.

BODY simple and cespitose from base, columnar, 8 cm. high, 4 cm. wide. TUBERCLES arranged in 8 and 13 spirals, semi-firm in texture, dark grayish green, conic, blunt at apex, keeled ventrally, with watery sap, 5 mm. long, 5 mm. wide at base. AREOLES oval, with very scant wool only in the youngest. AXILS naked. CENTRAL SPINES 8-11, 8-10 mm. long, stout acicular, straight to slightly curved, smooth, stiff, not enlarged at base, dark reddish brown, divergent porrect, one mid center. RADIAL SPINES 30, 5-6 mm. long, acicular, straight to slightly recurved, smooth, semi-stiff, not enlarged at base, pale yellow, darker at base and forming a ring around the areole, slightly ascending and recurved. FLOWERS funnelform, 10 mm. long, 15 mm. wide. *Outer perianth-segments* greenish yellow, greener mid-stripe, lanceolate, tip acute, margins entire. *Inner perianth-segments* paler greenish yellow, linear-lanceolate, tip acute, margins entire. *Filaments* pale yellow. *Anthers* orange-yellow. *Stigma-lobes* 4, greenish yellow. FRUIT yellow-tan, clavate, 20x5 mm., with dried perianth persisting. SEEDS light brown, glossy, curved pyriform with lateral hilum at base, very finely pitted, less than 1 mm. long.

Distribution: Queretaro, Mexico.

Type locality: None given but reported from Sierra de San Moran (Schmoll).

Illustration: Fig. 266 is from a photograph of a plant sent to us by Sr. F. Schmoll of Cadereyta, Qro.

FIG. 267. *Mammillaria microheliopsis* x 0.8

237. Mammillaria microheliopsis Werdermann

Mammillaria microheliopsis Werdermann, Notizbl. Bot. Gart. Mus. Berl., 11:278, 1931.
Mamillaria microhelia microheliopsis Backeberg, Blatter Kakteen, 131, sp. 8, 1937.

BODY simple, short cylindric, to 12 cm. high, 4-5 cm. wide, slightly sunken and woolly at apex. TUBERCLES arranged in 8 and 13 spirals, light green, faintly glossy, finely pitted, short conic, not angled, with watery sap, 8-9 mm. long, 8-9 mm. wide at base. AREOLES eliptical, 3-4x2 mm., with white wool in youth. AXILS with scant wool in youth, later becoming naked. CENTRAL SPINES 6-8, 6-11 mm. long, acicular, straight to recurved, stiff, enlarged at base, smooth, in youth flesh-color with black tip, later brownish to dark brown, spreading porrect. RADIAL SPINES 30-40, 4-8 mm. long, upper shorter, fine acicular, straight, smooth, whitish to greenish yellow, at base forms a golden yellow to brownish circle around the areole. FLOWERS wide funnelform, lateral, 15-16 mm. long, 11-12 mm. wide. *Outer perianth-segments* light green at base, bright rust-red to pale violet mid-stripe, lighter margins, lanceolate, tip acuminate, margins mostly entire. *Inner perianth-segments* pale violet, lighter margins, darker mid-line, oblong to lanceolate, tip acute, margins entire. *Filaments* whitish. *Anthers* yellow. *Style* greenish at base, to pale tan to whitish at top. *Stigma-lobes* 4-6, pale yellowish to pale green, overtop anthers. FRUIT whitish to dirty pale green, clavate, 15-18x5-6 mm., with dried perianth persisting. SEEDS bright brown, faintly glossy, curved pyriform, with small lateral oval hilum, very finely pitted, 1 mm. long.

Distribution: Queretaro, Mexico.

Type locality: None given but reported from Sierra de San Moran.

Illustration: Fig. 267 is from a photograph of a plant obtained from Sr. F. Schmoll of Cadereyta, Qro.

Fig. 268. *Mammillaria pottsii* x 1

238. **Mammillaria pottsii** Scheer

Mamillaria pottsii Scheer in Salm-Dyck, Cact. Hort. Dyck., 1849, 104, 1850.
Mamillaria leona Poselger, Allg. Gartenz., 21:94, 1853.
Cactus pottsii Kuntze, Rev. Gen. Pl., 1:261, 1891.
Neomammillaria pottsii Britton & Rose, The Cactaceae, 4:136, 1923.
Chilita pottsii Orcutt, Cactography 2, 1926.
Coryphantha pottsii Berger, Kakteen. 279, 1929.

BODY simple and cespitose from base and body, cylindric, rounded at apex, 15 cm. high, 4 cm. wide. TUBERCLES arranged in 8 and 13 spirals, bluish-green, conic, with watery sap, 5 mm. long, 5 mm. wide at base. AREOLES nearly round to slightly oval, with scant wool only in youngest. AXILS with very scant wool. CENTRAL SPINES 7-9, 4-10 mm. long, upper one longest, all stout acicular, straight or upper recurved, smooth, somewhat enlarged at base, yellowish to chalky purple with darker to brown tip, more or less ascending. RADIAL SPINES to 35, 4-5 mm. long, fine acicular, straight, weak, smooth, enlarged at base, white, horizontal interlacing. FLOWERS funnelform, lateral, 10 mm. long. *Outer perianth-segments* reddish rose mid-stripe near apex, cream margins, ovate-lanceolate, tip acute, margins serrate. *Inner perianth-segments* reddish rose mid-stripe, orange-rose margins, eliptical, tip acute, margins entire. *Filaments* pale purple above. *Anthers* whitish. *Style* pink. *Stigma-lobes* 4, orange-yellow. FRUIT red, clavate, with dried perianth persisting. SEEDS dark brown, nearly black, deeply pitted.

Distribution: Texas, U. S.; Nuevo Leon, Coahuila, Chihuahua, Durango, Zacatescas; Mexico.

Type locality: None given but reported from Big Bend country along the Rio Grande, Texas; Rinconda and Saltillo, N. L., Villa Juarex, D.

Illustration: Fig. 268 is from a photograph of a plant we obtained in Texas.

Mamillaria leona similis Hildmann (Verzeichnis 4, 1888) is only a name.

Examination of Salm-Dyck's description in regards to the tubercles will clear up the uncertainty. It calls for a very slight groove and many wrinkles, and this together with the material in the Missouri Botanical Garden, would indicate that the material used by Salm-Dyck was doubtlessly in a very shrunken atypical condition.

UNCLASSIFIED SPECIES

The following species are proposed but due to the fact that some of the important data is not available, we are uncertain as to their exact relationship in the key to the species.

Some of them were plants that died before the full description was obtained while others are recent imports on which we have not the full description of flowers, fruit, and seeds.

FIG. 269. *Mammillaria alamensis* x 1

Mammillaria alamensis sp. nov.

Corpus simplex, mamillis ad 5 et 8 seriebus ordinatis, conicis, suco aquario, axillis nudis; spinis centralibus 1, 9 mm., hamatis, acicularibus, porrectis; spinis radialibus 9, 6 mm., rectis, acicularibus tenuibus; flores et fructus ignota; semina nigra.

BODY simple, conical, 45 mm. high, 40 mm. wide. TUBERCLES arranged in 5 and 8 spirals, somewhat firm in texture, dark grayish green, conic, blunt apex, with watery sap, 5 mm. long, 5 mm. wide at base. AREOLES round, naked. AXILS naked. CENTRAL SPINES 1, 9 mm. long, hooked, acicular, smooth, stiff, base not enlarged, dark purplish brown to black, becoming lighter, porrect to somewhat erect. RAIDAL SPINES 9, 6 mm. long, nearly equal, fine acicular, straight, smooth, stiff, white, dark brown tip, horizontal. FLOWER and FRUIT unknown. SEEDS black, glossy, globular, obovate, smooth or some very faintly pitted, 0.7x.5 mm.

Distribution: Sonora, Mexico.

Type locality: Near Alamos.

Illustration: Fig. 269 is from a photograph of a plant collected by Mr. Howard Gentry in 1936 as No. 625.

Unclassified Species

Fig. 270. *Mammillaria armatissima* x 0.6

Mammillaria armatissima sp. nov.

Corpus simplex, globosus, mamillis pyramidis, angulis, suco lacteo, axillis nudis; spinis centralibus 0-1, 10-14 mm., subulatissimis, porrectis; spinis radialibus ad duo seriebus ordinatis, externa: 6, inferior 18-20 mm., alia 5-10 mm., subulatissima, interna: 4-6, 2-5 mm., aciculare; flores, fructa et semina ignota.

BODY simple, globose, 15 cm. high, 12 cm. wide. TUBERCLES medium light green, pyramidal, sharply angled especially in youth, with a very milky sap, 15 mm. long, 11-12 mm. wide at base. AREOLES oval, with white wool in youth. AXILS naked. CENTRAL SPINES 0-1, 10-14 mm. long, very heavy subulate, straight to curved, base cream to reddish brown to brown or black at tip, becoming horn colored, porrect. RADIAL SPINES in two rows, outer row: 6, lower one 18-20 mm. long, others 5-10 mm. long, as much as 1.6 mm. in diameter, all very heavy subulate, straight to curved to somewhat tortuous, color same as central, horizontal to slightly ascending, inner row: 4-6, 2-5 mm., acicular, straight to curved, same color, horizontal. FLOWERS, FRUIT and SEEDS unknown.

Distribution: Mexico (?)

Type locality: Not known.

Illustration: Fig. 270 is from a photograph of a plant obtained from a local dealer.

Mammillaria auriareolis Tiegel

Mammillaria (*Neomammillaria*) *auriareolis* Tiegel in Moellers, Deut. Gart. Zeitung 48:412, 1933.

BODY simple and cespitose by dichotomous branching, flattened globular, with sunken apex, 65 mm. high, to 80 mm. wide. TUBERCLES arranged in 13 and 21 spirals, firm in texture, bluish green, 4-angled below, cylindric to slightly conic at apex, keeled ventrally, with milky sap, 6-8 mm. long, 4-6 mm. wide at base. AREOLES with scant white wool

Unclassified Species

FIG. 271. *Mammillaria auriareolis*

in youth. AXILS with white wool and with bristles longer than tubercles. CENTRAL SPINES 4, 3-6 mm. long, lower ones longer, all straight, stiff acicular, smooth, base enlarged, whitish, reddish brown tip, bright golden yellow base (species charactenic), somewhat ascending. RADIAL SPINES 24, (35-40)* 3-5 mm. long, setaceous, straight, smooth, glossy white, somewhat recurved. FLOWERS unknown. FRUIT bright carmine red, clavate, 20 mm. SEEDS dull yellowish brown, curved pyriform with lateral hilum, 1.2x0.7 mm., somewhat rugose.

Distribution: Guanajuato, Queretaro, Mexico.

Type locality: Boundary between the two states.

Illustration: Fig. 271 is a reproduction of an illustration in Moeller Deut. Gar. Zeit. 48:412, 1933, as *Mammillaria* (*Neomammillaria*) *auriareolis* and a copy of same in Cact. Succ. Journ. 7:47, 1935, as *Neomammillaria auriareolis* Trez. (?)

Mammillaria auricantha sp. nov.

Mammillaria sp. No. 509, Gentry, Rio Mayo Plants 196, 1942. (Mention of distribution, without description).

Corpus simplex, columnarus, mamillis ad 8 et 13 seriebus ordinatis, conoidis ad cylindratis, suco lacteo, 5-6 mm. per longitudinem, areolis ovatis, axillis lanatis parvis, setis paucis; spinis centralibus 2-4 (fere 2), 5-14 mm., acicularibus, rectis, flavis; spinis radialibus 12-15, 3-7 mm., acicularibus tenuibus, rectis; flores, fructus et semina ignota.

BODY simple, cylindric, with apex slightly sunken and woolly, 40 mm. wide, 60 mm. high. TUBERCLES arranged in 8 and 13 spirals, firm in texture, light green, conic to nearly cylindric, nearly terete, somewhat keeled ventrally, with milky sap, 5-6 mm. long, 5-6 mm. wide at base. AREOLES short oval, with cream wool in youth soon becoming naked. AXILS with very scant wool and a few white bristles. CENTRAL SPINES 2-4 (mostly 2, few 3, the 4th is uppermost and much smaller), 5-14 mm. long, lower longest, all acicular, straight, stiff, smooth, golden yellow, divergent porrect. RADIAL SPINES 12-15, 3-7 mm. long, upper ones shorter, all very fine acicular, straight, smooth, semi-flexuous, white to pale yellow, horizontal. FLOWERS, FRUIT and SEEDS unknown.

Distribution: SW. Chihuahua and SE. Sonora, Mexico.

Type locality: Sierra Canelo, Rio Mayo.

Illustration: Fig. 272 is from a photograph of a plant collected by Mr. Howard S. Gentry while on botanical survey of Rio Mayo, Sonora, in 1936.

(*) Shurly (Journal British 11, 1935) reports the spine count as 40 while Tiegel reported only 24 (original description) and our specimens had 35-40. All material came from the same source (Sr. F. Schmoll, Cadereyta, Qro.)

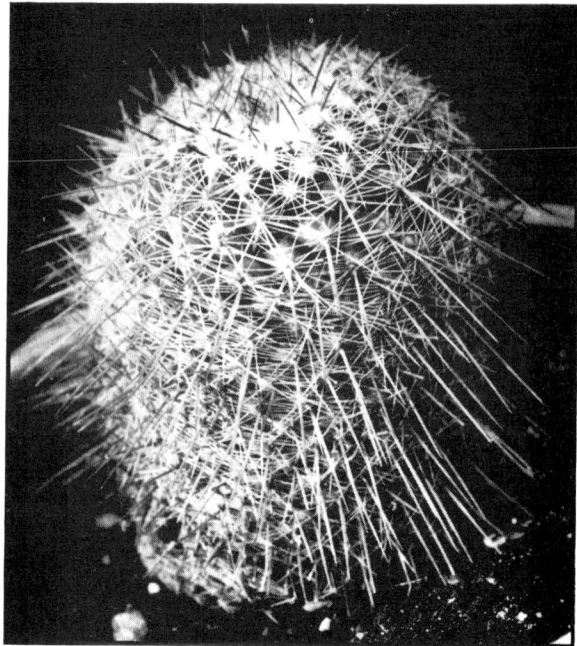

Fig. 272. *Mammillaria auricantha* x 1

Fig. 273 *Mammillaria auritricha* x .8

Mammillaria auritricha sp. nov.

Mammillaria sp. No. 510 Gentry, Rio Mayo Plants 196, 1942. (Mention of distribution, without description.)

Corpus simplex, ovatus, mamillis ad 13 et 21 seriebus ordinatis, cylindratis, suco lacteo, 4 mm. per longitudinem, areolis ovatis, axillis lanatis albis et setis; spinis centralibus 4-6, 10-12 mm. longis, acicularibus tenuibus, rectis, flavis; spinis radialibus ad 25, 1-5 mm. longis, setis, rectis, albis; flores, fructus et semina ignota.

BODY simple, ovoid, with apex slightly sunken and woolly, 70 mm. long, 50 mm. wide. TUBERCLES arranged in 13 and 21 spirals, hidden by spines, firm in texture, light green, cylindric, nearly terete, with milky sap, 4 mm. long and wide at base. AREOLES oval, with white wool in youth, soon becoming naked. AXILS with white wool and bristles. CENTRAL SPINES 5-7, 10-12 mm. long, slender acicular, straight, smooth, stiff, base slightly enlarged, golden, divergent ascending. RADIAL SPINES 25, 1-5 mm. long, upper ones shorter, very fine acicular, straight, smooth, white, horizontal. FLOWERS, FRUIT and SEEDS unknown.

Distribution: SW. Chihuahua and SE. Sonora, Mexico.

Type locality: Sierra Canelo, Rio Mayo in pine oak belt.

Illustration: Fig. 273 is from a photograph of a plant collected by Mr. Howard S. Gentry while on botanical survey of Rio Mayo, Sonora, in 1936.

FIG. 274. *Mammillaria bellacantha* x 0.8

Mammillaria bellacantha sp. nov.

Mammillaria No. 593 Gentry, Rio Mayo Plants 196, 1942 (Mention of distribution only).

Corpus simplex, globosus, mamillis ad 8 et 13 seriebus ordinatis, quadrangulatis autem sine angulis, suco lacteo, 9 mm. per longitudinem, areolis ovatis, axillis lanatis albis et setis; spinis centralibus 4, 5-7 mm. longis, acicularibus tenuibus, rectis, luteis; spinis radialibus 15, 3-8 mm. longis, setis, albis; flores, fructus et semina ignota.

BODY simple, depressed globular, with apex strongly sunken, 10 cm. wide, 8 cm. high. TUBERCLES arranged in 8 and 13 spirals, firm in texture, dark gray-green, 4-sided, not angled, areole facing more or less ventrally, with dorsal surface forming lip over areole, with milky sap, 9 mm. long, 9 mm. wide at base. AREOLES oval, sunken, with white wool in youth, later becoming naked. AXILS with white wool, white bristles nearly as long as tubercles. CENTRAL SPINES 4, 5-7 mm. long, slender acicular, straight, stiff, smooth, light reddish tan, darker tip, soon becoming horn-colored, cross formation divergent from porrect. RADIAL SPINES 15, 3-8 mm. long, upper ones shortest, all setaceous, straight, stiff, smooth, white with tip of some of them dark brown to black, markedly ascending. FLOWER, FRUIT, and SEED unknown.

Distribution: SW. Chihuahua and SE. Sonora, Mexico.

Type locality: Sierra Canelo, Rio Mayo.

Illustration: Fig. 274 is from a photograph of a plant collected by Mr. Howard S. Gentry in 1936 while on botanical survey of Rio Mayo.

FIG. 275. *Mammillaria bellisiana* x 0.6

Mammillaria bellisiana sp. nov.

Mammillaria sp. No. 567 Gentry, Rio Mayo Plants 197, 1942 (Mention of distribution only).

Corpus simplex et cespitosus, globosus, mamillis ad 13 et 21 seriebus ordinatis, conoidis, suco lacteo, areolis rotundis, axillis lanatis sed non seta; spinis centralibus 1, 15 mm., subulatissimis, rectis; spinis radialibus 6-9, 3-5 mm. rectis, acicularibus ad subulatis, rectis; flores, fructus, et semina ignota.

BODY simple, also branching from body and base, globular, with apex rounded and densely woolly, 13 cm. wide and 15 cm. high. TUBERCLES widely separated in 13 and 21 spirals, firm in texture, dull light green, conic, nearly terete, somewhat keeled ventrally, with milky sap, 14 mm. long, 12-14 mm. wide. AREOLES nearly round, with white wool in youth, soon becoming naked. AXILS with tufts of white wool nearly to top of tubercles, but no bristles. CENTRAL SPINES 1, 15 mm. long, stout subulate, over 1 mm. thick, smooth, mostly straight to slight curve, enlarged base, pinkish tan to purplish brown in youth, later becoming horn-colored, darker tip, porrect. RADIAL SPINES 6-9, upper ones 3-5 mm. long, five lower 9-15 mm. long, upper fine acicular, lower heavy acicular to subulate, all straight, stiff, smooth, base enlarged, tannish pink, later becoming horn-colored, somewhat ascending. FLOWERS, FRUIT and SEEDS unknown.

Distribution: SW. Chihuahua and SE. Sonora, Mexico.

Type locality: Sierra Canelo, Rio Mayo.

Illustration: Fig. 275 is from a photograph of a plant collected by Mr. Howard S. Gentry in 1936 while on botanical survey of Rio Mayo.

This species is named in honor of Mr. C. E. Bellis, the inspector for the U. S. Department of Agriculture of Nogales, Arizona.

FIG. 276. *Mammillaria cadereytensis* x 0.8

Mammillaria cadereytensis sp. nov.

Corpus simplex, globosus, mamillis ad 13 et 21 seriebus ordinatis, pyramidalis, autem sine angulis, suco laceto, axillis lanatis et 6-10 setis; spinis centralibus ad 6, 4-10 mm., rectis; spinis radialibus 30, 3-4 mm., albis; flores ignota; fructus coccineus; semina spadicia.

BODY simple and dichotomous branching, globose to short sylindric, 110 mm. high, 85 mm. wide. TUBERCLES arranged in 13 and 21 spirals, separated, firm in texture, dull gray green, pyramidal but not angled, keeled ventrally, with milky sap, 12 mm. long, 8 mm. wide at base. AREOLES wide oval, with dirty white wool in youth. AXILS with some wool in flowering area, 6-10 short tortuous bristles. CENTRAL SPINES to 6, 4-10 mm. long, subulate, straight, stiff, smooth, chalky white, black tip, nearly horizontal to ascending. RADIAL SPINES 30, 3-5 mm. long, slender acicular, straight, smooth, flexuous. FLOWERS unknown. FRUIT scarlet, slender clavate, 15x4 mm. with dried perianth persisting. SEEDS reddish brown, pyriform with lateral hilum near base, 1.3x0.5 mm., slightly rugose.

Distribution: Queretaro, Guanjuato, Mexico.

Type locality: Cadereyta, Qro.

Illustration: Fig. 276 is from a photograph of a plant obtained from Sr. F. Schmoll in 1941 as sp. nov. No. 1009.

Mammillaria tiegeliana is a name used by F. Schmoll and illustrated in Cact. Succ. Journ. 7:21, 1935, but it was not described as far as we know and it is probably referrable here.

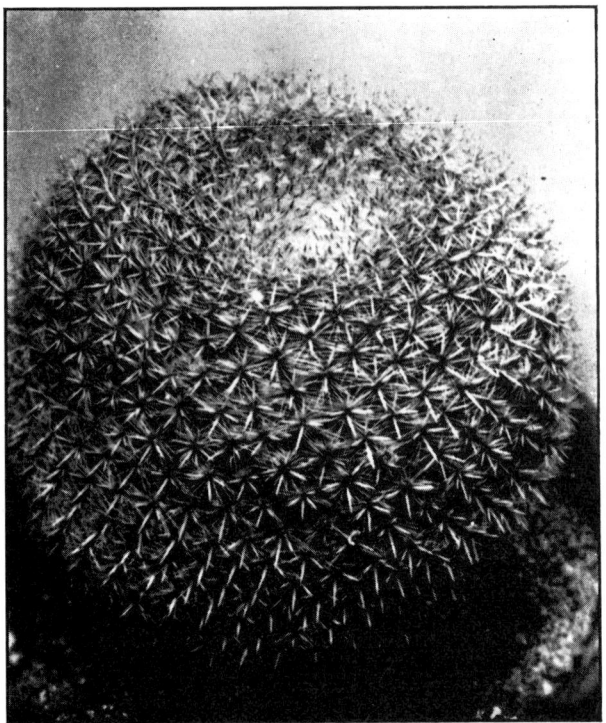

Fig. 277. *Mammillaria cadereytensis* var. *quadrispina* x 1.

a. var. **quadrispina** var. nov.

Spinis centralibus 4, 3-4 mm.

Sentral spines 4, 3-4 mm., acicular.

Type locality: Tarajeas, Qro.

Illustration: Fig. 277 is from a photograph of a plant received from Sr. F. Schmoll.

Mammillaria caerulea sp. nov.

Corpus simplex, clavatus; mamillis ad 13 et 21 seriebus ordinatis, viridibus caerulis pyramidatis, apicis teretibus, suco lacteo; axilis lanatis; spinis centralibus 3-6, ferme 4, 6-10 mm. longis, rectis, acicularibus, fuscis nigris; spinis radialibus 18-20, 1-5 mm. longis, acicularibus tenuibus, albis; flores, fructus et semina ignota.

BODY simple, clavate, 12 cm. high, 6 cm. wide. TUBERCLES closely arranged in 13 and 21 spirals, firm in texture, dark bluish green, quadrangular at base, pyramidal to terete above, with milky sap, 7 mm. long, 5 mm. wide at base. AREOLES oval, with slight white wool in youngest. AXILS white wool in flowering area only (occasional bristle but not typical). CENTRAL SPINES 3-6, mostly 4, 8-10 mm. long, acicular, straight to upper one recurved, smooth, base enlarged, dark brown in youth, later reddish chalky brown, spreading porrect, mostly in cross formation. RADIAL SPINES 18-20, 1-5 mm. long, lateral longest, all very fine acicular to bristle-like, straight, smooth, flexuous, white, somewhat ascending. FLOWERS, FRUIT and SEEDS unknown.

Distribution: Coahuila, Mexico.

Type locality: Near Saltillo.

Illustration: Fig. 278 is from a photograph of a plant received from Sr. F. Schmoll of Cadereyta de Montes, Qro., in December, 1944, as sp. nov.

Unclassified Species

Fig. 278. *Mammillaria caerulea* sp. nov.

Fig. 279. *Mammillaria canelensis* x 0.8

Mammillaria canelensis sp. nov.

Mammillaria sp. No. 613, 614 Gentry, Rio Mayo Plants 197, 1942 (Mention of distribution only).

Corpus simplex, globosus, mamillis ad 13 et 21 seriebus ordinatis, globosis, suco lacteo, areolis rotundis ad ovatis, axillis lanatis densis et setis; spinis centralibus 2-4, 30 mm., acicularibus robustis, rectis ad curvatis, luteis; spinis radialibus 22-25, 5-15 mm., acicularibus tenuibus, rectis, albis; flores, fructus et semina ignota.

BODY simple, globular with apex somewhat sunken and white woolly, 11 cm. wide. TUBERCLES arranged in 13 and 21 spirals, firm in texture, glossy yellowish green, globular, blunt apex, with milky sap, 8 mm. long, 8 mm. wide at base. AREOLES oval to round, very dense white wool in youth, soon becoming naked. AXILS with very dense white wool to apex of tubercle and numerous long white bristles. CENTRAL SPINES 2-4, 30 mm. long, heavy acicular, straight to variously curved, smooth, semi-flexuous, orange-yellow, spreading divergent to overtop apex. RADIAL SPINES 22-25, 5-15 mm. long, very fine acicular, smooth, straight, flexuous, white, horizontal. FLOWERS, FRUIT and SEEDS unknown.

Distribution: SW. Chihuahua and SE. Sonora, Mexico.

Type locality: Sierra Canelo.

Illustration: Fig. 279 is from a photograph of a plant collected by Mr. Howard S. Gentry.

This species may be near *Mammillaria standleyi* (Britton and Rose) Orcutt but the radial spines count is slightly more and the central spines are longer and very orange yellow.

FIG. 280. *Mammillaria crispiseta* x 0.8

Mammillaria crispiseta sp. nov.

Corpus simplex, mamillis ad 13 et 21 seriebus ordinatis, pyramidalis, angulis ad apece, suco lacteo, axillis setis tortuousis; spinis centralibus 4-5, 7-13 mm., acicularibus; spinis radialibus 9, 4-7 mm., acicularibus tenuibus, albis, flores ignota, fructus coccineus, semina subfulva.

BODY simple, flattened globular, 85 mm. wide. TUBERCLES arranged in 13 and 21 spirals, dark green, firm in texture, pyramidal, angled to tip, with milky sap, 10 mm. long, 8 mm. wide at base. AREOLES nearly round with very pale tannish white wool in only youngest. AXILS with long tortuous white bristles. CENTRAL SPINES 4-5, 7-18 mm. long, acicular, recurved, smooth, stiff, base slightly enlarged, color variable from cream to pinkish, brown to dark brown to black, darker tips, strongly ascending nearly porrect. RADIAL SPINES 8, 4-7 mm. long, upper shorter, all slender acicular, smooth, slight recurved white, tip brown, somewhat ascending. FLOWERS unknown. FRUIT crimson, clavate, 15 mm. long with dried perianth persisting. SEEDS light tan, curved pyriform with lateral hilum near base, 1 mm., slightly roughened.

Distribution: Mexico, possibly Queretaro?

Type locality: Unknown.

Illustration: Fig. 280 is from a photograph of a plant sent to us by Sr. F. Schmoll as sp. nov. No. 718.

FIG. 281. *Mammillaria fera rubra* x 1

Mammillaria fera rubra Schmoll sp. nov.

Corpus simplex, globosus, mamillis ad 13 et 21 seriebus ordinatis, pyramidalis autem sine angulis, suco lacteo, axillis lanatis brevis; spinis centralibus 6 (7), 12 mm., acicularibus magnis, flavis ad fuscis in diversas; spinis radialibus 15-18, 3-7 mm., acicularibus tenuibus albis, horizontaliter radiantes; flores ignoti; fructus coccineus, clavaeformus; semina fusca.

BODY simple, globular to short cylindric with sunken apex, 10 cm. high, 9 cm. wide. TUBERCLES arranged in 13 and 21 spirals, firm in texture, dull yellowish gray-green, 4-sided but not angled, blunt apex, with watery sap, 9 mm. long, 7 mm. wide at base. AREOLES rounded oval, with white wool in very youngest. AXILS with short wool. CENTRAL SPINES 6 occasionally 7, 12 mm. long, heavy acicular, smooth, straight to slightly recurved, base yellowish tan, above orange-brown, 6 divergent, seventh when present is mid-center and porrect. RADIAL SPINES 15-18, 3-7 mm. long, upper shorter, all slender acicular but upper more so, smooth, straight, white, horizontal. FLOWERS unknown. FRUIT scarlet, clavate, 20x6 mm., with dried perianth persisting. SEEDS brown, pyriform with lateral hilum at base, nearly smooth, 1.2x.7 mm.

Distribution: Quertaro, Mexico.

Type locality: San Lazara.

Illustration: Fig. 281 is from a photograph of a plant sent to us by Sr. F. Schmoll of Cadereyta, Qro.

FIG. 282. *Mammillaria laneusumma* x 1

Mammillaria laneusumma sp. nov.

Mammillaria sp. No. 575 Gentry, Rio Mayo Plants 197, 1942 (Mention of distribution only).

Corpus simplex, globosus, mamillis ad 13 et 21 seriebus ordinatis, conicis, suco lacteo, 5 mm. per longitudinem, areolis ovatis, axillis lanatis albis et setis; spinis centralibus 2-3, 10-20 mm., acicularibus, rectis, aureis; spinis radialibus 13-15, 5-15 mm., setis, albis; flores, fructus, et semina ignota.

BODY simple, flattened globular with apex markedly sunken. TUBERCLES arranged in 13 and 21 spirals, firm in texture, light green, short conic, blunt apex, with milky sap, 5 mm. long, 7-9 mm. wide at base. AREOLES narrow oval, with some wool, soon becoming naked. AXILS abundant white wool in upper part of plant, nearly to top of tubercles and long white bristles, longer than tubercles. CENTRAL SPINES 2, occasionally 3, 10-12 mm. long, acicular, straight, smooth, stiff, brownish orange, porrect and slightly divergent. RADIAL SPINES 13-15, 5-15 mm. long, upper ones shorter, all fine acicular, straight, more or less stiff, smooth, white, horizontal. FLOWERS, FRUIT and SEEDS unknown.

Distribution: SW. Chihuahua and SE. Sonora, Mexico.

Type locality: Sierra Canelo, Rio Mayo.

Illustration: Fig. 282 is from a photograph of a plant collected by Mr. Howard S. Gentry while on botanical survey of Rio Mayo in 1936.

Fig. 283. *Mammillaria mexicensis* x1

Mammillaria mexicensis sp. nov.

Corpus simplex, globosus, mamillis ad 8 et 13 ceriebus ordinatis, pyramidalis, suco lacteo, axillis lanatis exiguis albis; spinis centralibus 2-3, 8-11 mm., acicularibus, rectis, subalbis; spinis radialibus, 15-16, 8-12 mm., acicularibus, restis, subalbis; flores, fructa et semina ignota.

BODY simple, globose, 60 mm. wide and high. TUBERCLES arranged in 8 and 13 spirals, light gray-green, pyramidal but not sharply angled, with milky sap, 9 mm. high, 11 mm. laterally, 8 mm. dorso-ventrally. AREOLES round to slightly elongate, with white wool, soon becoming naked. AXILS with very small tuft of white wool, soon becoming naked. CENTRAL SPINES 2-3, 8-11 mm. long, acicular, straight, smooth, creamy white, light brown tip, spreading porrect. RADIAL SPINES 15-16, 8-12 mm. long, acicular, straight, smooth, creamy white, some with light brown at very tip, slightly ascending. FLOWERS, FRUIT and SEED unknown.

Distribution: Mexico.

Type locality: Unknown.

Illustration: Fig. 283 is from a photograph of a plant sent to us by Schwarz & Georgi of San Luis Potosi, S.L.P. in 1936.

Mammillaria montensis sp. nov.

Mammillaria sp. No. 504 Gentry, Rio Mayo Plants 196, 1942 (Mention of distribution only).

Corpus simplex, globosus, mamillis ad 8 et 13 seriebus ordinatis, conicis, suco lacteo, 6 mm. per longitudinem, areolis ovatis, axillis lanatis albis parvis et setis; spinis centralibus 2, 10-15 mm., rectis, fuscis; spinis radialibus 20, 3-7 mm., setis, rectis, albis; flores, fructus et semina ignota.

BODY simple, flattened globular with apex slightly sunken and woolly, 50 mm. wide, 30 mm. high. TUBERCLES arranged in 8 and 13 spirals, hidden by spines, firm in texture, dark green and slightly glossy, more or less terete, conic, with milky sap, 6 mm. long, 6 mm. wide at base. AREOLES oval, with white wool in youth, very soon becoming naked. AXILS with scant white wool and white bristles longer than tubercles.

FIG. 284. *Mammillaria montensis* x 1

CENTRAL SPINES 2, occasionally 3, 10-15 mm. long, lower longer, all acicular, straight to slightly recurved, stiff, smooth, dark purplish brown, markedly divergent dorsally and ventrally. RADIAL SPINES 20, 3-7 mm. long, upper ones shorter, all fine acicular, straight, stiff, smooth, white, the very tip on some brown, mostly horizontal. FLOWERS, FRUIT and SEED unknown.

Distribution: SW. Chihuahua and SE. Sonora, Mexico.

Type locality: Sierra Cajurichi, Rio Mayo.

Illustration: Fig. 284 is from a photograph of a plant collected by Mr. Howard S. Gentry in 1936, while on botanical survey of Rio Mayo.

a. var. **monocentra** var. nov.

Spinis centralibus 1, 4 mm., spinis radialibus 20, 3 mm.

Central spines 1, 4 mm., radial spines 20, 3 mm.

Type locality: Sierra Charuco, Rio Mayo.

b. var. **quadricentra** var. nov.

Spinis centralibus 4-5, ad 25 mm., spinis radialibus 20-22, 5-8 mm.

Central spines 4-5, to 25 mm., radial spines 20-22, 5-8 mm.

Type locality: Sierra Canelo, Rio Mayo.

Mammillaria movensis sp. nov.

Corpus simplex, globosus, mamillis ad 13 et 21 seriebus ordinatis, conoidis, suco lacteo, areolis pyriformis, axillis lanatis et setis; spinis centralibus 1-4, 5-20 mm., acicularibus robustis, rectis, spadicibus; spinis radialibus 10-13, 3-15 mm., acicularibus tenuibus, rectis, albis ad spadicibus; flores ignota; fructus ruber, clavaeformus; semina fulva.

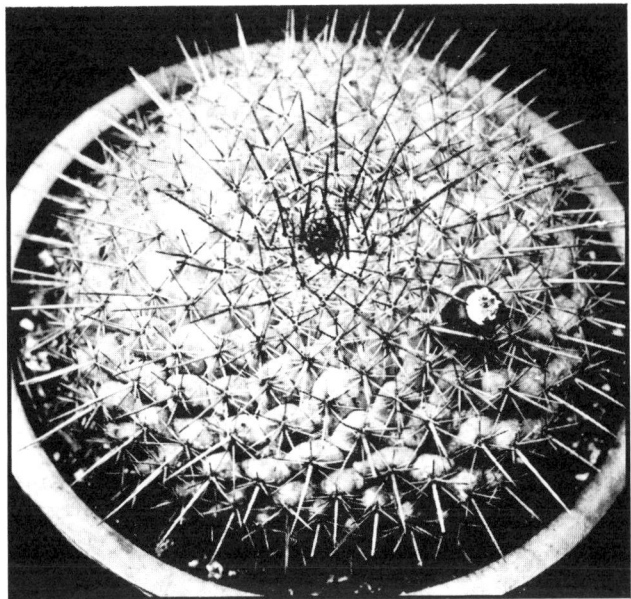

FIG. 285. *Mammillaria morensis* x 0.7

BODY simple, flattened globular, 10 cm. wide and 5 cm. high, with apex sunken and only slightly woolly. TUBERCLES arranged in 13 and 21 spirals, firm in texture, light yellowish gray-green, 4-sided base, above ovoid-conic, keeled ventrally, rounded dorsally, blunt apex, with milky sap, 8 mm. long, 5-9 mm. wide at base. AREOLES somewhat ventral from apex, inverted pyriform, continuing in a groove on the ventral side for 1-2 mm., with light tan wool in youngest, soon becoming naked. AXILS with 4-8 white tortuous bristles, some short and others longer than tubercles. CENTRAL SPINES 1-4 (usually 4), 5-20 mm. long, lower longest and heavier, upper acicular, all straight, smooth, stiff, enlarged base, reddish brown, upper ascending, lower nearly porrect. RADIAL SPINES 10-13, 3-15 mm. long, upper 3-4 shortest and fine acicular, others acicular, all straight, smooth, somewhat flexuous, upper white, lateral with brown tips, lower reddish brown with black tips, horizontal to slightly ascending. FLOWERS unknown. FRUIT red, clavate, 10-20 mm. long, with dried perianth persisting. SEED tan, dull, curved pyriform with lateral hilum near base, 0.8x.4 mm., rugose. ROOTS fiberous.

Distribution: SE. Sonora, Mexico.

Type locality: Movas.

Illustration: Fig. 285 is from a photograph of a plant from type locality, as No. 689.

This species (679, 689, 691, 693) was collected for us by Howard S. Gentry in the spring of 1937, but all of the specimens died before they flowered, so we are referring it to the uncertain group until more material can be collected to complete the description.

Mammillaria multicentralis sp. nov.

Corpus cespitosus; mamillis mollis, suco aquario; axillis nudis; spinis centralibus 12-15, 4 mm. longis, medio uno hamato et item alteris rectis; spinis radialibus 30-40, 12 mm. longis, setis; flores infundibuliformis, sepalis serratis, petalis subissimispuniceis, stigmatibus 3, flavoviridis; fructus coccineus; semina spadix.

BODY cespitose, forming low spreading mats, globular, 20 mm. wide. TUBERCLES flabby in texture, dark green, cylindric, terete, with watery sap, 5 mm. long, 3 mm. wide at base. AREOLES round, with scant deep yellow mat only in youngest. AXILS naked. CENTRAL SPINES 12--15, 4 mm. long, two types: mid center one hooked, fine acicular, pale yellow to yellowish tan, porrect; others straight, very fine acicular, white with very tip tannish yellow, subcentral, some ascending, others nearly horizontal. RADIAL SPINES 30-40, to 12 mm. long, very fine hair-like, tortuous, flexuous, smooth, white, nearly horizontal. FLOWERS wide funnelform, 6 mm. wide, 8 mm. long, April. *Outer perianth-segments* very pale yellowish green, linear-lanceolate, tip acute, margins serrate. *Inner perianth-segments* very pale pink, slightly darker mid-line, linear-lanceolate, tip acuminate, deeply serrate at tips but less so below. *Filaments* and *style* very pale yellow. *Anthers* bright yellow. *Stigma-lobes* 3, greenish yellow, overtop anthers. FRUIT reddish, oval, 5x13 mm., with dried perianth persisting. SEEDS dark reddish brown, globular curved pyriform with lateral hilum near base, lightly pitted, 1.2x0.9 mm. ROOTS tuberous "carrot shaped" flabby tap roots.

Distribution: Queretaro, Mexico.

Type locality: Probably Tarajeas (Schmoll).

We received this species from Sr. F. Schmoll of Cadereyta, Qro., in 1936, but unfortunately we did not obtain a photograph of it before it died and we have not been able to obtain additional material for further study.

FIG. 286. *Mammillaria ocotillensis* var. *longispina* x 0.8

Mammillaria ocotillensis sp. nov.

Corpus simplex, globosus; mamillis ad 12 et 21 seriebus ordinatis, suco lacteo, axillis nudis; spinis centralibus (1)-2, superis 3-15 mm. longis, inferis 12-35 mm. longis, acicularibus, rectis; spinis radialibus 3 (2-4), 3-10 mm. longis, flores ignoti; fructus coccineus; semina spadix.

BODY simple, flattened globular, deep seated, 5-6 cm. wide, 6-7 cm. high. TUBERCLES arranged in 13 and 21 spirals, firm in texture, dull light green, 4 sided but not angled, keeled ventrally, blunt at apex, with milky sap, 5-8 mm. long, 4-6 mm. wide at base, becoming corky in age. AREOLES oval, very small, ventrally from apex, with very little white wool only in the youngest. AXILS naked. CENTRAL SPINES 1 to mostly 2 to occasionally 3, upper 3-15 mm. long, lower 12-35 mm. long (see varieties), stout acicular, straight to curved ventrally in lower, smooth, stiff, base not to slightly enlarged, dark purplish brown to nearly black, especially at tip, divergent dorsally and ventrally but nearly porrect. RADIAL SPINES mostly 3, occasionally 2-4, 3-10 mm. long (see varieties), lower longer, slender acicular, straight, smooth, stiff, chalky white, nearly horizontal extending laterally and ventrally. FLOWER (from dried specimen, cream with reddish mid-stripe). FRUIT carmine, clavate, 17x5 mm., with dried perianth persisting. SEEDS dull reddish brown, irregular pyriform, with basal hilum, faintly pitted, 1.3x0.7 mm.

Distribution: Queretaro, Mexico.

Type locality: Ocotillo in Sierra de San Moran.

Illustrations: Fig. 286 and 287 are from photographs of plants sent to us by Sr. F. Schmoll.

FIG. 287. *Mammillaria ocotillensis* var. *brevispina* x 1

a. var. brevispina

Mamillis 5 mm. longis; spinis centralibus superis 3-4 mm. longis, inferis 12 mm. longis; spinis radialibus 3-4 mm. longis.

Tubercles shorter, 5 mm. long. Central spines upper 3-4 mm. long, lower to 12 mm. long. Radial spines 4-10 mm. long.

b. var. longispina

Mammillis 8 mm. longis; spinis centralibus superis 15 mm. longis, inferis 35 mm. longis; spinis radialibus 4-10 mm. longis.

Tubercles longer, 8 mm. long. Central spines upper to 15 mm. long, lower to 35 mm.

long. Radial spines 4-10 mm. long.

Inasmuch as we have not the complete information on the flower, other than from the dried remnants, its association in the key is uncertain. In some ways it appears to be close to M. *hamiltonhoytae* but other characteristics are quite different.

FIG. 288. *Mammillaria parensis* x 1.

Mammillaria parensis sp. nov.

Corpus cespitosus; mamillis ad 8 et 13 seriebus ordinatis, suco aquario; axillis setis; spinis centralibus 4-5, 5-8 mm. longis, acicularibus, luteis; spinis radialibus 18-20, 3-4 mm. longis, acicularibus, bellis, albis; flores, fructus et semina ignota.

BODY deep seated, cespitose from body and by dichotomous branching, globose to short columnar, 25 mm. wide, 35 mm. long (individual head). TUBERCLES closely set in 8 and 13 spirals, firm in texture, dull gray-green, conic, terete to slightly compressed 4 sided at base, with watery sap, 3-4 mm. long, 4 mm. wide at base. AREOLES oval with scant dirty white wool in youth only. AXILS with several white bristles. CENTRAL SPINES 4-5, 5-8 mm. long, lower shorter, acicular, somewhat tortuous, smooth, stiff, orange-yellow, ascending, spreading, recurved, fifth one often porrect. RADIAL SPINES 18-20, 3-4 mm. long, slender acicular, somewhat tortuous, smooth, stiff, white, with yellow base so as to form a yellow ring at the areole, horizontal. FLOWER, FRUIT and SEEDS unknown.

Distribution: Chihuahua and Coahuila; Mexico.

Type locality: Between Parras, Coahuila and Chihuahua.

Illustration: Fig. 288 is from a photograph of a plant obtained from Sr. F. Schmoll of Cadereyta de Montes, Qro.

Mammillaria queretarica sp. nov.

Corpus simplex, globosus; mamillis ad 13 et 21 seriebus ordinatis, suco lacteo; axillis lanatis parvissimis; spinis centralibus 4, 4-7 mm. longis, acicularibus crassis, rectis; spinis radialibus 30, 3-4 mm. longis, setis, rectis, albis; flores, fructus, et semina ignota.

FIG. 289. *Mammillaria queretarica* x 0.8

BODY simple, globose, 7 cm. wide, 6 cm. high. TUBERCLES arranged in 13 and 21 spirals, firm in texture, dull dark green, 4 sided but not angled, keeled ventrally, with milky sap, 4-5 mm. long, 5-6 mm. wide at base. AREOLES oval, with dirty white wool only in the youngest. AXILS with very scant wool. CENTRAL SPINES 4, 4-7 mm. long, lower longer, stout acicular, straight, smooth, stiff, with base only slightly enlarged, tan with dark brown to black tip, later becoming chalky, somewhat ascending in cross formation. RADIAL SPINES 30, 3-4 mm. long, setaceous, straight to slightly curved, flexuous, white, horizontal. FLOWER, FRUIT and SEEDS unknown.

Distribution: Queretaro, Mexico.

Type locality: Rio del Infernillo, Qro.

Illustration: Fig. 289 is from a photograph of a plant obtained from Sr. F. Schmoll.

We obtained this plant in 1942 but as yet it has not flowered for us, so its association is uncertain.

Mammillaria rosensis sp. nov.

Corpus simplex et dichotomos, mamillis ad 13 et 21 seriebus ordinatis, suco lacteo, axillis lanatis; spinis centralibus 4, supris et infris 12-17 mm., lateralis 4-6 mm., rectis, subulatis; spinis radialibus 25, 3-4 mm., acicularibus albis; flores ignota; fructus coccineus; semina subfulva.

BODY simple and dichotomous branching, flattened globular, to 10 cm. wide, 4 cm. high. TUBERCLES closely set in 13 and 21 spirals, firm in texture, gray-green, conic, not angled, with milky sap, 8 mm. long, 7-8 mm. wide at base. AREOLES oval, with scant white wool. AXILS with dense white wool in flowering area, but no bristles. CENTRAL SPINES 4, occasionally 5, upper and lower 12-17 mm. long, lateral 4-6 mm. long, upper and lower stout subulate, lateral stout acicular, all mostly straight, smooth, stiff, enlarged base, in youth dark purplish brown to black, later chalky tan with dark brown tips, lateral nearly horizontal, upper and lower ascending, upper often porrect, 5th one is lateral. RADIAL SPINES 25, 3-4 mm. long, slender acicular, mostly straight, smooth, stiff, white, horizontal. FLOWERS unknown. FRUIT carmine, curved clavate, 14x5 mm.,

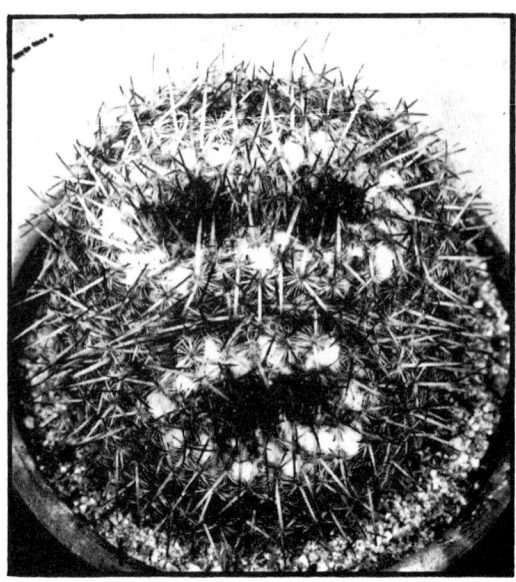

FIG. 290. *Mammillaria rosensis* x 0.7

with dried perianth persisting. SEEDS light brown, and somewhat curved globular pyriform, with small white lateral hilum at base, 1x.8 mm., nearly smooth to very slightly rugose. ROOTS large tap and fiberous.

Distribution: Queretaro, Mexico.

Type locality: San Juan de las Rosas.

Illustration: Fig. 290 is from a photograph of a plant obtained from Sr. F. Schmoll in 1942 as sp. nov. No. 1018.

Mammillaria tolimensis sp. nov.

Corpus simplex et cespitosus; mamillis ad 8 et 13 etiam 13 et 21 seriebus ordinatis, pyramidalis tenuibus, suco lacteo, 6-11 mm. longis, 8-10 mm. laxitibus; axillis setis tortuosis; spinis centralibus 5-7, 5-65 mm. longis, subulatis, rectis at tortuosis; spinis radialibus 5-10, 1-5 mm. longis, acicularibus, tortuosis; flores incertis (sepalis serratis, petalis rubris); fructus coccineus, clavatus; semina fulva.

BODY simple and cespitose, flattened globose, with sunken apex, to 10 cm. wide. TUBERCLES closely set in 8 and 13 also 13 and 21 spirals, firm in texture, dull gray-green to yellowish green, rounded pyramidal, flattened dorsally, keeled ventrally, with milky sap, 6-11 mm. long, 8-10 mm. wide at base. AREOLES sunken, oval, very scant white woolly mat. AXILS with strong tortuous white bristles, white wool. CENTRAL SPINES 5-7, usually 6, 5-65 mm. long, (see varieties), upper slender subulate, lower stout subulate, straight to curved to tortuous, angular in cross section, smooth, chalky white, dark brown to black tip, strongly spreading, one sometimes more central. RADIAL SPINES 5-10, 1-5 mm. long, lower longer, slender acicular to acicular, tortuous, smooth, stiff, chalky white, horizontal to strongly ascending, lower 1-3 in groove in lower part of areole. FLOWERS (from dried specimen. *Outer perianth-segments* lanceolate, tip acute, margins serrate. *Inner perianth-segments* reddish, lanceolate, tip acuminate, margins

FIG. 291. *Mammillaria tolimensis* var. *brevispina* x 0.8

entire). FRUIT carmine red, columnar clavate, 20x8 mm., with dried perianth persisting. SEEDS dull light brown, curved pyriform with lateral hilum near base, nearly smooth, 1.2x0.8 mm.

Distribution: Queretaro, Mexico.

Type locality: Toliman.

Illustrations: Figs. 291 and 292 are photographs of plants sent to us by Sr. F. Schmoll of Cadereyta, Qro.

a. var. brevispina

Mamillis ad 8 et 13 seriebus ordinatis, 11 mm. longis, 8 mm. laxitibus; spinis centralibus 6, 5-18 mm. longis, subulatis; spinis radialibus ad 9, 1-5 mm. longis, acicularibus tenuis, tortuosis.

Tubercles arranged in 8 and 13 spirals, 11 mm. long, 8 mm. wide at the base. Central spines 6, 5-18 mm. long, upper shorter and slender subulate, lower stout subulate, somewhat curved. Radial spines to 9, 1-5 mm. long, tortuous, slender acicular.

b. var. longispina

Mamillis ad 8 et 13, 13 seriebus ordinatis, 7-8 mm. longis, 10-12 mm. laxitibus; spinis centralibus 6, superis 4-12 mm. longis, inferis ad 65 mm. longis, subulatis, inferis cassioribus et tortuosis; spinis radialibus 5, 1-8 mm. longis, acicularibus, tortuosis.

Tubercles arranged in 8 and 13 spirals, 7-8 mm. long, 10-12 mm. wide at base. Central spines 6, uppers 4-12 mm. long, lower one to 65 mm. long, subulate, lower stouter and tortuous. Radial spines 5, 1-8 mm. long acicular.

Fig. 292. *Mammillaria tolimensis* var. *longispina* x 0.3

c. var. **subuncinata**

Mamillis ad 12 et 21 seriebus ordinatis, 6 mm. longis, 8 mm. laxitibus; spinis centralibus 5, 7-15 mm. longis, acicularibus crassis, tortuosis ad hamatis imperfecte; spinis radialibus 5, 1-4 mm. longis, acicularibus.

Tubercles arranged in 13 and 21 spirals, 6 mm. long, 8 mm. wide at base. Central spines 5, 1-4 mm. long, acicular, tortuous to incompletely hooked. Radial spines 5, 1-4 mm. long, acicular.

This species presents some of the characteristics of both *M. compressa* and *M. mystax* but not all of either of them and probably represents an intermediate form. It was sent to us by Sr. F. Schmoll as *M. cirrhifera* but the original description of the latter species does not justify that classification.

Mammillaria yaquensis sp. nov.

Corpus caespitosus cylindricus, mamillis ad 5 et 8 seriebus ordinatis, conicis, suco aquario, axillis subnudis; spinis centralibus 1, 7 mm., acicularibus, hamatis, spadicibus; spinis radialibus 18, 5-6 mm. acicularibus tenuibus, rectis; flores ignoti; fructus coccineus, globosus ad clavatus; semina nigra, globosa-pyriforma.

BODY very cespitose, forming mats under the bushes, joints easily detached from body, clindric to 70 mm. high, 15 mm. wide. TUBERCLES arranged in 5 and 8 spirals, semi-flabby in texture, purplish green, short conic, terete, with watery sap, 3 mm. long, 5 mm. wide at base. AREOLES round, with practically no wool. AXILS with only very faintest mat of white wool. CENTRAL SPINES 1, 7 mm. long, acicular, strongly hooked, stiff, smooth, reddish brown becoming dark brown, porrect. RADIAL SPINES 18, 5-6 mm. long, fine acicular, straight, stiff, smooth, cream, light brown tip, horizontal.

FIG. 293. *Mammillaria yaquensis* x 1

FLOWER unknown. FRUIT scarlet, elongate globular to short clavate, 9x5 mm. SEEDS black, glossy, globular pyriform with ventral hilum, pitted, 1 mm.

Distribution: Sonora, Mexico.

Type locality: Near Fort Pithaya, Rio Yaqui.

Illustration: Fig. 292 is from a photograph of a plant collected by Mr. and Mrs. John Hilton and the author in 1937 in the flat lowlands of the lower Rio Yaqui a few miles from Fort Pithaya. Mr. George Lindsay recollected material for us in 1940 at the same locality. It appears to be very limited in its distribution.

Type material has been deposited in the Dudley Herbarium at Stanford University, California.

LITTLE KNOWN SPECIES

The following species were very briefly described but usually without flower, fruit, or seed data. The type locality was seldom given so that recollection is very improbable. As a consequence, few of them can ever be identified. We are listing them here for the sake of reference as no effort has been made to edit them.

Mammillaria acicularis Lemaire

Mammillaria acicularis Lemaire, Cact. Gen. Nov. Spec. Mon., 34, 1839.
Cactus acicularis Kuntze, Rev. Gen. Pl., 1:261, 1891.

BODY subglobose with sunken apex. TUBERCLES pale green, ovate, conic, 4-sided base, 8 mm. long and 8-10 mm. wide at base. AREOLES small, round, with long wool. AXILS with white wool. CENTRAL SPINES 1, aprox. 25 mm. long, acicular, golden yellow, later yellowish, porrect. RADIAL SPINES 6, nearly equal, 8-10 mm. long, straight, golden yellow, later yellowish, horizontal.
Distribution: Mexico (Deitrich).

Cactus acicularis Kuntze, Rev. Gen. Pl., 1:260, 1891, as based on some name by Lehmann, is unknown to us.

Mammillaria actinoplea Ehrenberg

Mamillaria actinoplea Ehrenberg, Allg. Gartenz., 16:266, 1848.
Mammillaria crebrispina v. *nitida* Labouret, Monog. Cact., 75, 1853.
Cactus actinopleus Kuntze, Rev. Gen. Pl., 1:260, 1891.

BODY flattened-globular, branching, 75 mm. wide. TUBERCLES green to gray-green, conic to ovoid, 6 mm. long. AREOLES oval, with short wool in youth. AXILS with short wool and single short bristle. CENTRAL SPINES 10, 8-10 mm. long, acicular, middle one stronger and hooked, all dark reddish brown, lighter at base. RADIAL SPINES 20, 6-8 mm. long, lateral longer, bristle-like, light red tip.
Distribution: Mexico.

Mammillaria albiseta Foerster

Mammillaria albiseta Hort. in Foerster, Handb. Cact., ed. 2, 1:354, 1886.

BODY globular, somewhat flattened. TUBERCLES upper surface is curved ventrally so the areole is somewhat under a lip. AXILS naked. CENTRAL SPINES 1, somewhat longer. RADIAL SPINES 10, short, setaceous, grayish white, with tip darker in youth. FLOWER rose colored.
Distribution: Probably Mexico.
Britton and Rose referred this to *M. actinoplea* but it is different.

Mammillaria amabilis Ehrenberg

Mammillaria amabilis Ehrenberg, Allg. Gartenz., 17:326, 1849.
Cactus amabilis Kuntze, Rev. Gen. Pl., 1:260, 1891.

BODY globose to cylindric, 75 mm. high and wide. TUBERCLES bright green, elongate, ovoid conic, blunt 4-angled, 4-sided below, 8 mm. long, 5-6 mm. wide at base. AREOLES oval, with abundant reddish yellow wool. AXILS with little wool. CENTRAL SPINES 6-8, 8-10 mm. long, acicular, straight, stiff, one more central and longer, all are cream, with tip brown. RADIAL SPINES 22-24, 4-8 mm. long, yellow to white, horizontal, upper and lower more ascending. FLOWER and FRUIT unkonwn.
Distribution: Mexico.
Britton and Rose list this as a synonym of *M. actinoplea* but it is not to be regarded as such because of the lack of axillary bristles, the difference in spine count and the lack of hooked central spines.

Mammillaria anancistra Lemaire

Mammillaria anancistra Lemaire, Cact. Gen. Nov. Spec. Mon., 39, 1839.
Mammillaria inuncinata Lemaire, Cact. Gen. Nov. Spec. Mon., 39, 1839 (sine scriptore).
Mammillaria ancistra Walper, Repert. Bot. Syst., 2:296, 1843.
Cactus ancistrius Kuntze, Rev. Gen. Pl., 1:261, 1891.

BODY somewhat globose. TUBERCLES light green. CENTRAL SPINES 1, 10-12 mm. long, stout-acicular, reddish. RADIAL SPINES 16-19, 6-7 mm. long, 3 lateral longer, all acicular, recurved, 1-2 upper stronger, all reddish, recurved toward body.
Distribution: Unknown.
This species has usually been referred to *M. decipiens* but it calls for too many radial spines.

Mammillaria argentea Fennel

Mammillaria argentea Fennel, Allg. Gartenz., 15:66, 1847.
Cactus argenteus Kuntz. Rev. Gen. Pl., 1, 260. 1891.

BODY globose, 75 mm. wide. TUBERCLES closely set, bright green, conic. AXILS naked. CENTRAL SPINES 2, 10 mm. long, upper bent, lower straight, all horn colored, with darker tip, divergent dorsally and ventrally. RADIAL SPINES 13-15.
Distribution: Unknown.

Mammillaria atrorubra Ehrenberg

Mamillaria atrorubra Ehrenburg, Allg. Gartenz., 17:327, 1849.
Cactus atroruber Kuntze, Rev. Gen. Pl., 1:260, 1891.

BODY globular to cylindric, 85 mm. high, 75 mm. wide, with apex little sunken. TUBERCLES dark green, conic, blunt-angled, 10 mm. long, 5-6 mm. wide at base. AREOLES oval, with short white wool, later becoming naked. AXILS with short white wool. CENTRAL SPINES 4, 8-10 mm. long, irregular, subulate, straight or bent, dark red, later dark gray, in cross formation. RADIAL SPINES 16, 2-4 mm. long, very fine hair-like, straight, white. FLOWERS *outer perianth-segments* dark carmine-red, greenish mid-stripe. *Inner perianth-segments* carmine-red, very wide, 12-16 mm. long. *Filaments* and *style* red. *Anthers* bright yellow. *Stigma-lobes* dark red.
Distribution: Unknown.

Mammillaria atrosanguinea Ehrenberg

Mammillaria atrosanguinea Ehrenberg, Allg. Gartenz., 17:270, 1849.
Cactus atrosanguineus Kuntze, Rev. Gen. Pl., 1:260, 1891.

BODY globular to cylindric, 125 mm. high, 75 mm. wide. TUBERCLES dark green, ovoid-conic, above truncate, 6 mm. long, 5 mm. wide at base. AREOLES oval, golden yellow wool in youth, then white, later naked. AXILS with wool and many bristles. CENTRAL SPINES 6 seldom 4, 5 upper ones 6-7 mm. long, lower one 14 mm. long, all thin, acicular, stiff, straight or hooked, dark blood-red, then dark brownish red, spreading. RADIAL SPINES 24-30, upper 2 mm. long, lower to 6 mm. long, all bristle-like, fine, golden yellow, later whitish, horizontal, lower ascending.
Distribution: Mexico.

Mammillaria aureiceps Lemaire

Mammillaria aureiceps Lemaire, Cact. Aliq., Nov. 8, 1838.
Mammillaria rhodantha aureiceps Salm-Dyck, Hort. Dyck. 1844, 7, 1845.
Cactus aureiceps Kuntze, Rev. Gen. Pl. 1:260, 1891.
Neomammillaria aureiceps Britton and Rose, The Cactaceae, 4:114, 1923.
Mammillaria aureiceps v. *sanguinea* Neale, Cact. and other Succ., 83, 1935.
Mammillaria aureiceps v. *pfeifferi* Borg, Cacti, 331, 1937.

BODY simple, seldom branching, globose, apex mostly rounded. TUBERCLES somewhat dark green, conic, base nearly oval, with watery sap, 10-12 mm. long. AREOLES round, with short wool, soon naked. AXILS white wool in flowering area, short pale saffron-yellow tortuous spreading bristles. CENTRAL SPINES 6, rarely 7, 14-18 mm. long, slender subulate, dark golden, upward recurved. RADIAL SPINES 30 or more, all acicular, golden yellow, horizontal, interwoven. FLOWERS unknown (dark red to cream). FRUIT green, SEEDS unknown.
Distribution: Mexico. Queretaro (Schmoll). State of Mexico (Boedeker).
Type locality: None but reported from near Rancho del Sabino (Schmoll).

Blanc reports the flowers as cream. Britton and Rose confused this species with one of the variations of *M. rhodantha* and so changed both the central and radial spine count and also reported the color of the flower as dark red which is to be questioned. Lemaire (1840) gives it as synonym of *M. pfeifferi*. Salm-Dyck (1845) gives it as synonym of *M. rhodantha aureiceps*.

Mammillaria badispina Foerster

Mammillaria badispina C. F. Foerster, Hamburger Gartenz. Blumenz., 17:159, 1861.

BODY simple, globular, 50 mm. high and wide. TUBERCLES green, ovoid. AREOLES white woolly only in youth. AXILS naked. CENTRAL SPINES 2-4, 16-18 mm. long, stronger, acicular, ascending. RADIAL SPINES 11-12, 2-6 mm. long, upper ones shorter, all bristle-like, white somewhat ascending.
Distribution: Unknown.

Mammillaria bellatula Foerster

Mamillaria bellatula Foerster, Allg. Gartenz., 15:51, 1847.
Cactus bellatulus Kuntze, Rev. Gen. Pl., 1:259, 1891.

BODY globular, somewhat compressed. TUBERCLES bright green, wide conic, 4 mm. long, 6 mm. wide at base. AREOLES with white wool in youth. AXILS naked. CENTRAL SPINES 2, 12-16 mm. long, straight, black, later grayish brown, divergent dorsally and ventrally. RADIAL SPINES 12-16, 6-8 mm. long, bristle-like, whitish, spreading. FLOWERS unknown. (Schlumberger say large yellow.)

Distribution: Brazil?

"This species is said to have been grown from Brazilian seed; if this were true, it would exclude it from this genus and for this reason Schumann questioned whether it might not be an Echinocactus. Judging from the description, we believe that it is closely related to *Neomammillaria elegans* and is probably of Mexican origin." Britton and Rose, Cact. 4:165, 1923.

Mammillaria bergeana (Hildm.) Quehl

Mammillaria bergeana Hildmann, Verzeichnis, 3, 1888 (nomen).
Mamillaria bergeana (Hildm.) Quehl, Monat. Kak., 79, 1894.

BODY flattened globular, 25 mm. high, 35 mm. wide. TUBERCLES cylindric, with watery sap, 6-8 mm. long. AXILS naked. CENTRAL SPINES 4-5 mm. long, acicular, 1 hooked, 3 straight, all white below, reddish above. RADIAL SPINES 25, 10 mm. long, setaceous, white. FLOWERS white. FRUIT and SEEDS unknown.

Distribution: Unknown.

Various authors have referred this species to *M. glochidiata* but it has too many radial spines, and inasmuch as the description is so incomplete it will have to be referred to the uncertain species as no type locality is given so that more material could be collected to complete the description.

Mammillaria bergenii Ehrenberg

Mammillaria bergenii Ehrenberg, Allg. Gartenz., 17:326, 1849.

BODY globular, 100 mm. high and wide, with apex sunken. TUBERCLES light to later dark green, pyramidal conic, 4-sided below, a little rounded above. AREOLES oval, short white wool. AXILS with long white bristles. CENTRAL SPINES 2, seldom 3, very seldom 4, 10-12 mm. long, subulate, straight, horn color, divergent dorsally and ventrally. RADIAL SPINES 12-13, 5-6 mm. long, all bristle-like, stiff, whitish, brown tip, later gray, horizontal to little ascending.

Distribution: Probably Mexico.

Description is from seedling plant obtained from German nursery.

Mammillaria bifurca Dietrich

Mamillaria bifurca Dietrich, Allg. Gartenz., 18:188, 1850.

BODY obovate, 50 mm. high, 90 mm. wide. TUBERCLES dark green, pitted, 4-sided, not angled, thick. AXILS woolly, later naked. CENTRAL SPINES 2, 6 mm. long, subulate, upper straight, lower recurved, all yellowish brown, tip dark brown to black, divergent dorsally and ventrally. RADIAL SPINES 12, 4-6 mm. long, setaceous, straight, white, horizontal. FLOWERS funnelform, 12 mm. long. *Outer perianth-segments* greenish red, whitish margins, lanceolate, tip acute. *Inner perianth-segments* whitish red, rose-red mid-line, lanceolate, tip acuminate. *Filaments* rose-red. *Anthers* yellow. *Stigma-lobes* 6, whitish to rose-red. FRUIT and SEEDS unknown.

Distribution: Mexico.

Mammillaria breviseta Ehrenberg

Mammillaria breviseta Ehrenberg, Allg. Gartenz., 17:251, 1849.
Cactus brevisetus Kuntze, Rev. Gen. Pl., 1:260, 1891.

BODY globose to cylindric, 75-100 mm. high, 50-75 mm. wide. TUBERCLES dark green, ovoid conic, base 4-sided, 4 mm. long, 5 mm. wide at base. AREOLES oval, with short white wool, then golden yellow. AXILS short wool, white bristles. CENTRAL SPINES 8-15, 6-12 mm. long, setaceous to acicular, stiff, straight, whitish yellow, tip dark red. RADIAL SPINES 20-22, 2-4 mm. long, bristle to hair-like, white.

Distribution: Mexico.

Mammillaria caespititia DeCandolle

Mammillaria cespititia DeCandolle, Mem. Mus. Hist. Nat. Paris, 17:112, 1828.
Mammillaria caespitata Don, Gen. Syt. Gard. Bot., 159, 1834.
Mammillaria nitida Scheidweiler, Allg. Gartenz., 9:42, 1841.
Cactus caespititius Kuntze, Rev. Gen. Pl., 1:260, 1891.

BODY globose, cespitose from base. TUBERCLES ovate, small, (8 mm.) AREOLES naked. AXILS naked. CENTRAL SPINES 1-2, longer, straight, rigid, yellow to white, gray in age. RADIAL SPINES 9-11, straight, yellow to white, gray in age, radiating. FLOWERS unknown.

Distribution: Mexico. (Coulter) Mineral del Monte, Hidalgo, and Oaxaca (Foerster).

Pfeiffer (Allg. Gartenz. 8:250, 1840) gives as synonym of *M. horripila.* Salm-Dyck (Cact. Hort. Dyck, Cult. 7, 1850) gives as synonym of *M. multiceps.* Labouret (Des Cactees 75, 1853) refers as synonym *M. crebrispina* v. *nitida.* Schumann (Gesamtb. Kakteen, 444, 1898) gives as synonym of *Echinocactus horripilus* (ibid 527) synonym of *M. pusilla.* Quehl (Monat. Kakteenk. 28, 1896) refers to *M. pusilla caespititia.* Britton and Rose lists under little known species. The plant was collected in Mexico by Thomas Coulter but he reported no type locality. The description is too short and indefinite to refer it to any specific species.

FIG. 294. *Mammillaria cerralboa* (Britton & Rose) Orcutt

Mammillaria cerralboa (Britton & Rose) Orcutt

Neomammillaria cerralboa Britton & Rose, The Cactaceae, 4:116, 1923.
Mamillaria cerralboa Orcutt, Cactography, 7, 1926.

BODY simple, cylindric, with apex rounded, 100-150 mm. high, 50-60 mm. wide. TUBERCLES closely set, yellowish green, obtuse conic, terete, with watery sap. AREOLES nearly round. AXILS naked. CENTRAL SPINES 1, 20 mm. long, acicular, straight, yellowish, porrect. RADIAL SPINES 10, 10 mm. long, slender acicular, straight, yellowish, ascending. FLOWERS lateral, 10 mm. long.

Distribution: Islands in Gulf of California, Baja California, Mexico.

Type locality: Cerralboa Island.

Illustration: Fig. 294 is a reproduction of Fig. 121a in Britton & Rose, The Cactaceae, 4:116, 1923 as *Neomammillaria cerralboa.*

Mammillaria cirrosa Poselger

Mamillaria cirrosa Poselger, Allg. Gartenz., 21:94, 1853.
Mamillaria cirrhosa Schumann, Gesamtb. Kakteen, 582, 1898.
Mamillaria centricirrha cirrhosa Schelle, Handb. Kakteen, 266, 1907.
Neomammillaria magnimamma (syn. of), Britton & Rose, The Cactaceae, 4:78, 1923.

BODY subglobose, 75 mm. wide, 60 mm. high. TUBERCLES closely set, bluish green, 4-sided base, pyramidal, 4-angled, keeled ventrally. AREOLES with some wool. AXILS naked. CENTRAL SPINES 2-3, 40-60 mm. long, lower often more elongate, tortuous, reddish gray, apex black. RADIAL SPINES 6, 2-4 mm. long, white. FLOWER, FRUIT and SEED unknown.

Distribution: Uncertain but probably Mexico.

Type locality: San Agostin de Palmar.

The presence of the central spines excludes this species from the *M. magnimamma* group but the incomplete description makes its definite relationship uncertain.

Mammillaria citrina Scheidweiler

Mammillaria citrina Scheidweiler in Foerster, Handb. Cact., 254, 1846, (Ref. only). 518,1886.

BODY globular, 40 mm. high and wide. TUBERCLES blue green, conic, 4-8 mm. long. AREOLES scant white wool. AXILS yellow-white wool. CENTRAL SPINES 4, 16-20 mm. long, 3 upper ones somewhat bent, all citron-yellow, spreading. RADIAL SPINES 18-20, 4-6 mm. long, yellow to citron-yellow.

Distribution: Unknown.

Britton & Rose refers to *Nomen nudem* but they failed to take into consideration the description in Foerster's second hand book of 1886.

Mammillaria conica Haworth

Mammillaria conica Haworth, Suppl. Pl. Succ., 71, 1819.
Cactus conicus Kuntze, Rev. Gen. Pl., 1:259, 1891.

TUBERCLES large, conic. SPINES less than 10, all radial, red but paler at base. FLOWERS and FRUIT unknown.

Distribution: Unknown—Index Kewensis refers to South America.

Britton & Rose suggest that it might be *M. mammillaris* or *Discocactus placentiformis?*

Mammillaria corollaria Ehrenberg

Mamillaria corollaria Ehrenberg, Allg. Gartcnz., 17:194, 1849.
Cactus corollarius Kuntze, Rev. Gen. Pl., 1:260, 1891.

BODY globular to cylindric, 75-125 mm. high, 75 mm. wide. TUBERCLES light to yellowish green, ovoid conic to long ovate or also 4-sided, 4-5 mm. long, 4-5 mm. wide at base. AREOLES white wool in youth, later naked and yellowish. AXILS short white wool and white bristles longer than tubercles. CENTRAL SPINES 5-8, 8-12 mm. long, upper and lower ones longer, all acicular, fine, straight, at base golden yellow, tip red. RADIAL SPINES 24-38, upper ones 2-3 mm. long, lower 6 mm. long, fine setaceous, yellowish to white, lateral horizontal, upper and lower ones ascending. FLOWER unknown.

Distribution: Mexico.

Mammillaria coronaria (Willdenow) Haworth

Cactus coronatus Willdenow, Enum. Pl. Hort. Berol. Suppl., 30, 1813. Not Lamark 1783.
Mammillaria coronaria Haworth, Rev. Pl. Succ., 69, 1821.

This plant was described as five feet tall and one foot wide and had one of the central spines hooked. It is doubtful if this would come within the limits of this genus because as a general condition this genus does not have sufficient internal fiberous skeletal structure to support a plant of such size.

M. cylindracea (DeCandolle) and *Cactus cylindricus* (Ortega) are often confused with this species but they are to be referred elsewhere.

Schumann redescribed a species under this name which has since been referred to *M. neocoronaria* (Knuth).

Mammillaria coronata Scheiweiler

Mammillaria coronata Scheidweiler, Allg. Gartenz. 8:338, 1840.

BODY cespitose, cylindric, 75 mm. high, 25 mm. wide. TUBERCLES green, conic to somewhat recurved. AXILS naked. CENTRAL SPINES 4 (6-8), 16-20 mm. long, rigid, straight, golden, later purplish, divergent. RADIAL SPINES 20-25, setaceous, white, radiating, interlacing. FLOWER unknown.

Distribution: Unknown.

Cactus coronatus Lamarck, (Encyclo. Method., 537, 1783) and *Cactus coronatus* Willdenow (Enumerat. Pl. Hort. Reg. Bot. Berol., 30, 1809) are not referrable here.

Salm-Dyck (Hort. Dyck., 7, 1845), gives it as synonym of *Mammillaria crebrispina* but the latter has fewer radial spines. Later Salm-Dyck (Cact. Hort. Dyck., 10, 1850), gives as synonym of *M. coronaria*. Various authors have referred it first to one then to the other of these two synonyms.

Mammillaria crebrispina DeCandolle

Mammillaria crebrispina DeCandolle, Mem. Mus. Hist. Nat. Paris, 17:111, 1828.
Cactus crebrispinus Kuntze, Rev. Gen. Pl., 1:260, 1891.

BODY cespitose from base, ovate, 50 mm. high, 35 mm. wide. TUBERCLES ovate, compressed, short. AREOLES naked. AXILS naked. CENTRAL SPINES 3, straight, brown, porrect spreading. RADIAL SPINES 16-17, straight, white, interlocking. FLOWERS and FRUIT unknown.

Distribution: San Luis Potosi, Mexico, (Foerster).
Type locality: None.
It was thought by Pfeiffer (Enumerat., 35, 1837), to be related to *M. conoidea* (Now *Neolloydia conoidea*). In. Allg. Gartenz., 247, 1840, it was given as a synonym of *M. diaphanacantha* (Neol. conoidea). Foerster gives as synonyms *M. coronata* and *M. polychlora*, without the flower and other more definite data, it will have to remain as uncertain.

Mammillaria cylindrica (DC) Don

Mammillaria cylindracea DeCandolle, Rev. Fam. Cact., 17:111, 1829.
Mammillaria cylindrica Don, Gen. Syst. Gard. Bot., 157, 1834.
Mammillaria cylindrica flavispina Labouret, Des Cactees, 88, 1853.
Cactus cylindraceus Kuntze, Rev. Gen. Pl., 1:260, 1891.

BODY simple, cylindric, 125 mm. high, 25 mm. wide. TUBERCLES deep green, ovate, AREOLES nearly naked. AXILS little setose. CENTRAL SPINES 2, 6-8 mm. long, rigid, divergent. RADIAL SPINES 25-30, 3-4 mm. long, white.

Distribution: Mexico. Type locality: None.

This species has been referred by various authors to both *M. coronaria* and *M. eriacantha* but it differs from both.

Cactus cylindricus Ortega (Nov. Rar. Plant., 128, 1797) is not referrable here as the radial spine count is less and it has milky sap.

Mammillaria divaricata Forbes

Mammillaria divaricata Forbes, Journ. Hort. Tour., 150, 1837.

BODY cylindric, forked. CENTRAL SPINES yellowish pink. RADIAL SPINES white.
This brief description cannot be associated with any known plant.

Mammillaria divaricata Dietrich

Mamillaria divaricata Dietrich, Allg. Gartenz., 16:210, 1848.
Cactus divaricatus Kuntze, Rev. Gen. Pl., 1:260, 1891.

BODY globose, with sunken apex, 50 mm. wide and high. TUBERCLES bluish green, white pitted, compressed, rounded 4-sided. AREOLES round, naked. AXILS with wool. CENTRAL SPINES 2, 6 mm. long, acicular, dark, divergent dorsally and ventrally. RADIAL SPINES 16, 5 mm. long, setaceous, stiff, white, dark brown tip, spreading. FLOWERS lateral, small, 8 mm. long, 10 mm. wide. *Outer perianth-segments* brownish red, wide lanceolate, acute. *Inner perianth-segments* carmine-red, lanceolate, tip acute, margins entire. *Filaments* below white, above dark red. *Anthers* yellow. *Style* carmine-red. *Stigma-lobes* 6, small, yellow, slender.

Distribution: Unknown.

Mammillaria divaricata Hort in Foerster

Mamillaria divaricata Hort. in Foerster, Handb. Kakteenk., 370, 1892.

BODY flattened globose, apex sunken. TUBERCLES grayish green, many sided, rounded dorsally. AREOLES and AXILS with wool. CENTRAL SPINES 2-3, 30 mm. long, lower ones bent, all silver gray, brown tip, spreading. RADIAL SPINES 3-4, much shorter, lateral. FLOWERS or FRUIT unknown.

Distribution: Unknown.

Schumann does not recognize the species but gives reference of both Dietrich and Foerster. Schelle (Handb. Kakteenk., 266, 1907), gives the Dietrich reference but lists it as synonym of *M. centricirrha*. Britton & Rose evidently followed the mistake of Schelle in confusing the Dietrich description with that of Foerster and refers it to *M. magnimamma* with reference to Dietrich's type. The Dietrich description could certainly not be confused with the *M. magnimamma* type of plant as it calls for 16 radial spines and red flowers while the Foerster description calls for 3-4 radials and 2-3 centrals but no flower description which might be associated with some of the various forms of the magnimamma type.

Mammillaria eborina Ehrenberg

Mamillaria eborina Ehrenberg, Allg. Gartenz., 17:309, 1849.
Cactus eborinus Kuntze, Rev. Gen. Pl., 1:260, 1891.

BODY columnar, 75-100 mm. high, 60 mm. wide. TUBERCLES yellowish green, ovoid conic, bluntly keeled, rounded above, 6-8 mm. long, 4-6 mm. wide at base. AXILS with wool and white bristles. CENTRAL SPINES 4-7 (mostly 4), 3 upper ones 10 mm. long, lower 14-16 mm. long, strong, straight, stiff, white, brown tip, cross formation. RADIAL SPINES 20-22, 6-10 mm. long, lower ones longer, all white. FLOWER and FRUIT unknown.

Distribution: Mexico. Type locality: None.

Mammillaria echinops Scheidweiler

Mammillaria echinops Scheidweiler, l'Hort. Belg., 95, 1838.
Neomammillaria echinops Fosberg. Bull. So. Calif. Acad. Sci., 5, 1931.

BODY simple, globose, 80 mm. wide. TUBERCLES light green, obscurely 4-sided, with milky sap. AXILS with wool and bristles. CENTRAL SPINES 4, subulate, white, apex reddish, later gray, cross formation. RADIAL SPINES 12-13, 3 upper ones shorter, all setaceous, white. FLOWERS unknown. FRUIT red, clavate, 8 mm. long.
Distribution: Mexico. Type locality: Unknown.
Illustration: Hort. Belg., 5: pl. 5 as *Mammillaria echinops*.
Britton & Rose refers to it as related to *M. polygona*. Dietrich (Syn. Plant. 92, 1843), refers it to *M. oothele*.

Mammillaria emundtsiana Hort. in Foerster

Mammillaria emundtsiana Hort. in Foerster, Handb. Cact. ed. 2, 341, 1885.

BODY globular, with sunken apex. TUBERCLES bluish green, blunt 4-angled to conic, with milky sap, 6-8 mm. long. AREOLES abundant white wool. AXILS abundant white wool, later naked. CENTRAL SPINES none. RADIAL SPINES 3-5 (6), lateral longer, to 15 mm. long, strong, silver gray, darker tip, recurved. FLOWERS and FRUIT unknown.
Distribution: Unknown.
Britton & Rose lists as known by name only. We have plants that answer this short description but they have not flowered as yet.
Mentioned as *M. edmundtsiana* in Cact. Suc. Journ., 86, 1939.

Mammillaria erectacantha Foerster

Mammillaria erectacantha Foerster, Allg. Gartenz., 15:50, 1847.
Cactus erectacanthus Kuntze, Rev. Gen. Pl., 1:60, 1891.

BODY globular, 35 mm. high and wide. TUBERCLES dark green, conic, keeled ventrally, length and width at base 4 mm. AREOLES with white wool in youth. AXILS naked. CENTRAL SPINES 1, 6-8 mm. long, rigid, brownish with darker tip, erect. RADIAL SPINES 10-11, 2-6 mm. long, lateral longer, white, brown tip, radiating. FLOWERS and FRUIT unknown.
Distribution: Unknown.

Mammillaria eschanzieri Coulter

Cactus eschanzieri Coulter, Contr. U. S. Nat. Herb., 3:104, 1894.
Neomammillaria eschausieri Britton & Rose, Contr. U. S. Nat. Herb., 23:1678, 1926.

BODY depressed globose, 30 mm. wide, simple. TUBERCLES broader at base, 6-8 mm. long. AXILS naked. CENTRAL SPINES 1, 15-20 mm. long, pubescent, slender, usually hooked, somewhat twisted upwards, color reddish. RADIAL SPINES 15-20, lateral 10-12 mm. long, upper shorter, lower weaker, all pubescent, dusky tips, lower curved. FLOWERS red (?). FRUIT reddish. SEEDS reddish, 1.2 mm., oblique obovate, subventral hilum, pitted.
Distribution: San Luis Potosi, Mexico. Type locality: None.
Coulter says that it resembles *Cactus grahamii* but the spine count, the pubescence of the spines and the habitat would place it with some of the more recently described species from San Luis Potosi but it cannot be certain due to lack of definite data on the flower.
Orcutt changed the spelling of this species and referred it to his new genus *Chilita* as *eschauzieri* to which he referred such unlike species as *M. cephalaphora, M. painteri, M. erythrosperma, M. camptotricha, M. trichacantha*.

Mammillaria estanzuelensis Berger

Mammillaria estanzuelensis Hort. in Berger, Kakteen, 287, 1929.

BODY globose to cylindric, 70-80 mm. high or more, 60 mm. wide. TUBERCLES bright green, conic, upper and lower sides somewhat compressed, with watery sap, 8 mm. long. AREOLES oblong, woolly, later naked. AXILS nearly naked. CENTRAL SPINES none. RADIAL SPINES numerous, 5-6 mm. long, inner row shorter, all acicular, glossy white, at base yellowish, horizontal, more or less pectinate. FLOWERS whitish. FRUIT faded red, clavate. SEEDS unknown.
Distribution: Durango?, Mexico. Type locality: None.

Mammillaria euchlora Ehrenberg

Mamillaria euchlora Ehrenberg, Allg. Gartenz. 16:266, 1848.
Cactus euchlorus Kuntze, Rev. Gen. Pl., 1:260, 1891.

BODY cylindric, 50-100 mm. high, 35-65 mm. wide. TUBERCLES dark green, conic, 6-10 mm.

long, 4-6 mm. wide at base. AREOLES oval, woolly in youth. AXILS with short wool. CENTRAL SPINES 4, seldom 5, 4-8 mm. long, lower 2 mm. long, acicular, lower straight or hooked, stronger than radials, all dark reddish brown, irregular, 3 upper, and 1 lower. RADIAL SPINES 16-18, 4-8 mm. long, bristle-like, brownish yellowish or white with brown tip, horizontal. FLOWERS and FRUIT unknown.

Distribution: Mexico.

Mammillaria fellneri Ehrenberg

Mamillaria fellneri Ehrenberg, Allg. Gartenz., 17:261, 1849.
Cactus fellnerii Kuntze, Rev. Gen. Pl., 1:260, 1891.

BODY columnar, with apex somewhat sunken, 125 mm. high, 60-75 mm. wide. TUBERCLES light to yellowish green, ovate to conic, robust, 6 mm. long, 5-6 mm. wide at base. AREOLES oval, with short yellowish wool in youth. AXILS yellow to white wool, numerous yellow bristles, one or more longer than tubercles to 14-26 mm. long. CENTRAL SPINES 4-6, longer than radials, upper ones 8-10 mm. long, lower ones 16-18 mm. long, all acicular, stronger than radials, straight or hooked, brown-yellow, to reddish, base lighter, irregular cross formation. RADIAL SPINES 24-26, upper ones 4 mm. long, lower ones 8 mm. long, all setaceous, fine, straight or somewhat bent, with sulphur-yellow color, transparent, horizontal. FLOWERS and FRUIT unknown.

Distribution: Mexico.

Mammillaria flavicoma Foerster

Mammillaria flavicoma Hort. in Foerster, Handb. Kakteenk., 2 ed. 1:298, 1892.

BODY globular, apex sunken. TUBERCLES bright green, conic. AREOLES somewhat felted. AXILS with white wool. CENTRAL SPINES 6, 12-15 mm. long, upper ones longer, somewhat bent, stiff, in youth yellow, base brownish. RADIAL SPINES 22-24, fine bristle-like, yellow, later white. FLOWERS in more than 1 row in top, red.

Distribution: Unknown.

"This species was described from garden plants of unknown origin" Britton & Rose 4:166, 1923.

Mammillaria geminata Scheidweiler

Mammillaria geminata Scheidweiler, Allg. Gartenz., 9:42, 1841.
Cactus geminatus Kuntze, Rev. Gen. Pl., 1:260, 1841.

BODY dichotomous branching, with apex sunken. TUBERCLES green, 4-sided pyramidal, with milky sap, 8 mm. long. AREOLES woolly in youth, later naked. AXILS woolly. CENTRAL SPINES 1, 6 mm. long, stronger, curved, black. RADIAL SPINES 6, 5 mm. long, blackish, star shaped. FLOWERS and FRUIT unknown.

Distribution: Oaxaca, Mexico. Type locality: Unknown. Altitude 5000 ft.

Illustration: Moellers Deut. Gart. Zeit., 25:475, f. 8, No. 20.

This might be associated with the plant that we have designated as *M. confusa* var. *centrispina* but inasmuch as there is no flower data it cannot be definite.

Mammillaria glabrata Salm-Dyck

Mamillaria glabrata Salm-Dyck, Cact. Hort. Dyck. Cultae. 1849, 109, 1850.
Cactus glabratus Kuntze, Rev. Gen. Pl., 1:260, 1891.

BODY hemispheric. TUBERCLES spreading, glaucous green, 4-sided base, attenuate, rounded dorsally, apex obtuse, with milky sap. AREOLES oval, often with groove ventrally, little elongated, woolly, later naked. AXILS frequently naked. CENTRAL SPINES 1-3, very short to missing, color yellow to brown, porrect. RADIAL SPINES 12-14, lower ones longer, rigid, somewhat recurved, white to pale yellowish brown, lower ones in groove. FLOWERS and FRUIT unknown.

Distribution: Mexico. Potts reports from Chihuahua, Mexico.

Salm-Dyck in previous article compares this species with *M. caput medusa* and *M. heyderi,* so we are assuming that the tubercle has milky sap.

a. var. *leucacantha* Regel and Klein

M. glabrata (Salm-Dyck) var. *leucacantha* Regel & Klein, Ind. Sem. Petro.

Centrals 2-4, 10 mm.

Mammillaria glareosa Boedeker

Mammillaria glareosa Boedeker, Mammill. Vergl. Schluss., 59, 1933.

BODY flattened globular, very small. TUBERCLES 4-angled, obtuse above, with milky sap. AXILS woolly. CENTRAL SPINES 1, heavier acicular. RADIAL SPINES 9, to 6 mm. long, 3 upper ones thinner, and whitish, lower ones yellow. FLOWERS unknown.

Distribution: Baja California?, Mexico.

Mammillaria granulata Meinshausen

Mamillaria granulata Meinshausen, Wochenschr. Garteneri, 264, 1858.
Cactus granulatus Kuntze, Rev. Gen. Pl., 1:260, 1891.

BODY globose, branching laterally, 50-60 mm. high, 40-50 mm. wide. TUBERCLES ovate-oblong, dark green, 8-12 mm. long, 4-6 mm. wide. AREOLES oval, nearly naked. AXILS with wool and 8-15 very thin white bristles. CENTRAL SPINES 6, 12-14 mm. long, straight, rigid, yellow, purplish tip. RADIAL SPINES 18-20, 12-14 mm. long, setaceous, straight, flexuous, pubescent, white. FLOWERS and FRUIT unknown.

Distribution: Mexico. Type locality: La Escondida.

There is some confusion regarding this species because Meinshausen reported that the habitat was similar to *M. pusilla* and as a result some authors have referred it to that species. Later he compared it with *M. glochidiata* which has hooked central spines while this one is reported as straight. The type locality is rather indefinite, so recollection is very uncertain.

Mammillaria grisea Salm-Dyck

Mammillaria grisea Salm-Dyck, Cact. Hort. Dyck. 1849, 110, 1850.
Cactus griseus Kuntze, Rev. Gen. Pl., 1:260, 1891.

BODY thick to columnar, 100-125 mm. high, 75 mm. wide. TUBERCLES glaucous green, compressed, 4-sided ovate, oblique truncate. AREOLES small, naked. AXILS with wool and bristles. CENTRAL SPINES 4-6, 50 mm. long or more, stronger, recurved, white, apex red, later gray, ascending. RADIAL SPINES 10-12, short, rigid, white, interlacing. FLOWERS and FRUIT unknown.

Distribution: Mexico.

"This is perhaps different from *M. grisea galeotti* Foerster (Handb. Cact. 219, 1846), which was never described." Britton & Rose refers it to little known species.

Mammillaria grusonii Runge

Mammillaria grusonii Runge, Gartenflora, 38:105, 1889.

BODY globular, later cylindric, mostly simple, to 250 mm. wide. TUBERCLES light green, 4-angled, with milky sap, 6-8 mm. long. AREOLES little wool, later naked. CENTRAL SPINES 2, 4-6 mm., stronger, straight, reddish in youth, later snow white, divergent dorsally and ventrally. RADIAL SPINES 14, upper ones 6-8 mm. long, lower ones shorter, straight, all reddish in youth, later white. FLOWERS 25 mm. long and wide, yellow. FRUIT scarlet (like *M. applanata*).

Distribution: Coahuila, Mexico. Type locality: Sierra Bola.

Britton & Rose compares it with *M. scheeri* and *Escobaria chihuahuensis* but the milky sap of this species definitely separates them.

Mammillaria haematactina Ehrenberg

Mamillaria haematactina Ehrenberg, Allg. Gartenz. 16:266, 1848.
Cactus haematactinus Kuntze, Rev. Gen. Pl., 1:260, 1891.

BODY globular to cylindric, 100 mm. high, 60 mm. wide. TUBERCLES dark green, ovoid to conic, 6 mm. long, 8 mm. wide at base. AREOLES short wool in youth. AXILS with short wool, 2-4 bristles 24 mm. long. CENTRAL SPINES 12, 6-8 mm. long, upper ones longer, acicular, stouter than radials, all blood-red in youth, later white with red tip. RADIAL SPINES 20-22, 4-6 mm. long, upper ones shorter, all setaceous, white, some with red tip.

Distribution: Mexico.

Mammillaria hamata Lehmann

Cactus cylindricus Ortega, Nov. Rar. Pl., 128, 1797. Not Lamarck, 1783.
Mammillaria hamata Lehmann in Pfeiffer, Enum. Cact., 34, 1837.
Mamillaria hamata v. *longispina* Salm-Dyck, Cact. Hort. Dyck., 10, 1850.
Mamillaria hamata v. *brevispina* Salm-Dyck, Cact. Hort. Dyck., 10, 1850.
Mammillaria hamata v. *principis* Labouret, Des Cactees, 34, 1853.
Cactus hamatus Kuntze, Rev. Gen. Pl., 1:260, 1891.
Neomammillaria hamata Britton & Rose, The Cactaceae 4:140, 1923.

BODY simple, elongata globular, sometimes branching from the base, to 600 mm. high. TUBERCLES conic, pyramidal, compressed base. AREOLES oval, woolly, later naked. AXILS nearly naked, rarely setose. CENTRAL SPINES 3-4, (sometimes listed as 4-6), strong, elongate, lower one hooked, brownish, spreading. RADIAL SPINES 15-20, unequal, lower ones longer, acicular, white with yellowish brown tip, radiating. FLOWERS small, from near top of plant. *Inner perianth-segments* scarlet-red, lanceolate, tip acute. *Filaments* white. *Anthers* yellow. *Stigma-lobes* 4, yellow. FRUIT red, slender clavate. SEEDS brown, minute.

Distribution: Mexico. Type locality: Unknown.

As far as is known this incompletely described species has not been recollected and as no type locality is given, it is therefore being referred to the group of uncertain species. Britton & Rose and followed by Boedeker and Bravo, report the species as having milky sap, but we find no previous mention of it. Schumann refers it to *M. coronaria* which has a watery sap.

The size of the plant gives a hint that it might be similar to the plants we collected in Zapilote Canyon, near Rio Balsas, Guerrero, No. 1026, which we have tentatively referred to as *M. zapilotensis*.

M. hamata	*M.* No. 1026
Sap?	Sap watery
Axils nearly naked	Axils with bristles
Radial spines 15-20	Radial spines 30
Central spines 3-4	Central spines 2-4
Flowers scarlet red	Flowers unknown
Seeds brown	Seeds brown

Mammillaria principis Monville in Labouret (Monogr. Cact., 34, 1853) is given as a synonym of *M. hamata longispina*.

Mammillaria haynii Ehrenberg

Mammillaria heinii Ehrenberg, Botanische Zeitung, 2:833, 1844.
Mamillaria digitalis Ehrenberg, Allg. Gartenz., 16:267, 1848.
Mamillaria haynii v. *viridula* Salm-Dyck, Cact. Hort. Dyck., 10, 1850.
Mamillaria haynii v. *minima* Salm-Dyck, Cact. Hort. Dyck., 10, 1850.
Cactus heinei Kuntze, Rev. Gen. Pl., 1:260, 1891.

BODY cespitose, cylindric, apex somewhat sunken, 100 mm. high, 60 mm. wide. TUBERCLES green, thick, 4-sided, blunt, above rounded and sloping, 6 mm. long, 4 mm. wide at base. AREOLES elongate, woolly. AXILS woolly. CENTRAL SPINES 2-4, 8-16 mm. long, lower one longer, somewhat stouter, stiff, lower one hooked, all reddish brown. RADIAL SPINES 18-20, 2-8 mm. long, straw-yellow, upper horizontal, lower ones ascending. FLOWERS reddish.
Distribution: Mexico.

Ehrenberg (Bot. Zeitung, 833, 1844) published this species as *M. heinei* but later (Allg. Gartenz. 402, 1844), corrected it to *M. haynii* as it was named after Herrn Hayn of Waldenburg, Schlesien.

This species has been referred to *M. umbrina* by many authors possibly as a result of an error by Salm-Dyck. In his description of the latter species he did not follow the original in that he recorded the bristles in the axils.

Mammillaria helicteres DeCandolle

Cactus helicteres Mocino & Sesse, Fl. Mex. (drawing only).
Mamillaria helicteres (1) DeCandolle, Prodr. Syst. Nat., 460, 1828.
Mammillaria convoluta (2) St. Lager, Ann. Soc. Bot. Lyon, 7:130, 1880.

BODY simple, obovate, glabrous. TUBERCLES disposed in numerous, nearly vertical, regularly spiral series, bearing straight spines at the apex. FLOWERS rose colored, a little longer than the tubercles.
Distribution: Mexico.

Mammillaria hexacantha Salm-Dyck

Mammillaria hexacantha Salm-Dyck, Hort. Dyck, 344, 1834.
Cactus hexacanthus Kuntze, Rev. Gen. Pl., 1:260, 1891.

BODY simple, depressed. TUBERCLES light green, compressed. AREOLES oval, white wool in youth. CENTRAL SPINES 6, lateral ones 8 mm. long, lower ones 18 mm. long, all strong, straight, brown. RADIAL SPINES 25-30, 4 mm. long, white, radiating. (FLOWERS reddish purple) Labouret.
Distribution: Mexico. (Type locality: Chihuahua desert) Harshberger.

Schumann refers to *M. coronaria,* but Britton & Rose say that it "has nothing to do with that plant."

Mammillaria irregularis DeCandolle

Mammillaria irregularis DeCandolle, Mem. Mus. Hist. Nat. Paris, 17:111, 1828.
Cactus irregularis Kuntze, Rev. Gen. Pl., 1:260, 1891.

BODY cespitose, ovate, base somewhat tuberous, 50 mm. high, 25 mm. wide. TUBERCLES oblong. AREOLES naked. AXILS naked. CENTRAL SPINES none. RADIAL SPINES 20-25, 4 mm. long, somewhat reflexed, white.
Distribution: Mexico.

This plant was collected by Coulter but its identity is very uncertain.

---

(1) Index Kewensis gives this reference as appearing in DeCandolle's other publication Mem. Mus. Hist. Nat. Paris, 17:31, 1828. Both were published in the same year, so it is uncertain as to which has priority.

(2) The reference of *M. convoluta* as reported by Index Kewensis is in error as it does not appear at the place cited according to Mr. Shurly.

Mammillaria joossensiana Quehl

Mamillaria joossensiana Quehl, Monat. Kakteenk., 95, 1908.

BODY simple, globose to cylindric, apex little sunken, 50 mm. high, 40 mm. wide. TUBERCLES light green, conic, occasionally angled, with watery sap, 10 mm. long. AREOLES with white wool in youth. CENTRAL SPINES 4 or more, to 15 mm. long, stouter than radials, one hooked, all transparent white, spreading. RADIAL SPINES over 20, to 10 mm. long, upper ones shorter, all setaceous, white transparent, horizontal. FLOWERS yellow. *Stigma-lobes* 6, yellow. FRUIT and SEED unknown.

Distribution: Mexico.

Without the type locality and with only a very incomplete description, it is very uncertain as to the true nature of this species.

Mammillaria jucunda Ehrenberg

Mamillaria jucunda Ehrenberg, Allg. Gartenz., 17:250, 1849.
Cactus jucundus Kuntze, Rev. Gen. Pl., 1:260, 1891.

BODY columnar, 100-150 mm. high, 75 mm. wide. TUBERCLES yellowish green, later gray green, ovoid to conic, or blunt 4-sided with blunt angles, 4-6 mm. long, 4-5 mm. wide at base. AREOLES oval, long thick white wool, later naked. AXILS with long wool and numerous bristles which overtop the tubercles. CENTRAL SPINES 8-12, 1 or 2 in the middle somewhat longer than the others, one of them sometimes hooked, 8-10 mm. long, all acicular, straight, yellowish white, to cream colored with red tip, radiating. RADIAL SPINES 20-30, 4-6 mm. long, setaceous, almost hair-like, yellowish white.

Distribution: Mexico.

Mammillaria kleinii Regel

Mammillaria kleinii Regel, Ind. Sem. Hort. Petrop., 47, 1860.
Cactus kleinii Kuntze, Rev. Gen. Pl., 1:260, 1891.

BODY simple, clavate to cylindric, 80 mm. high, 45 mm. wide. TUBERCLES separated, grayish green, base oblong ovate, apex oblique, grooved above at apex, 10 mm. long, 8 mm. wide at base. AREOLES white woolly in youth, later naked. AXILS woolly in youth, later naked. CENTRAL SPINES 1, 6 mm. long, stronger, hooked, red. RADIAL SPINES 12, nearly equal, upper ones longer, 8-12 mm. long, somewhat recurved, white, tip reddish, interlocking, horizontal.

Distribution: Mexico.

The description of this species is somewhat confusing in that it calls for a groove on the upper side of the tubercle which would take it out of *Mammillaria* and would suggest possibly *Coryphantha*, *Neobesseya* or *Escobaria* but on the other hand it calls for a central spine that is hooked. The later two genera are all straight spined. The *Coryphantha* has curved central spines but not usually hooked. We did find a single specimen of *Coryphantha recurvata* in the upper Sonora river country south of Cananea that did have a hooked central spine but this is a rather unusual condition.

Mammillaria lamprochaeta Jacobi

Mamillaria lamprochaeta Jacobi, Allg. Gartenz., 24:82, 1856.

BODY simple, cylindric, apex sunken, 23 cm. high, 8 cm. wide. TUBERCLES grayish green, white pitted, conic, compressed. AREOLES round, woolly in youth, later naked. AXILS white wool in youth, tortuous hair-like bristles. CENTRAL SPINES 4, upper 12-14 mm. long, lateral 8 mm. long, lower ones 10-12 mm. long, all recurved, subulate, reddish at base in youth, apex darker, later dirty yellow, apex brownish, later grayish, cross formation and ascending. RADIAL SPINES 20, upper ones 4 mm. long, lateral and lower ones 7 mm. long, acicular, rigid, recurved, transparent white. FLOWERS small, rose colored.

Distribution: Unknown.

Mammillaria leucocentra Berg

Mammillaria leucocentra Berg, Allg. Gartenz., 8:130, 1840.
Cactus leucocentrus Kuntze, Rev. Gen. Pl., 1:260, 1891.

BODY simple, ovate, 12 cm. wide and high. TUBERCLES closely set, green, ovate, little compressed, with milky sap. AREOLES white woolly, later naked. AXILS with white wool. CENTRAL SPINES 5-6, 8-12 mm. long, lower ones longer, subulate, straight, white, apex brown, spreading. RADIAL SPINES numerous, nearly equal, white, interlacing. FLOWERS reported as carmine red or orange-rose.

Distribution: Mexico.

Ehrenberg reports it from Zimapan, Britton & Rose reports it from Oaxaca, Schumann refers it to *M. parkinsonii*, Berg says it is near *M. supertexta*.

Mammillaria leucodictia Linke

Mammillaria leucodictia Linke, Allg. Gartenz., 16:330, 1848.
Cactus leucodictyus Kuntze, Rev. Gen. Pl., 1:260, 1891.

BODY simple, globular, slightly sunken apex, 75 mm. wide and high. TUBERCLES crowded, green, pyramidal, 8 mm. long, 4-5 mm. wide at base. AREOLES snow white wool in youth. AXILS woolly, later naked. CENTRAL SPINES 3-4, 12-14 mm. long, stouter, little recurved, base rose to horn-color with dark brown tip. RADIAL SPINES 24-28, upper ones 6 mm. long, lateral and lower 10 mm. long, all setaceous, white, rose at base, interlacing. FLOWERS unknown.
Distribution: Unknown.
Hamsley (Bio. Cent. Amer. 513, 1879) spells it *leucodictya* Walper (Anal, Bot. Syst. 2:674, 1851), refers to it as near *M. micans*.

Mammillaria livida Fennel

Mamillaria livida Fennel, Allg. Gartenz., 15:66, 1847.
Cactus lividus Kuntze, Rev. Gen. Pl., 1:260, 1891.

BODY simple, long globose, 10 cm. high, 8 cm. wide. TUBERCLES gray, crowded conic, 7 mm. long, 4 mm. wide at base. AXILS white woolly. CENTRAL SPINES 4, 6 mm. long, stiff, in youth brownish, later white with black tip, cross formation. RADIAL SPINES 24-26, 4 mm. long.
Distribution: Mexico.
M. farinosa is referred here by Fennel but was never described.

Mammillaria loricata Martius

Mammillaria heteracantha Martius, Verz. Konig. Bot. Gard. Munchen, 127, 1829, Nomen. Hort. Reg. Monac, 127, 1829. Nomen.
Mammillaria loricata Martius in Pfeiffer, Enum. Cact., 13, 1837.
Coryphantha loricata Lemaire, Cactees, 35, 1868.
Cactus loricatus Kuntze, Rev. Gen. Pl., 1, 260, 1891.

BODY simple, nearly globose, 40-50 mm. wide. TUBERCLES glaucous-green, ovate, base 4-sided, 5 mm. long. AREOLES with dense wool. AXILS woolly. CENTRAL SPINES 2, 8-10 mm. long, straight, stronger than radials, apex black, lower ones downward curved. RADIAL SPINES 12, 6-8 mm. long, rigid, yellow, horizontal. FLOWERS, FRUIT and SEEDS unknown.
Distribution: Mexico.
M. heteracantha was mentioned but not described by Martius and later listed as synonym of above species in Pfeiffer, Enum. Cact., 13, 1837.

Mammillaria macracantha DeCandolle

Mammillaria macracantha DeCandolle, Mem. Mus. Hist. Nat. Paris, 17:113, 1828.
Mammillaria centricirrha macracantha Schelle, Handb. Kakteenk., 267, 1907.
Neomammillaria macracantha Britton & Rose, The Cactaceae, 4:79, 1923.
Mammillaria macracantha v. *retrocurva* Keller, Kakteenk., 6:86, 1937.

BODY simple, flattened globular, 4-5 cm. high, 10-15 cm. wide. TUBERCLES oval to somewhat 4-sided. AXILS with some dense wool. CENTRAL SPINES 1-2, to 5 cm. long, somewhat angled, white to brownish. RADIAL SPINES none. FLOWERS uncertain.
Distribution: Mexico, Central plateau? Type locality: None.
Illustration: DeCandolle Mem. Cact. pl. 9, and Britton & Rose (Cact. 4:80), Fig. 72a but not 72.
Schumann refers this species to *M. centricrrha* (see *M. magnimamma*) but it does not appear to us to be referrable. Rumpler refers it to *M. zuccarinii* but that species has radial spines. Britton & Rose restores it to specific status but we believe that they confused it with another species as their listing of the flower characters were drawn from a plant that has a different spine arrangement.
We have not been able to obtain any plants that we feel are true representatives of this species.
Walpers (Repert. 1843) and Labouret (Monogr. Cact. 1853) misspelled this name *macrantha*.

Mammillaria microthele (Sprengel) Muhlenpfordt

Cactus microthele Sprengel, Syst. Veg. 494, 1825.
Mammillaria microthele Martius, Hort. Reg. Monac. 127, 1829, Nomen.
Mammillaria microthele Muhlenpfordt, Allg. Gartenz 16:11, 1848.
Mamillaria microthele brongniartii Salm-Dyck, Cact. Hort. Dyck. 1849, 9, 1850.
Cactus bispinus Coulter, Contr. U. S. Nat. Herb. 3:101, 1894.

BODY cespitose, globose, small. TUBERCLES grayish green, ovoid to cylindric, compressed, 7 mm. long. AXILS naked. CENTRAL SPINES 2, 2 mm. long, straight, acicular, white, divergent dorsally and ventrally. RADIAL SPINES 22-24, 3-4 mm. long, setaceous, white. FLOWERS white, reddish yellow mid-stripe. FRUIT red, clavate, 10 mm. long. SEEDS probably black, rather large.
Distribution: Mexico probably. Type locality: Unknown.

The Sprengel plant may belong elsewhere but because of the very short and indefinite description it is not certain.

Engelmann and Coulter compare this species with *Mammillaria micromeris* (*Epithelantha micromeris*) but Britton & Rose believe that it was related to *M. elegans*. The presence of the central spines excludes it from the first reference and the color of the flower excludes it from the second. "Coulter renamed *Mammillaria microthele* because of an older *Cactus microthele*. Martius used the name *M. microthele* in 1829 (Hort. Reg. Monac. 129) but without description. The names *M. brongniartii* Hortus, *M. microthele brongniartii* and *M. compacta* Hortus (not Engelmann, 1848) have been used (Salm-Dyck, Cact. Hort. Dyck. 1849, 9, 1850) but without descriptions." Britton & Rose (Cact. 4:109, 1923).

Monville ex Foerster, (Handb. Cact. ed., 1:335, 1896), refers it to *M. caracassana*.

Mammillaria mucronata Ehrenberg

Mamillaria mucronata Ehrenberg, Allg. Gartenz., 17:294, 1849.
Cactus mucronatus Kuntze, Rev. Gen. Pl., 1:260, 1891.

BODY globular to cylindric, 125-150 mm. high, 75-100 mm. wide. TUBERCLES glossy dark green, ovate conic, above truncate, 6-8 mm. long, 4-6 mm. wide at base. AREOLES oval, white, then golden yellow, later gray short wool. AXILS white wool and bristles. CENTRAL SPINES 6-9, 5-6 mm. long, acicular, stiff, red-brown. RADIAL SPINES 26-28, upper ones 2-3 mm. long, lower ones 5 mm. long, setaceous, straight, golden yellow, then white, horizontal, lower somewhat ascending. FLOWER, FRUIT and SEED unknown.

Distribution: Mexico Type locality: Unknown.

Mammillaria multiseta Ehrenberg

Mamillaria multiseta Ehrenberg, Allg. Gartenz., 17:242, 1849.
Cactus multisetus Kuntze, Rev. Gen. Pl., 1:261, 1891.

BODY globular, clavate to cylindric, simple and 2-headed, 75-125 mm. high, 50-75 mm. wide. TUBERCLES light green, base 4-sided, pyramidal, 6-7 angled, with milky sap, 6mm. long, 5 mm. wide at base. AREOLES round to somewhat 4-sided, short yellowish wool, later naked. AXILS white wool, 20-30 white bristles 12-16 mm. long. CENTRAL SPINES one. RADIAL SPINES 4-6, often 1 more centrally placed, stronger, 8-10 mm. long, darker in color, others 6-7 mm. long, all acicular, straight, white, dark brown tip. FLOWERS, FRUIT and SEED unknown.

Distribution: Mexico. Type locality: Unknown.

Mammillaria nervosa cristata

Mammillaria nervosus cristatus, Journ. Hort. Home Farm., 3:60 (?), 7, 1910.
Mammillaria nervosa cristata Britton & Rose, The Cactaceae, 4:168, 1923.

"We know this plant only from a brief description and illustration on pages 7 and 8 of the Journal cited:

" '*Mammillaria nervosus cristatus* grows in convoluted masses like a great brain mass. The growths are covered with spiny mamillae(whence the name of the genus) and are of a dull olive-brownish hue. It, too, is Mexican.' " Britton & Rose 4:168, 1923.

Mammillaria nigra Ehrenberg

Mamillaria nigra Ehrenberg, Allg. Gartenz., 17:287, 1849.
Cactus niger Kuntze, Rev. Gen. Pl. 1:261, 1891.
Mamillaria nigra euchlora Rebut, Catal. 7, 1896.

BODY hemispheric to columnar, 50-100 mm. high, 50-75 mm. wide. TUBERCLES dark green, ovate-conic, 8-12 mm. long. AREOLES oval, scant woolly felt golden yellow, later gray. AXILS little white wool. CENTRAL SPINES 4-7, acicular, upper ones straight, lower ones hooked, all dark red. RADIAL SPINES 16-18, upper 2-3 mm. long, lower to 6 mm, all setaceous, stiff, upper ones thinner, brownish, later white, tip black to dark red. FLOWER, FRUIT and SEED unknown.

Distribution: Mexico. Type locality: Unknown.

Mammillaria nuda DeCandolle

Mammillaria nuda DeCandolle, Prodr., 3:460, 1828.
Cactus nudus Kuntze, Rev. Gen. Pl., 1:261, 1891.

BODY simple cylindric, TUBERCLES without spines. FLOWERS rose.
Distribution: Mexico.

The description is based on a drawing as *Cactus nudus* by Mociño and Sessé, Pl. Mex. So *ined*. It has never been subsequently identified and very probably is not of this genus.

Mammillaria obliqua Ehrenberg

Mamillaria obliqua Ehrenberg, Allg. Gartenz., 17:250, 1849.
Cactus obliquus Kuntze, Rev. Gen. Pl., 1:261, 1891.

BODY globular to elongate, with sunken apex. TUBERCLES gray-green, ovate to conic, 4-6 mm. long, 4-6 mm. wide at base. AXILS long white wool, long bristles. CENTRAL SPINES 9-12, 2-3 upper ones 6-8 mm. long, 7-8 lower ones 10-12 mm. long, middle one 16-18 mm. long, all acicular, strong, stiff, upper ones thinner, center one stronger and hooked, reddish, tip dark brownish-red, later chalky white, spreading. RADIAL SPINES 20-22, 2-6 mm. long, upper ones shorter, all setaceous, white, brown tip. FLOWER, FRUIT and SEED unknown.

Distribution: Mexico. Type locality: Unknown.

Mammillaria olorina Ehrenberg

Mamillaria olorina Ehrenberg, Allg. Gartenz., 17:326, 1849.
Cactus olorinus Kuntze, Rev. Gen. Pl., 1:261, 1891.

BODY globular to cylindric. TUBERCLES yellowish green, ovoid conic, below 4-sided, above truncate, 6-8 mm. long, 5-6 mm. wide at base. AREOLES oval, white wool in youth. AXILS short white wool, 6-8 long white bristles. CENTRAL SPINES 4-5, 3 upper ones 12-14 mm. long, lower one 24-28 mm. long, stronger, all straight or hooked, reddish, brown tip, later white. RADIAL SPINES 24, upper ones 4-6 mm. long, lower ones 8-10 mm. long, all setaceous, white, upper and lower ascending, lateral horizontal, interlocking. FLOWER, FRUIT and SEED unknown.

Distribution: Mexico. Type locality: Unknown.

Mammillaria oothele Lemaire

Mammillaria oothele Lemaire, Cact. Gen. Nov. Sp., 37, 1839.
Mammillaria ovimamma Lemaire, Cact. Gen. Nov. Sp., 49, 1839.
Mamillaria ovimamma brevispina Salm-Dyck, Cact. Hort. Dyck., 108, 1850.
Mammillaria ovimamma oothele Labouret, Monogr. Cact., 85, 1853.
Cactus oothele Kuntze, Rev. Gen. Pl., 1:261, 1891.
Cactus ovimamma Kuntze, Rev. Gen. Pl., 1:261, 1891.

BODY globose to cylindric, with apex sunken and woolly, 100 mm. high, 85 mm. wide. TUBERCLES ovate, conic, angled below, with milky sap, 10-12 mm. long, 11 mm. wide at base. AREOLES oval, woolly in youth. AXILS dense white wool to top of tubercles in youth. CENTRAL SPINES 3-4, 1-2 mm. long, subulate, straight to curved, reddish black, tip black, porrect. RADIAL SPINES 6-7, 2-8 mm. long, upper ones shorter, lower ones 12-14 mm. long, all straight, horny, ascending. FLOWER, FRUIT and SEED unknown.

Distribution: Unknown.

M. ovimamma is usually referred to this species but it should probably have at least varietal rank as it is described as having only 1 central spine and 8-9 radial spines.

Mammillaria palmeri Jacobi

Mamillaria palmeri Jacobi, Allg. Gartenz., 24:82, 1856.

BODY simple, globose to cylindric, 120 mm. high, 60-70 mm. wide. TUBERCLES dark green, wide to cylindric conic, compressed, with watery sap. AREOLES round, 3-4 mm. long, with white wool. AXILS thick wool. CENTRAL SPINES 4, 4-6 mm. long, lower ones longer, all straight to little bent, at base amber-yellow, above reddish brown, recurved. RADIAL SPINES 24-26 (18-22), 4-7 mm. long, upper ones shorter, all straight, setaceous, transparent white, more laterally, interlocking and covering plant. FLOWER funnelform, 15-17 mm. long, 20 mm. wide. *Outer perianth-segments* bright greenish yellow, wide reddish yellow mid-stripe, scale shape, 2-4 mm. long, 2-3 mm. wide, ciliate. *Inner perianth-segments* bright greenish yellow, fine yellowish-red mid-stripe at tip, lanceolate, 8-11x3-4 mm., acute to acuminate, tip split. *Filaments* white. *Anthers* chrome-yellow. *Style* white, tip bright carmine. *Stigma-lobes* 4, bright green, 2.5 mm.

Distribution: None reported.

Jacobi received the plant from Fennel who is sometimes given credit for this species. He compares it with *M. haageana* and *M. phaeacantha*. The original description is supplemented with flower data by Gurke (Monat. Kakteen. 16:174, 1906) who also reports the fewer number of radial spines.

Mammillaria persicina Ehrenberg

Mamillaria persicina Ehrenberg, Allg. Gartenz., 17:250, 1849.
Cactus persicianus Kuntze, Rev. Gen. Pl., 1:261, 1891.

BODY globular to cylindric. TUBERCLES grayish green, robust, oval conic, above truncate, 6 mm. long, 6 mm. wide at base. AREOLES oval, woolly in youth. AXILS with wool and bristles. CENTRAL SPINES 6-16, 8-10 mm. long, one more centrally placed, lower ones 12-14 mm. long, strong acicular, stiff, lower ones hooked, bright red, spreading. RADIAL SPINES 22-26, lower ones 5-6 mm. long, upper ones 2 mm. long, setaceous, white. FLOWER, FRUIT and SEED unknown.

Distribution: Mexico. Type locality: Unknown.

Mammillaria phaeotricha Monville

Mamillaria phaeotricha Monville, Catal., 1846. Nomen.
Mamillaria phaeotricha Monville, in Labouret, Mongr. Cact., 39, 1853.
Cactus phaeotrichus Kuntze, Rev. Gen. Pl., 1:261, 1891.

BODY globose. TUBERCLES conic, somewhat compressed below, above truncate. AREOLES round, with white wool, later naked. AXILS naked. CENTRAL SPINES 4-6, somewhat recurved, white to red above, spreading porrect. RADIAL SPINES 18-20, upper ones shorter, fine, white.
Distribution: Mexico.

Mammillaria plecostigma Meinshausen

Mamillaria plecostigma Meinshausen, Wochenschr. Gartn. Pflanz., 1:27, 1858.
Mamillaria plecostigma v. *minor*, Ibid.
Mamillaria plecostigma v. *major*, Ibid.
Cactus plecostigma Kuntze, Rev. Gen. Pl., 1:261, 1891.

BODY cespitose, cylindric. TUBERCLES cylindric, apex obtuse, truncate, with watery sap. AREOLES little woolly. AXILS at first naked, then flexuous setose. CENTRAL SPINES 3-4, 2-3 upper ones straight, lower one hooked, all yellow then reddish brown. RADIAL SPINES 16-20 setaceous, white, horizontal. FLOWERS campanulate. *Outer perianth-segments* purplish brownish brick-red, yellow margins, lanceolate. *Inner perianth-segments* flesh-white, dirty brown mid-line, oblong linear, apex rounded obtuse to cuspidate. *Filaments* numerous. *Style* exserted. *Stigma-lobes* 4-5, short, ovate, lateral margins reflexed. FRUIT and SEEDS unknown.
Distribution: Mexico.

The variety minor has tubercles more slender and the flower is only half the size. The author compared it with *M. schelhasei* and *M. wildii*.

Mammillaria pleiocephala Regel & Klein

Mamillaria pleiocephala Regel & Klein, Ind. Sem. Hort. Petrop., 47, 1860.
Cactus pleiocephalus Kuntze, Rev. Gen. Pl., 1:261, 1891.

BODY cespitose. TUBERCLES grayish green, below 4-sided pyramidal, compressed, above truncate, with milky sap (?), length and width 12 mm. AREOLES somewhat oval, woolly in youth. AXILS wooly in youth. CENTRAL SPINES 1-2, rarely 3, 2-6 mm. long, reddish in youth. RADIAL SPINES 10, 2-4 mm. long, lower ones longer to 8 mm. long, reddish tip in youth.
Distribution: Mexico.

Mammillaria procera Ehrenberg

Mamillaria procera Ehrenberg, Allg. Gartenz., 17:241, 1849.
Cactus procerus Kuntze, Rev. Gen. Pl., 1:261, 1891.

BODY simple, cylindric, 100-125 mm. high, 25 mm. wide. TUBERCLES widely separated, bright green, elongated conic, apex terete, 10 mm. long. AREOLES white woolly, later naked. AXILS naked. CENTRAL SPINES 1, 8-12 mm. long, strong, straight, purplish brown, porrect. RADIAL SPINES 9, acicular, rigid, in youth reddish yellow, later white with brown tip, radiating. FLOWERS 8 mm. long, 12 mm. wide, dark purplish red, margins ciliate, stigma-lobes yellowish green.
Distribution: Mexico. Type locality: Unknown.

Mammillaria pugionacantha Foerster

Mamillaria pugionacantha Foerster, Allg. Gartenz., 15:50, 1847.
Cactus pugionacanthus Kuntze, Rev. Gen. Pl., 1:261, 1891.

BODY cylindric, 85 mm. high, 60 mm. wide. TUBERCLES grayish green, conic, 6 mm. long, 4 mm. wide at base. AREOLES oval, dense wool in youth only. AXILS white wool in youth only. CENTRAL SPINES 4, upper ones 6-10 mm. long, lower ones 25-35 mm. long, upper 3 straight, lower ones bent downward, all acicular, snow white, lower ones with brown tip. RADIAL SPINES 30-36, 2-6 mm. long, lateral longer, white, black tip, horizontal to ascending. FLOWER, FRUIT and SEED unknown.
Distribution: Mexico. Type locality: Unknown.

Mammillaria punctata Labouret

Mamillaria punctata Labouret in Foerster, Hand. Cact. ed. 2, 293, 1885.

BODY simple and cylindric. TUBERCLES grayish green, with many small white pits, conic. AREOLES and AXILS with white wool. CENTRAL SPINES 6, 6-7 mm. long, stronger, brownish yellow. RADIAL SPINES 20, 4 mm. long, bristle-like, white, horizontal.
Distribution: Unknown.

Mammillaria purpurascens Ehrenberg

Mamillaria purpurascens Ehrenberg, Allg. Gartenz., 17:260, 1849.

BODY globular, 50 mm. high, 65 mm. wide. TUBERCLES dark green, conic, ovate, below 4-sided, 4 mm. long. AREOLES oval, with short white wool in youth. AXILS nearly naked, with single short white bristle. CENTRAL SPINES 9, 7 lateral ones nearly equal to 12 mm. long, upper one more central 18-20 mm. long, lower one 24-26 mm. long, acicular, straight, dark voilet, spreading. RADIAL SPINES 26, 4-5 mm. long, setaceous, white, horizontal. FLOWER funnelform, 12 mm. long *Outer perianth-segments* 12, smudgy purple-red, linear, tip acute, margins fine ciliate. *Inner perianth-segments* bright rose-purple-red, linear, tip acute-caudate, margins entire or fine ciliate. *Filaments* purple-red, bent. *Anthers* bright yellow. *Style* above rose-red. *Stigma-lobes* 5, greenish yellow, slender. FRUIT and SEED unknown.

Distribution: Mexico. Type locality: Unknown.

The flower description was in a subsequent publication but without the knowledge of the type of sap it is uncertain as to its accurate association.

Mammillaria purpurea Ehrenberg

Mamillaria purpurea Ehrenberg, Allg. Gartenz., 17:270, 1849.
Cactus purpureus Kuntze, Rev. Gen. Pl., 1:261, 1891.

BODY cylindric, 75-125 mm. high, 50 mm. wide. TUBERCLES dark green, oval conic, above truncate, 3-4 mm. long, 4-5 mm. wide at base. AREOLES oval, long white wool. AXILS with white wool and white bristles. CENTRAL SPINES 6, upper ones 8-12, lower ones 12-16 mm. long, all fine acicular, straight and hooked, purple-red, spreading. RADIAL SPINES 20-22, 3-4 mm. long, fine hair-like, straight, transparent white, horizontal. FLOWERS, FRUIT and SEED unknown.

Distribution: Mexico. Type locality: Unknown.

Mammillaria regia Ehrenberg

Mamillaria regia Ehrenberg, Allg. Gartenz., 17:269, 1849.
Cactus regius Kuntze, Rev. Gen. Pl., 1:261, 1891.

BODY columnar, apex sunken, 100 mm. high, 60 mm. wide. TUBERCLES grayish green, ovate conic, base 4-sided, apex oblique truncate, 6 mm. long, 5 mm. wide at base. AREOLES oval, with short wool in youth. AXILS short white wool,long white bristles. CENTRAL SPINES 6-8, 8-16 mm. long, acicular, hooked and straight, blood-red, base yellowish. RADIAL SPINES 24-26, 3-6 mm. long, upper ones shorter, all setaceous, very fine nearly hair-like, white. FLOWER, FRUIT and SEED unknown.

Distribution: Mexico. Type locality: Unknown.

Mammillaria rosea Scheidweiler

Mammillaria rosea Scheidweiler, l'Hortic. Belge., 5:118, 1838.
Mammillaria rhodeocentra Lemaire, Cact. Gen. Nov. Sp., 52, 1839.
Mammillaria discolor nigricans Salm-Dyck in Walpers, Repert. Bot., 2:271, 1843.
Mammillaria rhodeocentra gracilispina Salm-Dyck, Cact. Hort. Dyck, 1849, 14, 1850.
Cactus roseus Kuntze, Rev. Gen. Pl., 1:261, 1891.
Cactus rhodeocentrus Kuntze, Rev. Gen. Pl., 1:261, 1891.

Body smiple, cylindric with sunken apex. TUBERCLES light to grayish green, obtuse, 4-sided, with milky sap, 6-8 mm. long, 6-8 mm. wide at base. AREOLES round, woolly, later naked. AXILS long wool, hairy, covering tubercles. CENTRAL SPINES 3-5, straight, stiff. RADIAL SPINES 15 (16-18), unequal, acicular to setaceous, straight, white, tip dark brown, radiating. FLOWERS (by Dietrich) tubular, 12 mm. long. *Outer perianth-segments* few, some as long as bloom, some shorter, dark brown, margins lighter, oblong, tip acute, margins ciliate. *Inner perianth-segments* 12, bright rose-red, or light purple, darker mid-stripe, lanceolate, tip acute, margins serrated. *Filaments* purplish. *Anthers* flesh to whitish. *Style* purplish. *Stigma-lobes* 5, bright rose-red.

Distribution: Unknown. Probably Mexico. (Labouret.)

Scheidweiler compares it with *M. galleotti* and *M. vandermaelen*. Deitrich compares it with *M. oothele*.

Illustration: l'Hortic. Belge, 5, pl. 7, as *Mammillaria rosea*.

Mammillaria roseocentra Boedeker & Ritter

Mammillaria roseocentra Boedeker & Ritter, Mamm. Verg. Schlus., 21, 1933.

BODY cespitose, globose. TUBERCLES short, with watery sap. AXILS naked. CENTRAL SPINES none. RADIAL SPINES 25, in 2 rows, inners shorter, white, upper ones rose to red. FLOWERS unknown. ROOT heavy tap.

Distribution: Coahuila, Mexico. Type locality: Viesca.

Mammillaria rufidula Ehrenberg

Mamillaria rufidula Ehrenberg, Allg. Gartenz., 17:295, 1849.
Cactus rufidulus Kuntze, Rev. Gen. Pl., 1:261, 1891.

BODY globular to cylindric, 75-100 mm. high, 50-75 mm. wide. TUBERCLES dark green, ovoid conic, 8-10 mm. long, 6 mm. wide at base. AREOLES oval, white woolly, later naked. AXILS with long white wool, white bristles to 12 mm. long. CENTRAL SPINES 4-6, upper ones 8-12 mm. long, lower one 18-20 mm. long thin acicular, straight, reddish to brownish. RADIAL SPINES 24-26, upper ones 5 mm. long, lower ones 8 mm. long, all setaceous, thin hair-like, yellowish, horizontal.

Distribution: Mexico.

Mammillaria saxatilis Scheer

Mamillaria saxatilis Scheer, Botany of Voyage of Hearald, 286, 1852.
Cactus saxatilis Kuntze, Rev. Gen. Pl., 1:261, 1891.

BODY small cylindric. SPINES brownish to straw-colored.

Distribution: Chihuahua, Mexico.

Two small plants were collected in crevices of rocks near Chihuahua by Mr. Potts who reports "it is as difficult to see as to extract from its site". Scheer reports that it has a "great affinity for *M. humboltii* which grows in the meadows of Guanajuato".

The incomplete description makes it nearly impossible to identify. Howard Gentry sent me a small plant from western Chihuahua that might be referred here, but it like Scheer's plant died very soon.

Mammillaria schmerwitzii Haage

Mammillaria schmerwitzii Haage in Foerster, Handb. Cact. ed. 2, 270, 1885.
Mammillaria schmerwitziana Blanc, Hints on Cacti, 74, 1894.

BODY globular, somewhat flattened, strongly sunken apex, 100 mm. wide, 90 mm. high. TUBERCLES light green, short conic. AREOLES with white wool, later naked. AXILS naked. CENTRAL SPINES upper ones longer to 15 mm. long, yellowish white, brown in youth, recurved over apex. RADIAL SPINES 16-18, 4-5 mm. long, yellow. FLOWERS June and July. *Outer perianth-segments* red, lanceolate. *Inner perianth-segments* brilliant dark red. *Filaments* crimson. *Anthers* whitish. *Stigma-lobes* 4, crimson-red.

Distribution: Mexico.

Mammillaria seemannii Scheer

Mamillaria seemannii Scheer in Seemann, Botany of Voyage of H.M.S. Herald, 288, 1852.
Cactus seemannii Kuntze, Rev. Gen. Pl., 1:261, 1891.

BODY simple, hemispheric, 100 mm. wide, 75 mm. high. TUBERCLES greenish, minutely pitted, somewhat ovoid, elongate. AXILS with white wool. CENTRAL SPINES 1, 4-5 mm. long, subulate, straight, blackish purple in youth becoming white, porrect. RADIAL SPINES 11-13, nearly equal, less than 6 mm. long.

Distribution: Sonora or Durango, Mexico.

"This plant, with *M. salm-dyckiana* et *spinauea*, come probably from Sonora or from Durango, in which latter place Dr. Seemann has seen it, and says that, together with other species, it is used for culinary purposes." Scheer, ibid.

Britton & Rose lists under little known species and thought that it might be a *Coryphantha*.

Mammillaria seidelii Terscheck

Mammillaria seidelii Terscheck, Suppl. Cact. Verz. 1 (Walpers Repert. Bot. Sys. 2:301, 1843).
Cactus seidelii Kuntze, Rev. Gen. Pl., 1:261, 1891.

BODY globose. TUBERCLES short conic, length twice width. AREOLES are round, no wool. AXILS naked. CENTRAL SPINES 4, 20 mm. long, strong, somewhat arched erect. RADIAL SPINES 9, 18 mm. long, horizontal interlocking.

Distribution: Mexico.

Labouret (Des Cactees 136, 1846) gives it as synonym of *M. scepontocentra* which has been referred to *Coryphantha pycnacantha*. Britton & Rose lists it under known by name only.

Mammillaria severini Regel & Klein

Mammillaria severini Regel & Klein, Index Sem. Hort. Bot. Imp. Petro., 46, 1860.
Cactus severinii Kuntze, Rev. Gen. Pl., 1:261, 1891.

BODY depressed hemispheric. TUBERCLES dark grayish green, compressed, angled pyramidal, many sided, 12 mm. long, 10 mm. wide at base. AREOLES deep seated, with short gray wool, later naked. AXILS naked. SPINES 4, lateral 6 mm. long, upper ones 12 mm. long, lower ones 14 mm. long, all slender, base yellow, white, tip brown, cross formation.

Distribution: Mexico.

Mammillaria sororia Meinshausen

Mamillaria sororia Meinshausen, Wochenschr. Gartn., 1:28, 1858.
Cactus sororius Kuntze, Rev. Gen. Pl., 1:261, 1891.

BODY depressed globose, simple. TUBERCLES yellowish green, large pyramidal, compressed conic, above rounded, with milky sap, keeled ventrally. AREOLES woolly, later naked. AXILS naked. CENTRAL SPINES 1, 4 mm. long, horny, apex brownish, porrect. RADIAL SPINES 6, lower ones longer, subulate, horny, apex brownish, horizontal, interlacing, less so below. FLOWERS *Outer perianth-segments* below green, tip purplish, distinct whitish margin. *Inner perianth-segments* white, greenish purplish mid-line.

In young specimens there are only 4-5 radial spines and no centrals. Tubercles broad to narrow, more or less acute angled, lighter green, similar to *M. subpolyedra* but distinct.

Distribution: Unknown. Type locality: Jaumave, Santa Barbara, Katorza.

Britton & Rose lists this under little known species. Inasmuch as the author says that it is similar to *M. subpolyedra* which I have referred to *M. polyedra* which is found in the southern states of Puebla, Guerrero and Oaxaca, it is very likely that this might be its distribution. If the complete description was available and if that is its distribution, it might be synonomous with either *M. collinsii* or *M. confusa* v. *centrispina*.

Mammillaria speciosa DeVriese

Mammillaria speciosa DeVriese, Tijdschr. Nat. Geschr., 6:52, 1839. Not Gilles 1830.
Cactus vrieseanus Kuntze, Rev. Gen. Pl. 1:261, 1891.

BODY cylindric, robust, 150 mm. high, 60 mm. wide. TUBERCLES green, small depressed conic, base 4-sided, short. AREOLES woolly in youth, later naked. AXILS with some wool. CENTRAL SPINES 5-6-8, elongate, white, tip reddish. RADIAL SPINES 22, short, setaceous, white.

Distribution: Mexico. Type locality: Unknown.

Schumann did not know it. Britton & Rose lists it as probably belonging to some species of *Coryphantha* but the description gives no evidence of such.

Mammillaria spinaurea Salm-Dyck

Mamillaria spinaurea Salm-Dyck, Allg. Gartenz., 18:394, 1850.
Cactus spinaureus Kuntze, Rev. Gen. Pl., 1:261, 1891.

BODY globose to depressed. TUBERCLES green, 4-sided ovate, below keeled, apex sloping obtuse. AREOLES oval, ventrally frequently a small groove that is absent to elongate. AXILS flocculent with white wool in older ones. CENTRAL SPINES 5-6, twice the length of radials, heavier, golden, recurved. RADIAL SPINES 12, lower ones longer, stiff, thin, golden, spreading, lower ones in groove. FLOWERS and FRUIT unknown.

Distribution: Mexico.

Salm-Dyck thought that it was near *M. glabrata*. The original material was supposed to have been sent from Chihuahu by John Potts. Scheer thought that it might have been collected in Sonora or Durango. Walpers gives the distribution as Rio del Norte, Mexico, but this is very vague.

Mammillaria suaveolens Poselger

Mammillaria suaveolens Poselger in Rebut, Catalogue, 8, 1886. Nomen. Hort. ex. Foerster, Handb. Cact. ed 2, 1:297, 1886.

BODY globular, 40 mm. high and wide. TUBERCLES very compact, dark green, ovoid, small. AREOLES woolly in youth. AXILS white woolly. CENTRAL SPINES 4, bent above, brown. RADIAL SPINES 13-15, small, setaceous, whitish, later brownish yellow.

Distribution: Mexico.

Britton & Rose credit the author of this species to Rumple in Foerster (Handb. Cact. ed. 2, 297, 1885) but both Rebut in his 1896 Catalogue and Hildmann in his 1888 catalogue give the author as Poselger in above reference. Britton & Rose lists it under little known species, and states that it was grown in Germany from Mexican seed.

Mammillaria subulifera Ehrenberg

Mamillaria subulifera Ehrenberg, Allg. Gartenz., 17:242, 1849.
Cactus subulifer Kuntze, Rev. Gen. Pl. 1:261, 1891.

BODY flattened and globular, with sunken apex, 100 mm. high, 150 mm. wide. TUBERCLES dark green, pyramidal, 4-5-sided, 10-12 mm. long, 8 mm. wide at base. AREOLES oval, very small, long white wool in youth. AXILS with thick white wool. CENTRAL SPINES 1, 4-8 mm. long, all acicular, little bent, black, reddish base, later silver gray. RADIAL SPINES none. FLOWERS July, red, lighter margins.

Distribution: Mexico. Type locality: San Toro.

It is reported found in white marble and basalt with *M. webbiana.*

Mammillaria tacubayensis (Heese) Fedde

Mammillaria stella de tacubaya Heese, Gartenflora, 214, 1905.
Mammillaria tacubyensis Fedde, Nov. Gen. Sp. Ind., 443, 1905.
Neommmillaria tacubayensis Britton & Rose, The Cactaceae, 4:164, 1923.

BODY simple, globose, with apex a little sunken. TUBERCLES arranged in 13 and 21 spirals, bright green, cylindric, sap? (watery Britton & Rose), 8 mm. long, 3-4 mm. wide at base. AREOLES oval, 3-4 mm. long, with white wool. AXILS little wool. CENTRAL SPINES 1, 5-6 mm. long, acicular, hooked, black, and porrect. RADIAL SPINES 35-40, 3-5 mm. long, setaceous, white in color, horizontal, interlocking, covering plant. FLOWERS 15 mm. long. *Outer perianth-segments* 20-25, reddish white, darker mid-stripe. *Inner perianth-segments* reddish white, rounded, margins entire. *Stigma-lobes* 6, greenish. FRUIT reddish, 20 mm. long. SEED unknown.

Distribution: Federal District of Mexico (Bravo). Type locality: Rancho de Tacubaya.

Illustration: Gartenflora 53:214, f. 33, as *Mammillaria stella de tacubaya.*

Mammillaria tecta Miquel

Mammillaria tecta Miquel, Linnaea, 12:12, 1838.
Cactus tectus Kuntze, Rev. Gen. Pl., 1:261, 1891.

BODY globose to ovate cylindric, with apex sunken. TUBERCLES green, ovate conic, compressed. AXILS with white wool. CENTRAL SPINES 2, short, nearly equal, lower ones little longer, 5 mm. long, white, tip dark red, divergent dorsally and ventrally. RADIAL SPINES 25, 5-7 mm. long, horizontal, interlocking. FLOWERS unknown. FRUIT red, glossy, dried perianth not persisting, cylindric clavate, obtuse.

Distribution: Mexico.

Mammillaria tomentosa Ehrenberg

Mamillaria tomentosa Ehrenberg, Allg. Gartenz, 17:262, 1849.
Mamillaria flava Ehrenberg, Allg. Gartenz, 17:261, 1849.
Mamillaria tomentosa v. *flava* Salm-Dyck, Cact. Hort. Dyck. 1849, 12, 1850.
Cactus tomentosus Kuntze, Rev. Gen. Pl., 1:261, 1891.

BODY columnar, 100-125 mm. high, 65-75 mm. wide. TUBERCLES yellowish green, ovate to conic, above truncate, 6 mm. long and wide at base. AREOLES oval, yellowish wool. AXILS with white wool and white bristles. CENTRAL SPINES 4-6, mostly 4, upper ones 8-10 mm. long, lower one 12-20 mm. long, all acicular, stout, sometimes hooked, yellow, reddish tip. RADIAL SPINES 20-22, upper ones 4 mm. long, lower ones 8 mm. long, all bristle-like, straight, pale yellow, horizontal. (FLOWERS rose carmine) Foerster.

Distribution: Mexico.

Mammillaria uniseta Quehl

Mammillaria uniseta Quehl, Monatsschr. Kakteenk., 14:128, 1904.

BODY simple, globular, 50 mm. wide, 70 mm. high, sunken apex. TUBERCLES dark green, 4-angled. AREOLES with white wool, later naked. AXILS naked. CENTRAL SPINES 1, 2-3 mm. long. RADIAL SPINES 5, 2-3 mm. long, gray. FLOWERS unknown.

Described from a plant in the Botanical Garden at Halle. It is of unknown origin but probably Mexico.

Mammallaria variamamma Ehrenberg

Mamillaria variamamma Ehrenberg, Allg. Gartenz., 17:242, 1849.
Cactus variamamma Kuntze, Rev. Gen. Pl., 1:261, 1891.

BODY globular to cylindric, 150 mm. high, 50-75 mm. wide. TUBERCLES grass-green, 4-5-6-7-sided, base 4-sided, sharply angled, with milky sap, 6-8 mm. long, 4-5 mm. wide at base. AREOLES oval, 4-angled, sunken, with short yellow wool, later naked. AXILS long white wool, very long, straight or tortuous white bristles. CENTRAL SPINES none. RADIAL SPINES 5-6, 6-8 mm. long, equal length or lower somewhat longer, all acicular, straight or somewhat bent, stiff, whitish, transparent, tip dark brown, spreading.

Distribution: Mexico.

Mammillaria viridula Ehrenberg

Mamillaria viridula Ehrenberg, Allg. Gartenz., 16:267, 1848.
Mamillaria viridula v. *minima* Salm-Dyck, Cact. Hort. Dyck, 1849, 10, 1850.

BODY cylindric, 50-100 mm. high, 25-50 mm. wide. TUBERCLES light green, ovate to conic to pyramidal with blunt angles, 4-6 mm. long and wide at the base. AREOLES oval, with short wool. AXILS with short wool, later naked. CENTRAL SPINES 2-3-4, acicular, stiff, reddish brown to light brown with darker tip. RADIAL SPINES 12-14, 4-8 mm. long, bristle-like, brown to yellow to white with brown tip.

Distribution: Mexico.

This species has been referred to *M. heinei* (haynii) as a variety by Salm-Dyck (Cact. Hort. Dyck., 10, 1850) but the latter calls for a hooked central spine and there is no mention of any in the original description of this species.

Britton & Rose lists it under *M. heinei*.

Mammillaria wegeneri Ehrenberg

Mammillaria wegeneri Ehrenberg, Botanische Zeitung, 1:738, 1843, and Allg. Gartenz., 11:1843.
Cactus wegeneri Kuntze, Rev. Gen. Pl., 1:261, 1891.

BODY simple, flattened globular. TUBERCLES short and wide, conic, with 4-sided base, compressed, 4-6 mm. long, 8-10 mm. wide at base. AREOLES oval with white wool. AXILS with white wool. CENTRAL SPINES 4-5-6, 10-14 mm. long, yellow. RADIAL SPINES 22-24, 10-12 mm. long, yellow to gray-white with brownish red tip. FLOWERS unknown.

Distribution: Oaxaca, Mexico.

Foerster (Handb. Cact. 190, 1846) gives it as synonym of *M. castenoides*. Haage lists (Cact. Kult. Ed. 2, 1900) a var. *cristata*.

Mammillaria zegschwitzii Terscheck

Mammillaria zegschwitzii Terscheck, Supplem. Cact. Verz., 1.
Cactus zegschwitzii Kuntze, Rev. Gen. Pl., 1:261, 1891.

BODY ovate to cylindric. TUBERCLES green opaque, 4-angled. AREOLES woolly in youth. CENTRAL SPINES 1, 8 mm. long, straight, porrect. RADIAL SPINES 8, upper ones 4 mm. long, 3 lateral 8-10 mm. long, lower one 6 mm. long, slender. FLOWER unknown.

Distribution: Unknown.

Mammillaria zepnickii Ehrenberg

Mammillaria zepnickii Ehrenberg, Botanische Zeitung, 2:834, 1844. Allg. Gartenz., 12:402, 1844.
Cactus zepnickii Kuntze, Rev. Gen. Pl. 1:261, 1891.

BODY cespitose, globular to cylindric. TUBERCLES dark green, conic, below blunt angled, above terete, 8 mm. long, 6 mm. wide at base. AXILS woolly. CENTRAL SPINES 2-4, 8-14 mm. long, upper ones longer, straight to little bent, yellow with brown tip, in youth violet. RADIAL SPINES 16-20, unequal, 4-6 mm. long, bristle-like, whitish transparent. (FLOWERS very small, carmine red. Jacobi).

Distribution: Mexico.

KNOWN BY NAME ONLY

The following names have been recorded in the literature, some only mentioned without description or at the most only a few descriptive words. Some are names used in commercial catalogues and some are names proposed but not formally or sufficiently described.

Mammillaria acutangula Monat. Kakteenk., 172, 1895. Ref. to Grus. Cat.
M. adunca Scheidweiler in Pfeiffer & Otto, Abbild. Bluhend. Cact., 19, 1839, as a synonym of *M. uncinata*, but not described.
M. alamoensis. Trade name of probably *M. sonorensis*.
M. alba Cact. Succ. Journ. (Amer.), 137, 1938. Mention as *M. alba minor*.
M. albiseta Schlumberger, Rev. Hort., IV, 5:404, 1856. Reports only flowers like *M. spinosissima*.
M. aljibensis. Name used in F. Schmoll Catalogue. Illustration and mention in Cact. Succ. Journ., 7:59, 1935.
M. amarilla Schenkel, Seed Cat., 1936.
M. ancistrohamata Cact. Journ. (Great Brit.), 17, 1934-5; 75, 1936. Mention only.
M. ancistroides Lehmann, Del. Sem. Hort. Hamb., 1832. This has been referred to *M. glochidiata* but we have not seen the description.
M. angulier Kaktusarske Listy, 25, 1929. Probably a misprint of *M. angularis*.
M. argyphaea Walton, Cactus Price List, 19, 1893. Mention only.
M. armata Monatsschr. Kakteenk., 168, 1900. Misprint of *M. armillata*.
M. asteriflora Cels in Förster, Handb. Kakteen., 254, 1846. Name only.
M. aulacantha is not to be found in DeCandolle, Mem. Mus. Hist. Nat. Paris, 17:113, 1828, as reported by Schumann. *Cactus aulacantha* Kuntze, Rev. Gen. Pl., 1:260, 1891, probably belongs here.
M. aurata DeCandolle, Prodr. Syst. Nat., 460, 1828. Ref. only Hort. Berol. Later referred to *M. rhodantha* but never described.
M. aurea Walper, Repert. Bot. Syst., 2:270, 1843. Ref. Pfeiffer. Salm-Dyck, Cact. Hort. Dyck., 11, 1850, refers to *M. rhodantha aurea*.
M. baselluta Monatsschr. Kakteenk., 176, 1909. Mention only.
M. binops Förster, Handb. Kakteenk., 254, 1846. Ref. Haage. Name only.
M. bocasiana F. Schlumberger, Revue Hort., 404, 1846. Clear yellow flower.
M. boregui Schmoll, Catalogue. Probable *Escobaria*. Mention only.
M. borhei Rebut, Catalogue, 6, 1886. Mention only.
M. brandi Blanc, Hints on Cact., 67, 1894. Spines straw. Flowers cream.
M. brandtii Walton, Cactus Price List, 19, 1898. Mention only. Haage, Cact. Kult., 130, 1900, long straw-yellow spines, bloom cream white.
M. brongniartii Salm-Dyck, Cact. Hort. Dyck., 9, 86, 1850. Variation of *M. microthele*. No description.
M. bruennowii Rebut, Catalogue, 7, 1896. Mention only.
M. bruennowiana Monatsschr. Kakteenk., 10:22. Mention only.
M. brunispina Schmoll, Catalogue. Mention only.
M. buchhalziana Monatsschr. Kakteenk., 15:146. Mention only.
M. buchiana Rebut, Catalogue, 6, 1886. Ref. Hort. Germ.
M. californica Monatsschr. Kakteenk., 6:51. Ref. to "*M. grahamii* var. *arizonica*."
M. cantera Walper, Repert. Bot. Syst., 2:273, 1843. Mention only.
M. carycina Rebut, Catalogue, 6, 1886; 7, 1896. Mention only.
M. centa Gartenwelt, 9:249, 1905. Mentioned by C. A. Purpus.
M. cerroprieto Neal, Cact. Other Succ., 84, 1935. White spines.
M. chrysantha Link & Otto, Uber. Gat. Melo. Echino., 21, 1827. Ref. Berl. Bot. Gard.
M. circumtexta Martius, Hort. Reg. Monac., 127, 1829. Name only.
M. columnaroides Loudon, Gard. Mag., 22:314, 1841. Mention only.
M. concigera Walton, Cactus Price List, 20, 1898. Mention only.
M. conopea Monatsschr. Kakteenk., 3:78; 6:58. Mention only.
M. contacta Walper, Repert. Bot. Syst., 2:271, 1843. Ref. Wendland.
M. coryphides Forbes in Förster, Handb. Kakt., 254, 1846. Ref. name only.
M. costarica Blanc, Catalogue, 16, 1893. No author, scant description.
M. crassihamata Cact. Journ. (Brit.), 18, 1936. Mention only.
M. crinigera Walper, Repert. Bot. Syst., 2:270, 1843. Ref. Salm-Dyck.
M. crinigera Otto in Foerster, Handb. Kakteen. 254, 1846. Name only.
M. cuneiflora Hitchen in Forbes, Journ. Hort. Tour., 150, 1837. Six yellow centrals, white radials, red flowers.
M. cunendstiana Schlumberger, Revue Hort., 404, 1856. Ref. to *M. clillifera* (misprint of *M. cirrhifera*) as to flower only.
M. daedalea viridis Fennel in Förster. Name only.

M. dechlora Schlumberger, Revue Hort., IV, 5:404, 1856. Small red flower.
M. degrandi Rebut, Catalogue, 7, 1896. Ref. only.
M. degrandii Walton, Cact. Price List, 22, 1898. Ref. only.
M. deleuli Rebut, Catalogue, 7, 1896. Ref. only.
M. desnoyersii Monatsschr. Kakteenk., 3:78; 6:58. Mention.
M. destorum Hildmann in Schumann, Gesamtb. Kakteen, 582, 1898. Ref. only.
M. desertorum Walton, Cactus Price List, 20, 1898. Mention only. Schelle, Handb. Kakteenk., 266, 1907 refers to *M. magnimamma*.
M. dichotoma Forbes, Journ. Hort. Tour., 150, 1837. Cylindric, yellow central spines.
M. dispina Walton, Cactus Price List, 20, 1898. Mention.
M. dolichocantha Förster, Handb. Kakteenk., 213, 1846. Not described but often referred to *M. dolichocentra* (*M. tetracantha*).
M. donkeleari Hildmann in Haage, Cact. Kult., 131, 1900. Mention only.
M. dubia Rebut, Catalogue, 7, 1896. Mention only.
M. dyckii Monatsschr. Kakteenk., 3:78; 6:58. Mention only.
M. ebenacantha F. Schmoll, Catalogue.
M. elyii Texas Cactus Co. Catalogue. Possible *Neobesseya* or *Coryphantha*.
M. enneacantha Otto In Förster, Handb. Kakt., 254, 1846. Name only.
M. eriantha Pfeiffer, Enum. Cact., 32, 1837. Ref. Hort. Syn. of *M. eriacantha* but no description.
M. erythrocarpa Cact. Succ. Journ. (Amer.), 86, 1939. Mention only.
M. eschausieri Succulenta, 15, 1925. No author, comment.
M. esshaussieri Kaktusarske Listy, 35, 1931. Mention.
M. filipendula Bailey, Cyclo. Amer. Hort., 977, 1900. Name only.
M. flavispina Walton, Cactus Price List, 21, 1898. Mention only. Neal, Cact. Other Succ., 87, 1935. No author. Dark green, few short gray-yellow radials, 2 fairly long curved yellow centrals, flowers pinkish white.
M. fulvescens Salm-Dyck, Hort. Dyck., 7, 1845. Mention only.
M. fulvolanata Haage, Cact. Kult. ed. 2, 1900. Name only. Ref. to *M. malletiana* (Cels) in Monatsschr. Kakteenk., 3: 1905. Schelle ref. to Hildmann as author as syn. of *M. malletiana*, also as var. of *M. pyrrhocephala fulvolanata*. Kaktusarske Listy gives as syn. of *M. bicolor*.
M. geminiflora Haage, Cact. Kult., ed. 2, 1900. Name only.
M. gigantothele Otto in Förster, Handb. Kakt., 183, 1846. Mention.
M. glabrescens Walton, Cactus Price List, 21, 1898. Mention only.
M. goeringii Haage, Cact. Kult. ed. 2, 1900. Name only.
M. grandis Hitchen in Forbes, Journ. Hort. Tour, Germ. 149, 1837. Handsome, subglobose, central spines 6, yellow, radials white.
M. grusonii similis Haage, Cact. Kult. ed. 2, 1900. Name only.
M. guebwilleriana Haage, Cact. Kult. ed. 2, 1900. Name only.
M. handsworthii Walton, Cactus Price List, 21, 1898. Mention only.
M. heldii Walton, Cactus Price List, 21, 1898. Mention only.
M. herrerai Fric, Ceskoslov. Zahrod. Listii. (Kakt. Sukk.), 140, 1924. Name only. Probably a missspelling of *M. herrerae*.
M. heteracentra Otto ex Förster, Handb. Kakteen., ed. 1, 254, 1846. Name only.
M. hevernickii Senke in Rebut, Catalogue, 7, 1886. Ref. var. *M. discolor?*
M. hexacentra Otto ex Förster, Handb. Kakteen., ed. 1, 183, 1846. Ref. to *Dolichothele longimamma* but not described.
M. hildemanniana Cact. Succ. Journ. (Amer.), 86, 1939. Mention.
M. hochderferi Gartenwelt, 21:249, 1905. Mentioned by C. A. Purpus.
M. humilis Donn, Hort. Cantabrigiensis, Fig. 158, 1815. Dwarf Indian Fig.
M. hystricina Loudon, Gard. Mag., 22:314, 1841. Mention only.
M. ignota Loudon, Gard. Mag., 22:314, 1841. Mention only.
M. inermis Cact. Succ. Journ. (Amer.), 9:177, 1938. Name only.
M. intricata Otto ex Förster, Handb. Kakteenk, ed. 1, 254, 1846. Ref. Otto.
M. inuncinata Lemaire, Cact. Monovil., 39, 1839. Ref. to *M. anancistra* without description.
M. karwinskii Loudon, Gar. Mag., 22:315, 1941. Mention only.
M. klenneirii Schlumberger, Rev. Hort., IV, 5:404, 1856. Rose colored flower.
M. lapaixi Rebut, Catalogue, 7, 1896. Name only.
M. leucospina Walton, Cactus Price List, 21, 1898. Mention only.
M. lewisiana unpublished name suggested by H. Gates for one of the variations of *M. brandegeei*.
M. liebneriana Schumann, Monatsschr. Kakteenk., 175, 1894. Mention only.
M. lindbergii Hildmann, Verzeichnis, 4, 1888. Mention only.
M. lophothele Manchester, Collect. Cacti, 44, 1908. Mention only.
M. lorenzii Rebut, Catalogue, 7, 1896. Ref. Berge.
M. louisae Cact. Succ. Journ. (Amer.), 86, 1939. Mention only.
M. luevedoi Cact. Succ. Journ., 12:145, 1940. Mention only. Misprint (?) of *M. quevedoi*.
M. lutescens Forbes, Journ. Hort. Tour, 150, 1837. Mention only.
M. mieckleyi Schelle, Handb. Kakteenk. 274, 1907. Orange colored flower.
M. microdasys Rebut in Haage, Cact. Kult., ed. 2, 1900. Mention only.
M. miqueliana Pfeiffer in Förster, Handb. Kakt., ed. 1, 254, 1895. Mention.
M. mitis DeCandolle, Prodr., 3:460, 1828. Ref. Miller.
M. moelleri Zeitsschr. Suk., 37:89. Mention only.
M. monothele Haage, Cact. Kult. ed. 2, 1900. Mention only.
M. morini ibid.

M. multicolor Sieber in Haage, Cact. Kult., ed. 2, 1900.
M. multidigitata Cact. Succ. Journ. (Amer.), 5, 1940. Large clusters, central spines white, tipped with color, radial spines numerous, white.
M. multiradiata Martius, Hort. Reg. Monac., 127, 1829. Name only.
M. nicholsoni Hort in Sieber, Journ. Hort. Home Farm. Ser., 3:1, 7, 1910. Mention only.
M. nidulata Neal, Cact. Other Succ. 91, 1935. Ref. Tiegel. Long straw colored spines, bristles in axils.
M. nickelsi Sieber in Haage, Cact. Kult. ed. 2, 1900. Mention only.
M. nigerrima ibid.
M. numina ibid.
M. obducta Scheidweiler in Walper, Repert. Bot. Syst. 2:273, 1843. Mention only.
M. palmeri Fennel in Förster, Handb. Kakt.
M. pfersdorfii Cact. Journ. Brit., 17, 1934. Mention of fruit only.
M. pinispina trade name of London, England, dealer.
M. polia Sieber in Haage, Cact. Kult. ed. 2, 1900. Mention only.
M. polycantha Loudon, Gard. Mag., 22:313, 1841. Mention only.
M. porphyracantha Jacobi, Allg. Gartenz., 24:81, 1856.
M. potoniensis Walton, Cact. Price List 22, 1898. Mention only.
M. preinreichiana Monat. Kakteenk., 5:14; 20:136. Mention only.
M. pyrrhacantha Pfeiffer in Förster, Handb. Kakt. ed. 2, 254, 1885. Name only.
M. pyrrhacantha pallida ibid.
M. quehlii Walton, Cactus Price List, 23, 1898. Mention only.
M. quevedoi Schmoll, Catalogue. Cact. Succ. Journ. (Amer.) 145, 1940. Mention only.
M. rebuti Morin in Rebut, Catalogue, 7, 1896. Mention only.
M. rigida Schenkel, Seed Catalogue of 1936. Name only.
M. ritteri Monatsschr. Deut. Kakteen. Ges. 6, 1930. Mention only.
M. roematactina Schlumberger, Revue Hort., 404, 1856. Small red flowers.
M. roessingii is only a name used by Rebut but without description.
M. roii Rebut, Catalogue, 7, 1896. Mention only.
M. rubra Forbes, Journ. Hort. Tour. 151, 1837. Red flowered. Syn. *M. crinita?*
M. rueshiana (*M. rushiana*) Regel, Ind. Sem. Hort. Turic., 4, 1830. Name only.
M. saillardi Rebut, Catalogue, 1896. Ref.
M. salmiana Fennel in Förster, Handb. Kakt. ed. 1, 254, 1895. Name only.
M. saluciana F. Schlumberger, Revue Hort., 404, 1856. Flower flesh color.
M. schelhasii sericata Gartenflora, 6:207. Illustration.
M. scheideana Pfeiffer, Enum. Cact., 14, 1837. Name referred to *M. magnimamma* as synonym but not described.
M. schniedeana Cact. Succ. Journ. (Amer.), 10:132, 1938. Name only.
M. schulzeana Boedeker in Berger, Kakteen, 308, 1929. Mention only.
M. semigloba Walton, Cactus Price List, 23, 1898. Mention only.
M. semilonia Ehrenberg in Haage, Cact. Kult. ed. 2, 1900. Mention only.
M. seminolia Monatsschr. Kakteenk., 47, 1897. Mention only.
M. simonis Ehrenberg in Walton, Cactus Price List, 23, 1898. Mention only.
M. speciosissima Walton, Cactus Price List, 23, 1898. Mention only.
M. speciosissima brunea Salm-Dyck in Haage, Cact. Kult., 139, 1900. Mention only.
M. sphaeroides Salm-Dyck, Cact. Hort. Dyck., 85, 1850. Related to *M. humboldtii* but not described.
M. spinisfuscis Allnutt, The Cactus, 55, 1877. Mention only.
M. spirocentra Denter, Alphabet. Cact. Pls. LaMotola, 37, 1897. Mention.
M. stephani Walper, Repert. Bot. Syst. 2:273, 1843. Ref. H. Vind.
M. suberecta Pfeiffer in Förster, Handb. Kakt., 254, 1846. Ref.
M. supertexta caespitosa Monville in Salm-Dyck, Hort. Dyck., 6, 1845. Name only.
M. supertexta tetracantha Lemaire in Salm-Dyck, Hort. Dyck., 6, 1845. Name only.
M. tellii Walton, Cactus Price List, 24, 1892. Mention only.
M. tetragona Cactus Journ. (Brit.), 77, 1936. Pink flower. Trade name.
M. tiegeliana Neale, Cact. Other Succ., 94, 1935. New white species with fairly long centrals tipped slightly with brown, Ref. Schmoll. Cact. Succ. Journ. (Amer.), 31, 1936, as registered but not published, discovered by F. Schmoll.
M. tournefortii Dinter, Alphabet. Cact. Pls. LaMortola, 37, 1897. Mention.
M. trigonia Rebut, Catalogue, 8, 1886.
M. trigoniana Dams, Monatsschr. Kakteenk., 20, 1904, as samilar to *M. rhodantha?*
M. vicina Manchester, Collect. Cact., 48, 1908. Ref. Brandegee, Baja California.
M. villa-lerdo Monatsschr. Kakteenk., 140, 149, 1908. Mention only.
M. villosa Fennel in Förster, Handb. Kakt., ed. 1, 255, 1846. Mention only.
M. virentis Salm-Dyck, Cact. Hort. Dyck., 118, 1850. Ref. Scheidweiler.
M. viridescens Monatsschr. Kakteenk., 130, 1904. Ref. to *M. centricirra.* Cact. Succ. Journ. (Amer.), 145, 1940. Mention of plants sent out by F. Schmoll but not related to first reference.
M. vivida Monatsschr. Kakteenk., 105, 1918. Undescribed variation of possibly *M. erythrosperma.*
M. werdermanniana Schmoll, Catalogue. We refer it to *M. hahniana.*
M. werdermannii Weiss, Kakteenk., 3:18, 1934. Name only.
M. witurna Allg., Gartenz., 274, 1838. Bloom white, filaments red. Ref. *M. schelhasii.* Monatsschr. Kakteenk., 71, 1917. Ref. *M. glochidiata.*
M. xanthispina Walton, Cactus Price List, 25, 1898. Mention only.
Neomammillaria minuta Bartlett in Beard, Bot. Gaz., 99:14, 1937. Name only.

The following species has been described but as yet we have not been able to obtain the original description:

Neomammillaria sinaloensis Rose, Apuntes Fl. Indig. Sinaloa Fam. Cact., 5, No. 3, 1929.

NAMES TO BE EXCLUDED FROM THE GENUS MAMMILLARIA

M. acanthostephes Lehmann, Allg. Gartenz., 3:228, 1835. (*Coryphantha pycnacantha?*)
M. acanthostephes recta Hort. in Labouret Monogr., Cact., 138, 1853. (*Coryphantha pycnacantha.*)
M. aggregata Engelmann in Emory, Mil. Reconn., 157, 1848. (*Coryphantha aggregata.*)
M. albina a misprint of *M. alpina.* (See below.)
M. aloidaea pulvilligera Monville, Cat. 1846. (*Ariocarpus?*)
M. aloides Monville, Cat. 1846. (*Ariocarpus retusus.*)
M. alpina Martius in Salm-Dyck, Cact. Hort. Dyck. 1849, 79, 1850. (*Coryphantha?*)
M. alversonii Zeissold, Monatsschr. Kakteenk., 5:70, 1895. (*Coryphantha alversonii.*)
M. ambigua G. Don, Loudon, Hort. Brit., 194, 1830. (*Echinocactus?*)
M. ancistracantha Lemaire, Cact. Gen. Nov. Sp., 36, 1839. (*Neolloydia clavata.*)
M. andreae J. A. Purpus & Boedeker, Zeitschr. Sukkulentenk., 251, 1928. (*Coryphantha andreae.*)
M. areolosa Hemsley, Biol. Centr. Amer. Bot., 1:503, 1880. (*Ariocarpus retusus.*)
M. arietina Lemaire, Cact. Aliq. Nov., 10, 1838. (*Coryphantha pycnacantha.*)
M. arietina spinosior Monville in Salm-Dyck. Hort. Dyck. 14, 1845. (*Coryphantha pycnacantha.*)
M. arizonica Engelmann, Bot. Calif. 1:124, 1876. (*Coryphantha arizonica.*)
M. aselliformis Watson? Cact. Cult., 188. 1889. Monville? 1843. (*Pelecyphora aselliformis.*)
M. asperispina Boedeker, Mamm. Vergl. Schl., 14, 1933. (*Neobesseya asperispina.*)
M. asterias Cels in Salm-Dyck, Cact. Hort. Dyck. 1849, 129, 1850. (*Coryphantha exsudans* by Britton & Rose, *Coryphantha asterias* by Boedeker.)
M. atrata Mackie, Curtis's Bot. Mag., 65:pl. 364, 1839. (*Neoporteria subgibbosa.*)
M. aulacothele Lemaire, Cact. Aliq. Nov. 8, 1838. (*Coryphantha octacantha.*)
M. aulacothele flavispina Salm-Dyck, Cact. Hort. Dyck. 1844. 13, 1845. (*Coryphantha octacantha.*)
M. aulacothele multispina Scheidweiler, Bull. Acad. Sci. Bruz., 6:92, 1839. (*Coryphantha octacantha.*)
M. aulacothele spinosior Monville in Lemaire, Cact. Gen. Nov. Sp. 93, 1839. (*Coryphantha octacantha.*)
M. aulacothele sulcimamma Pfeiffer in Walpers Bot. Repert. 2:302, 1843. (*Coryphantha octacantha.*)
M. aylostera Werdermann, Kakteenk., Sept., 1938. (*Dolichothele?*)
M. barlowii Regel & Klein, Ind. Sem. Hort. Petrop., 1860. 46, 1860. (*Dolichothele uberiformis.*)
M. beguinii Weber in Bois, Hort. Dict., 466, 1896. (*Neolloydia beguenii.*)
M. bella Hort. (*Escobaria bella* Britton & Rose The Cactaceae, 4:56, 1923.)
M. bergeriana Boedeker, Monatsschr. Deutsch. Kakt. Ges., 191, 1929. (*Coryphantha bergeriana.*)
M. besleri Link & Otto in Forster, Handb. Cact. ed., 2, 1020, 1885. (*Melocactus besleri.*)
M. biglandulosa Pfeiffer, Allg. Gartenz., 6:274, 1838. (*Coryphantha octacantha.*)
M. bisbeeana Orcutt, Hort. (*Escobaria bisbeeana.*)
M. borealis Engelmann, Cact. Mex. Bound., 68, 1859. (*Coryphantha neomexicana.*)
M. borwigii Purpus in Berger. Kakteen, 272, 1929. (*Coryphantha borgwigii.*)
M. brachyelphis Schumann in Just, Bot. Jahresb., 26:343, 1898. (*Maihuenia brachyelphis.*)
M. brevimamma Zuccarini in Pfeiffer, Enum. Cact., 34, 1837. (*Coryphantha exsudans.*)
M. brevimamma exsudans Salm-Dyck, Cact. Hort. Dyck., 1849. 19, 1850. (*Coryphantha exsudans.*)
M. brownii Toumey, Bot. Gaz., 22:253, 1896. (*Coryphantha robustispina.*)
M. bumamma Ehrenberg, Allg. Gartenz., 17:243, 1849. (*Coryphantha bumamma.*)
M. bussleri Mundt in Schumann, Monatsschr. Kakteenk. 12:47, 1902. (*Coryphantha ottonis.*)
M. caespitosa Gray, First Lessons in Botany, 96, 1857. (*Neobesseya similis, Echinocereus reichenbachii.*)
M. calcarata Engelmann, Bost. Journ. Nat. Hist., 6:195, 1850. (*Coryphantha ?*)
M. callipyga Lemaire, l'Hort. Universel 256, 1839. Ref. *M. elephantidens.* (*Coryphantha elephantidens.*)
M. calochlora Hort., Monatsschr. Kakteenk., 26:167, 1916; 27:133, 1917. (*Coryphantha calochlora* Boedeker Mamm. Vergl. Schl. 7, 1933.)
M. canescens DeCandolle, Prodr., 3:460, 1828. (*Neolloydia conoidea?*)
M. caudata Gillies in Sweet, Hort. Brit. ed., 3:285, 1839, as *Cereus.*
M. cephalophora Salm-Dyck, Cact. Hort. Dyck., 1849. 137, 1850. (*Coryphantha pycnacantha?*)
M. ceratites Quhel, Monatsschr. Kakteenk., 19:155, 1909. (*Neolloydia ceratites.*)
M. ceratocentra Berg, Allg. Gartenz., 8:130, 1840. (*Coryphantha erecta.*)
M. chaffeyi Hort. (*Escobaria chaffeyi* Britton & Rose, 4:55, 1923.)
M. chihuahensis Hort. (*Escobaria chihuahuaensis* Britton & Rose, 4:55, 1923.)
M. childsii Blanc, Illustr. Cat. 14, 1894. (*Cereus childsii, Coryphantha? Echinomastus erectocentrus.*)
M. chlorantha Engelmann in Rothrock; Rep. U. S. Geogr. Surv. 6:127, 1878. (*Coryphantha chlorantha.*)
M. clava Pfeiffer, Allg. Gartenz. 8:282, 1840. (*Coryphantha clava.*)
M. clavata Scheidweiler, Bull. Acad. Sci. Brux. 5:494, 1838. (*Neolloydia clavata.*)
M. clunifera Lemaire, Cact. Aliq. Nov. 1, 1838. (*Coryphantha elephantidens.*)
M. coccinea G. Don in Loudon, Hort. Brit. 194, 1830. (*Denmoza rhodacantha.*)
M. communis? Steudel, Nom. ed. 2, 1:245, 1840. (*Melocactus communis.*)
M. compacta Engelmann in Wislizenus, Mem. Tour. North. Mex. 105, 1848. (*Coryphantha compacta.*)

M. congesta (H. Berlin) Hort. in Forster, Handb. Cact., 183, 1846. (*Dolichothele longimamma.*)
M. conimamma Linke, Allg. Gartenz., 25:239, 1857. (*Coryphantha sulcolanata.*)
M. conoidea DeCandolle, Mem. Mus. Hist. Nat. Paris, 17:112, 1828. (*Neolloydia conoidea.*)
M. corbula Herrera, Rev. Univ. Cuzco., 8:61, 1919. (*Lobivia corbula.*)
M. cordigera Heese, Gartenflora, 59:445, 1910. (*Coryphantha ?*)
M. corioides Bosch in Sweet, Hort. Brit. ed., 3:281, 1839. (*Echinocactus ?*)
M. cornifera DeCandolle, Mem. Mus. Hist. Nat. Paris, 17:112, 1828. (*Coryphantha cornifera.*)
M. cronifera impexicoma Salm-Dyck, Cact. Hort. Dyck, 20, 1850. (*Coryphantha radians.*)
M. cornimamma Brown, Gard. Chron., III. 2:186, 1887. (*Coryphantha sulcolanata.*)
M. cornuta Hildmann in Schumann, Gesamtb. Kakteen, 496, 1898. (*Coryphantha cornuta* Berger.)
M. crebrispina DeCandolle, Mem. Mus. Hist. Paris, 17:111, 1828. (*Neolloydia conoidea?*)
M. cubensis Zuccarini in Labouret. Monogr. Cact., 59, 1853. (*Coryphantha cubensis.*)
M. curvata Pfeiffer, Enum. Cact., 15, 1837. (*Coryphantha exsudans?*)
M. dactylithele Labouret, Monogr. Cact., 146, 1853. (*Coryphantha macromeris.*)
M. daimonoceras Lemaire, Cact. Aliq. Nov., 5, 1838. (*Coryphantha radians.*)
M. dasyacantha Engelmann, Proc. Amer. Acad., 3:268, 1856. (*Escobaria dasyacantha, Coryphantha ?*)
M. decora Forster, Hamb. Gartenz., 17:159, 1861. (*Coryphantha?*)
M. delaetiana Quehl, Monatsschr. Kakteenk., 18:59, 1908. (*Coryphantha salm-dyckiana*)
M. densispina Hort. (*Coryphantha desispina* Werdermann in Boedeker, Mamm. Vergl. Schl. 9, 1933.)
M. deserti Engelmann, Bot. Calif., 2:449, 1880. (*Coryphantha deserti.*)
M. diaphancantha Lemaire, Cact. Aliq. Nov., 39, 1838. (*Neolloydia conoidea.*)
M. difficilis Quehl, Monatsschr. Kakteenk., 18:107, 1908. (*Coryphantha poselgeriana.*)
M. diguetii Weber, Bull. Mus. Hist. Nat. Paris, 10:383, 1904. (*Mamillopsis diguetii.*)
M. disciformis DeCandolle, Mem. Mus. Hist. Nat .Paris, 17:114, 1828. (*Strombocactus disciformis.*)
M. durangensis Runge in Schumann, Gesamtb. Kakteen, 478, 1898. (*Coryphantha durangensis.*)
M. echinocactoides Pfeiffer, Allg. Gartenz., 8:281, 1940. (*Neolloydia conoidea.*)
M. echinoidea Quehl, Monatsschr. Kakteenk., 23:42, 1913. (*Coryphantha echinoidea.*)
M. echinus Engelmann, Proc. Amer. Acad., 3:267, 1856. (*Coryphantha echinus.*)
M. elephantidens Lemaire, Cact. Aliq. Nov., 1, 1838. (*Coryphantha elephantidens.*)
M. elephantidens bumamma Schumann, Keys Monogr. Cact., 43, 1903. (*Coryphantha humamma.*)
M. elongata Hemsley, Biol. Centr. Amer. Bot. 1:509, 1880. Not DeCandolle, 1828. (*Ariocarpus retusus.*)
M. emskotteriana Quehl, Monatsschr. Kakteenk., 20:139, 1910. (*Neolloydia sp.*)
M. erecta Lemaire in Pfeiffer, Allg. Gartenz., 5:370, 1837. (*Coryphantha erecta.*)
M. evanescens Forster, Handb. Cact. 243, 1846. (*Coryphantha erecta?*)
M. evarescens Forster, Handb. Cact. 243, 1846. (*Coryphantha erecta?*)
M. evarescentis Lemaire, Cact. Aliq. Nov. 4, 1838. (*Coryphantha erecta?*)
M. exsudans Zuccarini in Pfeiffer, Enum. Cact. 15, 1837. (*Coryphantha exsudans.*)
M. fissurata Engelmann, Proc. Amer. Acad. 3:270, 1856. (*Ariocarpus fissuratus.*)
M. floribunda Hooker in Curtis's Bot. Mag. 56: pl. 3647, 1839. (*Neoporteria subgibbosa.*)
M. fobei Hort. (*Escobaria fobei* [*Fobea viridiflora*, Fric].)
M. furguracea Watson, Proc. Amer. Acad. 25:150, 1890. (*Ariocarpus retusus.*)
M. georgii Boedeker, Monatsschr. Kakteen-Gesell. 163, 1931. (*Coryphantha georgii.*)
M. gibbosa Salm-Dyck, Hort. Dyck. 343, 1834. (*Echinocactus exsculptus?*)
M. gielsdorfiana Werdermann, Monatsschr. Kakteenk. 215, 1929. (*Thelocactus gielsdorfianus.*)
M. gigantothele Foerster, Handb. Cact. 183, 1846. (*Dolichothele longimamma?*)
M. gladiispina Boedeker, Zeitschr. Sukkulentenk. 120, 1925-1926. (*Coryphantha gladispina.*)
M. glanduligera Otto & Dietrich, Allg. Gartenz. 16:298, 1848. (*Coryphantha exsudans.*)
M. globosa Link, Allg. Gartenz. 25:240, 1857. (*Dolichothele longimamma.*)
M. goeringii Haage, Cact. Kult. ed. 2, 1900. (*Coryphantha sulcata.*)
M. golziana Haage, Jr., Monatsschr. Kakteenk. 19:100, 1909. (*Coryphantha ottonis.*)
M. grandiflora Otto in Pfeiffer, Enum. Cact. 33, 1838. (*Neolloydia grandiflora.*)
M. greggii Safford, Ann. Rep. Smiths. Inst. 1908: 531. pl. 4, f.l. 1909. (*Epithelantha greggii.*)
M. guerkeana Boedeker, Monatsschr. Kakteenk. 24:52, 1914. (*Coryphantha guerkeana.*)
M. halei Brandegee, Proc. Calif. Acad. II. 2:161, 1889. (*Cochemiea halei.*)
M. haseloffii Ehrenberg, Allg. Gartenz. 17:303, 1849. (*Mammillopsis senilis.*)
M. hesteri Hort. (*Coryphantha hesteri* [Wright] Cact. Succ. Journ. 4:274, 1932.)
M. heteromorpha Scheer in Salm-Dyck, Cact. Hort. Dyck. 1849. 128, 1850. (*Coryphantha macromeris.*)
M. hexacentra Otto, Forster, Handb. Cact. 183, 1846. (*Dolichothele longimamma.*)
M. hirschtiana Haage, Monatsschr. Kakteenk. 6:127, 1896. (*Coryphantha vivipara.*)
M. hoffmannseggii Salm-Dyck, Hort. Dyck. 3:43, 1834. (*Neoporteria subgibbosa.*)
M. hookeri Hort. (*Coryphantha?*)
M. horripila Lemaire, Cact. Aliq. Nov., 7, 1838. (*Neolloydia horripila.*)
M. humilior Forster in Schumann's Index Gesamtb. Kakteen. 1824, 1898. (*Neolloydia clavata.*)
M. impexicoma Lemaire, Cact. Aliq. Nov., 5, 1838. (*Coryphantha radians.*)
M. inconspicua Scheidweiler, Bull. Acad. Sci. Bruz. 5:495, 1838. (*Neolloydia conoidea.*)
M. incurva Scheidweiler, Bull. Acad. Sci. Brux. 6:92, 1839. (*Coryphantha pallida?*)
M. knuthianus Boedeker, Monatsschr. Kakteenk. 139, 1930. (*Thelocactus knuthianus.*)
M. kotschoubeyoides Hort. (*Ariocarpus kotschubeyoides.*)
M. laeta Rumpler in Forster, Handb. Cact. ed. 2:247, 1885. (*Dolichothele uberiformis.*)
M. latimamma DeCandolle, Mem. Mus. Hist. Nat. Paris, 17:114, 1828. (*Coryphantha pycnacantha?*)
M. latispina Tate in Loudon, Gar. Mag. 16:26, 1840. (*Ferocactus latispina.*)

M. leei Rose. (*Escobaria leei.*)
M. lehmanni Otto in Pfeiffer, Enum. Cact. 23, 1837. (*Coryphantha octacantha.*)
M. lehmannii sulcimamma Miquel, Linnaea 12:9, 1838. (*Coryphantha clava.*)
M. leucacantha DeCandolle, Mem. Mus. Hist. Nat. Paris, 17:113, 1828. (*Coryphantha octacantha.*)
M. leucodasys Salm-Dyck in Scheer, Seemann, Bot. Herald, 286, 1856. (*Epithelantha micromeris?*)
M. lewinii Karsten, Deutsche Fl. ed. 2. 2:457, 1895. (*Lophophora williamsii.*)
M. linkei Ehrenberg in Salm-Dyck, Cact. Hort. Dyck. 1849, 8, 1850. (*Mamillopsis senilis.*)
M. lloydii Hort. (*Escobaria lloydia* Britton & Rose, The Cactaceae 4:57, 1923.)
M. longicornis Boedeker, Monatsschr. Deutsch. Kakt. Ges. 249, 1931. (*Coryphantha longicornis.*)
M. longihamata Engelmann—Coulter, Contr. U. S. Nat. Herb. 3:105, 1849. (*Cochemia poselgeri.*)
M. longimamma DeCandolle, Mem. Mus. Hist. Nat. Paris 17:113, 1828. (*Dolichothele longimamma.*)
M. longimamma congesta Hort in Forster, Handb. Cact. 183, 1846. (*Dolichothele longimamma.*)
M. longimamma gigantothele Berg in Forster, Handb. Cact. 183, 1846. (*Dolichothele longimamma.*)
M. longimamma globosa Schumann, Gesamtb. Kakteen 508, 1898. (*Dolichothele longimamma.*)
M. longimamma hexacentra Berg, Allg. Gartenz. 8:130, 1840. (*Dolichothele longimamma.*)
M. longimamma luteola Hort in Forster, Handb. Cact. ed. 2, 246, 1885. (*Dolichothele longimamma.*)
M. longimamma sphaerica K. Brandegee in Bailey, Cyl. Amer. Hort. (*Dolichothele sphaerica.*)
M. longimamma uberiformis Schumann, Gesamtb. Kakteen 508, 1898. (*Dolichothele uberiformis.*)
M. macromeris Engelmann in Wislizenus, Mem. Tour North Tex. 97, 1848. (*Coryphantha macromeris.*)
M. macromeris longispina Schelle, Handb. Kakteenk. 237, 1907. (*Coryphantha macromeris.*)
M. macromeris v. *nigrispina* Schelle, Handb. Kakteenk. 237, 1907. (*Coryphantha macromeris.*)
M. macrotheli Martius in Pfeiffer, Enum. Cact. 24, 1837. (*Coryphantha octacantha.*)
M. macrotheli biglandulosa Salm-Dyck, Cact. Hort. Dyck. 19, 1850. (*Coryphantha octacantha.*)
M. macrotheli lehmanii Salm-Dyck, Cact. Hort. Dyck. 1850. (*Coryphantha octacantha.*)
M. macrotheli nigrispina Schelle, Handb. Kakteenk. 243, 1903. (*Coryphantha octacantha.*)
M. maelenii Salm-Dyck, Hort. Dyck. 1844, 14, 1845. (*Thelocactus leucacanthus.*)
M. magnimamma Otto, (*latimamma: Coryphantha pycnacantha.*)
M. magnimamma var. *arietina* Salm-Dyck in Forester, Handb. Cact. 235, 1846. (*Coryphantha pycnacantha.*)
M. magnimamma var. *lutescens* Salm-Dyck, Cact. Hort. Dyck. 1849, 17, 121, 1850. (*Coryphantha pycnacantha.*)
M. magnimamma spinosior Lemaire, Cact. Gen. Nov. Sp. 94, 1839. (*Coryphantha pycnacantha.*)
M. mammillariaeformis Foerster, Handb. Cacteenk. 252, 1846, Syn. *M. cephalophora* (*Coryphantha pycnacantha?*)
M. martiana Pfeiffer, Linnaea, 12:140, 1838. (*Coryphantha octocantha.*)
M. melaleuca Karwinsky in Salm-Dyck, Cact. Hort. Dyck. 1849. 108, 1850. (*Dolichothele melaleuca.*)
M. microceris var. *greggii* Engelmann, Proc. Amer. Acad. 3:261, 1856. (*Epiphilantha micromeris.*)
M. micromeris Engelmann, Proc. Amer. Acad. 3:260, 1856. (*Epiphilantha micromeris.*)
M. missouriensis Sweet, Hort. Brit. 171, 1826. (*Neobesseya missouriensis.*)
M. missouriensis caespitosa S. Watson, Bibl. Index. 1:403, 1878. (*Neobissya similis.*)
M. missouriensis nutallii Schelle, Handb. Kakteen. 241, 1907. (*Neobessya missouriensis.*)
M. missouriensis robustior Engelmann, Proc. Amer. Acad. 3:265, 1856. (*Neobesseya wissmannii.*)
M. missouriensis similis Schumann, Gesamtb. Kakteen. 498, 1898. (*Neobesseya similis* Engelmann, 1898.)
M. missouriensis viridescens Briton & Brown, Illustr. R. 2:2525. (*Neobessya missouriensis.*)
M. monoclova Schumann-Hort, Gesamtb. Kakteen. 495, 1898. (*Coryphantha radians.*)
M. montana Blanc in Hints on Cacti, p. 72; Darel. Illustr. Handb. Kakteen. 96 f. 81. (*Coryphantha vivipara.*)
M. muehlbaueriana Boedeker, Mamm. Vergl. Schl. 15, 1933. (*Neobesseya muehlbaueriana.*)
M. muehlenpfordtii Hort. (*Coryphantha muehlenpfordtii,* Poselger Allg. Gartenz. 21:102, 1853.)
M. neo-mexicana Engelmann, Cact. Mex. Bound. 64, 1859. (*Coryphantha neo-mexicana.*)
M. nickelsae Brandegee, Zoe 5:31, 1900. (*Coryphantha nickelsae.*)
M. nogalensis Runge in Schumann, Gesamtb. Kakteen, 495, 1898. (*Coryphantha recurvata.*)
M. notesteinii Britton, Bull. Torr. Club. 18:367, 1891. (*Neobesseya notesteinii.*)
M. nutallii Engelmann, Pl. Fendl. 1:49, 1849. (*Neobesseya missouriensis.*)
M. nutallii borealis Engelmann, Proc. Amer. Acad. 3:264, 1856. (*Neobesseya missouriensis.*)
M. nutallii caespitosa Engelmann, Proc. Amer. Acad. 3:265, 1856. (*Neobesseya similis.*)
M. nutallii robustior Engelmann, Proc. Amer. Acad. 3:265, 1856. (*Neobesseya wissmannii.*)
M. obscura Boedeker, Monatsschr. Kakteen-Gesell. 25, 1930. (*Coryphantha obscura.*)
M. octocantha DeCandolle, Mem. Mus. Hist. Nat. Paris, 17:113, 1828. (*Coryphantha octocantha.*)
M. odorata Boedeker, Monatsschr. Kakteen-Gesell. 169, 1930. (*Coryphantha odorata.*)
M. ottonis Pfeiffer Allg. Gartenz. 6:274, 1838. (*Coryphantha ottonis.*)
M. ottonis tenuispina Pfeiffer, Hort. (*Coryphantha ottonis.*)
M. pallida Hort. (*Coryphantha pallida,* Britton & Rose 4:40, 1923.)
M. palmeri Hort. (*Coryphantha palmeri* Britton & Rose 4:39, 1923.)
M. papyracantha Engelmann, Pl. Fendl. 49, 1849. (*Toumeya papyracantha.*)
M. pectinata Engelmann, Proc. Amer. Acad. 3:266, 1856. (*Coryphantha pectinata.*)
M. pectinata cristata Hort in Forster, Handb. Cact. ed. 2, 403, 1885. (*Coryphantha pectinata.*)
M. pectinefera Weber, Dict. Hort. Bois. 804, 1898. (*Solisia pectinata.*)
M. pfeifferiana De Vriese, Tydschr. Nat. Geschr. 6:51, 1839. (*Coryphantha cornifera.*)
M. pilispina Purpus, Monatsschr. Kakteenk. 22:150, 1912. (*Neolloydia pilispina.*)
M. plaschnickii Otto in Pfeiffer, Enum. Cact. 24, 1837. (*Coryphantha octacantha.*)
M. plaschnickii straminea Salm-Dyck, Cact. Hort. Dyck. 19, 1850. (*Coryphantha octacantha.*)

M. polychlora Scheidweiler in Forster, Handb. Cact. 205, 1846. (*Neolloydia conoidea?*)
M. polymorpha Scheer in Muhlenpfordt, Allg. Gartenz. 14:373, 1846. (*Coryphantha octacantha.*)
M. pondii Greene, Pittonia, 1:268, 1889. (*Cochemia pondii.*)
M. poselgeri Hildmann, Garten-Zeitung. 4:559, 1885. (*Cochemiea poselgeri.*)
M. poselgeriana Hort. (*Coryphantha poselgeriana* Britton & Rose Cact. 4:28, 1923.)
M. potosiana Jacobi, Allg. Gartenz. 24:92, 1856. (*Coryphantha* sp., *Neolloydia clavata?*)
M. prismatica Hemsley, Biol. Centr. Amer. Bot. 1:519, 1880. (*Ariocarpus retusus.*)
M. pseudechinus Boedeker, Monatsschr. Kakteen. Gesell. 18, 1929. (*Coryphantha pseudechinus.*)
M. purpuracea Reported in Index Kew. Suppl. 1:263, 1906, for *M. furfuracea.* (*Ariocarpus retusus.*)
M. purpusii Schumann, Monatsschr. Kakteenk. 4:165, 1894. (*Pediocactus simpsonii.*)
M. pulvilligera Monville in Foerster, Handb. Cact. ed. 2:231, 1885. (*Ariocarpus retusus.*)
M. pycnacantha Martius, Nov. Act. Nat. Cur. 16:325, 1832. (*Coryphantha pycnacantha.*)
M. quadrata Don in Loudon, Hort. Brit. 194. (*Echinocactus?*)
M. radians DeCandolle, Mem. Mus. Hist. Nat. Paris 12:111, 1828. (*Coryphantha radians.*)
M. radians var. *daemonoceras* Schumann, Gesamtb. Kakteen, 496, 1898. (*Coryphantha radians.*)
M. radians var. *echinus* Schumann, Gesamtb. Kakteen, 496, 1898. (*Coryphantha echinus.*)
M. radians var. *globose* Scheidweiler, Bull. Acad. Sci. Brux. 5:494, 1838. (*Coryphantha radians.*)
M. radians var. *impexicoma* Schumann, Gesamtb. Kakteen, 495, 1898. (*Coryphantha radians.*)
M. radians var. *sulcata* Schumann, Gesamtb. Kakteen, 496, 1898. (*Coryphantha sulcata.*)
M. radicantissima Quehl, Monatsschr. Kakteenk. 22:164, 1912. (*Neolloydia clavata.*)
M. radiosa Engelmann, Bost. Journ. Nat. Hist. 6:196, 1850. (*Coryphantha vivipara.*)
M. radiosa arizonica Schumann, Gesamtb. Kakteen, 481, 1898. (*Coryphantha neo-mexicana.*)
M. radiosa alversonii Schumann, Gesamtb. Kakteen 481, 1898. (*Coryphantha deserti.*)
M. radiosa borealis Engelmann, Cact. Mex. Bound. 68, 1859. (*Coryphantha neo-mexicana.*)
M. radiosa chlorantha Schumann, Gesamtb. Kakteen, 481, 1898. (*Coryphantha chlorantha.*)
M. radiosa deserti Schumann, Gesamtb. Kakteen, 481, 1898. (*Coryphantha desertii.*)
M. radiosa neo-mexicana Engelmann, Cact. Mex. Bound. 64, 1859. (*Coryphantha neo-mexicana.*)
M. radiosa texana Engelmann, Cact. Mex. Bound. 68, 1859. (*Coryphantha neo-mexicana.*)
M. radliana Quehl, Monatsschr. Kakteenk. 2:104, 1892. (*Cochemiea poselgeri.*)
M. ramosissima Quehl, Monatsschr. Kakteenk. 18:127, 1908. (*Coryphantha* Sp.)
M. raphidacantha Lemaire, Cact. Gen. Nov. Sp. 34, 1839. (*Neolloydia clavata.*)
M. recurvata Engelmann, Trans. St. Louis Acad. 2:202, 1863. (*Coryphantha recurvata.*)
M. recurvens Rebut, Nomen. (*Coryphantha recurvata?*)
M. recurvispina Engelmann Proc. Amer. Acad. 3:266, 1856. (*Coryphantha recurvata.*)
M. recurvispina DeVriese, Tijdsehr. Geschr. 6:53, 1839. (*Coryphantha?*)
M. reduncuspina Boedeker, Kakteenk. 153, 1933. (*Coryphantha reduncuspina.*)
M. retusa Pfeiffer, Allg. Gartenz. 5:369, 1837. (*Coryphantha retusa*, Pfeiffer; *Ariocarpus retusus*, Scheidweiler.)
M. rhaphidacantha Lemaire, Cact. Gen. Nov. Sp. 34, 1839. (*Neolloydia clavata.*)
M. rhaphidacantha var. *ancistracantha* Schumann, Gesamtb. Kakteen, 506, 1898. (*Neolloydia clavata.*)
M. rhaphidacantha var. *humilior* Salm-Dyck in Forster, Handb. Cact. 244, 1846. (*Neolloydia clavata.*)
M. robustispina Schott in Engelmann, Proc. Amer. Acad. 3:265, 1856. (*Coryphantha robustispina.*)
M. roederiana Boedeker, Monatsschr. Kakteen-Gesell. 153, 1929. (*Coryphantha roederiana.*)
M. roseana Brandegee, Zoe, 2:19, 1891. (*Cochemiea poselgeri.*)
M. runyonii Hort. (*Coryphantha. Escobaria* Britton & Rose The Cactaceae 4:55, 1923.)
M. salm-dyckiana Scheer in Salm-Dyck, Cact. Hort. Dyck. 1849, 134, 1850. (*Coryphantha salm-dyckiana.*)
M. salm-dyckiana brunnea Salm-Dyck, Allg. Gartenz. 18:394, 1850. (*Coryphantha salm-dyckiana.*)
M. scepontocentra Lemaire, Cact. Gen. Nov. Sp. 43, 1839. (*Coryphantha pycnacantha.*)
M. scheerii Muhlenpfordt, Allg. Gartenz. 13:346, 1845. (*Coryphantha poselgeriana, Lemaire Neolloydia conoidea.*)
M. scheeri var. *valida* Engelmann, Proc. Amer. Acad. 3:265, 1856. (*Coryphantha muehlenpfordtii.*)
M. schlechtendalii Ehrenberg, Linnaea 14:377, 1840. (*Coryphantha clava.*)
M. schlechtendalii levior Salm-Dyck, Cact. Hort. Dyck. 127, 1850. (*Coryphantha clava.*)
M. schumannii Hildmann, Monatsschr. Kakteenk. 1:125, 1891. (*Bartschella schumannii.*)
M. schwartziana Boedeker, Mamm. Vergl. Schlus. 1933. (*Coryphantha schwartziana.*)
M. scolymoides Scheidweiler, Allg. Gartenz. 9:44, 1841. (*Coryphantha cornifera.*)
M. scolymoides longiseta Salm-Dyck, Cact. Hort. Dyck. 132, 1850. (*Coryphantha cornifera.*)
M. scholymoides nigricans Salm-Dyck, Cact. Hort. Dyck. 132, 1850. (*Coryphantha cornifera.*)
M. scolymoides raphidacantha Salm-Dyck, Cact. Hort. Dyck. 1849, 132, 1850. (*Neolloydia clavata.*)
M. senilis Loddiges in Salm-Dyck, Cact. Hort. Dyck. 1849, 82, 1850. (*Mamillopsis senilis.*)
M. senilis diguetii Weber, Bull. Mus. Hist. Nat. Paris. 10:383, 1904. (*Mamillopsis diguetii.*)
M. setispina Engelmann in K. Brandegee, Erythea 5:117, 1897. (*Cochemiea setispina.*)
M. similans Hort. (*Neobesseya similis?*)
M. similis Engelmann, Bost. Journ. Nat. Hist. 5:246, 1845. (*Neobesseya similis,* Engelmann.)
M. similis robustior Engelmann, Bost. Journ. Nat. Hist. 6:200, 1850. (*Neobesseya wissmanii.*)
M. similis caespitosa Engelmann, Bost. Journ. Nat. Hist. 6:200, 1850. (*Neobesseya similis.*)
M. simplex Torrey & Gray, F. N. Amer. 1:553, 1840. (*Neobesseya missouriensis.*)
M. simpsonii Jones, Zoe. 3:302, 1893. (*Pediocactus simpsonii.*)
M. sneedii Hort. (*Escobaria sneedii* Britton & Rose, The Cactaceae 4:56, 1923.)
M. solitaria Don in Loudon, Hort. Brit. 194. (*Echinocactus?*)
M. speciosa De Vriese. (Tijdschr. Nat. Geschr. 6:52, 1839. Not Don, 1830.)
M. speciosa Boedeker, Monatsschr. Deutsch. Kak. Ges. 23, 1930. (*Coryphantha speciosa.*)

M. speciosa Gillies, some Chilean plant.
M. sphaerica Dietrich in Poselger, Allg. Gartenz. 21:94, 1854. (*Dolichothele sphaerica.*)
M. spaethiana Schumann by Spath, Cat. 1894-1895. (*Pediocactus simpsonii.*)
M. spinosa Don is reported to have come from Chile so is not of this genus.
M. spinosissima Hort. (*Echinocactus spinosissimus,* Forbes, Journ. Hort. Tour. Ger. 152, 1837.)
M. stipitata Scheidweiler, Bull. Acad. Sci. Bruz. 5:495, 1838. (*Neolloydia clavata.*)
M. strobiliformis Scheer in Salm-Dyck, Cact. Hort. Dyck. 104, 1850. (*Escobaria tuberculosa.*)
M. strobiliformis Muhlenpfordt, Allg. Gartenz. 16:19, 1848. (*Coryphantha sulcata.*)
M. strobiliformis Engelmann in Wislizenus, Mem. Tour. North Mex. 113, 1848. (*Neolloydia conoidea.*)
M. strobiliformis pubescens Quehl, Monatsschr. Kakteenk. 17, 87, 1907. (*Escobaria tuberculosa.*)
M. strobiliformis durispina Quehl, Monatsschr. Kakteenk. 17, 87, 1907. (*Escobaria tuberculosa.*)
M. strobiliformis rufispina Quehl, Monatsschr. Kakteenk. 17, 87, 1907. (*Escobaria tuberculosa.*)
M. strobiliformis caespititia Quehl, Monatsschr. Kakteenk. 19, 173, 1909. (*Escobaria tuberculosa.*)
M. subulata Mühlenpfordt refers to *Pereskia subulata.*
M. sulcata Engelmann, Bost. Journ. Nat. Hist. 5:246, 1845. (*Coryphantha sulcata, Ariocarpus kotschoubeyanus?*)
M. sulcimamma Pfeiffer, Allg. Gartenz. 6:274, 1838. (*Coryphantha octacantha.*)
M. sulcoglandulifera Jacobi, Allg. Gartenz. 24:92, 1856. (*Neolloydia clavata.*)
M. sulcolanata Lemaire, Cact. Aliq. Nov. 2, 1838. (*Coryphantha sulcolanata.*)
M. texensis Hort. (*Neolloydia texensis* Britton & Rose, The Cactaceae 4:18, 1923.)
M. thelocamptos Lehmann in Linnaea, 13:Litt. 101, 1839. (*Coryphantha octacantha?*)
M. trigona? Index Kewensis, (Suppl. 2:16, 1904.)
M. tuberculata Hort. (*Escobaria tuberculosa?*)
M. tuberculosa Engelmann, Proc. Amer. Acad. 3:268, 1856. (*Escobaria tuberculosa.*)
M. turbinata Hooker in Curtis's Bot. Mag. 69. pl. 3984, 1843. (*Strombocactus disciformis.*)
M. uberiformis Zuccarini in Pfeiffer, Enum. Cact. 23, 1837. (*Dolichothele uberiformis.*)
M. uberiformis gracilior, Meinshausen, Wochenschr. Gartn. Pflanz. 1:26, 1858. (*Dolichothele longimamma.*)
M. uberiformis hexacentra Salm-Dyck, Cact. Hort. Dyck. 1849, 6, 1850. (*Dolichothele longimamma.*)
M. uberiformis major Hort. in Forster, Handb. Cact. ed. 2, 244, 1885. (*Dolichothele uberiformis.*)
M. uberiformis variegata Hort. in Forster, Handb. Cact. ed. 2. 244, 1885. (*Dolichothele uberiformis.*)
M. unicornis Boedeker, Zeitschr. Sukkulentenk. 205, 1928. (*Coryphantha unicornis.*)
M. urbaniana Vaupel, Monatsschr. Kakteenk. 22:65, 1912. (*Coryphantha cubensis.*)
M. utahensis Hildmann by Schumann, Gesamtb. Kakteen. 481, 1898. (*Coryphantha chlorantha.*)
M. valdezianus Moeller, Moellers. Deutscher, Gaertener Zeitung 21. (*Pelecyphora valdeziana.*)
M. valida Purpus, Monatsschr. Kakteenk. 21:97, 1911. (*Coryphantha poselgeriana.*)
M. varicolor Hort. 1932. (*Escobaria varicolor* Teigel.)
M. vaupeliana Boedeker, Zeitschr. Sukkulentenk. 206, 1928. (*Coryphantha vaupeliana.*)
M. venusta Brandegee, Zoe 5:8, 1900. (*Bartschella schumannii.*)
M. vivipara Nutall in Frasier's Cat. No. 22, 1813, as *Cactus viviparus.* (*Coryphantha vivipara.*)
M. vivipara radiosa neo-mexicana Engelmann, Proc. Amer. Acad. 3:269, 1856. (*Coryphantha neo-mexicana.*)
M. vivipara radiosa Engelmann, Proc. Amer. Acad. 3:269, 1856, also Cact. Mex. Bound. 15, 1859, as subspecies. (*Coryphantha vivipara.*)
M. vivipara vera Engelmann, Proc. Amer. Acad. 3:269, 1856. (*Coryphantha vivipara.*)
M. voghtherriana Werdermann & Boedeker, Monatsschr. Deutsch. Kak.-Ges. 32, 1932. (*Coryphantha vogtherriana.*)
M. werdermannii Boedeker, Monatsschr. Deutsch. Kak.-Ges. 155, 1929. (*Coryphantha werdermannii.*)
M. williamsii Coulter, Contr. U. S. Nat. Herb. 2:129, 1891. (*Lophophora williamsii.*)
M. winkleri Forster, Allg. Gartenz. 15:50, 1847. (*Coryphantha pycnacantha.*)
M. wissmannii Hildmann in Schumann, Gesamtb. Kakteen, 498, 1898. (*Neobesseya wissmannii.*)
M. zilziana Boedeker, Mamm. Vergl. Schl. 14, 1933. (*Neobesseya zilziana.*)
Neomammillaria
N. longimamma Fosberg, Bull, So. Calif. Acad. Sci. 30:58, 1931. (*Dolichothele longimamma.*)
N. pectinata Fosberg, Bull. South. Calif. Acad. Sci. 30:58, 1931. (*Pelecyphora pectinata.*)
N. schwarzii Rose ex Fric, Zivot. Prirode, 29, Kakt. & Succ. 69, 1925. (*Haagea schwarzii* in Symom.)
N. sphaerica Fosberg, Bull. South. Calif. Acad. Sci. 30:58, 1931. (*Dolichothele sphaerica.*)
N. uberiformis Fosberg, Bull. South. Calif. Acad. Sci. 30:58, 1931. (*Dolichothele uberiformis.*)

ASSOCIATED GENERA

A large number of widely variable species have at different times been included in the genus *Mammillaria*. Various groups of these species, that exhibit distinctive sets of characteristics, have been separated and placed in new genera. Some of these segregations have been on strong and consistent division lines while others have been made on rather weak and not too distinctive separation criteria. The monotypic genus *Phellosperma*, as erected by Britton & Rose, was based solely on the corky base of the seeds but there are several other *Mammillarias* that possess more or less this same characteristic but not to the same extent. *Phellosperma* has been herein returned to the genus *Mammillaria* as originally described by Engelmann. The genera *Dolichothele* and *Bartschella*, although quite closely related to and which superficially appear to possibly belong to the genus *Mammillaria*, have been herein treated as associated but separate genera.

DOLICHOTHELE

Schumann (Gesamtb. Kakteen, 506, 1891) first used the name *Dolichothele* for one of the sections of his "*Mamillarieae*." Under the latter section he grouped many of the genera that Britton & Rose (Cact. 4:3, 1923) later grouped under the "*Coryphanthanae*" but he still used the generic name of *Mamillaria* for all of them. Britton & Rose (Cact., 4:61) later used the name with generic rank to cover this small group of plants.

The members of this genus have been referred by some authors back to the genus *Mammillaria* but they differ from it in several respects while at the same time they shade into it through several of the species of the latter genus. The size of the flower has often been used as a differenciating factor but it is not a distinctive characteristic as there are several species of the genus *Mammillaria* that possess large blooms. The size of the tubercles in both length and width is quite a distinguishing factor especially in certain of the species. The elongated shape of the tubercle of *M. camptotricha* shows a tendency toward association with this genus but it does not possess the other characteristics of the genus. A strong distinction between the two genera appears to be in the fact that the tubercles in the genus *Dolichothele* are not consistently arranged in regular series of spiral rows that cross in both clockwise and counter clockwise directions as they are universally found in the genus *Mammillaria*.

Dolichothele sphaerica (Dietrich) Britton & Rose

Mamillaria sphaerica Dietrich in Poselger, Allg. Gartenz., 21:94, 1853.
Cactus sphaericus Kuntze, Rev. Gen. Pl., 1:261, 1891.
Mamillaria longimamma sphaerica K. Brandegee in Bailey, Cyclo. Amer. Hort., 2:975, 1900.
Dolichothele sphaerica Britton & Rose, The Cactaceae, 4:61, 1923.
Neomammillaria sphaerica Fosberg, Bull. So. Calif. Acad. Sci., 30:58, 1931.

BODY cespitose, low spreading, to 50 cm. wide. TUBERCLES semi-flabby in texture, pale green, conic-cylindric to somewhat ovate, terete, with watery sap, 12-16 mm. long, 5-6 mm. wide at base. AREOLES round, small, with scant wool in only the youngest. AXILS with scant wool. CENTRAL SPIINES 1, 3-4 mm. long, slender subulate, straight, stiff, smooth, chalky yellowish, porrect. RADIAL SPINES 12-15, 6-9 mm. long, acicular, straight, stiff, smooth, base not enlarged, white to pale yellow, darker at base, horizontal to little recurved. FLOWERS from top of plant but from axils of old tubercles, 6-7 cm. wide. *Inner perianth-segments* bright yellow, wide spreading, oblong lanceolate, tip acute to apiculate. *Anthers* yellow. *Stigma-lobes* 8, yellow, linear. FRUIT greenish white to purplish, 10-15 mm. long, short oblong, juicy, with pleasant odor. SEEDS black, flattened,

Fig. 295. *Dolichothele sphaerica*

with straight ventral face, rounded on back, with subventral hilum. ROOTS large, thick tuberous.

Distribution: Texas, U.S.; Nuevo Leon, Tamaulipas, Mexico.

Type locality: Near Corpus Christi, Texas; also reported southward to Victoria, Tam.

Illustration: Fig. 295 is reproduction of an illustration, Plate I, fig. 2, in Britton & Rose (Cact., *4*: frontispiece).

Dolichothele uberiformis (Zuccarini) Britton & Rose

Mammillaria uberiformis Zuccarini in Pfeiffer, Enum. Cact., 23, 1837.
Mamillaria uberiformis gracilior Meinshausen, Wöchenschr. Gartn. Pflanz., 1:26, 1858.
Mammillaria uberiformis major Hort. in Foerster, Handb. Cact. ed. 2, 244, 1885.
Mammillaria uberiformis variegata Hort. in Foerster, Handb. Cact. ed. 2, 244, 1885.
Mammillaria laeta Rümpler in Foerster, Handb. Cact. ed. 2, 247, 1885.
Cactus uberiformis Kuntze, Rev. Gen. Pl., 1:261, 1891.
Mamillaria longimamma uberiformis Schumann, Gesamtb. Kakteen, 508, 1898.
Dolichothele uberiformis Britton & Rose, The Cactaceae, 4:63, 1923.
Neomammillaria uberiformis Fosberg, Bull. So. Calif. Acad. Sci., 30:58, 1931.

BODY cespitose, low growing, somewhat flattened globose, 10 cm. wide, 7 cm. high. TUBERCLES widely separated, semi-flabby in texture, bright dark green, conic-cylindric, obtuse at tip, with watery sap, 25-35 mm. long, 12-16 mm. wide at base. AREOLES round, nearly naked. AXILS naked. CENTRAL SPINES none. RADIAL SPINES 4-5, rarely 3, occasionally 6, 7-16 mm. long, slender acicular, upper one sometimes slightly heavier, straight to somewhat tortuous, pubescent, not enlarged at base, horn colored, red-

FIG. 296. *Dolichothele uberiformis*

dish brown at very base forming a ring around the base, somewhat ascending in nearly cross formation. FLOWERS somewhat tubular, 35 mm. wide. *Tube* slender, white. *Outer perianth-segments* reddish. *Inner perianth-segments* in 2 series, yellow, tip acuminate, margins serrate. *Anthers* yellow. *Filaments* white, slender, tortuous. *Style* yellow. *Stigma-lobes* 5-6, recurved. FRUIT and SEEDS unknown. ROOTS large thick tuberous.

Distribution: Hidalgo, Mexico.

Type locality: Pachuca.

Illustration: Fig. 296 is from a photograph of a plant collected near Pachuca in 1941.

Dolichothele melaleuca (Karwinsky) Britton & Rose

Mamillaria melaleuca Karwinsky in Salm-Dyck, Cact. Hort. Dyck. 1849, 108, 1850.
Mammillaria longimamma melaleuca Graessner, Kakteen, 1912 & 1914. Nomen.
Dolichothele melaleuca (Hort.) Boedeker, Mammill. Vergl. Schluss., 19, 1933.

BODY cespitose. TUBERCLES semi-flabby in texture, bright green, conic to cylindric, with watery sap, 20-25 mm. long, 12-15 mm. wide at base. AREOLES small oval, naked. AXILS naked. CENTRAL SPINES none to 1, present in only about one third of areoles, 10 mm. long, straight, acicular, stiff, pubescent, chalky horn color, porrect.

Fig. 297. *Dolichothele melaleuca*

RADIAL SPINES 6-7 (9), 12-14 mm. long, acicular, straight to slight recurve, semi-flexuous, chalky horn color, horizontal. FLOWERS funnelform. *Inner perianth-segments* yellow, spatulate, tip obtuse and often split. *Anthers* yellow. *Filaments* pale yellow. *Stigma-lobes* 5, pale greenish yellow.

Distribution: Oaxaca, Mexico.

Type locality: None reported.

Illustration: Fig. 297 is from a photograph of a plant obtained from Sr. F. Schmoll of Cadereyta, Qro.

Mammillaria longimamma pseudo-melaleuca Haage & Schmidt, Catalogue, 1922. Nomen. May be referrable here.

Mamillaria centricirrha flaviflora Hort. Cat. was referred to *M. melaleuca* by Schumann and to *Dolichothele longimamma* by Britton & Rose may belong here.

This species represents an intermediate form between *D. longimamma* and *D. uberformis* because the central spines are present only in part of the areoles.

FIG. 298. *Dolichothele longimamma*

Dolichothele longimamma (DeCandolle) Britton & Rose

Mammillaria longimamma DeCandolle, Mem. Mus. Hist. Nat. Paris, 17:113, 1828.
Mammillaria longimamma hexacantha Berg, Allg. Gartenz., 8:130, 1840.
Mammillaria longimamma congesta Hort. in Foerster, Handb. Cact., 183, 1846.
Mammillaria longimamma luteola Hort. in Foerster, Handb. Cact., ed. 2, 246, 1885.
Cactus longimammus Kuntze, Rev. Gen. Pl., 1:260, 1891.
Dolichothele longimamma Britton & Rose, The Cactaceae, 4:62, 1923.
Neomammillaria longimamma Fosberg, Bull. So. Calif. Acad. Sci., 30:58, 1931.

BODY simple and cespitose from base. TUBERCLES widely separated, semi-flabby in texture, ovate-oblong, somewhat smooth, with watery sap, 3-5 cm. long, 1-1.5 cm. wide at base. AREOLES round to oval, slightly woolly in youth. AXILS woolly. CENTRAL SPINES usually 1, occasionally to 3, to 25 mm. long, acicular, straight, pubescent, flexuous, only slightly enlarged at base, light brown with black tip, porrect. RADIAL SPINES 9-10, 5-20 mm. long, acicular, straight, pubescent, somewhat flexuous, white to pale yellow, darker at base becoming dark grayish, horizontal. FLOWERS funnelform, 5-6 cm. wide, 4-6 cm. long. *Outer perianth-segments* greenish yellow, brownish yellow, brownish back stripe, lanceolate, tip acute. *Inner perianth-segments* bright yellow, serrated (?) at tip. *Anthers* yellow. *Stigma-lobes* 5-8, yellowish green. FRUIT yellowish (?), oblong. SEEDS dark brown, finely pitted.

Distribution: Hidalgo, Mexico.

Type locality: None stated but reported from Metztitlan.

Illustration: Fig. 298 is a reproduction of an illustration in Berger (Kakteen 285, 1929) as *Mamillaria longimamma*.

The following names have been used in the literature, mostly without description, but may be referrable here:

Mammillaria longimamma compacta (Rümpler) Schelle, Handb. Kakteenk., 1907.
Mammillaria longimamma congesta Krook, Handb. Cact., 41, 1855.
Mammillaria longimamma laeta Schelle, Handb. Kakteenk., 1907.
Mammillaria longimamma ludwigii Graessner, Kakteen, 1912 and 1914.

Mammillaria longimamma major Schelle, Handb. Kakteenk., 1907.
Mammillaria longimamma malaena Schelle, Handb. Kakteenk., 1907.
Mammillaria longimamma spinosior Link, Catalogue.

a. var. gigantothele Berg

Mammillaria longimamma gigantothele Berg. in Foerster, Handb. Cact., 183, 1846.
Mammillaria gigantothele Foerster, Handb. Cact., 183, 1846. Nomen.

Tubercles to 5 cm. long. Central spines 1, 25 mm. long. Radial spines 9, 20 mm. long.

b. var. globosa (Link) Schumann

Mamillaria globosa Link, Allg. Gartenz., 25:240, 1857.
Mamillaria longimamma globosa Schumann, Gesamtb. Kakteen, 508, 1898.

Body globose. Central spines 1-2, rarely 3. Radial spines 11-13.

The original description of *M. longimamma* states only the number of spines and does not differenciate as to whether they are all radial spines or whether some of them are central spines. Subsequent authors have all considered this species as having both types so we are doing likewise.

BARTSCHELLA

Hildmann originally described this plant as a *Mammillaria* without any information on the flower, fruit, seed, nor any reference as to its distribution. K. Brandegee later described as *M. venusta* a more typical form of the species and added the missing data not reported by Hildmann. Britton & Rose erected the monotype genus of *Bartschella* and pointed out several distinctions. Marshall & Bock (Cact., 175, 1941) have returned this species to the genus *Mammillaria* but the distinctions between the two genera are quite definite. The circumscissile characteristic of the fruit is not found in any of the *Mammillariae*. The arrangement of the tubercles into incomplete ridges appears to lean toward the mamillated *Echinocacti* but the position of the flowers excludes it from that genus.

Bartschella schumannii (Hildmann) Britton & Rose

Mamillaria schumanni Hildmann, Monatsschr. Kakteenk., 1:125, 1891.
Mamillaria venusta K. Brandegee, Zoe, 5:8, 1900.
Bartschella schumannii Britton & Rose, The Cactaceae, 4:58, 1923.

BODY cespitose from base, to as many as 40 heads, individuals 6 cm. high and wide. TUBERCLES incompletely arranged with adjacent ones in ridges (as in some species of *Echinocactanae*), grayish green to grayish violet, rounded, quadrangular at base, with blunt sunken apex, with watery sap, short and thick. AREOLES round, with white wool in youth, later naked. AXILS with short wool, later naked. CENTRAL SPINES 1-2 (occasionally to 3-4) 10-15 mm. long, stout acicular, lower longer and usually strongly hooked, all white at base, black above, lower porrect, upper erect. RADIAL SPINES 12 (9-15), 6-12 mm. long, stout acicular, straight, pubescent, white with black tip, somewhat ascending. FLOWERS campanulate, with very short tube, from near the top of the plant but not in the apex, 4 cm. wide, September. *Outer perianth-segments* lanceolate, tip acute to acuminate. *Inner perianth-segments* (10) rose-colored, oblanceolate, tip acuminate, recurved spreading. *Anthers* yellow. *Style* pale, slender. *Stigma-lobes* 5-6, green to rosy brown. FRUIT scarlet, 15-20 mm. long, nearly dry, linear, circumscissile some distance from base leaving most of the seeds in an axillary cup. SEEDS dull black,

FIG. 299. *Bartschella schumannii*

less than 1 mm. long, basal part on one half as long and nearly as wide as the upper portion, minutely pitted, pits much obscured by delicate intervening striae, with large triangular basal and slightly depressed hilum.

Distribution: Baja California, Mexico.

Type locality: None stated but reported from San Jose del Cabo (Cape District).

Illustration: Fig. 299 is a reproduction of an illustration, Fig. 55, in Britton & Rose (Cact., 4:58) as *Bartschella schumannii*.

Schumann (Gesamtb. Kakteen, 545, 1898) report the height of the plant as 10 cm. but this very probably was from hothouse cultivated plants.

Mamillaria schumanniana (Monatsschr. Kakteenk., *12*:178, 1903) is very probably a misprint.

Britton & Rose (Cact. 4:3) in the "Key to Genera" under *Coryphanthae* give the position of the flowers as being central but the illustration shows them as being lateral but near the top of the plant.

The color of the stigma-lobes is sometimes changeable as in *M. blossfeldiana* in that they change color after the flower has been open for a day or so.

APPENDIX

Additional plants have been recently received that presented some additional information that we were not able to include in the main text which had already gone to press, so we are presenting it here.

FIG. 300. Species No. 1095

Species No. 1095

Body globular. Tubercles widely separated in 8 and 13 spirals, blunt ovoid, with watery sap, 5 mm. long, 9 mm. wide at base. Axils naked. Central spines 2-4, 12 mm. long, acicular, straight, pale lemon yellow, spreading. Radial spines 18-20, 12-15 mm. long, acicular, semi-flexuous, pale lemon yellow. Flower (from dried perianth only, outer perianth somewhat reddish). Fruit unknown. Seeds very dark brown, glossy, pitted, 1.3x0.8 mm. curved globular pyriform with nearly basal hilum.

This species has some of the characteristics of M. *densispina* but the central spine count is less. It might be related to M. *vetula* (which we known only from description) but the radial spine count is less but the reddish nature of the flower suggests a possible association. The radial spine count suggests a possible association with M. *esperanzaensis* but inasmuch as the flower data is incomplete the true relationship is uncertain.

Illustration: Fig. 300 is from a photograph of a plant received from Sr. F. Schmoll of Cadereyta de Montes, Qro., in December, 1944.

Type locality: Near Hacienda Ocotilla in Sierra de San Moran, Queretaro.

Fig. 301. *Mammillaria echinaria* variation No. 1092

Mammillaria echinaria variation No. 1092

This plant has most of the characteristics of *M. echinaria* but the growth is so much more robust that it gives the appearance of *M. densispina* or probably *M. vetula* but as the number of radial spines is only 18-19, it eliminates these species. The central spines are sometimes missing or at the most 1 which might suggest *M. elongata* but the spiral arrangement of the tubercles and the general structure of the plant certainly does not justify that association. No data are available on the flower characteristics so its association is uncertain but on the strength of the information available it will be considered as a possible variation of *M. echinaria*. This may have been the plant on which was based *M. densa* (Link & Otto) or *M. echinata densa* (Pfeiffer).

Illustration: Fig. 301 is from a photograph of a plant received from Sr. F. Schmoll of Cadereyta de Montes, Qro., in December, 1944.

Mammillaria ritteriana

A variation of this species from Parras, Coahuila, differs a little from the type in that the tubercles are arranged in 21 and 34 spirals and the central spines are mostly 2 or occasionally 3-4.

Mammillaria viperina

A variation of this species shows a straight central spine, which is 2 mm. long, subulate, but only very rarely present. The spine coloration differs from the type in that it is bright reddish brown.

Mammillaria mystax

Further study of collected material of this species discloses that the tubercles are more sharply angled than reported in the "Key to the Species." Plants that have been grown under cultivation with more vigorous growth sometimes do not display the angulation of the tubercles because of a more swollen condition.

Mammillaria eichlamii

Plants received from Guatemala showed a variation in the central spines in that some were very slender acicular of 3-4 mm. long to others that were heavy acicular to 25 mm. long. All of them had 1 to occasionally 2 central spines.

Mammillaria elongata

Mammillaria subcrocea anguinea and *M. subcrocea rutila* (Walpers, Repert. Bot., 2:272, 1843) are only names without descriptions.

Mammillaria tenuis arrecta, M. tenuis coerulescens and *M. tenuis derubescens* are only garden names in the Berlin Botanical Garden and listed by Foerster (Handb. Cact., 240, 1846).

Mamillaria subcrocea rufescens Salm-Dyck (Cact. Hort. Dyck. 1849, 100, 1850) should probably be referred to *M. elongata*.

FIG. 302. *Mammillaria pachyrhiza*

This illustration is a reproduction of an illustration accompanying the original description.

Fig. 303.

Mammillaria zuccariniana

This illustration is from a photograph of a plant obtained from Mrs. Bakkers of the Knickerbocker Nursery of San Diego, California.

Rumpler uses the name *Mammillaria zuccarinii* but it was probably intended to be *M. zuccariniana*.

Fig. 304

Mammillaria uncinata var. *biuncinata*

This illustration is from a photograph of a plant received from Sr. F. Schmoll of Cadereyta de Montes, Qro., in December, 1944.

Mammillaria obvallata

Mammillaria obvallata Otto, Dietrich, Allg. Gartenz. 14:308. 1846.
Cactus obvallatus Kuntze, Rev. Gen. Pl. 1:261. 1891.

BODY ovate, with sunken apex. TUBERCLES blue green, conic, with unknown sap, 10 mm. long. AREOLES with white wool later becoming naked. AXILS with wool. CENTRAL SPINES 4, 25 mm. long, lower longest to 35 mm. long, and heavier, all stronger and curved, deep yellow in youth, brown in age, in cross formation. RADIAL SPINES 16, fine bristle-like, white. FLOWERS 10 mm. long. *Outer perianth-segments* 8-10, brownish red, lanceolate, tip acute. *Inner perianth-segments* 15, purplish red, lanceolate, tip acuminate, margins entire. *Anthers* yellow. *Filaments* purplish red. *Style* purplish red. *Stigma lobes* 3-4, bright purplish red, short and heavy. FRUIT and SEEDS unknown.

Distribution: Mexico.

Mammillaria magnimamma

The following garden names were listed by Schumann (Gesamtb. Kakteen 582, 1898) as belonging to this species:

boucheana	jorderi	obconella
destorum	lehmannii	posteriana
detampico	longispina	spinosior
grandicornis	monstii	tetracantha
grandidens	moritziana	viridis
hystrix	nordmannii	zooderi

ERRATA AND OMISSIONS

Page 23. The reference should read *Cactus subpolyedrus* instead of *Cactus polyedrus*.

Page 35. *Mammillaria centricirrha polygona* and *M. centricirrha pulchra* Schelle (Handb. Kakteenk., 267, 1907) are omitted from the synonymy.

Page 47. *Cactus stenocephalus* Kuntze (Rev. Gen. Pl., 1:261, 1891) is omitted from the synonymy.

Page 98. Caption under the illustration should read *Mammillaria johnstonii sancarlensis*.

Page 141. *Mammillaria fragilis centrispina* (Hort) is the form which occasionally displays a central spine on the older clusters.

Page 210. The reference should read *Mammillaria bocasana glochidiata* instead of *M. bocasana flavispina*.

Page 222. The reference should read *Mammillaria phitauiana* instead of *Nammillaria phautauiana*.

Page 228. The reference should read *Mammillaria dolichocentra* instead of *Mammillaria dolichochentra*.

Page 233. *Cactus picta* Kuntze (Rev. Gen. Pl. 1:261, 1891) is omitted from the synonymy.

Page 253. *Cactus subechinus* Kuntze (Rev. Gen. Pl., 1:261, 1891) is omitted from the synonymy.

Page 259. The reference should read *Mammillaria esperanzaensis* instead of *Mammillaria esperanzaenis*.

On pages 260 and 261, the references should read *Mammillaria pulchella* instead of *Mammillaria puchella*.

Page 278. The reference should read *Mammillaria schaeferi* instead of *Mammillaria shaeferi*.

Page 351. *Dolichothele spaerica* should read *D. sphaerica*.

BIBLIOGRAPHY
Books and Bulletins

Aiton, Wm.—*Hortus Kewensis.* London, 1789. Second Edition 1810-13.
Allnut, Henry—*The Cactus and Other Tropical Succulents,* etc. London, 1877.
Armer, Laura A.—*Cactus.* London, 1934.
Backeberg, Curt—*Blatter fur Kakteenforschung* (Bulletin of Cactus Research). Hamburg, Germany, 1934-38.
Backeberg, Curt, and Knuth, F. M.—*Kaktus A.B.C.* Copenhagen, Denmark, 1935.
Backeberg, Curt, and Werdermann, Erich—*Neue Kakteen.* Frankfurt, Germany, 1931.
Bailey, Liberty H.—*Cyclopedia of American Horticulture.* New York, 1900-02.
Bailey, Liberty H.—*Standard Cyclopedia of Horticulture.* New York, 1916. 1927.
Baxter, Edgar—*California Cactus.* Los Angeles, California, 1935.
Benson, Lyman—*Cacti of Arizona.* Tucson, Arizona, 1940.
Benson, Lyman—*Goal and Methods of Systematic Botany.* Pasadena, California, 1943.
Berger, Alwin—*Hortus Mortolensis.* London, 1912.
Berger, Alwin—*Kakteen.* Stuttgart, Germany, 1929.
Blanc, A.—*Hints on Cacti.* Philadelphia, 1886.
Blanc, A.—*Price List of Cacti.* 1893.
Boedeker, Fredrick—*Ein Mammillarien Vergleichs Schlüssel.* Neudamm, Germany, 1933.
Boerhaave, Herman—*Index Plantarum quae in Horto Academico Lugduno Batavo Peperiunter.* London, 1710. Second Edition, 1727.
Bois, Desire—*Dictionaire d'Horticulture* (Cactaceae by Weber). Paris, France, 1893-99.
Boissevain, Charles—*Colorado Cacti.* Pasadena, California, 1940.
Borg, J.—*Cacti.* London, 1937.
Bradley, Richard—*History of Succulent Plants.* London, 1739.
Bravo, Helia—*Las Cactaceas de Tehuacan.* Chapultepec, D.F., Mexico, 1931.
Bravo, Helia—*Cuatro Especias Nuevas del Genero Neomammillaria,* in Anales del Instituto de Biologia, Mexico, D.F., Mexico, 1931.
Bravo, Helia—*Las Cactaceas del Cañon del Zapilote,* in Anales del Instituto de Biologia. Mexico, D.F., Mexico, 1932.
Bravo, Helia—*Las Cactaceas de Mexico.* Mexico, D.F., Mexico, 1937.
Breyne, Jakob—*Prodromi Fasciculi Rariorum Plantarum.* Holland, 1676.
Breyne, Jakob—*Exoticarum aliarumque minus cognitarum plantarum centuria prima.* Holland, 1678.
Britton, N. L., and Brown, Addison—*Illustrated Flora of North U.S.* New York, 1896-98.
Britton, N. L., and Rose, J. N.—*The Cactaceae.* Washington, D.C., 1923
Burmann, Jahan (Plumber, Charles in)—*Plantarum Americanarum Pasciculus Primus.* Amsterdam, 1758.
Candolle, Augustin de—*Plantarum Succulentarum.* Paris, 1799.
Candolle, Augustin de—*Catalogus Plantarum Horti Botanici Monspeliensis.* Paris, 1813.
Candolle, Augustin de—*Memoires Museum d'Histore Naturalis.* Paris, 1828-38.
Candolle, Augustin de—*Memoire sur Quelques Especes des Cactees.* Paris, 1834.
Candolle, Augustin de—*Prodromous Systematis Naturalis Regni Vegetabilis.* Paris, 1828.
Candolle, Augustin de—*Revue de la Famille des Cactees.* Paris, 1829.
Candolle, Augustin de, and Redoute, A. J.—*Plantes Grasses.* Paris, 1799-1829.
Cassell—*Dictionary of Practical Gardening.* Ed. W. P. Wright. London, 1902.
Castle, Lewis—*Cactaceous Plants.* London, 1884.
Chase, Pearl (Ed.)—*Cacti and Other Succulents in Santa Barbara Region.* Santa Barbara, California, 1930.
Clements, Fredrick E.—*Rocky Mountain Flowers.* 1914.
Colla, Luigi—*l'Antolegista Botanico Opera.* Torino, 1813.
Colla, Luigi—*Hortus Ripulensis.* Torino, 1824.
Colla, Luigi—*Illustrationes et Icones Rariorum Stirpium que in equs Horto Ripulis Florebant,* Ano 1824. Torino, 1827-31.

Commelin, Jan—*Catalogus Plantarum Horti Medici Amstelodamensis*. Amsterdam, Holland, 1689.
Commelin, Jan—*Horti Medici Amstelodamensis*. Amsterdam, Holland, 1697.
Coulter, John M.—*Manual of Botany of Rocky Mountain Region*. Washington, D.C., 1885.
Coulter, John M.—*Preliminary Revision of the North American Species of Cactus*. Contribution from the U. S. National Herbarium. Washington, D.C., 1891-94.
Coulter, John M.—*Manual of the Phanerogam and Pteridophytes of West Texas*. Washington, D.C., 1891.
Coulter, John M., and Nelson—*New Manual of Botany of Central Rocky Mountains*. 1909.
Day, Harry A.—*Flowers of the Desert*. London, 1938.
de Candolle—see Candolle.
Desfontaines, R. L.—*Tableau de l'Ecole de Botanique de Museum d'Historire Naturelle*. Paris, 1804.
Desfontaines, R. L.—*Tableau de l'Ecole de Botanique du Jardin de Roi*. Paris, 1815.
de Tussac—see Tussac.
de Vriese—see Vriese.
Dietrich, Freidrich—*Neuer Nachtrag zum Vollstandigen Lexicon der Gartnerei und Botanik*. Berlin, 1826.
Dietrich, Freidrich—*Synopsis Plantarum seu Enumeratio Systematica*. Berlin, 1843.
Deguet, Leon—*Les Cactacees Utiles du Mexique*. Paris, 1928.
Dinter, Kurt—*Alphabetical Catalogue of Plants at La Mortola*. Leipzig, 1897.
Don, George—*General System of Gardening and Botany*. London, 1834.
Donn, James—*Hortus Cantabigiensis*. Cambridge, Massachusetts, 1796-1845. (1815.)
Eaton, Amos, and Wright, John—*North American Botany*. Troy, N.Y., 1840.
Ehrenberg, Karl—*Beitrag zur Geschichte einiger Mexicanisher Cacteen*. Berlin, 1846.
Eklund, G. M.—*Kaktusboken*. Stockholm, 1935.
Engelmann, George—*Additions to the Cactus Flora of the Territory of the United States*. St. Louis, Missouri, 1862.
Engelmann, George—*Cactaceae of the Boundary Survey*. Washington, D.C., 1859.
Engelmann, George—*Cactaceae of Emory's Reconnaissance*. Washington, D.C., 1848.
Engelmann, George—*Cactaceae of the Ives Exploration*. Washington, D.C., 1861.
Engelmann, George—*Cactaceae of Plantae Fendlerianae*. Cambridge, Massachusetts, 1849.
Engelmann, George—*Cactaceae of Simpson's Expedition*. Washington, D.C., 1876.
Engelmann, George—*Notes on the Cereus Giganteus of Southern California and Other Californian Cactaceae*. New Haven, Connecticut, 1852.
Engelmann, George—*Synopsis of the Cactaceae of the Territory of the United States and Adjacent Regions*. Cambridge, Massachusetts, 1856.
Engelmann, George, and Bigelow, J.—*Description of the Cactaceae Collected en Route Near the Thirty-fifth Parallel*. (Whipple) (Railroad Report). Washington, D.C., 1856.
Engelmann, George, and Gray, Asa—*Cactaceae of Plantae Lindheimerianae*. Boston, Massachusetts, 1845.
Engelmann, George, and Rothrock, J. T.—*Cactaceae of Wheeler's Exploration*. Washington, D.C., 1878.
Engler, Adolf—*Die Naturlichen Pflanzenfamilien*. Leipzig.
Engler, Adolf, and Prantl, Karl—*Die Naturichen Pflanzenfamilien*. Leipzig, 1890-1910.
Fedde, A. Friedrich (Ed.)—*Repertorium novarum specierum regni vegetabilis*. Berlin, 1906, 1909, 1931.
Forbes, James—*Journal of a Horticultural Tour Through Germany, Belgium and Parts of France*. London, 1837.
Foerster, Carl F.—*Handbuch der Kakteenkunde*. Leipzig, 1846. Second Edition. Rumpler, Theodore, (Ed.) 1885-1896.
Fournier, P.—*Les Cactees et Plantes Grasses*. Paris, 1935.
Gentry, Howard—*Rio Mayo Plants*. Washington, D.C., 1942.
Graebener, L.—*Kakteenzucht*. Berlin, 1930.
Gray, Asa—*Introduction to Structural and Systematic Botany*. New York, N.Y., 1868.
Griffiths, David, and Thompson, Charles—*Cacti*. Washington, D.C., 1930.

Guillaumin, Andre—*Les Cactees Cultivetes*. Paris, 1931.
Haage, Friedrich A., Jr.—*Kakteen im Heim*. Erfurt, Germany, 1928.
Haage, Walther—*Kakteenzimmerkultur*. Erfurt, Germany, 1928.
Haage, Walther—*Kakteen*. Erfurt, Germany, 1930.
Harshberger, John Wm.—*Phyteographic Survey of North America*. Leipzig, 1911.
Haselton, Scott—*Cacti for the Amateur*. Pasadena, California, 1938.
Haworth, A. H.—*Synopsis Plantarum Succulentarum*. London, England, 1812.
Haworth, A. H.—*Supplementum et Revisiones Plantarum*. London, England, 1819.
Helms, A. T.—*Desert Cacti*. Phoenix, Arizona, 1931.
Hemsley, William B.—*Biologia Centrali Americana*. London, England, 1879.
Hermann, Paul—*Horti Academici Lugduno Batava Catalogus*. 1689.
Hermann, Paul—*Paradisus Batavus*. 1898.
Hertrich, William—*Guide to the Desert Plant Collection of Huntington Botanical Garden*. Pasadena, California, 1938.
Higgins, Ethyl—*Our Native Cacti*. New York, 1931.
Higgins, Vera—*The Study of Cacti*. London, England, 1933.
Hildmann—*Verzeichnis*. 1888.
Hirscht, Karl—*Bilder dem Kakteen*. Berlin, Germany, 1903.
Hirscht, Karl—*Der Kakteen und Succulenten Zimmergarten*. Neudamm, 1922.
Hoffmannsegg, Johann—*Preiss Verzeichnis*. Berlin, 1833.
Houghton, A. D.—*The Cactus Book*. New York, 1930.
Ives, J. C.—*Report on the Colorado River of the West*. Washingtn, D.C., 1861.
Jepson, Willis L.—*A Manual of the Flowering Plants of California*. San Francisco, California, 1925-36.
Jussieu, Antoine—*Genera Plantarum*. Paris, 1789.
Karsten, Herman—*Deutsche Flora*. Berlin, Germany, 1882.
Karsten, Herman—*Flora von Deutschland*. Berlin, Germany, 1895.
Karsten, Herman—*Mexikanische Cacteen*. Berlin, Germany, 1903.
Kearney, Thomas, and Peebles, R. H.—*Flowering Plants and Ferns of Arizona* Washington, D.C., 1942.
Knuth, F. M.—*Kaktusbogen,* Copenhagen, Denmark, 1928.
Knuth, F. M.—*Stora Kaktusboken*. Copenhagen, Denmark, 1931.
Knuth, F. M.—*Nye Kaktusbog*. Copenhagen, Denmark, 1930.
Kreuzingers—*Verzeichnis*. 1935.
Krook, J. J.—*Kenntness und Behandlung der Cacteen*. Amsterdam and Leipzig, 1855.
Kuntze, Otto—*Reviso Generum Plantarum*. Leipzig, 1891.
Kupper, Walter—*Das Kakteenbuch*. Berlin, Germany, 1927.
Labouret, J.—*Monographie de la Famille Des Cactees*. Paris, 1853.
Lamark, Jean Baptiste—*Encyclopedie Methodique*. Paris, 1783.
Lawson, H. C.—*Book of Cacti*. San Antonio, Texas, 1935.
Lehmann, Johann—*Delectus Seminum Horto Hamburgensium Botanico*. Hamburg, Germany, 1833.
Lemaire, Charles—*Cactearum Aliquot Novarum en Horto Monvill*. Paris, 1838.
Lemaire, Charles—*Cactearum Genera Nova et Species Nova en Horto Monvill*. Paris, 1839.
Lemaire, Charles—*l'Horticulture Universel*. Paris, 1839.
Lemaire, Charles—*Iconographie des Cactees*. Paris, 1853.
Lemaire, Charles—*Les Cactees*. Paris, 1868.
Link, Heinrich F.—*Enumeratio Plantarum Horti Regni Botanici Beralinensis*. Berlin, Germany, 1821-22.
Link, H., and Otto, Friedrich—*Icones Plantarum Rariorum Horti Regii Botanici Berolinensis*. Berlin, Germany, 1828.
Link, H., and Otto, Friedrich—*Uber die Gattungen Melocactus und Echinocactus*. Berlin, Germany, 1827.
Linne', Carl (Linnaeus)—*Hortus Cliffortianus*. Upsala, 1737.
Linne', Carl (Linnaeus)—*Upsaliensis*. Upsala, 1748.
Linne', Carl (Linnaeus)—*Species Plantarum*. Upsala, 1753. Second Edition, 1762.
Linne', Carl (Linnaeus)—*Genera Plantarum*. Upsala, 1764.

Loudon, John—*Hortus Britainnicus*. London, England, 1830.
Loudon, John—*Catalogue*. London, England, 1832.
Loudon, John—*Encyclopedia of Plants*. London, England, 1855.
Loddiges, Conrad—*Botanical Cabinet*. London, England, 1817.
Lynch, Robert—*Succulent Plants*. London, England, 1907.
Maasz, Harry—*Die Schoenheit Unserer Kakteen*. Frankfurt, Germany, 1928.
Martius, Karl von—*Nova Acta Physico Medica Academiae Caesareae Leopoldino Carolinae Naturae Curiosorum*. (Beschreibung einiger neuen nopaleen). Munchen, 1932.
—See also Schrank.
Miller, Philip—*Gardener's Dictionary*. London, England, 1768.
Miller, Philip—*Abridgement of the Gardener's Dictionary*. London, England, 1771.
Miller, Philip—*The Gardener's and Botanist's Dictionary*. London, England, 1807.
Miquel, F. A.—*Commentarii Phytographici*. Amsterdam, 1839.
Morrison, Robert—*Plantarum Historiae Universalis Oxoniensis*. Oxonii, 1715.
Neale, W. T.—*Cacti and Other Succulents*. New Haven, Sussex, England, 1935.
Nicholson, George—*Illustrated Dictionary of Gardening*. New York, 1886.
Nuttall, Thomas—*The Genera of North American Plants*. Philadelphia, Pennsylvania, 1818.
Ochoterena, Isaac—*Boletin de la Direccion de Estudios Biologicos*. Mexico, D.F., Mexico, 1918.
Ochoterena, Issac—*Las Cactaceas de Mexico*. Mexico, D.F., Mexico, 1922.
Orcutt, Charles—*American Plants*. San Diego, California, 1907-12.
Orcutt, Charles—*Cactography*. San Diego, California, 1926.
Ortega, Casimiro—*Novarum aut Rariorum Plantarum* (Hort. reg. botan. matrit, descript. decades). Madrid, 1797.
Paxton, Sir Joseph—*Botanical Dictionary*. London, England, 1868.
Palmer, F. T.—*Culture des Cactees* (suivie d'une description des principales especes et varieties). Paris, 1882.
Petersen, Elln—*Taschenbuch fer den Kakteenfruend*. Munchen, 1927.
Pfeiffer, Ludwig—*Beschriebung und Synonymik*. Berlin, Germany, 1837.
Pfeiffer, Ludwig—*Enumeratio Diagnostica Cactearum Hucusque Cognitarum*. Berlin, Germany, 1837.
Pfeiffer, L., and Otto, Freidrich—*Abbildbung und Beschreibung Bluhender Kacteen*. Berlin, Germany, 1843-50.
Plukenet, Leonard—*Opera Omnia Botanica*. London, England, 1720.
Poiret, Jean Louis—*Supplement to Lamark's Encyclopedie Methodique*. Paris, 1817.
Porter, Thomas, and Coulter, John—*Synopsis of the Flora of Colorado*. Washington, D.C., 1874.
Pursh, Fredrick—*Flora Americae Septentrionalis*. London, England, 1814.
Plumier, Charles—*Botanicum Americanarum M.S.S.* Paris, 1697.
Plumier, Charles—*Plantarum Americanarum Fasciculus Primus*. Paris, 1755-60.
Plumier, Charles—*Nova Plantarum Americanarum General*. Paris, 1702.
Rebut, P.—*Catalogue des cactees et plantes grasses diverses de le collection de P. Rebut*. 1886, 1896.
Regel, Edward—*Catalogus Plantarum quae in Horti Aksakoviano*. St. Petersburg, Russia, 1860.
Regel, Edward—*Acta Horti Petropolitani*. St. Petersburg, Russia, 1893.
Remark, Ferdinand—*Der Kakteenfreund*. Minden, 1890.
Roeder, W. von—*Fehlerbuch des Kakteenzuechters*. Stuttgart, Germany, 1929.
Roeder, W. von—*Kakteenzuecht Ltichtgemachet*. Stuttgart, Germany, 1929.
Rother, W. O.—*Praktischer Leitfaden fur die Anzucht und Pflege der Kakteen und andere Sukkulenten*. Frankfurt, Germany, 1928.
Rydberg, Per Axel—*Flora of the Rocky Mountains*. New York, 1917.
Safford, W. E.—*Cactaceae of Northeastern and Central Mexico*. (Smith. Rep. 19080.) Washington, D.C., 1909.
Salm-Dyck, Joseph von—*Dyckensis oder Verzeichniss der in dem Botanischen Garten zu Dyck*. Dusseldorf, 1834.
Salm-Dyck, Joseph von—*Horti Dyckensi*. Duseldorf, 1845.

Salm-Dyck, Joseph von—*Cacteae en Horto Dyckensi Cultae en anno 1849.* Bonn, 1850.
Sauvalle, Francisco—*Flora Cubana.* Havana, Cuba, 1873.
Seeman, Berthold—*Botany of the Voyage of H.M.S. Herald.* London, 1852-57.
Schelle, Ernest—*Handbuch der Kakteenkultur.* Stuttgart, 1907.
Schelle, Ernest—*Kakteen.* Tubingen, 1926.
Schrank, Franz, and Martius, Karl—*Hortus Regius Monacensis.* Munchen, 1829.
Schulz, Ellen D.—*Cactus Culture.* New York, 1932.
Schulz, Ellen D., and Runyon, Robert—*Texas Cacti.* San Antonio, Texas, 1930.
Schumann, Karl—*Bluhende Kakteen.* Neudamm, 1931.
Schumann, Karl—*Gesamtbeschreibung der Kakteen.* Neudamm, 1898.
Schumann, Karl—*Gesamtbeschreibung der Kakteen Nachtrag.* Neudamm, 1903.
Schumann, Karl—*Keys of the Monograph of the Cactaceae.* Neudamm, 1903.
Schumann, Karl—*Verbreitung der Cactaceae* (In Verhaltnis zu ihrer systematischen Gliederung). Berlin, 1899.
Schumann, K., and Guercke, M.—*Bluhende Kakteen.* Neudamm, 1898.
Shreve, Forrest—*The Cactus and Its Home.* Baltimore, Md., 1931.
Small, John—*Flora of the Southeast United States.* New York, 1903.
Sprengel, Kurt—*Systema Vegetabilium.* Halle, Germany, 1825.
Standley, Paul—*Trees and Shrubs of Mexico.* Washington, D.C., 1924-26.
Steudel, Ernest—*Nomenclatur Botanicus* (Enumerans ordine alphabetico nomina atque synonymatum generica tum specifica et a Linnaeo et recentioribus de ve botanica scriptoribus plantis phanerogamis imposita). Stuttgart, 1841.
Stockwell, Wr. P., and Breazeale, Lucretia—*Arizona Cacti.* Tucson, Arizona, 1933.
Sweet, Robert—*Hortus Britannicus.* London, 1839.
Thomas, Fritz—*Zimmerkultur der Kakteen,* Neudamm, 1935.
Thornbers, J., and Bonkers, Francis—*The Fantastic Clan.* New York, 1932.
Torrey, John, and Gray, Asa—*A Flora of North America.* Washington, D.C., 1838.
Torrey, John—*Collection of Rocky Mountain Plants.* Washington, D.C., 1826.
de Tussac—*Flore des Antilles.* 1818.
Tournefort, Joseph—*Institutiones Rei Herbariae.* Paris, 1719.
Thiebaut, P.—*Cactees et Plantes Grasses.* Paris, 1930.
Vaupel, Friedrich—*Guttungen und Arten aus der Familie der Cactaceae.* Neudamm, 1913.
Vaupel, Friedrich—*Cactaceae* (In Naturliche Pflanzenfamilien). Leipzig, 1913.
Vaupel, Friedrich—*Die Kakteen* (Monographie der Cactaceae). Berlin, 1926.
von Roeder—see Roeder.
Vriese, William de—*Tijdschrift Natuurlijke Gesch.* (Klein Byjdragen.) Amsterdam, 1838.
Watson, Serano—*Cactaceae of Clarence King's Expedition of the Fortieth Parallel.* Washington, D.C., 1871.
Watson, William—*Bibliographic Index.* London, 1878.
Watson, William—*Cactus Culture for Amateurs.* London, 1889, 1903, 1920.
Werdermann, Erich—*Bluhende Kakteen und andere Sukkulente Pflanzen.* Neudamm, 1931.
Werdermann, Erich, and Socnik, H.—*Mein Kakteen,* Frankfurt, 1937.
Willdenow, Karl—*Enumeratio Plantarum Horti Regii Botanici Berolinensis.* Berlin, 1809.
Wislizenus, Adolphus—*Memoir of a Tour of Northern Mexico in 1846-47.* Washington, D.C., 1848.
Wooten, Elmer, and Standley, Paul—*Flora of New Mexico.* Washington, D.C., 1915.
Weinmann, Johann—*Phytanthoza Iconographia.* Ratisbon, 1716.
Young, Maud—*Flora of Texas.* 1873.
Zuccarini, Joseph—*Planterum Novarum Horto Botanico Regio Monacensi.* Munchen, 1837.

Periodicals

Alianza Cientifica Universal (Boletin del Comite Regional del Estado de Durango).
    Durango, Mexico, 1910.
Allgemeinen Gartenzeitung. Berlin, 1835-56.
Anales del Instituto de Biologia. Mexico, D.F., Mexico, 1930-34.
Annals of the Missouri Botanical Garden. St. Louis, Mo., 1915.
Annals of the Lyceum of Natural History of New York. New York, 1828.
Annales de la Societe Botanique de Lyon. Lyon, France, 1880.
Baltimore Cactus Journal. Baltimore, Md., 1895-96.
Beitrage zur Sukkulentenkunde und Pflege. Berlin, 1938.
La Belgique Horticola. 1874.
Bluhende Kakteen. Neudamm, 1904-35.
Boletin de la Direccion de Estudios Biologicas. Mexico, D.F., Mexico, 1918.
Boston Journal of Natural History. Boston, Massachusetts, 1845-50.
Botanical Gazette (Univ. Chicago). Chicago, Ill., 1903.
Botanisches Zeitschrift. Berlin, 1843-1910.
Botanisches Zeitung. Berlin, 1843-48.
Bulletin de l'Academie Royal des Sciences et Belle Lettres de Bruxxelles. Brussels,
    Belgium, 1838-39.
Bulletin des Cactophiles Belges. Bruxelles, 1931.
Bulletin de Museum d'Histore Naturelle. Paris, 1828, 1904.
Bulletin de le Societe Botanique de France. 1855.
Bulletin of Southern California Academy of Science. Los Angeles, California, 1931.
Bulletin of the Torrey Botanical Club. New York, 1922.
Cactaceae (German Jahrbuch). 1937-38.
Cactus (Belgain Journal), 1931-37.
Cactus Journal. London, 1898-99.
Cactus Journal of Great Britain. London, 1932-40.
Cactus and Succulent Journal of America. Los Angeles and Pasadena, California, 1929-44.
Cactussen en Vetplanten. Holland.
Curtis Botanical Magazine. London.
Delectus Seminum Horto Hamburgensium Botanico. Hamburgh, 1833.
Desert Plant Life. Pasadena, California, 1930-37.
Desert. El Centro, California, 1937-44.
Deutsch Kakteen Geselschaft. Neudamm, 1938.
Edwards Botanical Register. London, 1815-47.
Erythea. Berkeley, California, 1897.
Gardeners Chronicle. London, 1887, 1907.
Gartenflora. Erlangen, Germany, 1852, 1862, 1885, 1889.
Gartenwelt. Berlin, 1905.
Gartenzeitung der Osterreich Gartenbau Gesellschaft. 1885, 1929.
Hamburg Gartenzeitung Blumenzeitung. Hamburg, 1861.
Hookers Journal of Botany and Kew Garden Miscellany. London, 1853.
l'Horticulteur Belge. Brussels, Belgium, 1837-38.
l'Horticulteur Universal. Paris, 1839-47.
l'Illustration Horticole. Gand, France, 1858.
Index Kewensis Plantarum Phanerogamarum. Brussels, 1893.
Index Kewensis Supplement. Brussels, 1901-35.
Index Seminum Hortus Botanicus Imperialis Petropolitanus. 1860.
Index Seminum Horti Botanici Turicensis. 1850.
Index Specierum Horti Botanici Berolinensis. Berlin, 1829.
Jahrbuch der Deutschen Kakteen Gesellschaft in der Deutschen Gesellschaft fur Gartenkultur. 1935-38.
Journal of Horticulture, Cottage Gardener and Home Farmer. London, 1903.
Journal of Royal Horticultural Society. London, 1908.

Just's Botanischer Jahresbericht. Leipzig, 1907 (1874).
Kakteen Almanac. Neudamm, 1939.
Kakteen Jahrbuch, Neudamm, 1935.
Kakteen und Andere Sukkulente. Neudamm, 1937.
Kakteenfreund. Neudamm, 1932-35.
Kakteenkunde, Neudamm, 1933-39.
Kakteenkunde und Kakteenfreund. Neudamm, 1937.
Kakteenkunde Veoffentlicht von der Deutschen Kakteen Gesellschaft. Neudamm, 1938.
Kaktusarske Listy. 1925-33.
Kew Bulletin of Miscellaneous Information. London, 1908.
Linaea. Berlin, 1838, 1840, 1847 (1826-82).
Meehan Monthly. Germantown, Pa., 1891.
Memorie Della Reale Accademiea di Torino. Torino, 1829.
Memoires of American Academy of Arts and Science. Cambridge, Massachusetts, 1849.
Memoires of New York Botanical Garden. New York, 1900.
Moeller Deutsche Garten Zeitung. Erfurt, Germany, 1856, 1926-33.
Monatsschrift der Deutschen Kakteen Gesellschaft. Berlin, 1929-32.
Monatsschrift fur Kakteenkunde. Berlin, 1891-1922.
National Horticultural Magazine. 1941.
Notizblatt des Botanischen Garten und Museums zu Berlin-Dahlem. Berlin, 1931.
Philosophical Magazine. London, 1824-30.
Pittonia. Berkeley, California, 1889-90.
Proceedings of American Academy of Arts and Science. Cambridge, Massachusetts, 1856-89.
Proceedings of California Academy of Science. 1889.
Revue Horticole. Paris, 1855-56.
Succulenta. Leeuwarden, Netherlands, 1919-39.
Transactions of the Academy of Science. St. Louis, Missouri, 1856-62.
Transactions of the American Philosophical Society. Philadelphia, Pa., 1863.
Walpers Annales Botanices Systematicae. Leipzig, 1851-56.
Walpers Reportorium Botanices Systematicae. Leipzig, 1843-45.
Walton Cactus Journal. London, 1898-99.
West American Scientist. San Diego, California, 1887, 1894, 1899, 1900, 1902.
Wiener Illustreierte Garten Zeitung. Wien, Austria, 1904.
Wochenschrift Gatenerei und Pflanzenkunde. Berlin, 1858-59.
Zeitschrift fur Sukkulentenkunde. Berlin-Dahlem, 1923-28.
Zoe. San Diego, California, 1891-1900.

INDEX

BARTSCHELLA
  schumannii 356
CACTUS
  acanthophlegmus 281
  acicularis 323
  aciculatus 261
  actinopleus 323
  aeruginosus 44
  affinis 47
  amabilis 323
  ancistrius 323
  ancistroides 219
  anguineus 253
  argenteus 324
  atroruber 324
  atrosanguineus 324
  aulacantha 343
  aureiceps 324
  auricomus 269
  auroreus 269
  barbatus 216
  bellatulus 325
  beneckei 174
  bergii 31
  bihamatus 43
  bispinus 334
  bocasanus 210
  bockii 34
  brandegeei 70
  brevisetus 325
  caespititius 325
  canescens 279
  carneus 44
  celsianus 278
  centricirrhus 34
  centrispinus 27
  chrysacanthus 235
  cirrhifer 18
  compressus 18
  conicus 327
  conopseus 34
  corollarius 327
  coronaria 327
  coronatus 327
  crassispinus 279
  crebrispinus 327
  crinitus 184
  crocidatus 37
  cruciger 125
  cylindraceus 328
  cylindricus 328, 331
  dealbatus 119
  decipiens 229
  densispinus 288

CACTUS—Continued
  depressus 42, 260
  diadema 34
  discolor 261
  divaricatus 34, 328
  divergens 34
  dolichocentrus 227
  dyckianus 281
  eborinus 328
  echinarius 253
  ehrenbergii 34
  elegans 281
  elongatus 141
  erectacanthus 329
  eriacanthus 286
  eschanzieri 329
  euchlorus 329
  eximius 269
  fasciculatus 173
  fellneri 330
  fischeri 26
  flavescens 67
  foersteri 34
  formosus 124
  foveolatus 20
  funckii 54
  fuscatus 279
  gabbii 71
  geminatus 330
  geminispinus 80
  glabratus 330
  gladiatus 34
  glaucus 34
  glochidiatus 163
  glomeratus 274
  goodridgii 181
  gracilis 253
  grahamii 208
  granulatus 331
  griseus 331
  guilleminianus 229
  gummifer 96
  gummiferus 96
  haageanus 246
  haematactinus 331
  hamatus 331
  haworthianus 275
  heinei 332
  helicteres 332
  hemisphaericus 94
  hexacanthus 332
  heyderi 122
  heyderi hemisphaericus 94
  humboldtii 142
  hystrix 34

CACTUS—Continued
  intertextus 141
  irregularis 332
  isabellinus 269
  jucundus 333
  karwinskianus 26
  kleinei 333
  klugii 281
  knuthii 281
  krameri 34
  lactescens 34
  lasiacanthus 150
  lasiacanthus denudatus 151
  leucocentrus 333
  leucodictyus 334
  leucotrichus 54
  linkeanus 269
  lividus 334
  longimammus 355
  longisetus 18
  loricatus 334
  ludwigii 75
  magnimamma 33
  mammillaris 106
  mammillaris glaber 106
  mammillaris languinosus 67, 106
  mammillaris prolifer 274
  maschalacanthus 54
  megacanthus 34
  meiacanthus 66
  meissneri 281
  microceras 34
  microthele 106, 334, 335
  minimus 142
  mirabilis 269
  mucronatus 335
  muehlenpfordtii 278
  multiceps 273
  multisetus 335
  mutabilis 54
  mystax 54
  neumannianus 34
  niger 335
  nigricans 238
  niveus 80
  nivosus 66
  nobilis 80
  nudus 335
  obconellus 227
  obliquus 335
  obvallatus 258
  odieranus 279
  olorinus 336
  oothele 336
  ovimamma 336
  pallescens 44
  palmeri 267
  parkinsonii 119
  parvimammus 106
  pazzanii 34

CACTUS—Continued
  pentacanthus 56
  persicianus 336
  phaeacanthus 238
  phaeotrichus 337
  phellospermus 196
  phymatothele 75
  picta 363
  plecostigma 337
  pleiocephalus 337
  polycentrus 269
  polyedrus 21, 363
  polygonus 50
  polythele 47
  polytrichus 21
  pomaceus 269
  pottsii 297
  praelii 29
  pretiosus 269
  pringlei 259
  procerus 337
  prolifer 275
  proliferus 274
  pseudomammillaris 260, 262
  pugionacanthus 337
  pulchellus 261
  pulcherrimus 269
  purpureus 338
  pusillus 274
  pyrrhocephalus 17
  pyrrhochracanthus 232
  quadrispinus 47
  recurvus 34
  regius 338
  rhodanthus 235
  rhodanthus sulphureospinus 279
  rhodeocentrus 338
  roseus 338
  ruficeps 235
  rufidulus 339
  rufocroceus 254
  rutilus 255
  saxatilis 339
  schaeferi 278
  scheidweilerianus 203
  schelhasii 189
  schiedeanus 149
  seemannii 339
  seidelii 339
  seitzianus 20
  sempervivi 58
  setosus 47
  severinii 339
  sororius 340
  spectabilis 136
  sphacelatus 240
  sphaericus 351
  sphaerotrichus 272
  spinaureus 340
  spinii 260

CACTUS—Continued
  spinosissimus 269
  squarrosus 18
  stella-auratus 141
  stellatus 274
  stellatus texanus 273
  stenocephalus 363
  stramineus 67
  stueberi 248
  subangularis 18
  subcroceus 141
  subcurvatus 34
  subechinus 363
  subpolyedrus 23, 363
  subtetragonus 44
  subulifer 340
  supertextus 282
  tectus 340
  tentaculatus 279
  tenuis 142
  tetracanthus 227
  tetracentrus 227, 228
  tetrancistrus 196
  texanus 273
  texensis 97
  tomentosus 341
  triacanthus 18
  uberiformis 352
  umbrinus 191
  uncinatus 42
  varimamma 341
  versicolor 34
  vetulus 289
  villifer 21
  virens 26
  viridis 30
  vrieseanus 340
  vulpinus 269
  webbianus 37
  wegeneri 342
  wildianus 152
  woburnensis 49
  wrightii 185
  xanthotrichus 54
  zegschwitzii 342
  zephyranthoides 178
  zepnickii 342

CHILITA
  albicans 262
  armillata 162
  barbata 216
  bocasana 210
  boedekeriana 212
  bombycina 191
  candida 272
  carretii 176
  decipiens 229
  denudata 151
  discolor 261
  echinaria 253

CHILITA—Continued
  elongata 141
  eriacantha 286
  eschausieri 133, 147, 149, 156, 164, 171, 211
  fertilis 232
  fordii 161
  fragilis 140
  fraileana 159
  glochidiata 163
  goodridgei 181
  grahamii 208
  hirsuta 198
  jaliscana 218
  kunzeana 200
  lasiacantha 150
  lenta 145
  longicoma 199
  longiflora 206
  mainae 172
  mazatlanensis 241
  mercadensis 205
  milleri 208
  multiceps 273
  multiformis 193
  multihamata 197
  nelsonii 174
  occidentalis 168
  oliviae 223
  palmeri 267
  plumosa 146
  pottsii 297
  prolifera 275
  saffordi 176
  scheidweileriana 203
  schelhasei 189
  schiedeana 149
  seideliana 202
  sheldonii 180
  slevinii 262
  sphacelata 240
  swinglei 170
  thornberi 173
  umbrina 191
  verhaertiana 160
  vetula 289
  viridiflora 188
  wilcoxii 186
  wildii 152
  wrightii 185
  xanthina 104
  zephyranthoides 178
  zuccariniana 73

DOLICHOTHELE
  longimamma 354, 355
  melaleuca 353
  sphaerica 351
  uberiformis 352

KRAINZIA
  longiflora 206

Roman type indicates recognized species. *Italic* type is used for synonyms and names not recognized. Figures followed by an asterisks (*) refer to the principal reference.

MAMMILLARIA

abdusta 281
acanthophlegma 115, 278, 280, 281
acanthophlegma abducta 281
acanthophlegma decandollii 280
acanthophlegma elegans 280
acanthophlegma leucocephala 281
acanthophlegma meissneri 280
acanthophlegma monocantha 281
acanthostephes 346
acanthostephes recta 346
acicularis 323
aciculata 48, 260, 261
actinoplea 323
acutangula 343
adunca 343
aeruginosa 44
affinis 46, 48
aggregata 346
alamensis sp. nov. 299
alamoensis 343
alba 343
alba minor 343
albescens 11, 134*
albiarmata 7, 40*
albicans 15, 262*, 263
albicoma 16, 270*, 271
albida 260, 261
albilanata 15, 251*, 252
albina 346
albiseta 323, 343
aleodantha 236
aljibensis 343
aloidaea pulviligera 346
aloides 346
alpina 346
alversonii 346
amabilis 323
amarilla 343
ambigua 346
amoena 15, 247*
anancistra 323, 344
ancistra 323
ancistracantha 346
ancistrata 219
ancistrina 219
ancistrohamata 343
ancistroides 219*, 343
andraea 346
andreae 235
angelensis sp. nov. 12, 165*
anguinea 253
angularis 18*, 20, 48, 343
angularis compressa 19
angularis fulvispina 18
angularis longiseta 18
angularis triacantha 19

anguliger 343
anisacantha 23
applanata 9, 94, 97*, 122, 246
areolosa 346
argyphaea 343
argentea 324
arida 9, 100*
arietina 33, 346
arietina spinosior 346
arizonica 346
armata 343
armatissima sp. nov. 300
armillata 12, 162*, 163, 343
aselliformis 346
asperispina 346
asterias 346
atrata 346
atrorubra 324
atrosanguinea 324
aulacantha 343
aulacothele 346
aulacothele flavispina 346
aulacothele multispina 346
aulacothele spinosior 346
aulacothele sulcimamma 346
aurata 237, 343
aurea 237, 343
aureiceps 324
aureiceps pfeifferi 324
aureiceps sanguinea 324
aureilanata 11, 147*
aureoviridis 198
auriareolis 300*, 301
auricantha sp. nov. 301*, 302
auricoma 268
aurihamata 11, 153*, 154
auritricha sp. nov. 302
aurorea 268
autumnalis 54, 55
aylostera 346
bachmannii 9, 74*
badispina 324
balsasensis 174, 175
balsasoides sp. nov. 5, 12, 158*
barbata 5, 14, 203, 216*
barlowii 346
baselluta 343
baumii 16, 291
baxteriana 8, 69*, 70, 100
beguinii 237, 346
bella 346
bellacantha sp. nov. 303
bellatula 325
bellisiana sp. nov. 304
beneckei 5, 12, 159, 174*, 175, 181
bergeana 325

MAMMILLARIA—Continued

bergenii 325
bergeriana 346
bergii 31
besleri 346
bicolor 80*, 81, 344
bicolor cristata 80
bicolor longispina 80
bicolor nivea 80
bicolor nobilis 80
bicorem 81
bifurca 325
biglandulosa 346
bihamata 43
binops 343
bisbeeana 346
blossfeldiana 12, 183*, 184, 357
blossfeldiana shurliana 184
bocasana 14, 210*
bocasana cristata 210, 211
(*bocasana flavispina*) 210, 211, 363
bocasana glochidiata 210, 211, 363
bocasana inermis 211
bocasana kunzeana 200
bocasana multihamata 210
bocasana sericata 211
bocasana splendens 210, 211
bocasiana 343
bocensis sp. nov. 8, 56*, 57
bockii 34
boedekeriana 14, 212*, 213
bogotensis 284, 285
bombycina 13, 201*, 202
borealis 346
boregui 343
borhei 343
borwigii 346
boucheana 363
brachyelphis 346
brandegeei 8, 70*, 71, 344
brandegeei gabbii 71
brandi 343
brandtii 343
brauneana 10, 109, 110
bravoae sp. nov. 10, 111, 112*, 113
brevimamma 346
brevimamma exsudans 346
breviseta 325
brevispina 79
brongniartii 343
brownii 346
bruennowii 343
bruennowiana 343
brunispina 343
bucareliensis sp. nov. 8, 61*
bucareliensis bicornuta var. nov. 62
buchalziana 343
buchheimiana 96, 97
buchiana 343

MAMMILLARIA—Continued

bullardiana 179
bumamma 346
bussleri 346
cadereytensis sp. nov. 305
cadereytensis quadrispina var. nov. 306
caerulea sp. nov. 306*, 307
caesia 268
caespititia 274, 325*
caespitosa 142, 346
calacantha 16, 276*
calacantha rubra 276
calcarata 346
californica 343
callipyga 346
calochlora 346
camptotricha 11, 133*, 135, 329, 351
candida 16, 143, 272*
candida rosea 272
candida sphaerotricha 272
canelensis sp. nov. 86, 307*
canescens 262, 346
cantera 343
capensis 11, 155, 156*
caput medusae 58*, 59, 330
caput medusae centrispina 58
caput medusae crassior 58
caput medusae tetracantha 59
caput medusae hexacantha 59
caracasana 106
caracassana 106*, 107, 335
carnea 7, 44*
carnea aeruginosa 44
carnea cirrosa 44
carnea robustispina var. nov. 45
carretii 12, 176*
carycina 343
castanaeformis 269
castanea 269
castaneoides 269, 342
cataphracta 48
caudata 48, 346
celsiana 16, 255, 278*
celsiana guatemalensis 234
celsiana longispina 278
centa 343
centricirrha 33*, 35, 73, 247, 328, 334
centricirrha amoena 34
centricirrha arietina 34
centricirrha bockii 34
centricirrha boucheana 34
centricirrha ceratophora 34
centricirrha cirrhosa 326
centricirrha conopsea 34
centricirrha cristata 34
centricirrha deflexispina 34
centricirrha destorum 34
centricirrha detampico 34
centricirrha diacantha 34
centricirrha diadema 34

MAMMILLARIA—Continued
   centricirrha diavaricata 34
   centricirrha divergens 34
   centricirrha ehrenbergii 34
   centricirrha falcata 34
   centricirrha flaviflora 354
   centricirrha foersteri 34
   centricirrha gebweileriana 34
   centricirrha gladiata 34
   centricirrha glauca 34
   centricirrha globosa 34
   centricirrha grandidens 34
   centricirrha guilleminiana 34
   centricirrha hopferiana 34
   centricirrha hystrix 34
   centricirrha hystrix grandicornis 34
   centricirrha hystrix longispina 34
   centricirrha jorderi 34
   centricirrha krameri 34
   centricirrha krameri longispina 34
   centricirrha krausei 34
   centricirrha lactescens 34
   centricirrha longispina 34
   centricirrha macracantha 334
   centricirrha macrothele 33
   centricirrha magnimamma 34
   centricirrha megacantha 34
   centricirrha microceras 34
   centricirrha monstii 35
   centricirrha moritziana 35
   centricirrha neumanniana 35, 36
   centricirrha nordmannii 35
   centricirrha nordmanniana 36
   centricirrha obconella 35
   centricirrha pachythele 65
   centricirrha pazzanii 35
   centricirrha pentacantha 56
   centricirrha polygona 363
   centricirrha pulchra 363
   centricirrha recurva 34
   centricirrha scheideana 35
   centricirrha schmidtii 35
   centricirrha spinosior 35
   centricirrha subcurvata 35
   centricirrha uberimamma 37
   centricirrha valida 35
   centricirrha versicolor 35
   centricirrha viridis 35
   centricirrha zooderi 35
   centricirrha zuccariniana 35
   centrispina 27
   cephalaphora 147*, 148, 329, 346
   ceratites 346
   ceratocentra 346
   ceratophora 33
   cerralboa 326
   cerroprieto 343
   cespititia 325
   chaffeyi 346

MAMMILLARIA—Continued
   chapinensis 49
   chapinensis rubescenes 49
   chihuahuaensis 346
   childsii 346
   chionocephala 10, 115*
   chlorantha 346
   chrysacantha 235
   chrysacantha fuscata 235
   chrysantha 343
   circumtexta 343
   cirrhifera 18*, 343
   cirrhifera angulosior 18
   cirrhifera longiseta 18
   cirrhosa 35, 326
   cirrosa 35, 326*
   citrina 327
   clava 346
   clavata 346
   clillifera 343
   closiana 27
   clunifera 346
   coccinea 346
   cochemoides 156
   collina 15, 249*, 250, 251
   collinsii 8, 25, 41, 53*, 340
   colonensis sp. nov. 14, 219*
   columbiana 16, 284*, 287
   columnaris 46
   columnaris minor 48
   columnaroides 343
   communis 346
   compacta 335, 346
   compressa 3, 7, 18*
   compressa compressa 19
   compressa fulvispina 19
   compressa longiseta 19
   compressa rubispina 19
   compressa triacantha 19
   concigera 343
   confinis 261
   confusa 7, 18, 23*, 27, 29, 41
   confusa conzattii var. nov. 24*
   confusa robustispina var. nov. 25*, 26
   confusa centrispina var. nov. 8, 25* 53, 330, 340
   congesta 347
   conica 327
   coniflora 262
   conimamma 347
   conoidea 328, 347
   conopea 343
   conopsea 33
   conopsea longispina 33
   conothele 248
   conspicua 16, 264*, 265
   contacta 343
   convoluta 332
   conzattii 24, 26

MAMMILLARIA—Continued
  corbula 347
  cordigera 202, 347
  corioides 347
  cornifera 347
  cornifera impexicoma 347
  cornimamma 347
  cornuta 347
  corollaria 327
  coronaria 175, 181, 255, 327*, 328, 332
  coronaria beneckii 181
  coronaria eugenia 181
  coronaria nigra 181
  coronaria nigra euchlora 181
  coronaria nigra euchlora cristata 181
  coronata 327, 328
  coryphides 343
  costarica 343
  craigiana 137
  craigii 8, 63*, 64
  crassihamata 343
  crassispina 237, 279, 280
  crassispina gracilior 279
  crassispina rufra 279
  crebrispina 327, 347
  crebrispina nitida 323, 326
  criniformis 11, 152*, 164
  criniformis rosei 152
  criniformis albida 152, 184
  crinigera 343
  crinita 13, 184*
  crinita pauciseta 153
  crinita rubra 152
  crispiseta sp. nov 308*
  crocidata 7, 37*
  crucigera 10, 125*, 126
  cubensis 347
  cuneiflora 343
  cuneiformis 152, 164
  cunnendstiana 343
  curvata 347
  curvispina 260
  curvispina parviflora 260
  cylindracea 287, 327, 328*
  cylindrica 287, 328*
  cylindrica flavispina 328
  dactylithele 347
  daedalea 80, 81
  daedalea viridis 56, 343
  dasyacantha 347
  dawsonii 8, 67*, 68
  dealbata 119, 120, 282
  decholara 344
  decipiens 14, 135, 229*, 324
  declivis 97
  decora 347
  deficiens 229
  deficum 229

MAMMILLARIA—Continued
  deflexispina 33
  degrandii 344
  delaetiana 347
  delenlii 295, 344
  densa 142, 253*, 360
  densispina 16, 288*, 347, 359
  denudata 151
  depressa 42, 261
  deserti 347
  desertorum 344
  desnoyersii 344
  destorum 344, 363
  de tampico 363
  diacantha 58, 59, 247
  diacantha nigra 246
  diacentra 74
  diacentra nigra 247
  diadema 33
  diamonoceras 347
  diaphanacantha 328, 347
  dichotoma 344
  dietrichae 119
  difficilis 347
  digitalis 332
  diguetii 347
  dioica 12, 161*, 182
  dioica insularis 267
  disciformis 347
  discolor 15, 48, 260*, 261, 262
  discolor aciculata 260
  discolor albida 260
  discolor breviflora 262
  discolor coniflora 261, 262
  discolor curvispina 261
  discolor droegeana 295
  discolor fulvescens 262
  discolor monstrosa 260
  discolor nigricans 338
  discolor nitens 261
  discolor prolifera 260, 262
  discolor pulchella 260
  discolor rhodacantha 260
  dispina 344
  divaricata 34, 328*
  divergens 33
  dolichocantha 228, 344
  dolichocentra 227, 228, 344, 363
  dolichocentra brevispina 227
  dolichocentra galeottii 228
  dolichocentra nigrispina 228
  dolichocentra phaeacantha 227, 228
  dolichocentra picta 228
  dolichocentra staminea 227
  dolichocentra straminea 228
  donatii 15, 250*
  donkeleari 344
  droegeana 16, 295*
  dubia 344

MAMMILLARIA—Continued
   dumetorum 149
   durangensis 347
   durispina 11, 137*, 138
   dyckiana 280
   dyckii 344
   ebenacantha 344
   eborina 328
   eburnea 80
   echinaria 15, 141, 142, 253*, 360
   echinaria rufo-crocea var. nov. 254
   echinata 142, 253
   echinata densa 253, 360
   echinocactoides 347
   echinoidea 347
   echinops 329
   echinus 347
   edmundtsiana 329
   ehrenbergii 33
   eichlamii 8, 52*, 361
   eichlamii alba 52
   ekmannii 106
   elegans 16, 81, 115, 118, 120, 140, 254, 265, 280*, 291, 335
   elegans aureispina 281
   elegans dealbata 281*
   elegans globosa 280
   elegans klugii 280
   elegans micracantha 280
   elegans minor 280
   elegans migrispina 281
   elegans schmollii var. nov. 283
   elegans supertexta 140, 282*
   elephantidens 347
   elephantidens bumamma 347
   elongata 11, 141*, 254, 347
   elongata anguinea 253
   elongata echinata 253
   elongata intertexta 141
   elongata minima 142
   elongata rufescens 141
   elongata rufocrocea 254
   elongata schmollii 141
   elongata stella aurata 141
   elongata subechinata 253
   elongata subcrocea 141
   elongata tenuis 142
   elongata viperiana 144
   elyii 344
   emskoetteriana 347
   emundstiana 329
   enneacantha 344
   erecta 347
   erectacantha 329
   erectohamata 13, 194*
   eriacantha 16, 284, 286*, 287, 328, 344
   eriantha 287, 344
   erinacea 64, 65, 237
   erythrocarpa 344

MAMMILLARIA—Continued
   erythrosperma 12, 156*, 157, 204, 329, 345
   erythrosperma similis 157
   esaussieri 289
   eschanzieri 329
   eschauzieri 344
   esperanzaensis 15, 259, 263*, 359, 363
   esseriana 9, 79*, 80
   estanzuelensis 329
   euchlora 329
   eugenia 255
   evanescens 347
   evarescens 347
   evarescentis 347
   evermanniana 9, 82*
   eximia 268
   exsudans 347
   falcata 34
   farinosa 334
   fasciculata 12, 173*
   fellneri 330
   fennelii 178
   fera rubra sp. nov. 309
   fertilis 14, 232*
   filipendula 344
   fischeri 26
   fissurata 347
   flava 341
   flavescens 67
   flaviceps 280
   flavicoma 330
   flavispina 344
   flavovirens 8, 55*, 56
   flavovirens cristata 55
   floccigera 237
   floccigera longispina 237
   floribunda 347
   fobeana 285, 286
   fobei 347
   foersteri 33
   fordii 161
   formosa 10, 124*, 127
   formosa dispicula 124
   formosa gracilispina 124
   formosa laevior 124
   formosa microthele 124
   formosa nigrispina 124
   fortispina 78
   foveolata 20
   fragilis 11, 140*, 141, 254
   fragilis centrispina 15, 363
   fraileana 12, 159*, 160
   fuliginosa 15, 257*, 258
   fuliginosa longispina 257
   fulvescens 344
   fulvispina 235, 237, 258
   fulvispina media 237
   fulvispina minor 237

MAMMILLARIA—Continued
  fulvispina pyrrhocentra 237
  fulvispina rubescens 235, 237
  fulvispina rubescentem 237
  fulvolanata 344
  funkii 54
  furfuracea 347
  fuscata 16, 279*, 280
  fuscata esperanza 263
  gabbii 71
  galeottii 228, 338
  gasseriana 14, 215*
  gatesii 8, 68*, 69
  gaumeri 10, 103*
  gebweileriana 34
  geminata 30
  geminiflora 344
  geminispina 9, 80*, 81, 118, 279, 280
  geminispina nivea 80
  geminispina tetracantha 280
  georgii 347
  gibbosa 347
  gielsdorfiana 347
  gigantea 10, 102*
  gigantothele 344, 347, 356
  gilensis 13, 198*
  giseliana 111
  glaborescens 344
  glabrata 330, 340
  glabrata leucacantha 330
  gladiata 33
  gladiata aculeis rectis 33
  gladiata aculeis minimis 33
  gladiata spinis longissimis 33
  gladispina 347
  gladuligera 347
  glareosa 330
  glauca 34
  globosa 347, 356
  glochidata 12, 142, 152, 163*, 164, 185, 201, 219, 325, 331, 343
  glochidiata crinita 184
  glochidiata inuncinata 229
  glochidiata prolifera 163
  glochidiata purpurea 189
  glochidiata sericata 190, 203
  glomerata 274, 275
  goeringii 344, 347
  golziana 347
  goodrichii 175, 181
  goodridgei 12, 181*, 182
  goodridgii 181
  gracilis 141, 253, 254
  gracilis fragilis 253
  gracilis monville 254
  gracilis pulchella 141
  gracilis pusilla 141
  graessneriana 15, 254*, 255

MAMMILLARIA—Continued
  grahamii 187, 208, 209, 217
  grahamii arizonica 208, 343
  grandicornis 33, 363
  grandidens 35, 363
  grandiflora 262, 290, 347
  grandis 344
  granulata 331
  greggii 347
  grisea 331
  grisea galeotti 331
  grusonii 331
  grusonii similis 344
  guanajuatensis 103
  guebwilleriana 344
  guelzowiana 5, 14, 217*, 218
  guelzowiana splendens 217
  guerkeana 347
  guerreronis 10, 129*, 130, 133, 266
  guerreronis recta var. nov. 130
  guerreronis subhamatam var. nov. 4, 130
  guilleminiana 229
  guirocobensis sp. nov. 14, 220*, 221
  gülzowiana 217
  gummifera 9, 96*
  haageana 15, 59, 246*, 247, 252, 336
  haageana validior 246
  haehneliana 14, 223*
  haematactina 331
  hahniana 10, 110*, 111, 345
  hahnina giseliana 111*, 112
  hahniana tarajensis 111
  hahniana werdermanniana var. nov. 112
  halbingeri 16, 263*, 264
  halei 347
  hamata 331
  hamata brevispina 331
  hamata longispina 331
  hamata principis 331
  hamiltonhoytae 9, 77*, 315
  hamiltonhoytae fulvaflora var. nov. 78
  hamuligera 172
  handsworthii 344
  haseloffii 268, 347
  haynii 332
  haynii minima 332
  haynii viridula 332
  heeseana 108
  heeseana brevispina 108
  heeseana longispina 108
  heinii 332, 342
  heldii 344
  helicteres 332
  hemisphaerica 9, 94*, 122
  hemisphaerica waltheri var. nov. 95
  hennisii 284, 285
  hepatica 268

MAMMILLARIA—Continued
  hermantiana 55
  hermantii 55
  herrerai 344
  herrerae 11, 148*, 149, 344
  herrerae albiflora 149
  herrerae intertexta 149
  herrmannii 268
  herrmanni flavicans 268
  hertrichiana sp. nov. 9, 92*, 93
  hertrichiana robustior var. nov. 93
  hesteri 347
  heteracantha 334
  heteracentra 344
  heteromorpha 347
  hevernickii 344
  hexacantha 59, 332
  hexacentra 344, 347
  heyderi 10, 94, 97, 122*, 246, 330
  heyderi hemisphaerica 94
  heyderi applanata 97
  hidalgensis 14, 136, 226*
  hildemanniana 344
  hirschtiana 347
  hirsuta 13, 198*, 199
  hochderferi 344
  hoffmanniana 15, 256*
  hoffmannseggii 347
  hookeri 347
  hopferiana 34
  horripila 326, 347
  humboldtii 11, 142*, 143
  humilior 347
  humilis 344
  hutchisoniana 12, 179*
  hybrida 237
  hystricina 344
  hystrix 33, 363
  icamolensis 12, 166*, 167
  ignota 344
  imbricata 235
  impexicoma 347
  inaiae 15, 239*
  incerta 161, 162
  inclinis 29
  inconspicua 347
  incurva 347
  inermis 344
  infernillensis sp. nov. 10, 123*
  insularis 13, 205
  intertexta 141
  intricata 344
  inuncinata 323, 344
  inuncta 235
  irregularis 332
  isabellina 268
  jalapensis 23
  jaliscana 14, 218*
  johnstonii 9, 98*

MAMMILLARIA—Continued
  johnstonii guaymensis var. nov. 10, 99*
  johnstonii sancarlensis var. nov. 98, 99*, 363
  johnstonii typica var. nov. 98, 99*
  joossensiana 333
  jorderi 363
  jucunda 333
  karstenii 106, 107
  karwinskiana 7, 23, 26*, 27, 29
  karwinskiana centrispina 27
  karwinskiana flavescens 26
  karwinskiana virens 26
  karwinskii 344
  kelleriana sp. nov. 14, 231*
  kewensis 11, 135*
  kwensis albispina 135
  kewensis craigiana var. nov. 136, 137*, 232
  kleinii 333
  kleinschmidtiana 19
  klenneirii 344
  klissingiana 10, 121*
  klugii 280
  knebeliana 13, 195*
  knippeliana 7, 28*
  knuthiana 347
  kotschubeyoides 347
  krameri 33
  krameri viridis 34
  krameri longispina 34
  krauseana 55
  krausei 34
  kunthii 280
  kunzeana 13, 200*, 201
  kunzeana flavispina 200
  kunzeana longispina 200
  kunzeana rubrispina 200
  lactescens 34
  laeta 347, 352
  lamprochaeta 333
  lamuligera 172
  lanata 11, 139*, 282
  laneusumma sp. nov. 310
  lanifera 262, 279
  lapaixi 344
  lasiacantha ix, 11, 147, 150*, 225
  lasiacantha denudata 151
  lasiacantha minor 150
  lasiacantha plumosa 147
  lasiandra denudata 151
  lassaunieri 246
  lassomeri 246
  lassonneriei 245, 246
  latimamma 347
  latispina 347
  leei 348
  lehmannii 348, 363
  lehmannii sulcimamma 348

INDEX

MAMMILLARIA—Continued
  lengdobleriana 151
  lenta 11, 145*, 146
  leona 297
  leona similis 298
  lesaunieri 15, 245*, 246
  leucacantha 155, 348
  leucantha 11, 154*, 155
  leucocarpa 55
  leucodasys 348
  leucocentra 121, 333*
  leucocephala 281
  leucodictia 334
  leucospina 344
  leucotricha 54
  lewinii 348
  lewisiana 344
  liebneriana 232, 344
  lindbergii 344
  lindheimeri 98
  lindsayi 9, 87*, 88
  lindsayi robustior 88
  linkeana 268
  linkei 348
  littoralis 241, 242
  livida 334
  lloydii 7, 39*, 348
  longicoma 13, 199*, 200
  longicornis 348
  longiflora 5, 14, 206*
  longihamata 348
  longimamma 348, 355*
  longimamma compacta 355
  longimamma congesta 348, 355
  longimamma gigantothele 348, 356
  longimamma globosa 348, 356
  longimamma hexacentra 348, 355
  longimamma ludwigii 355
  longimamma luteola 348, 355
  longimamma melaleuca 353
  longimamma pseudomelaleuca 354
  longimamma sphaerica 348, 351
  longimamma spinosior 355
  longimamma uberiformis 348, 352
  longiseta 18
  longispina 228, 363
  lophothele 344
  lorenzii 344
  loricata 334
  luevedoi 344
  louisea 344
  ludwigii 75
  ludwigii clavata 75
  lutescens 344
  macdougalii 9, 95*
  macdowellii 103
  macracantha 334
  macracantha retrocurva 334
  macrantha 334

MAMMILLARIA—Continued
  macromeris 348
  macromeris longispina 348
  macromeris nigrispina 348
  macrothele 348
  macrothele biglandulosa 348
  macrothele lehmanii 348
  macrothele nigrispina 348
  maelenii 348
  magallanii sp. nov. 14, 223*
  magnimamma 7, 33*, 56, 65, 150, 228, 247, 326, 328, 334, 345, 348, 363
  magnimamma arietina 33, 348
  magnimamma bockii 35
  magnimamma divergens 35
  magnimamma krameri 35
  magnimamma lutescens 348
  magnimamma recurva 35
  magnimamma spinosior 348
  mainae 12, 172*
  maletiana 17, 344
  maletiana pyrrhocephala 17
  maletiana pyrrhocephala fulvolanata 17
  mallettiana 17
  maltrata 44
  mammillaris 10, 67, 106*, 258, 285, 327
  mammillarieaformis 348
  marshalliana 69
  martiana 348
  martinezii 252
  maschalacantha 54
  maschalacantha dolichacantha 55
  maschalacantha leucotricha 54
  maschalacantha xanthotricha 54
  mayensis sp. nov. 10, 116*
  mazatlanensis 15, 241*, 242
  mazatlanensis monocentra var. nov. 242
  mazatlensis 242
  megacantha 34
  megacantha rigidior 34
  meiacantha 66, 122
  meiacantha longispina 66
  meissnerii 280
  melaleuca 348, 353*
  melanacantha 64
  melanocentra 66
  melanocentra 8, 64*, 65, 237
  melanocentra meiacantha var. nov. 65
  melanocentra runyonii var. nov. 65
  melanocentra typica var. nov. 64
  melispina 7, 45*
  mendeliana 7, 43*
  meonacantha 66
  mercadensis 13, 204, 205*
  meschalacantha 54
  mexicensis sp. nov. 311
  micans 280, 334

MAMMILLARIA—Continued
  mickleyi 344
  micrantha 107
  micracantha 107
  microcarpa 14, 181, 208*, 217, 224
  microceras 33
  microceris 348
  microdasys 344
  microhelia rubispina 294
  microhelia 16, 293*, 294
  microhelia albiflora 293
  microhelia microheliopsis 296
  microheliopsis 16, 296*
  micromeris 335, 348
  microthele 106, 107, 334*, 343
  microthele brongniartii 107, 334
  mieheana 288, 289
  mieheana globosa 288
  milleri 208
  minima 142
  miqueliana 344
  mirabilis 268
  missouriensis 142, 348
  missouriensis caespitosa 348
  missouriensis nutallii 348
  missouriensis robustior 348
  missouriensis similis 348
  missouriensis viridescens 348
  mitis 344
  moelleri 344
  moelleriana 14, 214*, 215
  mölleriana 214
  monancistra 157*, 204
  monocantha 279
  monocentra 74
  monoclava 348
  monothele 344
  monstii 363
  montana 348
  montensis sp. nov. 311*, 312
  montensis monocentra var. nov. 312
  montensis quadricentra var. nov. 312
  morganiana 10, 125*
  morini 344
  moritziana 363
  movensis sp. nov. 312*, 313
  mucronata 335
  muehlbaueriana 348
  muehlenpfordtii 278, 348
  multicentralis sp. nov. 313
  multiceps 16, 273*, 274, 326
  multiceps albida 274
  multiceps elongata 273
  multiceps grisea 273
  multiceps humilis 273
  multiceps perpusilla 273
  multiceps texana 273
  multicolor 345
  multidigitata **345**

MAMMILLARIA—Continued
  multiformis 13, 157, 193*
  multihamata 13, 197*, 198
  multimamma 47, 48
  multiradiata 345
  multiseta 335
  mundtii 15, 248, 249*
  mutabilis 54, 55
  mutabilis autumnalis 55
  mutabilis laevior 55
  mutabilis leucocarpa 54
  mutabilis xanthotricha 54
  mystax 8, 34, 54*, 360
  napina 11, 138*, 139
  napina centrispina 15, 139*
  nealeana 118
  neglecta 237
  nelsonii 174, 175
  neocoronaria 12, 181*, 327
  neomexicana 348
  neopotosina nom. nov. 10, 117*, 118
  neopotosina brevispina var. nov. 118
  neopotosina hexispina var. nov. 118
  neopotosina longispina var. nov. 118
  neopalmeri nom, nov. 16, 267*, 268
  nerispina 232
  nervosa cristata 335
  nervosus cristatus 335
  neumanniana 33
  nicholsoni 345
  nickelsea 348
  nickelsi 345
  nidulata 345
  nigerrima 345
  nigra 335
  nigra euchlora 335
  nigricans 238
  nitens 260
  nitida 325
  nivea 80
  nivea brevispina 80
  nivea cristata 81
  nivea daedalea 80
  nivea wendlei 81
  nivosa 8, 66*, 67
  nobilis 80
  nogalensis 348
  nolascana 82
  nordmanniana 35
  nordmannii 35, 363
  notesteinii 348
  nuda 335
  numina 345
  nunezii 16, 192, 265*
  nuttallii 348
  nuttallii borealis 348
  nuttallii caespitosa 348
  nuttallii robustior 348
  obavallata 363

MAMMILLARIA—Continued
  obconella 227, 363
  obconella glaeottii 227
  obducta 345
  obliqua 336
  obdusta 281
  obvallata 258
  obscura 8, 59*, 60, 62, 348
  obscura wagneriana tortulospina comb. nov. 60
  ocamponis 205
  occidentalis 12, 168*, 169, 242
  occidentalis patonii comb. nov. 169
  occidentalis sinalensis var. nov. 169
  ochoterenae 15, 258*, 259, 263
  ocotillensis sp. nov. 314
  ocotillensis brevispina var. nov. 315
  ocotillensis longispina var. nov. 315
  octacantha 348
  odieriana 279, 280
  odieriana aurea 237, 279
  odieriana rigidior 279
  odieriana rubra 279
  odorata 348
  oettingenii 19
  olivacea 280
  oliveriana 280
  oliviae 14, 223*, 224
  olorina 336
  oothele 329, 336*
  orcuttii 8, 46*
  ortega 7, 30*
  ortiz rubiona 16, 266*
  ortiz-rubiana 266
  ottonis 348
  ottonis tenuispina 348
  ovimamma 336
  ovimamma brevispina 336
  ovimamma oothele 336
  pachyrhiza 9, 97*, 361
  pachethele 65
  pacifica 69
  painteri 14, 157, 211*, 212, 329
  pallescens 44
  pallida 348
  palmeri 267, 268, 336*, 345, 348
  papyracantha 348
  parensis sp. nov. 316
  parkinsi 121
  parkinsonii 10, 119*, 121, 281, 333
  parkinsonii brevispina 120, 121*
  parmentieri 67
  parvimamma 106
  parvissima 274
  patonii 169
  pazzanii 34
  peacockii 282
  pectinata 388
  pectinata cristata 348

MAMMILLARIA—Continued
  pectinifera 348
  peninsularis 7, 36*
  pentacantha 56
  perbella 15, 244*, 245
  perbella fina 245
  perbella lanata 245
  perote 247
  perpusilla 274
  perringii 278
  persicina 336
  petrophila 8, 51*
  petterssonii 10, 108*, 109
  pfeifferi 279*, 324
  pfeifferi altissima 279
  pfeifferi dichotoma 279
  pfeifferi flaviceps 279
  pfeifferi fulvispina 279
  pfeifferi variabilis 279
  pfeifferiana 348
  pfersdorfii 345
  phaeacantha 15, 237, 238*, 336
  phaeotricha 337
  phellosperma 5, 13, 196*
  phitauiana 14, 222*, 363
  phoeotricha 337
  phymatothele 9, 75*
  phymatothele trohartii comb. nov. 76
  picta 14, 233*
  pilispina 348
  pinispina 345
  plaschnickii 348
  plaschnickii straminea 348
  plecostigma 337
  plecostigma minor 337
  plecostigma major 337
  pleiocephala 337
  plinthimorpha 18
  plumosa 11, 146*
  polia 345
  polyacantha 268
  polyactina 268
  polycantha 345
  polycentra 268
  polycephala 283
  polychlora 328, 349
  polyedra 7, 21*, 23, 78, 340
  polyedra anisacantha 23
  polyedra laevior 23
  polyedra scleracantha 23
  polygona 8, 23, 50*, 329
  polymorpha 348
  polythele 8, 46*, 48, 227, 257, 261
  polythele aciculata 48, 260, 261
  polythele affinis 47
  polythele columnaris 47
  polythele hexacantha 47
  polythele latimamma 47
  polythele quadrispina 47

MAMMILLARIA—Continued
  polythele setosa 47
  polytricha 21
  polytricha hexacantha 21
  polytricha laevior 21
  polytricha scleracantha 21
  polytricha tetracantha 21
  pomacea 268
  pondii 349
  porphyracantha 261, 345
  poselgeri 349
  poselgeriana 269, 349
  posseltiana 14, 207*, 208
  posteriana 363
  potoniensis 345
  potosiana 118, 349
  potosina 118
  pottsii 16, 297*
  praelii 7, 29*
  praelii viridis 30
  preinreichiana 345
  pretiosa 268
  pretiosa cristata 269
  principis 332
  pringlei 15, 259*, 260
  pringlei columnaris var. nov. 26
  prismatica 345
  procera 337
  prolifera 16, 274*, 275
  prolifera haitensis 275
  prolifera multiceps 273
  prolifera texana 273
  pruinosa 268
  pseudocrucigera sp. nov. 9, 101*
  pseudoechinus 349
  pseudofuscata 288
  pseudomammillaris 260
  pseudoperbella 16, 277*
  pseudoperbella rufispina 277
  pseudorekoi 131
  pubispina 12, 167*, 168
  pugionacantha 337
  pulchella 260, 363
  pulchella flore pallidiore 261
  pulchella nigricana 261
  pulcherrima 268
  pulchra 235
  pulvilligera 349
  punctata 337
  purpuracea 349
  purpurascens 338
  purpurea 338
  purpusii 349
  pusilla 142, 274*, 326, 331
  pusilla albida 275
  pusilla caespititia 274, 326
  pusilla elongata 275
  pusilla gemina 275
  pusilla haitensis 275

MAMMILLARIA—Continued
  pusilla humilis 275
  pusilla major 274
  pusilla mexicana 274
  pusilla multiceps 274
  pusilla neomexicana 274
  pusilla texana 273
  pycnacantha 349
  pygmaea 12, 164*
  pyramidalis 237
  pyrrhacantha 345
  pyrrhacantha pallida 345
  pyrrhocentra 237
  pyrrhocentra gracilior 237
  pyrrhocephala 7, 17*, 55
  pyrrhocephala confusa 17
  pyrrhocephala donkelaeri 17
  pyrrhocephala fulvolanata 17, 344
  pyrrhocephala maletiana 17
  pyrrhochrantha 232
  quadrata 349
  quadrispina 46
  quadrispina major 48
  quehlii 345
  queretrica sp. nov. 316*, 317
  quevedoi 113*, 344, 345
  radians 349
  radians daemonoceras 349
  radians echinus 349
  radians globosa 349
  radians impexicoma 349
  radians sulcata 349
  radiaissima sp. nov. 16, 292*
  radicantissima 349
  radiosa 349
  radiosa alversonii 349
  radiosa arizonica 349
  radiosa borealis 349
  radiosa chlorantha 349
  radiosa deserti 349
  radiosa neomexicana 349
  radiosa texana 349
  radliana 349
  radula 237
  ramosissima 349
  raphidacantha 349
  raphidacantha ancistracantha 349
  raphidacantha humilior 349
  rebsamiana 72
  rebuti 345
  recta 281
  recurva 33
  recurvata 349
  recurvens 349
  recurvispina 37, 349
  reduncuspina 349
  regia 330
  rekoi 4, 11, 128, 131*, 133
  rekoi pseudo rekoi comb. nov. 131

rekoiana nom. nov. 10, 128*, 129, 132
rettigiana 13, 190*, 191
retusa 349
rhaphidacantha 349
rhodacantha 261
rhodantha 14, 232, 235*, 236, 237, 248, 279, 280, 324, 343, 345
rhodantha andreae 235
rhodantha aurea 237, 343
rhodantha aureiceps 324
rhodantha callaena 235
rhodantha centrispina 235
rhodantha chrysacantha 235
rhodantha crassispina 235
rhodantha droegeana 262, 295
rhodantha esperanza 237
rhodantha fuscata 279
rhodantha fulvispina 235, 237
rhodantha gigantea 236
rhodantha inuncta 237
rhodantha isabelliana 237
rhodantha neglecta 235, 237
rhodantha odieriana 279
rhodantha pfeifferi 279
rhodantha prolifera 235
rhodantha pyramidalis 235, 237
rhodantha quadrispina 237
rhodantha rubens 235
rhodantha ruberrima 235
rhodantha rubescens 235
rhodantha rubra 235, 236, 237
rhodantha ruficeps 235, 236
rhodantha schochiana 237
rhodantha sulphurea 279
rhodantha stenocephala 47
rhodantha tentaculata 279
rhodantha wendlandii 235, 237
rhodeocentra 338
rhodeocentra gracilispina 338
rigida 345
rigidispina 227
ritteri 345
ritteriana 9, 88*, 89, 360
ritteriana quadricentralis var. nov. 89
robusta 235
robustispina 349
roederiana 349
roematactina 345
roessingii 345
roii 345
rosea 338
roseana 349
rosealeuca 38
rosensis sp. nov. 317, 318*
roseo-alba 7, 38*
roseocentra 338
rubra 345
rueshiana 345

ruestii 15, 234*
ruficeps 235, 236
rufidula 339
rufispina 20
rufocrocea 254
rungei 151
runyonii 65, 349
rüshiana 345
russea 235
rüstii 234
rutila 255
rutila octospina 255
rutila pallidior 255
saetigera 9, 83*
saetigera quadricentralis var. nov. 83*, 84
saffordii 176
saillardii 345
salmiana 345
salm-dyckiana 339, 349
salm-dyckiana brunea 349
saltillensis 64
saluciana 345
sanguinea 269
sartorii 9, 71*, 72
sartorii brevispina 72
sartorii longispina 72
saxatilis 339
scepontocentra 339, 349
schaeferi 278, 363
schaeferi longispina 278
scheeri 331, 349
scheeri valida 349
scheidweileriana 13, 203*, 204
schelhasii 13, 189*, 190, 337, 345
schelhasei lanuginosior 189, 211
schelhasii rosea 190
schelhasii sericata 190, 345
schelhasii triuncinata 190
schiedeana 11, 143, 149*, 150, 345
schiedeana denudata 151
schiedeana plumosa 147
schlechtendalii 349
schlechtendalii levior 349
schmidtii 34
schmerwitzii 339
schmerwitziana 339
schniedeana 345
schochiana 237
schmollii 16, 283, 290*
schmuckeri 232
schulzeana 254, 345
schumannii 349, 356*
schwartziana 349
scleracantha 23
scolymoides 349
scolymoides longiseta 349
scolymoides nigricans 349

MAMMILLARIA—Continued
  scolymoides raphidacantha 349
  scrippsiana 9, 78*
  seegeri 268
  seegeri gracilispina 268
  seegeri pruinosa 268
  seegeri mirabilis 269
  seemannii 57, 339*
  seideliana 13, 202*, 203
  seidelii 339
  seitziana 7, 20*
  semigloba 345
  semitonia 345
  seminolia 345
  sempervivi 8, 58*, 74, 247
  sempervivi laeteviridis 59
  sempervivi tetracantha 58, 59*
  senckei 21
  senckeana 21
  senilis 349
  senilis diguetii 349
  senkei 54, 55
  senkii 54
  sericata 149
  setispina 349
  setosa 46
  severini 339
  sheldonii 12, 180*, 209
  similans 349
  similis 349
  similis caespitosa 349
  similis flavescens 67
  similis robustior 349
  simonis 345
  simplex 106, 349
  simplex affinis 107
  simplex flavescens 67
  simplex parvimamma 106
  simpsonii 349
  sinistrohamata 14, 213*, 214
  slevinii 262, 263
  sneedii 349
  solissii 13, 192*, 266, 270
  solitaria 349
  sonorensis 9, 90*, 343
  sonorensis brevispina 90, 91*
  sonorensis gentryi 90, 91*
  sonorensis hiltonii 90, 92*
  sonorensis longispina 90, 91*
  sonorensis maccartyi 92
  sororia 340
  spaethiana 350
  speciosa 340, 349, 350
  speciosissima 345
  speciosissima brunea 345
  spectabilis 136
  sphacelata 15, 145, 240*
  sphaerica 350, 351*

MAMMILLARIA—Continued
  sphaeroides 345
  sphaerotricha 272
  sphaerotricha rosea 272
  spinaurea 339, 340*
  spinii 262
  spinisfuscis 345
  spinosa 350
  spinosior 363
  spinosissima 16, 192, 268*, 343, 350
  spinosissima auricoma 269
  spinosissima aurorea 269
  spinosissima brunnea 268
  spinosissima castenoides 270
  spinosissima eximia 269
  spinosissima flavida 268
  spinosissima haseloffii 269
  spinosissima hepatica 269
  spinosissima hermannii 269
  spinosissima isabelliana 269
  spinosissima linkeana 269
  spinosissima mirabilis 269
  spinosissima pruinosa 269
  spinosissima pulcherrima 270
  spinosissima pretiosa 270
  spinosissima rubens 268
  spinosissima sanguinea 269
  spinosissima seegeri 270
  spinosissima vulpina 270
  spirocentra 345
  splendens 280
  squarrosa 18
  standleyi 9, 85*, 308
  standleyi robustispina var. nov. 86
  staurotypa 59
  stella aurata 141, 142
  stella aurata gracilispina 141
  stella de tacubaya 341
  stellaris 274
  stellata 274
  stenocephala 47
  stephani 345
  stipitata 350
  straminea 67
  strobiliformis 350
  strobiliformis caespitatia 350
  strobiliformis durispina 350
  strobiliformis rufispina 350
  strobiliana 4, 7, 41*
  stueberi 237, 248*
  suaveolens 340
  subangularis 18
  subcirrhifera 20
  subcrocea 141
  subcrocea anguinea 361
  subcrocea intertexta 141
  subcrocea rufescens 361
  subcrocea rutila 361

MAMMILLARIA—Continued
subcurvata 33
subechinata 253
suberecta 345
subpolyedra 23*, 340
subpolygona 50
subtetragona 44
subulata 350
subulifera 340
sulcata 350
sulcimamma 350
sulco-glandulifera 350
sulcolanata 350
sulphurea 279
sulphurea longispina 279
supertexta 282*, 333
supertexta caespitosa 282, 283, 345
supertexta compacta 283
supertexta dichotoma 282
supertexta longioribus 283
supertexta rosea 142
supertexta rufa 142
supertexta tetracantha 282, 283, 345
surculosa 12, 177*
swinglei 12, 170*
tacubayensis 341
tarajensis 111
tecta 341
tellii 345
tenampensis 8, 50*
tentaculata 279, 280
tentaculata conothele 237, 248
tentaculata fulvispina 237
tentaculata longispina 237
tentaculata picta 279
tentaculata rubra 237
tentaculata ruficeps 237
tenuis 142
tenuis arrecta 361
tenuis coerulescens 361
tenuis media 141
tenuis derubescens 361
tenuis minima 142
tesopacensis sp. nov. 10, 58, 104*, 105
tesopacensis rubraflora var. nov. 105
tetracantha 14, 35, 59, 227*, 344, 363
tetracantha galeottii 228
tetracentra 35, 227, 228
tetragona 345
tetrancistra 196
texana 273
texensis 97, 98, 350
thelocamptos 350
thornberi 173
tigeliana 305, 345
toaldoae 80
tolimensis sp. nov. 318*-320*
tolimensis brevispina var. nov. 319
tolimensis longispina var. nov. 319

MAMMILLARIA—Continued
tolimensis subuncinata var. nov. 320
tomentosa 278, 341
tomentosa flava 341
tortolensis 67
tournefortii 345
triacantha 18
trichacantha 12, 171*, 329
trigona 350
trigoniana 345
trohartii 76
tuberculosa 350
tuberculata 350
turbinata 350
uberiformis 350, 352*
uberiformis gracilior 350, 352
uberiformis hexacentra 350
uberiformis major 350, 352
uberiformis variegata 350, 352
uhdeana 269
umbrina 13, 191*, 332
umbrina roessingii 191
uncinata 4, 7, 42*, 343
uncinata biuncinata 43*, 362
uncinata rhodacantha 42
uncinata spinosior 42
unicornis 350
unihamata 176
uniseta 341
urbaniana 350
utahensis 350
vagaspina sp. nov. 8, 62*
valdezianus 350
valida 64, 350
vandermaelon 338
varimamma 341
varicolor 350
vaupeliana 350
vaupelii 10, 127*, 128
vaupelii flavispina var. nov. 128
venusta 350, 356*
verhaertiana 12, 160*
versicolor 33
vetula 16, 289*, 290, 359
vetula major 290
vicina 345
viereckii 14, 230*
viereckii brunea 230
viereckii brunispina 230
villa-lerdo 345
villifera 21, 23
villifera aeruginosa 44
villifera carnea 44
villifera cirrosa 44
villosa 345
viperina 3, 11, 144*, 360
virens 26
virentis 345
viridescens 345

MAMMILLARIA—Continued
  viridiflora 188
  viridis 30
  viridis praelii 30
  viridula 342
  viridula minima 342
  vivida 345
  vivipara 350
  vivipara radiosa 350
  vivipara radiosa neomexicana 350
  vivipara vera 350
  voburnensis 49
  vogtherriana 350
  vulpina 268
  wagneriana 60
  wagneriana tortulospina 60
  waltheri 95
  waltonii 121
  webbiana 37, 341
  webbiana longispina 37
  wegnerii 342
  wegnerii cristata 342
  weingartiana 14, 209, 210*
  werdermannii 345, 350
  werdermanniana 345
  wiesingeri 15, 243*, 244
  wilcoxii 5, 13, 186*
  wilcoxia viridiflora 188
  wildiana 152
  wildiana aurea 153
  wildiana compacta 152
  wildiana cristata 152
  wildiana major 152
  wildiana rosea 152, 157, 203
  wildii 11, 152*, 153, 337
  wildii compacta 152
  wildii cristata 152
  wildii monstrosa 152
  wildii rosea 152
  wildii rosiflora 153
  williamsii 350
  winkleri 350
  winteriae 7, 31*, 32
  wissmannii 350
  witurna 345
  woburnensis 8, 49*
  woodsii 10, 111, 113*, 114
  wrightii 5, 13, 185*, 187, 188
  xanthina 10, 104*
  xanthispina 345
  xanthotricha 54
  xanthotricha aculeis axillaribus robustioribus 54
  xanthotricha laevior 55
  yaquensis sp. nov. 174, 320*
  yucatanensis 16, 285*, 286
  zahniana 7, 32*, 33
  zanthotricha 54
  zapilotensis sp. nov. 4, 11, 130, 132*, 332

MAMMILLARIA—Continued
  zegschwitzii 342
  zeilmanniana 13, 188*
  zephyranthiflora 178
  zephyranthoides 5, 12, 178*, 191
  zepnickii 342
  zeyeriana 10, 107*, 108
  zilziana 350
  zooderi appendix
  zuccariniana 9, 73*, 334, 362*
  zuccarinii 362

NEOMAMMILLARIA
  albicans 262
  amoena 247, 248
  applanata 97
  arida 100
  armillata 162
  auriarceolis 301
  aureiceps 324
  barbata 216
  baxteriana 69
  blossfeldiana 183
  bocasana 210
  boedekeriana 212, 213
  bombycina 201
  brandegeei 70
  bullardiana 179
  camptotricha 133
  candida 272
  capensis 156
  carnea 44
  carretii 176, 177
  celsiana 278
  cerralboa 326
  chinocephala 115
  collina 249, 251
  collinsii 25, 53
  compressa 19
  confusa 23
  conspicua 264
  conzattii 24
  crocidata 37
  dawsonii 67, 68
  dealbata 281, 282
  decipiens 229
  densispina 288
  denudata 151
  dioica 161
  discolor 261
  donatii 250, 251
  echinaria 253, 254
  echinops 329
  eichlamii 52
  elegans 281, 325
  elongata 141, 142
  eriacantha 286
  eschausieri 329
  evermanniana 82
  fasciculata 173

NEOMAMMILLARIA—Continued
  fertilis 232
  flavovirens 55
  formosa 124
  fragilis 140, 141
  fraileana 159
  galeottii 228
  gaumeri 103
  geminispina 80
  gigantea 102
  glochidiata 163
  goodridgei 181
  graessneriana 254
  guerreronis 129, 130
  gummifera 96
  haageana 246
  hahniana 111
  hamata 331
  hamiltonhoytae 77
  hemisphaerica 94
  heyderi 122
  hirsuta 198
  hoffmanniana 256
  hutchisoniana 179
  jaliscana 218
  johnstonii 98
  karwinskiana 26
  kewensis 135
  kunzeana 200
  lanata 139, 140
  lapacena 162, 163
  lasiacantha 150
  lenta 145, 146
  lloydii 39
  longicoma 199, 200
  longiflora 206
  longimamma 350, 355*
  macdougalii 95
  macracantha 334
  magnimamma 35*, 326
  mainae 172, 173
  mammillaris 106, 107
  marshalliana 69
  mazatlanensis 241
  meiacantha 66
  melanocentra 64
  mendeliana 43
  mercadensis 205
  microcarpa 208
  mieheana 288
  milleri 208
  minuta 345
  moelleriana 214
  morganiana 125
  multiceps 273, 274
  multiformis 193
  multihamata 197
  mundtii 249
  mystax 54

NEOMAMMILLARIA—Continued
  napina 138, 139
  nelsonii 174, 175
  nivosa 66, 67
  nunezii 265
  obscura 60
  occidentalis 168, 169
  ochoterenae 258, 259
  oliviae 223
  ortegae 30
  ortizrubiona 266
  pacifica 69
  painteri 157, 211, 212
  palmeri 267
  parkinsonii 119
  patonii 169
  pectinata 350
  peninsularis 36
  perbella 244
  petrophila 51
  petterssonii 108
  phaeacantha 238
  phitauiana 222
  phymatothele 75
  plumosa 146
  polyedra 21
  polygona 50
  polythele 47
  pottsii 297
  praelii 29
  pringlei 259
  prolifera 275
  pseudoperbella 277
  pygmaea 164
  pyrrhocephala 17, 18, 25
  rekoi 128, 129, 131
  rhodantha 235
  ruestii 234
  runyonii 65
  saffordii 176, 177
  sartori 71, 72
  scheidweileriana 203
  schelhasei 189
  schiedeana 149
  schmollii 290
  schwarzii 350
  scrippsiana 78, 79
  seideliana 202, 203
  seitziana 20
  sempervivi 58
  sheldonii 180
  sinaloensis 345
  slevinii 262, 263
  solisii 192
  sphacelata 240
  sphaerica 350, 351
  spinosissima 270
  standleyi 85
  subpolyedra 23

NEOMAMMILLARIA—Continued
  swinglei 170, 171
  swingleri 171
  tacubayensis 341
  tenampensis 50, 51
  tetracantha 227
  tetrancistra 196
  trichacantha 171
  uberiformis 350, 352
  umbrina 191
  uncinata 42
  vaupelii 127
  verhaertiana 160
  vetula 289, 290
  villifera 21
  viperina 144
  viridiflora 188
  wilcoxii 186
  wildii 152
  woburnensis 49
  wrightii 185, 186

NEOMAMMILLARIA—Continued
  xanthina 104
  yucatanensis 285, 286
  zephyranthoides 178
  zeyeriana 107
  zuccariniana 73
PHELLOSPERMA
  tetrancistra 196, 197
  Species No. 504, pg. 311
  Species No. 509, pg. 301
  Species No. 510, pg. 302
  Species No. 567, pg. 304
  Species No. 579, pg. 310
  Species No. 593, pg. 303
  Species No. 613, pg. 307
  Species No. 617, pg. 86
  Species No. 625, pgs. 221, 299
  Species No. 645, pg. 221
  Species No. 718, pg. 309
  Species No. 1018, pg. 318
  Species No. 1095, pg. 359